BIOLOGICAL AND MEDICAL PHYSICS, BIOMEDICAL ENGINEERING

BIOLOGICAL AND MEDICAL PHYSICS BIOMEDICAL ENGINEERING

Volumes Published in This Series:

The Physics of Cerebrovascular Diseases, Hademenos, G.J., and Massoud, T.F., 1997

Lipid Bilayers: Structure and Interactions, Katsaras, J., 1999

Physics with Illustrative Examples from Medicine and Biology: Mechanics, Second Edition, Benedek, G.B., and Villars, F.M.H., 2000

Physics with Illustrative Examples from Medicine and Biology: Statistical Physics, Second Edition, Benedek, G.B., and Villars, F.M.H., 2000

Physics with Illustrative Examples from Medicine and Biology: Electricity and Magnetism, Second Edition, Benedek, G.B., and Villars, F.M.H., 2000

Physics of Pulsatile Flow, Zamir, M., 2000

Molecular Engineering of Nanosystems, Rietman, E.A., 2001

Biological Systems Under Extreme Conditions: Structure and Function, Taniguchi, Y. et al., 2001

Intermediate Physics for Medicine and Biology, Third Edition, Hobbie, R.K., 2001

Epilepsy as a Dynamic Disease, Milton, J., and Jung, P. (Eds), 2002

Photonics of Biopolymers, Vekshin, N.L., 2002

Photocatalysis: Science and Technology, Kaneko, M., and Okura, I., 2002

E. coli in Motion, Berg, H.C., 2004

Biochips: Technology and Applications, Xing, W.-L., and Cheng, J. (Eds.), 2003

Laser-Tissue Interactions: Fundamentals and Applications, Niemz, M., 2003

Medical Applications of Nuclear Physics, Bethge, K., 2004

Biological Imaging and Sensing, Furukawa, T. (Ed.), 2004

Biomaterials and Tissue Engineering, Shi, D., 2004

Biomedical Devices and Their Applications, Shi, D., 2004

Microarray Technology and Its Applications, Muller, U.R., and Nicolau, D.V. (Eds), 2004

Emergent Computation: Emphasizing Bioinformatics, Simon, M., 2005

Molecular and Cellular Signaling, Beckerman, M., March 22, 2005

The Physics of Coronary Blood Flow, Zamir, M., May, 2005

The Physics of Birdsong Mindlin, G.B., Laje, R., August, 2005

Radiation Physics for Medical Physicists Podgorsak, E.B., September 2005

Neutron Scattering in Biology—Techniques and Applications Fitter, J., Gutberlet, T., Katsaras, J. (Eds.), January 2006

Forthcoming Titles

Topology in Molecular Biology: DNA and Proteins Monastyrsky, M.I. (Ed.), 2006

Optical Polarization in Biomedical Applications Tuchin, V.V., Wang, L. (et al.), 2006

Continued After Index

Ying Xu, Dong Xu, and
Jie Liang (Eds.)

Computational Methods for Protein Structure Prediction and Modeling

Volume 1: Basic Characterization

Springer

Ying Xu
Department of Biochemistry
and Molecular Biology
University of Georgia
120 Green Street
Athens, GA 30602
USA
email: xyn@bmb.uga.edu

Dong Xu
Department of Computer Science
Digital Biology Laboratory
University of Missouri–Columbia
201 Engineering Building West
Columbia, MO 65211
USA
email: xudong@missouri.edu

Jie Liang
Department of Bioengineering
Center for Bioinformatics
University of Illinois at Chicago
851 S. Morgan Street
Chicago, IL 60607
USA
email: jliang@uic.edu

Library of Congress Control Number: 2006929615

ISBN 10: 0-387-33319-3
ISBN 13: 978-0387-33319-9

Printed on acid-free paper.

9 8 7 6 5 4 3 2 1

springer.com

Preface

An ultimate goal of modern biology is to understand how the genetic blueprint of cells (genotype) determines the structure, function, and behavior of a living organism (phenotype). At the center of this scientific endeavor is characterizing the biochemical and cellular roles of proteins, the working molecules of the machinery of life. A key to understanding of functional proteins is the knowledge of their folded structures in a cell, as the structures provide the basis for studying proteins' functions and functional mechanisms at the molecular level.

Researchers working on structure determination have traditionally selected individual proteins due to their functional importance in a biological process or pathway of particular interest. Major research organizations often have their own protein X-ray crystallographic or/and nuclear magnetic resonance facilities for structure determination, which have been conducted at a rate of a few to dozens of structures a year. Realizing the widening gap between the rates of protein identification (through DNA sequencing and identification of potential genes through bioinformatics analysis) and the determination of protein structures, a number of large scientific initiatives have been launched in the past few years by government funding agencies in the United States, Europe, and Japan, with the intention to solve protein structures *en masse,* an effort called *structural genomics*. A number of structural genomics centers (factory-like facilities) have been established that promise to produce solved protein structures in a similar fashion to DNA sequencing. These efforts as well as the growth in the size of the community and the substantive increases in the ease of structure determination, powered with a new generation of technologies such as synchrotron radiation sources and high-resolution NMR, have accelerated the rate of protein structure determination over the past decade. As of January 2006, the protein structure database PDB contained ~34,500 protein structures.

The role of structure for biological sciences and research has grown considerably since the advent of systems biology and the increased emphasis on understanding molecular mechanisms from basic biology to clinical medicine. Just as every geneticist or cell biologist needed in the 1990s to obtain the sequence of the gene whose product or function they were studying, increasingly, those biologists will need to know the structure of the gene product for their research programs in this century. One can anticipate that the rate of structure determination will continue to grow. However, the large expenses and technical details of structure determination mean that it will remain difficult to obtain experimental structures for more than a small fraction of the proteins of interest to biologists. In contrast, DNA sequence determination has doubled routinely in output for a couple of decades. The genome projects have led to the production of 100 gigabytes of DNA data in Genbank, and as the cost of sequencing continues to drop and the rate continues to accelerate, the scientific community anticipates a day when every individual has the genes of their interest and the genomes of all related major organisms sequenced.

Structure determination of proteins began before nucleic acids could be sequenced, which now appears almost ironic. As microchemistry technologies continue to mature, ever more powerful DNA sequencing instruments and new methods for preparation of suitable quantities of DNA and cheaper, higher sequencing throughput, while enabling a revolution in the biological and biomedical sciences, also left structure determination way behind. As sequencing capacity matured in the last few decades of the twentieth century, DNA sequences exceeded protein structures by 10-fold, then 100-fold, and now there is a 1000-fold difference between the number of genes in Genbank and the number of structures in the PDB. The order of magnitude difference is about to jump again, in the era of metagenomics, as the analyses of communities of largely unculturable organisms in their natural states come to dominate sequence production. The J. Craig Venter Institute's Sargasso Sea experiment and other early metagenomics experiments at least doubled the number of known open reading frames (ORFs) and potential genes, but the more recent ocean voyage data (or GOS) multipled the number on the order of another 10-fold, probably more. The rate of discovery of novel genes and correspondingly novel proteins has not leveled off, since nearly half of new microbial genomes turn out to be novel. Furthermore, in the metagenomics data, new families of proteins are discovered directly proportional to the rate of gene (ORF) discovery.

The bottom line is quite simple. Despite the several fold reduction in cost in structure determination due to the structural genomics projects—the NIH Protein Structure Initiative and comparable initiatives around the world—and the steady increase in the rate of protein structure determination, the number of proteins with unknown structures will continue to grow vastly faster. At an early structural genomics meeting in Avalon, New Jersey, the experimental community voted in favor of experimentally solving 100,000 structures of proteins with less than 30% sequence identity to proteins with known structures. This seemed to some theoreticians at the time as solving "the protein structure problem" and removing the need for theory, simulation, and prediction. Now, while it appears that this goal is aiming too high for just the initiative alone, certainly, the structural community will have 100,000 structures in the PDB not long after the end of this decade—and probably sooner than expected as costs continue to go down and technologies continue to advance. Yet, those 100,000 structures will be significantly less than 1% of the known ORFs genes! The problem, therefore, is *not* about having structures to predict, but having robust enough methods to make predictions that are useful at deep levels in biology, from helping us infer function and directing experimental efforts to providing insight into ligand binding, molecular recognition, drug discovery, and so on. The kind of success in terms of "reasonable" accuracy for "most" targets has been the grand success of the CASP competition (see Chapter 1) but is completely inadequate for the biology of the twenty-first century and the expectations of both basic and applied life sciences. Prediction is not at the requisite level of comprehensive robustness yet, and therein is one of the features of critical importance of the discussions in this book.

Computational methods for predicting protein structure have been actively pursued for some time. Their acceptance and importance grew rapidly after the establishment of a blind competition for predicting protein structure, namely, CASP. CASP involves theoreticians predicting then-unknown protein structures and their verification and analysis following subsequent experimental determination. The validation of the general approach both enhanced funding and brought participants to the field and pointed to the limitations of current methods and the value of extensive research into advanced computational tools. Overall, the rapidly growing importance of structural data for biology fueled the emergence of a new branch of computational biology and of structural biology, an interface between the methods of bioinformatics and molecular biophysics, namely, *structural bioinformatics*. Similar to genomic sequence analysis, bioinformatic studies of protein structures could lead to both deep and general or broad insights about aspects such as the folding, evolution, and function of proteins, the nature of protein–ligand and protein–protein interactions, and the mechanisms by which proteins act. The success of such studies could have immense impacts not just on science but on the whole society through providing insight into the molecular etiology of diseases, developing novel, effective therapeutic agents and treatment regimens, and engineering biological molecules for novel or enhanced biochemical functions.

As one of the most active research fields in bioinformatics, structural bioinformatics addresses a wide spectrum of scientific issues, including the computational prediction of protein secondary and tertiary structures, protein docking with small molecules and with macromolecules (i.e., DNA, RNA, and proteins), simulation of dynamic behaviors of proteins, protein structure characterization and classification, and study of structure–function relationships. While proteins were viewed as essentially static three-dimensional structures up until the 1980s, the establishment of computational methods, and subsequent advances in experimental probes that could provide data at suitable time scales, led to a revolution in how biologists think about proteins. Indeed, over the past few decades, computational studies using molecular dynamics simulations of protein structure have played essential roles in understanding the detailed functional mechanisms of proteins important in a wide variety of biological processes. Within the applied life sciences, protein docking has been extensively applied in the drug discovery pipeline in the pharmaceutical and biotech industry.

Protein structure prediction and modeling tools are becoming an integral part of the standard toolkit in biological and biomedical research. Similar to sequence analysis tools, such as BLAST for sequence comparison, the new methods for structure prediction are now among the first approaches used when starting a biological investigation, conducted prior to actual experimental design. That computational analysis would become the first step for experimentalists represents a major paradigm shift that is still occurring but is clearly essential to deal with the maturation of the field, the large quantities of data, and the complexity of biology itself as reflected in the requirement for today's powerful experimental probes used to address sophisticated questions in biology. This paradigm shift was noted first by Wally Gilbert, in a prescient article fifteen years ago ("Toward a new paradigm for molecular biology,"

Nature 1991, 349:99), who asserted that biologists would have to change their mode of approach to studying nature and to begin each experimental project with a bioinformatics analysis of extant literature and other computational approaches. This paradigm shift is deeply interconnected with the increased emphasis on computational tools and the expectation for robust methods for structure prediction.

Similar to other fields of bioinformatics, structural bioinformatics is a rapidly growing science. New computational techniques and new research foci emerge every few months, which makes the writing of textbooks a challenging problem. While a number of books have been published covering various aspects of protein structure prediction and modeling, it is widely recognized that the field lacks a comprehensive and coherent overview of the science of "protein structure prediction and modeling," which span a range from very basic problems (around physical and chemical properties and principles), such as the potential function and free energies that determine the folded shape of a protein, to the algorithmic techniques for solving various structure prediction problems, to the engineering aspects of implementation of computer prediction software, and to applications of prediction capabilities for investigations focused on functional properties. As educators at universities, we feel that there is an urgent need for a well-written, comprehensive textbook, one that proverbially goes from soup to nuts, and that this requirement is most critical for beginners entering this field as young students or as experienced researchers coming from other disciplines.

This book is an attempt to fill this gap by providing systematic expositions of the computational methods for all major aspects of protein structure analysis, prediction, and modeling. We have designed the chapters to address comprehensively the main topics of the field. In addition, chapters have been connected seamlessly through a systematic design of the overall structure of the book. We have selected individual topics carefully so that the book would be useful to a broad readership, including students, postdoctoral fellows, research scientists moving into the field, as well as professional practitioners/bioinformatics experts who want to brush up on topics related to their own research areas. We expect that the book can be used as a textbook for upper undergraduate-level or graduate-level bioinformatics courses. Extensive prior knowledge is not required to read and comprehend the information presented. In other words, a dedicated reader with a college degree in computational, biological, or physical science should be able to follow the book without much difficulty. To facilitate learning and to articulate clearly to the reader what background is needed to obtain the maximum benefit from the book, we have included four appendices describing the prerequisites in (1) biology, (2) computer science, (3) physics and chemistry, and (4) mathematics and statistics. If a reader lacks knowledge in a particular area, he or she could benefit by starting from the references provided in the corresponding appendix.

While the chapters are organized in a logical order, each chapter in the book is a self-contained review of a specific subject. Hence, a reader does not need to read through the chapters sequentially. Each chapter is designed to cover the following material: (1) the problem definition and a historical perspective, (2) a mathematical or computational formulation of the problem, (3) the computational methods and algorithms, (4) the performance results, (5) the existing software packages, (6) the strengths, pitfalls, and challenges in current research, and (7) the most promising future directions. Since this is a rapidly developing field that encompasses an exceptionally wide range of research topics, it is difficult for any individual to write a comprehensive textbook on the entire field. We have been fortunate in assembling a team of experts to write this book. The authors are actively doing research at the forefront of the major areas of the field and bring extensive experience and insight into the central intellectual methods and ideas in the subdomain and its difficulties, accomplishments, and potential for the future.

Chapter 1 (A Historical Perspective and Overview of Protein Structure Prediction) gives a perspective on the methods for the prediction of protein structure and the progress that has been achieved. It also discusses recent advances and the role of protein structure modeling and prediction today, as well as touching briefly on important goals and directions for the future.

Chapter 2 (Empirical Force Fields) addresses the physical force fields used in the atomic modeling of proteins, including bond, bond-angle, dihedral, electrostatic, van der Waals, and solvation energy. Several widely used physical force fields are introduced, including CHARMM, AMBER, and GROMOS.

Chapter 3 (Knowledge-Based Energy Functions for Computational Studies of Proteins) discusses the theoretical framework and methods for developing knowledge-based potential functions essential for protein structure prediction, protein–protein interaction, and protein sequence design. Empirical scoring functions including single-body energy function, statistical method for pairwise interaction between amino acids, and scoring function based on optimization are addressed.

Chapter 4 (Computational Methods for Domain Partitioning of Protein Structures) covers the basic concept of protein structural domains and practical applications. A number of computational techniques for domain partition are described, along with their applications to protein structure prediction. Also described are a few, widely used, protein domain databases and associated analysis tools.

Chapter 5 (Protein Structure Comparison and Classification) discusses the basic problem of protein structure comparison and applications, and computational approaches for aligning two protein structures. Applications of the structure–structure alignment algorithms to protein structure search against the PDB and to protein structural motif search in the PDB are also discussed.

Chapter 6 (Computation of Protein Geometry and Its Applications: Packing and Function Prediction) treats protein structures as 3D geometrical objects, and discusses structural issues from a geometric point of view, such as (1) the union of ball models, molecular surface, and solvent-accessible surface, (2) geometric constructs such as Voronoi diagram, Delaunay triangulation, alpha shape, surface geometry (including cavities and pockets) and their computation, (3) local surface similarity measure in terms of shape and sequence, and (4) function prediction based on protein surface patterns. Also described are the application issues of these computational techniques.

Chapter 7 (Local Structure Prediction of Proteins) covers protein secondary structure prediction, supersecondary structure prediction, prediction of disordered regions, and applications to tertiary structure prediction. A number of popular prediction software packages are described.

Chapter 8 (Protein Contact Maps Prediction) describes the basic principles for residue contact predictions, and computational approaches for prediction of residue–residue contacts. Also discussed is the relevance to tertiary structure prediction. A number of popular prediction programs are introduced.

Chapter 9 (Modeling Protein Aggregate Assembly and Structure) describes the basic problem of structure misfolding and implications, experimental approach for data collection in support of computational modeling, computational approaches to prediction of misfolded structures, and related applications.

Chapter 10 (Homology-Based Modeling of Protein Structure) presents the foundation for homology modeling, computational methods for sequence–sequence alignment and constructing atomic models, structural model assessment, and manual tuning of homology models. A number of popular modeling packages are introduced.

Chapter 11 (Modeling Protein Structures Based on Density Maps at Intermediate Resolutions) discusses methods for constructing atomic models from density maps of proteins at intermediate resolution, such as those obtained from electron cryomicrosopy. Details of application of computational tools for identifying α-helices, ß-sheets, as well as geometric analysis are described.

Chapter 12 (Protein Structure Prediction by Protein Threading) describes the threading approach for predicting protein structure. It discusses the basic concepts of protein folds, an empirical energy function, and optimal methods for fitting a protein sequence to a structural template, including the divide-and-conquer, the integer programming, and tree-decomposition approaches. This chapter also gives practical guidance, along with a list of resources, on using threading for structure prediction.

Chapter 13 (*De Novo* Protein Structure Prediction) describes protein folding and free energy minimization, lattice model and search algorithms, off-lattice model and search algorithms, and mini-threading. Benchmark performance of various tools in CASP is described.

Chapter 14 (Structure Prediction of Membrane Proteins) covers the methods for prediction of secondary structure and topology of membrane proteins, as well as prediction of their tertiary structure. A list of useful resources for membrane protein structure prediction is also provided.

Chapter 15 (Structure Prediction of Protein Complexes) describes computational issues for docking, including protein–protein docking (both rigid body and flexible docking), protein–DNA docking, and protein–ligand docking. It covers computational representation for biomolecular surface, various docking algorithms, clustering docking results, scoring function for ranking docking results, and start-of-the-art benchmarks.

Chapter 16 (Structure-Based Drug Design) describes computational issues for rational drug design based on protein structures, including protein therapeutics based on cytokines, antibodies, and engineered enzymes, docking in structure-based drug design as a virtual screening tool in lead discovery and optimization, and ligand-based drug design using pharmacophore modeling and quantitative structure–activity relationship. A number of software packages for structure-based design are compared.

Chapter 17 (Protein Structure Prediction as a Systems Problem) provides a novel systematic view on solving the complex problem of protein structure prediction. It introduces consensus-based approach, pipeline approach, and expert system for predicting protein structure and for inferring protein functions. This chapter also discusses issues such as benchmark data and evaluation metrics. An example of protein structure prediction at genome-wide scale is also given.

Chapter 18 (Resources and Infrastructure for Structural Bioinformatics) describes tools, databases, and other resources of protein structure analysis and prediction available on the Internet. These include the PDB and related databases and servers, structural visualization tools, protein sequence and function databases, as well as resources for RNA structure modeling and prediction. It also gives information on major journals, professional societies, and conferences of the field.

Appendix 1 (Biological and Chemical Basics Related to Protein Structures) introduces central dogma of molecular biology, macromolecules in the cell (DNA, RNA, protein), amino acid residues, peptide chain, primary, secondary, tertiary, and quaternary structure of proteins, and protein evolution.

Appendix 2 (Computer Science for Structural Informatics) discusses computer science concepts that are essential for effective computation for protein structure prediction. These include efficient data structure, computational complexity and NP-hardness, various algorithmic techniques, parallel computing, and programming.

Appendix 3 (Physical and Chemical Basis for Structural Bioinformatics) covers basic concepts of our physical world, including unit system, coordinate systems, and energy surfaces. It also describes biochemical and biophysical concepts such as chemical reaction, peptide bonds, covalent bonds, hydrogen bonds, electrostatic interactions, van der Waals interactions, as well as hydrophobic interactions. In addition, this chapter discusses basic concepts from thermodynamics and statistical mechanics. Computational sampling techniques such as molecular dynamics and Monte Carlo method are also discussed.

Appendix 4 (Mathematics and Statistics for Studying Protein Structures) covers various basic concepts in mathematics and statistics, often used in structural bioinformatics studies such as probability distributions (uniform, Gaussian, binomial and multinomial, Dirichlet and gamma, extreme value distribution), basics of information theory including entropy, relative entropy, and mutual information, Markovian process and hidden Markov model, hypothesis testing, statistical inference (maximum likelihood, expectation maximization, and Bayesian approach), and statistical sampling (rejection sampling, Gibbs sampling, and Metropolis–Hastings algorithm).

Ying Xu
Dong Xu
Jie Liang
John Wooley

April 2006

Acknowledgments

During the editing of this book, we, the editors, have received tremendous help from many friends, colleagues, and families, to whom we would like to take this opportunity to express our deep gratitude and appreciation. First we would like to thank Dr. Eli Greenbaum of Oak Ridge National Laboratory, who encouraged us to start this book project and contacted the publisher at Springer on our behalf. We are very grateful to the following colleagues who have critically reviewed the drafts of the chapters of the book at various stages: Nick Alexandrov, Nir Ben-Tal, Natasja Brooijmans, Chris Bystroff, Pablo Chacon, Luonan Chen, Zhong Chen, Yong Duan, Roland Dunbrack, Daniel Fischer, Juntao Guo, Jaap Heringa, Xiche Hu, Ana Kitazono, Ioan Kosztin, Sandeep Kumar, Xiang Li, Guohui Lin, Zhijie Liu, Hui Lu, Alex Mackerell, Kunbin Qu, Robert C. Rizzo, Ilya Shindyalov, Ambuj Singh, Alex Tropsha, Iosif Vaisman, Ilya Vakser, Stella Veretnik, Björn Wallner, Jin Wang, Zhexin Xiang, Yang Dai, Xin Yuan, and Yaoqi Zhou. Their invaluable input on the scientific content, on the pedagogical style, and on the writing style helped to improve these book chapters significantly. We also want to thank Ms. Joan Yantko of the University of Georgia for her tireless help on numerous fronts in this book project, including taking care of a large number of email communications between the editors and the authors and chasing busy authors to get their revisions and other materials. Last but not least, we want to thank our families for their constant support and encouragement during the process of us working on this book project.

Contents

Contributors

Natasja Brooijmans

Chemical and Screening Sciences
Wyeth Research
Pearl River, New York 10965

Christopher Bystroff

Department of Biology
Rensselaer Polytechnic Institute
Troy, New York 12180

Liming Cai

Department of Computer Science
University of Georgia
Athens, Georgia 30602-7404

Orhan Camoglu

Department of Computer Science
University of California Santa Barbara
Santa Barbara, California 93106

Yang Dai

Department of Bioengineering
University of Illinois at Chicago
Chicago, Illinois 60607-7052

Haobo Guo

Department of Biochemistry and Cellular and Molecular Biology
University of Tennessee
Knoxville, Tennessee 37996

Hong Guo

Department of Biochemistry and Cellular and Molecular Biology
University of Tennessee
Knoxville, Tennessee 37996

Jun-tao Guo

Department of Biochemistry and Molecular Biology
University of Georgia
Athens, Georgia 30602-7229

Carol K. Hall

Department of Chemical and Biomolecular Engineering
North Carolina State University
Raleigh, North Carolina 27695

Jaap Heringa

Centre for Integrative Bioinformatics
Vrije Universiteit
1081 HV Amsterdam, The Netherlands

Xiche Hu

Department of Chemistry
University of Toledo
Toledo, Ohio 43606

Ling-Hong Hung

Department of Microbiology
University of Washington
Seattle, Washington 98195-7242

Xiang Li

Department of Bioengineering
University of Illinois at Chicago
Chicago, Illinois 60607-7052

Jie Liang

Department of Bioengineering
University of Illinois at Chicago
Chicago, Illinois 60607-7052

Guohui Lin

Department of Computing Science
University of Alberta
Edmonton, Alberta T6G 2E8, Canada

Zhijie Liu

Department of Biochemistry and Molecular Biology
University of Georgia
Athens, Georgia 30602-7229

Hui Lu

Department of Bioengineering
University of Illinois at Chicago
Chicago, Illinois 60607-7052

Jianpeng Ma

Department of Biochemistry and Molecular Biology
Baylor College of Medicine
Houston, Texas 77030
and
Department of Bioengineering
Rice University
Houston, Texas 77005

Alexander D. MacKerell, Jr.

Department of Pharmaceutical Chemistry
School of Pharmacy
University of Maryland
Baltimore, Maryland 21201

Shing-Chung Ngan

Department of Microbiology
University of Washington
Seattle, Washington 98195-7242

Ognjen Perišić

Department of Bioengineering
University of Illinois at Chicago
Chicago, Illinois 60607-7052

Brian Pierce

Department of Biomedical Engineering
Boston University
Boston, Massachusetts 02215

Kunbin Qu

Department of Chemistry
Rigel Pharmaceuticals, Inc.
San Francisco, California 94080

Ram Samudrala

Department of Microbiology
University of Washington
Seattle, Washington 98195-7242

Ilya Shindyalov

San Diego Supercomputer Center
University of California San Diego
San Diego, California 92093-0505

Victor A. Simossis

Centre for Integrative Bioinformatics
Vrije Universiteit
1081 HV Amsterdam, The Netherlands

Ambuj K. Singh

Department of Computer Science
University of California Santa Barbara
Santa Barbara, California 93106

Stella Veretnik

San Diego Supercomputer Center
University of California San Diego
San Diego, California 92093-0505

Zhiping Weng

Department of Biomedical Engineering
Boston University
Boston, Massachusetts 02215

Ronald B. Wetzel

Department of Structural Biology
Pittsburgh Institute for Neurodegenerative Diseases
University of Pittsburgh School of Medicine
Pittsburgh, Pennsylvania 15260

John C. Wooley

Associate Vice Chancellor for Research
University of California San Diego
San Diego, California 92093-0043

Zhexin Xiang

Center for Molecular Modeling
Center for Information Technology
National Institutes of Health
Bethesda, Maryland 20892-5624

Dong Xu

Computer Science Department
University of Missouri—Columbia
Columbia, Missouri 65211-2060

Ying Xu

Institute of Bioinformatics and Department of Biochemistry and Molecular Biology
University of Georgia
Athens, Georgia 30602-7229

Yuzhen Ye

Bioinformatics and Systems Biology Department
The Burnham Institute for Medical Research
La Jolla, California 92037

Xin Yuan

Department of Computer Science
Florida State University
Tallahassee, Florida 32306

1 A Historical Perspective and Overview of Protein Structure Prediction

John C. Wooley and Yuzhen Ye

1.1 Introduction

Carrying on many different biological functions, proteins are all composed of one or more polypeptide chains, each containing from several to hundreds or even thousands of the 20 amino acids. During the 1950s at the dawn of modern biochemistry, an essential question for biochemists was to understand the structure and function of these polypeptide chains. The sequences of protein, also referred to as their *primary structures*, determine the different chemical properties for different proteins, and thus continue to captivate much of the attention of biochemists. As an early step in characterizing protein chemistry, British biochemist Frederick Sanger designed an experimental method to identify the sequence of insulin (Sanger et al., 1955). He became the first person to obtain the primary structure of a protein and in 1958 won his first Nobel Price in Chemistry. This important progress in sequencing did not answer the question of whether a single (individual) protein has a distinctive shape in three dimensions (3D), and if so, what factors determine its 3D architecture. However, during the period when Sanger was studying the primary structure of proteins, American biochemist Christian Anfinsen observed that the active polypeptide chain of a model protein, bovine pancreatic ribonuclease (RNase), could fold spontaneously into a unique 3D structure, which was later called *native conformation* of the protein (Anfinsen et al., 1954). Anfinsen also studied the refolding of RNase enzyme and observed that an enzyme unfolded under extreme chemical environment could refold spontaneously back into its native conformation upon changing the environment back to natural conditions (Anfinsen et al., 1961). By 1962, Anfinsen had developed his theory of protein folding (which was summarized in his 1972 Nobel acceptance speech): "The native conformation is determined by the totality of interatomic interactions and hence, by the amino acid sequence, in a given environment."

Anfinsen's theory of protein folding established the foundation for solving the protein structure prediction problem, i.e., for predicting the native conformation of a protein from its primary sequence, because all information needed to predict the native conformation is encoded in the sequence. The early approaches to solving this problem were based solely on the thermodynamics of protein folding. Scheraga and his colleagues applied several computer searching techniques to investigate the free energy of numerous local minimum energy conformations in an attempt to find the global minimum conformation, i.e., the thermodynamically most stable conformation of the protein (Gibson and Scheraga, 1967a,b; Scott et al., 1967). The major challenge for an energy minimization approach to protein structure prediction is that proteins are very flexible; thus, their potential conformation space is too large to be enumerated. [Despite the huge space of possible conformations, that proteins fold reliably and quickly to their native conformation is known as "Levinthal's paradox" (Levinthal, 1968)]. To address this issue, one needs an accurate energy function to compute the energy for a given protein conformation and a rapid computer searching algorithm. The progress of peptide molecular mechanics enabled the development of molecular force fields that described the physical interactions between atoms using Newton's equations of motion. In general, the interactions considered in the force field include covalent bonds and noncovalent interactions, such as electrostatic interactions, the van der Waals interactions, and, sometimes, hydrogen bonds and hydrophobic interactions. The parameters used in these force fields were obtained through experimental studies of small organic molecules. On the other hand, many computational methods developed in the field of optimization theory and mechanics have been applied to the rapid conformation search. These fall into two categories: the molecular dynamics method and the Brownian dynamics (or stochastic dynamics) method. Both methods sample a portion of potential protein conformations and evaluate their free energy. Molecular dynamics samples the conformations by simulating the protein motion based on Newton's equation, starting from an arbitrarily chosen protein conformation. Brownian dynamics, instead, uses Monte Carlo random sampling technique or its derivatives to evaluate protein conformations. Combining various force fields and conformation searching methods, many software packages were developed, such as AMBER (Pearlman et al., 1995), CHARMM (Brooks et al., 1983) and GROMOS (van Gunsteren and Berendsen, 1990), all aimed at using computing simulations to predict the native conformation of proteins.

Despite the great theoretic interest in energy minimization methods, these have not been very successful in practice, because of the huge search space for potential protein conformations. In 1975, Levitt and Warshel used a simplified protein structure representation and successfully folded a small protein [bovine pancreatic trypsin inhibitor, (BPTI), 58 amino acid residues] into its native conformation from an open-chain conformation using energy minimization (Levitt and Warshel, 1975). Little progress, however, has been made since then; the simulation usually takes an unrealistic compute or run time, and the final prediction is not very satisfactory. For instance, in 1998, Duan and Kollman reported a simulation experiment of one small protein (the villin headpiece subdomain, 36 amino acid residues), running on a Cray T3D and then a Cray T3E supercomputer, that took months of computation with the entire machine dedicated to the problem (Duan and Kollman, 1998). Even though the resulting structure is reasonably folded and shows some resemblance to the native structure, the simulated and native structure did not completely match. Currently, energy minimization methods are largely used to refine a low-resolution initial structure obtained by experimental methods or by comparative modeling (Levitt and Lifson, 1969).

At nearly the same time as these energy minimization approaches were developed, computational biochemists were looking for practical approaches to the protein structure prediction problem, which need not and presumably does not "mimic" the protein folding process inside the cell. An important observation was that proteins that share similar sequences often share similar protein structures. Based on this concept, Browne and co-workers modeled the structure of α-lactalbumin using the X-ray structure of lysozyme as a template (Browne et al., 1969). This success opened the whole new area of protein structure prediction that came to be known as *comparative modeling* or *homology modeling*. Many automatic computer programs and molecular graphics tools were developed to speed up the modeling. The potential targets of homologous modeling were also expanded through the rapid development of homologous modeling software and approaches. New technologies, including threading or the assembly of minithreaded fragments, were proposed and have now been successfully applied to many cases for which the target modeled does not have a sequence similar to the template proteins.

In this chapter, we review the history of protein structure prediction from two different angles: the methodologies and the modeling targets. In the first section, we describe the historical perspective for predicting (largely) globular proteins. The specialized methodologies that have been developed for predicting structures of other types of proteins, such as membrane proteins and protein complexes and assemblies, are discussed along with the review of modeling targets in the second section. The current challenges faced in improving the prediction of protein structure and new trends for prediction are also discussed.

1.2 The Development of Protein Structure Prediction Methodologies

1.2.1 Protein Homology Modeling

The methodology for homology modeling (or comparative modeling), a very successful category of protein structure prediction, is based on our understanding of protein evolution: (1) proteins that have similar sequences usually have similar structures and (2) protein structures are more conserved than their sequences. Obviously, only those proteins having appropriate templates, i.e., homologous proteins with experimentally determined structures, can be modeled by homologous modeling. Nevertheless, with the increasing accumulation of experimentally determined protein structures and the advances in remote homology identification, protein homology modeling has made routine, continuing progress: both the space of potential targets has grown and the performance of the computational approaches has improved.

1.2.1.1 First Structure Predicted by Homology Modeling: α-Lactalbumin (1969)

The first protein structure that was predicted by the use of homologous modeling is α-lactalbumin, which was based on the X-ray structure of lysozyme. Browne and co-workers conducted this experiment (Browne et al., 1969), following a procedure that is still largely used for model construction today. It starts with an alignment between the target and the template protein sequences, followed by the construction of an initial protein model created by insertions, deletions, and side chain replacements from the template structure, and finally finished by the refinement of the model using energy minimization to remove steric clashes.

1.2.1.2 Homology: Semiautomated Homology Modeling of Proteins in a Family (1981)

Greer developed a computer program to automate the whole procedure of homologous modeling. Using this program, 11 mammalian serine proteases were modeled based on three experimentally determined structures for mammalian serine proteases (Greer, 1981). The prediction used in this work was based on the analysis of multiple protein structures from the same protease family. He observed that the structure of a protease could be divided into structurally conserved regions (SCRs) with strong sequence homology, and structurally variable regions (SVRs) containing all the insertions and deletions in order to minimize errors in the query–template alignments significantly. Next, SVRs of the eight structurally unknown proteins were constructed directly from the known structures, based on the observation that a variable region that has the same length and residue character in two different known structures usually has the same conformation in both proteins.

This successful modeling experiment demonstrated that mammalian serine proteases could be constructed semiautomatically from the known homologous structures; both the need for manual inspections using biological intuition and the use of energy force fields were greatly reduced. The whole modeling procedure from this exercise was later implemented in the first protein modeling program, Homology, and integrated into a molecular graphics package InsightII (commercialized by Biosym, now Accelrys). Several important features of Homology, including the identification of modeling template using pairwise sequence alignment in the same protein family, the layout of sequence alignment between target and template protein sequences, and the identification and distinct modeling of conserved and variable regions using multiple structural templates from the same family, have been included in more recently developed homology modeling programs.

1.2.1.3 Composer: High-Accuracy Homology Modeling Using Multiple Templates (1987)

Greer's homology modeling method used multiple protein structures from the same family to define the conserved and variable regions in the target protein. It, however, used only one protein structure as the template to model the target protein. Blundell and co-workers recognized that the structural framework (or the "average" structure) of multiple protein structures from the same family usually resembled the target protein structure more than any single protein structure did. Based on this concept, they implemented a program called Composer (Sutcliffe et al., 1987), which was later integrated into the protein modeling package Sybyl, which was commercialized by Tripos.

The framework-based protein modeling significantly increased the accuracy of model construction over the previous semiautomatic methods, and hence made modeled protein structures practically useful. However, Composer applies empirical rules for modeling SVRs and the structure of amino acid side chains. As a result, the accuracy of these regions is much lower than the backbone structures in the SCRs. Therefore, the modeling of SVRs (or loops) and side chain placement have become two independent research topics for protein modeling. Many different solutions have been proposed (see Section 1.2.4 for a detailed review).

1.2.1.4 Modeller: Automatic Full-Atom Protein Modeling (1993)

Before 1993, protein modeling was done through a semiautomatic and multistep fashion, including distinct modeling procedure for SCRs, SVRs, and side chains. MODELLER, developed by Sali and Blundell, was the first automatic computer program full-atom protein modeling (Sali and Blundell, 1993). MODELLER computes the structure of the target protein by optimally satisfying spatial restraints derived from the alignment of the target protein sequence and multiple related structures, which are expressed as probability density functions (pdfs) of the restrained structural features. MODELLER facilitates high-throughput modeling of protein targets from genome sequencing project (Sanchez et al., 2000) and remains one of the popular or widely used modeling packages.

1.2.1.5 Other Protein Modeling Programs

SWISS-MODEL is a fully automated protein structure homology-modeling server, which was initiated in 1993 by Manuel Peitsch (Peitsch and Jongeneel, 1993). SWISS-MODEL automates the complete modeling pipeline including homology template search, alignment generation and model construction. It uses ProMod (Peitsch, 1996) to construct models for protein query with an alignment of the query and template sequences. NEST (Petry et al., 2003) realizes model generation by performing operations of mutation, insertion, and deletion on the template structure finished with energy minimization to remove steric clashes. The minimization starts with those operations that least disturb the template structure (which is called an artificial evolution method). The minimization is done in torsion angle space, and the final structure is subjected to more thorough energy minimization. Kosinski et al. (2003) developed the "FRankenstein's monster" approach to comparative modeling: merging the finest fragments of fold-recognition models and iterative model refinement aided by 3D structure evaluation; its novelty is that it employs the idea of combination of fragments that are often used by *ab initio* methods.

1.2.2 Remote Homology Recognition/Fold Recognition

All homology-based protein modeling programs rely on a good-quality alignment of the target and the template (of known structure). The identification of appropriate templates and the alignment of templates and target proteins are two essential topics for protein modeling, especially when no close homologue exists for modeling. The power or accuracy of homology modeling benefits from any improvement in the homology detection and target–template(s) alignment. Initially, a sequence alignment algorithm was used to derive target–template(s) alignment. More complicated methods (considering structure information) were later developed to improve the target–template(s) alignment.

1.2.2.1 Threading

The process of aligning a protein sequence with one or more protein structures is often called *threading* (Bryant and Lawrence, 1993). The protein sequence is placed or threaded onto a given structure to obtain the best sequence–structure compatibility. Obviously, the problem of identifying appropriate templates for a given target protein sequence can also be formulated as a *threading problem*, in which the structure in the database that is most compatible to the target sequence will be discerned and distinguished from those that are sufficiently compatible. Evolutionary information has been introduced to improve the sensitivity of homology recognition and to improve the target–template alignment quality, resulting a series sequence–profile and profile–profile alignment programs.

The threading method is able to go beyond sequence homology and identify structural similarity between unrelated proteins; "fold recognition" might be a better term for such cases. Homology recognition is used to detect templates that are homologous to the target with statistically significant sequence similarity; however, with the introduction of the powerful *profile-based* and *profile–profile-based methods*, the boundary between homology and fold recognition has blurred (Friedberg et al., 2004).

The threading-based method is typically classified in a separate category that is parallel to the homology-based modeling and *ab initio* modeling; it can be further divided into two subclasses considering whether or not the target and template have sequence similarity (homology) for quality evaluation purposes (Moult, 2005). However, from a methodology point of view, most threading-based modeling packages borrow similar ideas or even the existing modules from homology-based methods, to model the structure of a template after deriving the target–template alignment.

The concept of the threading approach to protein structure prediction is that in some cases, proteins can have similar structures but lack detectable sequential similarities. Indeed, it is widely accepted that there exist in nature only a limited number of distinct protein structures, called *protein folds*, which a virtually infinite number of different protein sequences adopt. As a result, it is hopeful that it is more sensible comparing the template protein structures with the target protein sequence than comparing their sequences. Protein threading methods fall into two categories. One kind of method represents protein structures first as a sequence of symbolic *environmental features*, e.g., the secondary structures, the accessibility of amino acid residues, and so on; next, it aligns this sequence of features with the target protein sequence using the classical dynamic programming algorithm for sequence alignment with a special scoring function. The other kind of method is based on a *statistical potential*, i.e., the frequency of observing two amino acid residues at a certain distance, in order to evaluate the compatibility between a protein structure and a protein sequence. Threading approaches have three distinct applications in protein structure prediction: (1) identifying appropriate protein structure templates for modeling a target protein, (2) identifying protein sequences adopting a known protein fold, and (3) accessing the quality of a protein model.

1.2.2.2 3D-profile: Representing Structures by Environmental Features

The pioneering work of Bowie and co-workers on "the inverse protein folding problem" led to a simple method for assessing the fitness of a protein sequence onto a structure, thus laying the foundation of the first kind of protein threading approach. In their work, structural environments of an amino acid residue were simply defined in terms of solvent accessibility and secondary structure (Bowie et al., 1991; Luthy et al., 1992). Statistics of residue–structure environment compatibility (3D-profile) were then computed based on the statistics of the frequency of a particular type of amino acid appearing in a particular structural environment in the collection of known structures. Threading programs using 3D-profile include 123D (Alexandrov et al., 1996), 3D-PSSM (Kelley et al., 2000), and FUGUE (Shi et al., 2001).

1.2.2.3 Statistical Potential Models

An alternative approach to threading is to measure the protein structure–sequence compatibility by a statistical potential model, which represents the preference of two types of amino acids to be at some spatial distance. Sippl proposed the concept of "reverse Boltzmann Principle" to derive a statistical potential, which he called potential of mean force, from a set of unrelated known protein structures (Sippl, 1990; Casari and Sippl, 1992). The basic idea of this energy function is to compare the observed frequency of a pair of amino acids within a certain distance for known protein structures with the expected frequency of this pair of amino acid types in a protein. Bryant and Lawrence first used the term "threading" to describe the approach of aligning a protein sequence to a known structure when they reported a new statistical potential model (Bryant and Lawrence, 1993).

1.2.2.4 Algorithmic Development for Threading Using Statistical Potential

Unlike the 3D-profile approach, statistical potential-based threading approach cannot use the classical dynamic programming approach for structure–sequence comparison. In fact, if pairwise interaction between residues is considered in assessing the compatibility of sequence and structure, the problem becomes very difficult (specifically, it is an NP-hard problem).

Various algorithms have been developed to address this computational difficulty. Early threading programs used various heuristic strategies to search for the optimal sequence–structure alignment. For example, GenTHREADER (Jones, 1999) and mGenTHREADER (based on the original GenTHREADER method, but adding the PSI-BLAST profile and predicted secondary structure as inputs) adopted a double dynamic programming strategy, which did not treat pairwise interactions rigorously. New threading programs have come to use more rigorous optimization algorithms. For example, PROSPECT (Xu and Xu, 2000) introduced a divide-and-conquer technique, and RAPTOR (Xu et al., 2003) used linear programming.

1.2.2.5 Profile-Based Alignment

Threading is not the only way to improve the sensitivity of (remote) template identification and the quality of template–target alignment. The other kind of method to achieve this goal makes use of multiple sequences from the same protein families to improve the sensitivity of homology detection and to improve the quality of sequence alignment.

Sequence–profile alignment strategy was first used to increase the sensitivity of distant homology detection. The development of Position Specific Iterative BLAST (PSI-BLAST) (and of course the accumulation of protein sequences) boosted the development of profile-based database search for homologies. In PSI-BLAST, a profile (or Position Specific Scoring Matrix, PSSM) is generated by calculating position-specific scores for each position in the multiple alignment constructed from the highest scoring hits in an initial BLAST search. Highly conserved positions receive high scores and weakly conserved positions receive scores near zero. The profile is then used to perform a second BLAST search by performing a sequence–profile alignment and the results are used to refine the profile, and so forth. This iterative searching strategy results in significantly increased sensitivity. PSI-BLAST is now often used as the first step in many studies including the profile–profile alignments. Profile information is also employed in hidden Markov models (HMMs) (Krogh et al., 1994), as implemented in the SAM (Karplus et al., 1998) and HMMER (http://hmmer.wustl.edu), which have vastly improved the accuracy of sequence alignments and sensitivity of homology detection.

Several profile–profile alignment methods have been developed more recently, including FFAS (Rychlewski et al., 2000), COMPASS (Sadreyev et al., 2003), Yona and Levitt's profile–profile alignment algorithm (Yona and Levitt, 2002), a method developed in Sali's group (Marti-Renom et al., 2004), and COACH (using hidden Markov models) (Edgar and Sjolander, 2004). The FFAS program pioneered the profile–profile alignment; it is now used in many modeling pipelines and metaservers. Zhou and Zhou (2005) developed a fold recognition method by combining sequence profiles derived from evolution and from a depth-dependent structural alignment of fragments. A key process for this group of methods is the alignment of the profile of target and homologies and the profile of structural template and homologies. They share the basic idea of profile–profile alignment but differ in many details, such as the profile calculation, profile–profile matching score, and alignment evaluation. The application of profile-profile alignment in homology detection highly increases the sensitivity of homology detection, even to the level of fold recognition.

1.2.3 *Ab Initio* Protein Structure Prediction

Despite the great success of homology modeling and threading methods, there are still many important target proteins that have no appropriate template (the number of such proteins is expected to be reduced due to the efforts of Structural Genomics, which aims at experimentally determining protein structures from all families and thus with providing new folds). *Ab initio* methods (which predict structures from sequence without using any structural template) are more general in this sense. *Ab initio* approaches are in principal based on Anfinsen's folding theory (Anfinsen, 1973), according to which the native structure corresponds to the global free energy minimum. Successful *ab initio* protein structure prediction methods fall roughly into several broad categories: (a) approaches that start from random/open conformations and simulate the folding process or minimize the conformational energy, (b) segment assembly-based methods as represented by the Rosetta method, and (c) methods that combine the two types of approaches (Samudrala et al., 1999).

1.2.3.1 Protein Folding Simulation and *ab Initio* Structure Prediction

Protein folding simulation and protein tertiary structure prediction are two distinct yet closely coupled problems. The main goal of protein folding simulation is to help characterize the mechanism of protein folding and also the interactions that determine the folding process and serve to specify the native structure; the goal of protein structure prediction is to determine the native structure. The solution of both problems relies on the effectiveness of energy function and conformation search methods utilized. Folding simulation approaches can be applied to predict protein structure *ab initio*, as seen in examples in which "folded" states resembling the native structures were derived. But only very few folding simulation approaches have been widely adopted for protein structure prediction and applied to a large number of predictions.

Molecular dynamics (MD) simulation is a natural approach for simulating protein folding. This approach has a long history and is still widely used; this could be viewed as illustrated most dramatically by IBM's Blue Gene project (http://www.research.ibm.com/bluegene/). However, the computational cost of folding simulations requires that the proteins to be simulated are small and fold ultrafast, even when supported by powerful computing (Duan and Kollman, 1998). Besides, the inadequacy in current potential functions for proteins in solution complicates the problem. The folded state by simulation does not necessarily correspond to the native state of proteins; actually, for current simulations, folding to the stable native state has not (yet) occurred. Considering these two types of difficulties in fold simulation and *ab initio* prediction of protein structures, many researches have either adopted simplified representation of proteins (including lattice and off-lattice models) to alleviate computational complexity, and/or to apply some conformational constraints to reduce the conformational searching space (e.g., the application of local structures in segment assembly based methods). Doing so improves the efficiency of folding simulation and *ab initio* methods for protein structure prediction.

1.2.3.2 Reduced Models of Proteins and Their Applications

Reduced models of proteins are necessary for easy and unambiguous interpretation of computer simulations of proteins and to obtain dramatic reduction (by orders of magnitude) of the computational costs. Such reduced models are still very important tools for theorctical studies of protein structure, dynamics, and thermodynamics in spite of the enormous increase in computational power (Kolinski and Skolnick, 2004). Simplified representations of protein structures include lattice models, continuous space models (e.g., a protein structure is reduced to the Cα trace and the centroid of side chains), and hybrid models (in which some degrees of conformational freedom are locally discretized). The resolution of lattice models can vary from a very crude shape of the main chain to a resolution similar to that of good experimental structures. Usually, the protein backbone is restricted to a lattice. The side chain, if explicitly treated, could be restricted to a lattice or could be allowed to occupy off-lattice positions. The HP model, proposed by Lau and Dill (1989), is a type of simple lattice model, which only considers two types of residues, hydrophobic and polar in a simple cubic lattice. Lattice models of moderate to high resolutions were also designed to retain more details of actual protein structure, including SICHO (SIde CHain Only) model (Kolinski and Skolnick, 1998), CABS, and "hybrid" 310 lattice model (considering 90 possible orientations of the Cα-trace vectors with off-lattice side chains and multiple rotamers). Reduced representations of proteins were employed in many studies, for example in studies of the cooperativity of protein folding dynamics (Dill et al., 1993) and in the *ab initio* prediction of protein structures (Skolnick et al., 1993).

1.2.3.3 *Ab Initio* Methods Using Reduced Representation of Proteins

Levitt and Warshel made one of the very first attempts to model real proteins using a reduced representation of proteins in 1975 (Levitt and Warshel, 1975). They applied a simplified continuous representation of protein structures with each residue represented as two centers (Cα atom, and the centroid of the side chain) in the simulation of the folding of bovine pancreatic trypsin inhibitor (BPTI), in which BPTI was folded from an open-chain conformation into a folded conformation resembling the crystallographic structure, with a backbone RMSD in the range of 6.5 Å.

Skolnick et al. developed a hierarchical approach to protein-structure prediction using two cycles of the lattice method (the second on a finer lattice), in which reduced representations of proteins are folded on a lattice by Monte Carlo simulation using statistically derived potentials, and a full-atom MD simulation afterwards (Skolnick et al., 1993; Kolinski and Skolnick, 1994b). This procedure was applied to model the structures of the B domain of staphylococcal protein (60 residues) and mROP (120 residues) (Kolinski and Skolnick, 1994a). Skolnick's group also developed TOUCHSTONE, an *ab initio* protein structure prediction method that uses threading-based tertiary restraints (Kihara et al., 2001). This method employs the SICHO model of proteins to restrict the protein's conformational space and uses both predicted secondary structure and tertiary contacts to restrict further the conformational search and to improve the correlation of energy with fold quality.

Scheraga's group developed a hierarchical approach that is similar to Skolnick's hierarchical method, but uses *off-lattice* simplified representation of proteins in the first steps of the prediction process; namely, one based solely on global optimization of a potential energy function (Liwo et al., 1999). This global optimization method is called Conformational Space Annealing (CSA), which is based on a genetic algorithm and on local energy minimization. Using this method, Liwo et al. built models of RMSD to native below 6 Å for protein fragments of up to 61 residues. This method was further assessed through two blind tests; the results were reported in Oldziej et al. (2005).

In specialized cases, parallel computation allows protein fold simulations using all-atom representation of proteins, and even explicit solvents, at the microsecond level. As described in brief above, a representative example is the folding of HP35, which is a subdomain of the headpiece of the actin-binding protein villin (Duan and Kollman, 1998), which has only 36 residues and folds autonomously without any cofactor or disulfide bond. This simulation was enabled by a parallel implementation of classic MD using an explicit representation of water, and the folded state of HP35 significantly resembles the native structure (but is not identical). But all-atom simulations are still limited and only practical for small ultrafast folding proteins.

1.2.3.4 *Ab Initio* Methods by Segment Assembling

A significant progress in the development of *ab initio* methods was the introduction of conformational constraints to reduce the computational complexity. Several *ab initio* modeling methods have been developed based on this strategy (Zhang and Skolnick, 2004; Lee et al., 2005), which was pioneered in the implementation of the Rosetta method (Simons et al., 1997, 1999a).

The basic idea of Rosetta is to narrow the conformation searching space with local structure predictions and model the structures of proteins by assembling the local structures of segments. The Rosetta method is based on the assumption that short sequence segments have strong local structural biases, and the strength and multiplicity of these local biases are highly sequence dependent. Bystroff et al. developed a method that recognizes sequence motifs (I-SITES) with strong tendencies to adopt a single local conformation that can be used to make local structure predictions (Bystroff and Baker, 1998). In the first step of Rosetta, fragment libraries for each three- and nine-residue segment of the target protein are extracted from the protein structure database using a sequence profile–profile comparison method. Then, tertiary structures are generated using a Monte Carlo search of the possible combinations of likely local structures, minimizing a scoring function that accounts for nonlocal interactions such as compactness, hydrophobic burial, specific pair interactions (disulfides and electrostatics), and strand pairing (Simons et al., 1999b). A test of Rosetta on 172 target proteins showed that 73 successful structure predictions were made out of 172 target proteins with lengths below 150 residues, with an RMSD < 7 Å in the top five models (Simons et al., 2001). Rosetta has achieved the top performance in a series of independent, blind tests (Moult et al., 1999; Simons et al., 1999a), ever since those for CASP3 (see below for details about the CASP series of workshop). Rosetta has also been further refined and extended to related prediction tasks, namely, docking on predicted interactions (see below).

Zhang and Skolnick developed TASSER, a threading template assembly/refinement approach, for *ab initio* prediction of protein structures (Zhang and Skolnick, 2004). The test of TASSER on a comprehensive benchmark set of 1489 single-domain proteins in the Protein Data Bank (PDB) with length below 200 residues showed that 990 targets could be folded by TASSER with an RMSD < 6.5 Å in at least one of the top five models. The fragments used for assembly in TASSER are derived in a different way than in Rosetta. Specifically, the fragments or segments are excised from the threading results, and thus are generally much longer (about 20.7 residues on average) than the segments used by Rosetta (which are 3–9 residues).

1.2.4 Modeling of Side Chains and Loops

We review the modeling of side chains and loops as a separate section because these are two main problems that both homology modeling and *de novo* methods face, and because they differ more among protein homologues than do the backbone and protein cores. Yet, the conformation of side chains and loops may carry very important information for understanding the function of proteins.

There are mainly two classes of computational approaches to building the loop structures: knowledge-based methods and *ab initio* methods. Knowledge-based methods build the loop structures using the known structures of loops from all proteins in the structure database, whether or not they are from the same family as the target protein (Sucha et al., 1995; Rufino et al., 1997). This approach is based on the principle that the plausible conformations of loops within a certain length cannot be that many, i.e., must be limited. Assuming a sufficient variety of known protein structures, almost all plausible loop structures should be represented by at lease one protein structure in the database. In fact a library of plausible loop structures for a given loop size has been constructed (Donate et al., 1996; Oliva et al., 1997). Typically, for a given loop in the target protein, the selection of the optimal template structure usually relies on the similarity of the anchor regions (i.e., the flanking residues around the loop) between template loop structure and the modeled core structure of the target, and the compatibility of the template loop structure with the core structure as measured by a residue level empirical scoring function (van Vlijmen and Karplus, 1997). *Ab initio* methods build loop structures from scratch (Moult and James, 1986; Pedersen and Moult, 1995; Zheng and Kyle, 1996). Recently, methods that combine knowledge-based and *ab initio* methods for better loop modeling have been introduced (Deane and Blundell, 2001; Rohl et al., 2004). MODELLER (Sali and Blundell, 1993) uses a different methodology from the above, which builds both core and loop regions by optimally satisfying spatial restraints derived from the target–template alignment.

Similarly, side-chain conformations can be predicted from similar structures and from steric or energetic considerations (Vasquez, 1996). The construction of side-chain rotamers and the development of powerful conformation searching algorithms (such as Dead End Elimination, DEE) (Desmet et al., 1992) and the mean force field-based method (Lee, 1994; Koehl and Delarue, 1995) contributed to the success of side-chain conformation prediction. Rotamer libraries are generally defined in terms of side-chain torsional angles for preferred conformations of a particular side chain. Ponder and Richards set up the first rotamer library (Ponder and Richards, 1987). A backbone-dependent rotamer library was later constructed and used for side-chain prediction (SCWRL) (Dunbrack and Karplus, 1993; Canutescu et al., 2003). Wang et al. developed a rapid and efficient method for sampling off-rotamer side-chain conformations through torsion space minimization; this starts from discrete rotamer libraries supplemented with side-chain conformations taken from the unbound structures. This approach has been used to improve side chain packing in protein–protein docking.

1.2.5 Modeling Structural Differences

Mutation data are an important source of information in the study of the functions of proteins; similarly, analyzing the differences among protein families is one way to study their function and functional specificity. It is therefore very important to study the detailed structural differences associated with mutations and sequence differences among families. For example, homology modeling (Lee, 1995) and molecular dynamics (MD) were used for studying the consequences of mutations (see the section "Molecular Dynamics Simulations of Membrane Proteins").

Baker's group tried to model structural differences based on comparative modeling by free-energy optimization along principal components of natural structural variation, which serves to improve the accuracy of protein modeling (Qian et al., 2004). In comparative modeling, an issue has been that a given protein model is frequently more similar to the template(s) used for modeling than to the target protein's native structure. In principle, energy-based minimization might help to improve the resolution of models. However, in practice, energy-based refinement of comparative models generally leads to degradation rather than improvement in model quality. The work of Baker's group (Qian et al., 2004) led to an improved use of energy-based minimization, through restricting the search space along the evolutionarily favored direction and thereby avoiding the false attractors that might lead the minimization to wrong answers.

There are numerous limits within current efforts, and considerable effort is still required to improve the methods for predicting the structures resulting from mutations and the modeling of structural difference within families. The reasons underlying the difficulties include our inability to model protein structures in fine resolution despite the strict requirements for quality in modeling of the structural differences. Indeed, "modeling of the structure of a single mutation" and "modeling structure changes associated with specificity changes within protein families" were identified as two of the three modeling challenges as viewed by a community meeting in 2005 [see the summary from CASP6 (Moult et al., 2005), which is a summary from the sixth in a series of structure prediction meetings described below].

1.2.6 Novel Communitywide Activities to Improve Prediction and Demonstrate Value

CASP (Critical Assessment of Structure Prediction) is a communitywide experiment with the primary aim of assessing the effectiveness of modeling methods. CASP deserves special recognition in any consideration of the role of modeling/computational methods for biology, since the meeting/process has transformed the level of recognition (for modeling studies) coming from experimentalists; CASP has become a model for all computational biology communities and an exemplar for evaluating techniques or methods beyond software/the approaches of scientific computing. In light of these competitions and the overall efforts in the field, the general status for high-resolution refinement of protein structure models and overall progress in modeling has been reviewed in depth recently (Misura and Baker, 2005; Schueler-Furman et al., 2005b).

CASP was first held in 1994 and six CASP meetings were held through 2004; the most recent meeting was held in 2006 (as the 7th Community Wide Experiment on Critical Assessment of Techniques for Protein Structure Prediction). The key feature of CASP is that participants make *blind* predictions of structures. CASP has monitored since 1994 the progress of protein modeling (covering all categories of modeling methods). Also it provides a good arena for testing the performance of newly developed modeling methods. The prediction season, during a cycle, begins in spring and all predictions are due at the end of the summer. The essential aspect is that experimentalists make lists available of what they are likely to solve during this time period and agree not to release their structures, when obtained, until after the deadline for predictions. Establishing this clear process solved the longstanding assertions about structure prediction being based on previously known information.

How well one does in CASP has become important—some would say too important—as a metric for research in the field. As a consequence, as well as CASP, which is a manual method in which any amount of scientific knowledge and any collection of algorithms can be employed, an automated prediction approach has been added, to test the state of computational prediction schemes rather than the participants' insight into protein structure. This is the Critical Assessment of Fully Automated Structure Prediction (CAFASP). Besides using automated approaches for the competition, numerous protein prediction servers have been introduced for the community, including, for example, PROSPECT-PSPP (Guo et al., 2004) and Robetta (Kim et al., 2004). Other aspects of large-scale prediction servers are described below (Section 1.2.7). Interestingly, services, such as EVA, have also been created to monitor the quality or performance of the numerous prediction servers, and provide continuous, fully automatic, and statistically significant analysis of such servers (Koh et al., 2003).

CASP is now organized by the Protein Structure Prediction Center. The Center's goal is to help advance the methods of identifying protein structure from sequence. The Center has been organized to provide the means of objective testing of these methods via the process of blind prediction. In addition to support of the CASP meetings, their goal is to promote an objective evaluation of prediction methods on a continuing basis. Some of the recent successes in CASP have been described previously.

A very powerful related community scheme looks at the nature of macromolecular interactions or docking, Critical Assessment of PRedicted Interactions (CAPRI), which grew up directly from the successes of CASP, where this new chain of meetings was launched after 1996. While few of the proteins identified through major genome sequencing efforts will ever have their structure solved, since proteins actually carry out biological processes as larger, multimeric or even heterologous complexes, characterizing the structure of proteins in native complexes is more important, and even fewer of those complexes will ever be experimentally determined, due to the greater inherent difficulties in doing so. To test what are therefore essential computational methods, the starting points for predicting the structures of protein complexes ("docked" proteins) are the independently solved structures of the constituents of a protein complex, whose 3D structure is unknown, and against which the community's algorithms and approaches can be tested. For example, an in-depth evaluation of certain docking algorithms in early CAPRI rounds (3, 4, and 5) has been provided (Wiehe et al., 2005); of particular value has been the introduction of benchmarks for analysis, such the Protein-Protein Docking Benchmark 2.0 (Mintseris et al., 2005), which provides a platform for evaluating the progress of docking methods on a wide variety of targets. An extension of the very successful Rosetta approach to the challenges of predicting the structure of complexes is RosettaDock, which uses real-space Monte Carlo minimization on both rigid body and the side chain degrees of freedom in order to find the lowest free energy arrangement of two docked protein structures; more recently, this has been extended to take into account backbone flexibility and employed very successfully in more recent CAPRI competitions (Schueler-Furman et al., 2005a). More details about docking approaches in general are discussed below.

1.2.7 Protein Modeling Metaservers

Several protein modeling metaservers have appeared since 2001, including Pcon (a neural-network–based consensus predictor) (Lundstrom et al., 2001), Structure Prediction Meta Server (Bujnicki et al., 2001), 3D-Jury (Ginalski et al., 2003), GeneSilico protein structure prediction metaserver (Kurowski and Bujnicki, 2003), and 3D-SHOTGUN (Daniel, 2003). These automatic servers collect models from other servers and use that input to produce consensus structures. According to the assessment performed via CASP, protein modeling metaservers perform generally better than other single modeling methods; their performance is even close to that of human experts. [The noteworthy progress between CASP4 and CASP5 was partly due to the effective use of metaservers (Moult, 2005).] However, some CASP participants have worried that the increasing successes of metaservers might discourage researchers from developing new prediction methods. This seems a small worry, in light of the various objectives for improved modeling methods and the potential impact from delivering more accurate, high-throughput genome annotation to enhanced drug discovery. Of course, there is a community goal, to seek improved tools and validation of the overall approach in the eyes of experimentalists, and the many personal goals, to seek to make the best contribution possible. As a consequence, the larger worry, under the current environment for CASP itself, is that it is hard to dissect the individual computational contributions to prediction and ascertain progress and what tools to choose since considerable manual or intellectual intervention is inevitably involved in order to achieve the highest validated successes in prediction. This difficulty is among the factors that led to the introduction of automated approaches, including metaservers, in the first place.

1.3 A Shift in the Focus for Protein Modeling

In recent years, the efforts in genome sequencing have been enormously successful. Hundreds of whole or complete microbial genomes and dozens of eukaryotic plant and animal genomes have been sequenced, and many more genome projects are underway. In contrast to the quickly increasing number of predicted protein sequences (open reading frames or ORFs) that are deposited in the community database, Genbank, the number of proteins whose architecture has been solved increases much more slowly. This continues despite the advances in structure determination techniques and the effects of the (National Institutes of Health, NIH) Protein Structure Initiative in the United States and Structural Genomics Projects worldwide. Therefore, more modeling per se as well as improved computational modeling of protein structures is of crucial importance to keep pace with the advances of genome sequencing and functional genomics, that is, our ability to predict the structure of newly discovered or predicted proteins has to increase greatly in order for the community to be able to characterize and utilize fully the extraordinary delivery of new sequence information. Accordingly, the focus of modeling has shifted in recent years, from modeling of monomers to modeling of simple protein–protein complex and even the modeling of large protein assemblies; that is, the focus has moved from small-scale modeling to large-scale modeling (and even genome-scale efforts at comprehensive modeling). In this section, we will focus on a discussion of modeling of different targets. Also, we will discuss specific methods that have already been developed and those that are emerging to deal with the various requirements, which are different from the methods discussed above. (These methods are mostly for modeling soluble, single-domain globular proteins.)

1.3.1 Modeling of Membrane Proteins

Membrane proteins play a central role in many cellular and physiological processes. Any aspect of cell activity is regulated by extracellular signals that are recognized and transduced inside the cell via different classes of plasma membrane receptors. It is estimated that integral membrane or transmembrane (TM) proteins make up about 20–30% of the proteome (Krogh et al., 2001). They are essential mediators of material and information transfer across cell membranes. Identifying these TM proteins and deciphering their molecular mechanisms is of great importance for understanding many biological processes. In addition, membrane proteins are of particular importance in biomedicine, because they are the targets of a large number of pharmacologically and toxicologically active substances, and are directly involved in their uptake, metabolism, and clearance. Membrane proteins can be loosely associated on the surface of the lipid bilayer (peripheral membrane proteins) or embedded (integral membrane protein, e.g., bacteriorhodopsin). The prediction and analysis of membrane proteins largely involves a focus on integral membrane proteins.

Membrane proteins account for less than 1% of the known high-resolution protein structures (White, 2004), despite their importance in essential cellular functions. Solving the structure of a membrane protein remains challenging and no high-throughput methods, or even general methods, have been developed. In the first instance, structure determination of membrane proteins remains a challenge because of difficulties in expressing sufficient quantities of protein and in manipulating the protein *in vitro* with an artificial environment mimicking some attributes of the *in situ* environment. Even when these challenges are met, there are remaining difficulties in obtaining ordered crystals for analysis by X-ray crystallography. NMR remains the modality of choice for structural analysis of membrane proteins but cannot readily tackle larger proteins and requires substantive quantities of material. Given the challenges for crystallographic analysis, membrane proteins were inevitably listed as "lower priority" or "avoided" targets for the Structural Genomics Centers, during the early phase of the Protein Structure Initiative. Research funding has even included set-aside opportunities to address the challenges of characterizing the biophysical properties and structure of membrane proteins. However, no demonstrated method yet exists to deliver a pipeline for high-throughput structure determination of membrane proteins.

Given the relatively and absolutely (!) small number of known, high-resolution membrane protein structures, computational methods are very important in predicting the structures of membrane protein, and in this case especially, if a prediction could be said to "determine" the structure, computational methods would have a huge impact on fundamental biology and biomedicine, and on applied life sciences research around drug targets. Most of the tools used for analyzing and predicting the structure of soluble, nonmembrane proteins can also be used for this important class. That is, many secondary structure prediction methods from primary sequences based on statistical methods, physicochemical methods, sequence pattern matching, and evolutionary conservation can also be applied for modeling the structures of membrane proteins, as can the conventional 3D structure prediction methods, including homology modeling techniques. At the same time, due to the limited number of known structures of membrane proteins, the application of homology modeling in predicting membrane protein structures remains very limited.

In the absence of a high-resolution 3D structure (experimental or computational), an important cornerstone for the functional analysis of any membrane protein is an accurate topology model. A topology model describes the number of TM spans and the orientation of the protein relative to the lipid bilayer. The secondary structure of a membrane-spanning segment can be an α-helix or a β-strand, but a TM β-strand usually has fewer residues than an α-helix. Nearly all TM β-strand proteins are found in prokaryotes, and belong to only a few protein families. Generally, integral membrane transporters of the inner membrane consist largely of α-structures, and they traverse the membrane as α-helices, whereas those of the outer membranes consist largely of ß-barrels. Because of this, many methods have been developed to focus on the prediction of transmembrane α-helices. These methods are mainly based on the special properties of membrane proteins (Chen and Rost, 2002), such as differences in amino-acid compositions in cytoplasmic and extracellular regions (positive-inside rule) (Heijne, 1986), the hydrophobic/hydrophilic patterns of TM regions (Kyte and Doolittle, 1982), and the minimum length of TM regions.

1.3.1.1 Methods for Topology Model Prediction (α-Helix Membrane Proteins)

One of the earliest and still most widely practiced methods for identification of membrane regions is hydropathy analysis, which uses a sliding-window approach to calculate the average hydrophobicity of an amino-acid position. By definition, hydrophobicity is the property of being water-repellent. Rose first introduced the concept of hydrophobicity analysis as a means of identifying chain turns in soluble proteins in 1978 (Rose, 1978), and in 1982, Kyte and Doolittle developed the first hydropathy scale (KD hydropathy scale, or KD scale), which is widely used by many prediction programs for evaluating the hydrophobicity of a protein along the amino acid sequence (Kyte and Doolittle, 1982). In practice, there are multiple ways to quantify the hydrophobicity of amino acids. Indeed, to date, more than 100 hydrophobicity scales have been published in the literature. These were either derived experimentally based on the free energy of transfer or empirically calculated based on surface accessibility.

The use of more complex processing of the hydrophobicity scale (and in combination with other physicochemical parameters) helped to improve the performance of membrane protein prediction. An early effort used discriminant analysis to classify membrane proteins as integral or peripheral and to estimate the odds that the classification is correct (Klein et al., 1985). TopPred (von Heijne, 1992) combines hydrophobicity analysis with the positive-inside rule and achieves better performance than using hydrophobicity alone. The Dense Alignment Surface (DAS) method optimizes the use of hydrophobicity plots by assessing sequence similarities between segments of the query protein and known transmembrane segments (Cserzo et al., 1997). For making predictions, the SOSUI method combines four physicochemical parameters: KD scale, amphiphilicity, relative and net charges, and protein length (Hirokawa et al., 1998). TMFinder combines segment hydrophobicity and the nonpolar phase helicity to predict TM segments (Deber et al., 2001).

A more general strategy is to infer the statistical preference of amino acids in membrane proteins from unknown membrane proteins (since consecutive residues have preferences for certain secondary structure states), and then to use the derived preference (instead of hydrophobility) for prediction. This strategy can be used for general secondary structure prediction for globular proteins and, upon considering different states, for membrane proteins. Methods developed following this strategy include MEMSAT, SPLIT, TMAP, and TMpred (for a review see Chen and Rost, 2002).

Many advanced methods have been developed employing statistical preferences and machine learning methods, including neural networks (NN; e.g., PHDhtm), hidden Markov models [HMM; e.g., HMMTOP (Tusnady and Simon, 1998) and TMHMM (see below)], and SVM (e.g., SVMtm—see below) for membrane protein prediction. Rost et al. (1995) developed a neural network system for predicting the locations of TM helices in integral membrane proteins using evolutionary information as input. TMHMM (Krogh et al., 2001) embeds a number of statistical preferences and rules into a hidden Markov model to optimize the prediction of the localization of TM helices and their orientation. It incorporates hydrophobicity, charge bias, helix lengths, and grammatical constraints (i.e., cytoplasmic and noncytoplasmic loops have to alternate) into one model for which algorithms for parameter estimation and prediction already exist. TMHMM achieved highly accurate performance: it correctly predicts 97–98% of the TM helices, and discriminates between soluble and membrane proteins with both specificity and sensitivity better than 99% (but the accuracy drops when signal peptides are present). This high degree of accuracy makes it possible to use this method to predict integral membrane proteins reliably from numerous genomes. Based on this prediction across a wide collection of complete genomes, an estimate has been made that 20–30% of all genes in most genomes encode membrane proteins, which is in agreement with previous estimates. A more recent method SVMtm (Yuan et al., 2004) applies support vector machines to predict transmembrane segments; various sequence coding schemes (including three different hydropathy scales and 21-UNIT) (Rost et al., 1995) were tested.

1.3.1.2 Methods for Topology Model Prediction (β-Strand Membrane Proteins)

β-Strand TMs lack a clear pattern in their membrane-spanning strands, making them different from the α-helical membrane proteins, which have hydrophobic segments and the positive-inside rule. Predictions made for TM β-strands are currently less successful than those for TM α-helices. An early method developed in 1995 used Gibbs motif sampling to detect bacterial outer membrane protein repeats; these were then used in searching for outer membrane proteins (Neuwald et al., 1995).

One of the key structural determinants of β-barrel membrane proteins is a pattern of β-barrel dyad repeats. β-Barrel proteins of known 3D structure share two physicochemical properties (i.e., hydrophobicity and amphipathicity): most of the TM strands correspond to a peak of hydrophobicity, but the hydrophobic values of these peaks are generally not as high as those of the TM α-helices of cytoplasmic integral membrane proteins. Most of the TM β-strands exhibit peaks of amphipathicity caused by the alternating hydrophilic residues located inside the barrel and the hydrophobic residues located outside the barrel. These two physicochemical properties laid the basis for many software programs aimed at β-barrel TM proteins. The β-Barrel Outer Membrane protein Predictor (BOMP) program (Berven et al., 2004) combines two independent methods for identifying the possible integral outer membrane proteins and also a filtering mechanism to remove false positives; it was designed to predict whether a protein sequence specifically from Gram-negative bacteria is an integral β-barrel outer membrane protein (80% accuracy and 88% sensitivity achieved when applied to *E. coli* K12 and *S. typhimurium*). Similar to predictions for α-helix TM proteins, statistical preferences and machine learning methods (NN in BBF, OM_Topo_predict and TMBETA-NET; HMM in BETA-TM, BIOSINO-HMM, HMM-B2TMR, PRED-TMBB, and ProfTMB) have also been introduced to improve the prediction of β-barrel TM proteins. BBF, Beta-Barrel Finder (Zhai and Saier, 2002), is a program based on physicochemical properties (both hydropathy and amphipathicity), which uses NNs to identify TM β-barrel proteins in *E. coli*. TBBPred (Natt et al., 2004) uses both NNs and SVMs for predicting TM β-barrel regions.

1.3.1.3 Molecular Dynamics Simulations of Membrane Proteins

MD simulations are widely used in studying the structures of membrane proteins such as the conformational dynamics of the receptors, the functions (such as open or closed states) of ion channels (Giorgetti and Carloni, 2003), and the receptor and ligand interactions. The simulations enable us to extrapolate from the essentially static (time- and space-averaged) structure revealed by X-ray diffraction to a more dynamic picture of the behavior of a membrane protein in a more realistic environment that mimics a small patch of the membrane. The first MD simulation of a biological process was the 1976 simulation of the primary event in rhodopsin (Warshel, 1976). MD simulations were next applied to the earliest simulations of enzymatic reactions and electron transfer reactions and then simulations of proton translocations and ion transport in proteins (see the review by Warshel, 2002). MD simulations have been employed in a number of studies on outer membrane proteins, in order, for example, to probe protein and solvent dynamics in relationship to permeation mechanisms in porins (Tieleman and Berendsen, 1998), to explore possible pore-gating mechanisms in OmpA (Bond et al., 2002), and to examine the role of calcium binding and dimerization in the catalytic mechanism of OMPLA (Baaden et al., 2003). MD simulations can also be used to assess whether any *mutation* in a protein has an effect on the structure and function of the protein before more time-consuming experiments have been performed; for example, this has been done with the computational alanine scanning of human growth hormone-receptor complex (Huo et al., 2002) and the study of TM domain mutants of Vpu from HIV-1 along with the consequences of these mutations on its structure (Candler et al., 2005).

1.3.1.4 Modeling and Simulation of GPCR (G-Protein-Coupled Receptor)

GPCRs constitute the largest family of signal transduction membrane proteins, which mediate the cellular responses to a variety of bioactive molecules, including biogenic amines, amino acids, peptides, lipids, nucleotides, and proteins. The GPCRs play a crucial role in many essential physiological processes as diverse as neurotransmission, cellular metabolism, secretion, cell growth, immune defense, and differentiation. GPCRs are also (not surprisingly) the most common targets for the drugs currently used in clinics and for the wealth of drug candidates that high-throughput methods are expected to deliver in the immediate future. Extensive computational analysis (see the review by Fanelli and DeBenedetti, 2005), which includes predicting families and subfamilies of GPCRs from sequences, 3D structure modeling, and MD simulation of the consequences of mutants, has been done for GPCRs; a dedicated database was created for GPCRs, the G-protein-coupled receptor database (GPCRDB) at http://www.gpcr.org/7tm.

1.3.1.5 Global Topology Analysis of the *E. coli* Membrane Proteome

A study that deserves special mention is the global topology analysis of the *E. coli* inner membrane proteome by Daley et al. (2005). This is the first reported large-scale prediction of membrane proteins in combination with large-scale experiments. Their work exploits the observation that topology prediction can be greatly improved by constraining it with an experimentally determined reference point, such as the location of a protein's C-terminus; an estimate is that at least ten percentage points in overall accuracy in whole-genome predictions can be gained in this way (Melen et al., 2003). Using C-terminal tagging with the alkaline phosphatase and green fluorescent protein, they determined the locations of the C-termini (either periplasmic or cytoplasmic) for 601 inner membrane proteins. Then, by constraining topology predictor TMHMM with these data, they derived high-quality topology models for these proteins; this research provides a firm foundation for future functional studies of this and other membrane proteomes.

1.3.2 Modeling of Multiple-Domain Proteins

Domain fusion/shuffling is one of the most important events in the evolution of modern proteins (Patthy, 1999; Kriventseva et al., 2003). The majority of proteins, especially in higher organisms, are built from multiple domains (modules), which can be found in various contexts in different proteins. Such domains usually form stable three-dimensional structures, even if excised from a complete protein, and perform the same or similar molecular functions as parts of the protein.

The identification of domain boundaries is critical for both experimental and computational (including *ab initio* and comparative modeling) protein structure determination. NMR spectrometry has a length limitation in solving protein structures, and X-ray diffraction requires high-quality crystals and thus may fail or may have regions lacking in detail due to flexible linker regions between domains. As a consequence, there are inevitably fewer structures deposited in PDB that can be used as structural templates for modeling multiple-domain proteins. *Ab initio* prediction methods also encounter huge difficulties in predicting large, multidomain proteins due to the exceptional computational barrier in exploring conformational space and for the determination of domain–domain interactions. Consequently, current *ab initio* structure prediction methods can only model structures of relatively small size and do so at worse resolution than obtained by homology modeling. Both experimental and computational approaches to protein structure determination would benefit significantly from predicted domain assignments.

One way of dealing with the multidomain problems is to model the structures of domains of a protein separately and then, if possible, to assemble the domains together. Any computational methods for protein structure modeling can be applied to model structures of individual domains, including comparative modeling, threading, and *ab initio* methods. Special issues in modeling multidomain proteins involve the first step of domain dissection (Contreras-Moreira and Bates, 2002) and the last step of predicting spatial arrangement of constituent domains (Inbar et al., 2003), which will be discussed in this section.

1.3.2.1 Domain Assignment of Proteins

Although the concept of domains as structural components of proteins has been around for years, ever since studies conducted by Wetlaufer (1973) and Rossman and Liljas (1974), and is now well accepted, its definition is full of ambiguities. Caution in using current domain assignments was recommended by Veretnik et al., who systematically assessed the consistency of current domain assignments by investigating six methods [three "human-expert methods"—authors' annotation, CATH, and SCOP—and three "fully-automated methods"—DALI (Holm and Sander 1994), DomainParser (Guo et al., 2003), and PDP (Alexandrov and Shindyalov, 2003)].

Their survey of the consistency of domain assignment also indicated where additional work is needed in domain assignment, including the assignment of the domain boundaries and the assignment of small domains. Nevertheless, significant advances have been made in the domain assignment of proteins (with or without structure information), which hence will be discussed here.

For a protein whose structure is known, domain assignments can usually be done manually by experts, or by automatic programs, or by a combination of both. Earlier, when there were only a few known protein structures, simple visual inspection of protein structures was quite adequate (Wetlaufer, 1973). The SCOP database is one of the widely cited databases of protein domains; the database represents largely the results of human experts, who have been assisted by computer visualization, and in particular, considered evolutionary information (Murzin et al., 1995). Fully automatic methods, i.e., those that are run without intervention and are not affected by human subjectivity in terms of consistently following criteria, are becoming ever more important in order to keep pace with the current accumulation of experimental structures of proteins. One early automatic method was developed by Wodak and Janin, who used surface area measurements based on atomic positions to give a quantitative definition of the structural domains in proteins (Wodak and Janin, 1981). The incorporation of secondary structure information (DOMAK) (Siddiqui and Barton, 1995) or information on hydrophobic cores (DETECTIVE) (Swindells, 1995) has subsequently been shown to enhance the automatic domain assignment. Another program, PUU, was based on achieving/expecting maximal interactions within each unit but minimal interaction between units (or domains) (Holm and Sander, 1994). The CATH database (Orengo et al., 1997) uses three algorithms (DETECTIVE, PUU, and DOMAK) for domain decomposition as a first step in the assignment process, followed by an expert's inspection. The VAST algorithm, which is used for structure neighboring in the Entrez system, is a fully automatic method that splits protein chains at points between secondary structure elements (SSEs) when the ratio of intra- to interdomain contacts exceeds a certain threshold (Madej et al., 1995).

In the absence of a known protein structure (as would be the case for any protein structure modeling), algorithms developed to predict domain boundaries have been based on sequence information, multiple sequence alignments, and/or homology modeling. Early approaches to domain boundary prediction relied on information theory (Busetta and Barrans, 1984) and used statistical potentials (Vonderviszt and Simon, 1986). Later prediction methods took into account the information of (predicted) secondary structure (DomSSEA), sequence conservation (Guan and Du, 1998; George and Heringa, 2002a; Rigden, 2002), or both, in order to improve domain assignment from sequences. DomSSEA (Marsden et al., 2002) uses a fold recognition approach, based on aligning predicted secondary structure of a query protein sequence to the assigned secondary structure of known structures, and then transferring the SCOP assigned domains from the best fold match to the query protein. SnapDRAGON is a suite of programs developed to predict domain boundaries based on the consistency observed in a set of alternative *ab initio* 3D models generated for a given protein multiple sequence alignment (George and Heringa, 2002b).

Many of the Open Reading Frames (ORFs) or predicted protein sequences discovered along the genome of any fully sequenced bacterial organism have been found not to be conserved across organisms; indeed, nearly half of all new ORFs appear to be unique. In these cases, algorithms that do not rely on sequence conservation have been applied to assign domains. The Domain Guess by Size (DGS) algorithm makes predictions based on observed domain size distributions (Wheelan et al., 2000). Galzitskaya and Melnik (2003) developed a method based on the assumption that the domain boundary is conditioned by amino acid residues with a small value of side chain entropy, which correlates with the side chain size. The Armadillo program (Dumontier et al., 2005) uses an amino acid index, called the domain linker propensity index (DLI) and derived from the amino acid composition of domain linkers using a nonredundant structure data set, to convert a protein sequence to a smoothed numeric profile from which domains and domain boundaries may be predicted. In general, most approaches predict the number of domains accurately, but only a few predict the domain boundaries well; prediction of domain boundaries only has a moderate sensitivity of ~50–70% for proteins with single domains, and does considerably worse (~30%) for multidomain proteins. Multidomain proteins are also harder to study experimentally. Thus, the proteins from eukaryotes are more difficult for both experimental and computational analysis.

1.3.2.2 Modeling of Domain–Domain Interactions

Considering the similarity of domain–domain interaction (folding) and protein–protein interaction (binding), docking techniques that have been developed for modeling protein–protein complexes (see Section 1.3.3) have been applied to build the model of multidomain proteins by docking separate structures of domains together. Unfortunately, few advances have been made; this is especially the case for predicting the spatial arrangement of protein domains.

Xu et al. (2001) modeled the structure of vitronectin by first modeling its C-terminal and central domains and then modeling the interaction of these two domains using GRAMM, a docking technique. In this work, the threading program PROSPECT was used to find the structure template for modeling and to generate the sequence–structure alignment, which was used as input for the program MODELLER to create the models. Experimental data were also used to guide the docking of the central and C-terminal domains by GRAMM.

Inbar et al. developed CombDock, a combinational docking algorithm, for protein structure prediction via combinatorial assembly of substructural units (building blocks/domains) (Inbar et al., 2003). Three steps are involved in this algorithm to predict the structure of a protein sequence: a dissection into fragments and the assignment of their structures; the assembly of the fragments into an overall structure of the protein sequence; and the prediction of the spatial arrangement of the assigned structures and then the completion and refinement of highly ranked predicted arrangements. The combinatorial assembly of domains is formulated as the problem of finding the spanning tree in a graph (where each substructure is a vertex, and an edge between two vertices presents the interaction of the two substructures), and a heuristic polynomial solution to this computational hard problem has been provided.

Jones et al. used a similar strategy, i.e., domain docking and microdomain folding, to model complete chains of selected CASP6 targets. Their method, called FRAGFOLD-MODEL, generates models of a complete chain by "docking" domains together by searching possible linker peptide conformations. To this end, a genetic algorithm or simulated annealing can be used for the conformational search (Jones et al., 2005).

1.3.3 Modeling of Protein Complexes

Modeling of protein complexes is far less successful than the modeling of protein monomers. At the same time, obtaining accurate models for complexes is of increasing importance because of their functional importance—in general, proteins act as part of large macromolecular assemblies. Indeed, experimental work routinely extends the known scale (number of proteins) of interactions in any given functional pathway, and the impact on our thinking about molecular processes is now expanding rapidly with the advent of improved and larger-scale identification of protein–protein interactions. A few docking programs have been developed since the late 1970s for predicting the structures of protein–protein complexes. Recent developments include the usage of computational models for docking, the combination of experimental data in computational docking, and the combination of homology modeling and cryoEM data to model large complex structures; these are discussed below.

1.3.3.1 Modeling Protein Complexes by Docking

Most docking methods consist of a global (or stochastic) search of translational and rotational space followed by refinement of the best predictions. The relative performance depends on the conformational searching ability and on the efficiency of complex evaluation. Very often, docking programs treat proteins as a "rigid body" during the first step and use a simple and "soft" energy function to evaluate the potential complex, and subsequently use more fine evaluation in the second step of refinement, during which some programs also consider the flexibility of the side chains, but few consider the flexibility of the backbone as well.

The first computational protein docking tools were developed in the late 1970s. Greer and Bush (1978) introduced a grid-based measure of complementarity between molecules, and used it to score interfaces between hemoglobin subunits. An early docking study by Wodak and Janin (1981) used a simplified protein model with one sphere per amino acid, which they used to dock BPTI to trypsin. The search involved rotating BPTI and varying its center-of-mass distance with trypsin. Newer programs use more complex shape-complementarity and an energy function to evaluate the complex models, and use a more rigorous definition of conformational space to improve the docking performance. A significant improvement in the conformational space search has been the use of the fast Fourier transform (FFT) to perform correlations in grid-based translational searching (Katchalski-Katzir et al., 1992). FFT is employed by several commonly used docking programs, including GRAMM, FTDock, and ZDOCK. These programs use the same strategy for the conformational search (FFT), but may use different scoring functions and do use different details for overall operation: FTDock's scoring of complexes is based on shape complementarity and on favorable electrostatic interactions (Gabb et al., 1997); GRAMM (Vakser, 1995) implements docking at different resolutions to account for the inaccuracy of input structures; ZDOCK (Chen et al., 2003) combines pairwise shape complementarity with desolvation and electrostatics for complex scoring. Other techniques including geometric hashing (Fischer et al., 1995), stochastic searches such as Monte Carlo search [e.g., RossetaDock (Gray et al., 2003)], or a genetic algorithm (e.g., Gardiner et al., 2001) have also been used for the conformational search step in docking.

Most of the existing docking programs adopt the "rigid-body" strategy while neglecting the conformational changes during binding. For complexes of an enzyme with its inhibitor, the conformational changes might be small and can be compensated by using some "soft" scoring to evaluate the potential complex, which is a key consideration of evaluating the potential complex for unbound docking. (Bound docking uses the structures from the complex structure as input, which obviously has little predictive use, whereas unbound docking uses the structures from the individually crystallized subunits as input.) However, for other types of complexes, the conformational change may be greater. Due to the huge conformational space for protein structures, "flexible" protein–protein docking remains a challenge, despite the advances in incorporating the flexibility of receptors in protein–ligand docking (Jones et al., 1997; Alberts et al., 2005).

1.3.3.2 Data-Driven Docking Approaches

Applying proper constraints to the conformational space during docking can significantly improve the computation speed and the accuracy of docking (van Dijk et al., 2005). The constraints can be derived via many methods, both experimentally and computationally. NMR data have been used in combination with docking methods and in different ways in order to generate information about protein–protein complexes. For example, diamagnetic chemical shift changes and intermolecular pseudocontact shifts were combined with restrained rigid-body molecular dynamics to solve the structure of the paramagnetic plastocyanin–cytochrome f complex (Ubbink et al., 1998). Intermolecular NOEs and residual dipolar couplings (RDCs) were combined to solve the structure of the EIN–HPr complex (Clore, 2000). TreeDock (Fahmy and Wagner, 2002) enumerates the search space at a user-defined resolution subject to the condition that a pair of atoms, one from each molecule, are always in contact, which can be in principle derived from NMR chemical shift perturbation or mutagenesis data. Dominguez et al. (2003) developed an approach called HADDOCK (High Ambiguity Driven protein–protein Docking), which makes use of biochemical and/or biophysical interaction data such as chemical shift perturbation data resulting from NMR titration experiments or mutagenesis data. The data are transformed as Ambiguous Interaction Restraints (AIRs) between all residues shown to be involved in the interaction to drive the docking process.

1.3.3.3 Integrating Homology Modeling and EM Density Map for Modeling Protein Assemblies

With the advances in functional genomics, experimental methods allow us to determine on a large scale what the partners are for pairwise protein–protein interactions (by yeast two-hybrid system and protein chips) and what the constituents are among large protein assemblies [by tandem-affinity purification (TAP) and mass spectrometry]. Accordingly, the need to model protein–protein interactions and protein assemblies is increasing. The integration of high-resolution structures/models and the electron microscopy (EM) density map is an exciting advance for modeling a large protein–protein complex and assemblies. The basic idea is to fit known high-resolution structures into low-resolution structures of large complexes that are determined by EM to obtain the refined structure of large complexes. This technique has been applied in solving the structures of large biological machines/macromolecular complexes, such as viruses, ion channels, ribosomes, and proteasomes. In cases where no experimental high-resolution structures are available, computational models of the individual proteins may instead be used in fitting. In addition, intermediate-resolution cryo-EM density maps are helpful for improving the accuracy of comparative protein structure modeling in those cases for which no template for modeling can be found by a sequence-based search or a threading method. In fact, the application of EM density maps in structure modeling started quite early; for example, a model for the structure of bacteriorhodopsin was originally generated based on high-resolution electron cryomicroscopy (Henderson et al., 1990). An explosion of joint EM/crystallographic studies in the mid-1990s followed the development of strategies for generating pseudo-atomic-resolution models of macromolecular complexes by combining the data from high-resolution structures of components with lower-resolution EM data for the entire complex (Baker and Johnson, 1996).

Electron cryomicroscopy (cryo-EM) can image complexes in their physiological environment and does not require large quantities of the sample. Cryo-EM also provides a means of visualizing the membrane proteins *in situ*, as opposed to the usually artificial hydrophobic environments used for crystallizing membrane proteins. Structures of large macromolecular complexes can now be visualized in different functional states at intermediate resolution (6–9 Å) (Chiu et al., 2005). The corresponding cryo-EM maps are generally still insufficient for atomic structure determination on their own. However, one can fit atomic-resolution structures of the components of the assembly (e.g., protein domains, whole proteins, and any subcomplexes) into the lower-resolution density of the entire assembly. In early applications, researchers employed mainly “visual docking” to position the protein components in the envelopes derived from low-resolution data (Schroder et al., 1993). More recently, computational programs were developed to obtain quantitative means to fit the data. Wriggers et al. used topology-representing neural networks (TNN) to vector-quantize and to correlate features within the structural data sets to generate pseudo-atomic structures of large-scale protein assemblies by combining high-resolution data with volumetric data at lower resolution (Wriggers et al., 1998, 1999). Roseman et al. developed a fitting procedure that uses a real-space density-matching procedure based on local correlation of the density derived from the atomic coordinates of protein components and the density of the EM map (Roseman, 2000). Ceulemans and Russell developed 3SOM (Ceulemans and Russell, 2004) for finding the best fit through surface overlap maximization.

In cases where experimentally determined atomic-resolution structures of assembly components are not available, or the induced fit severely limits their usefulness in the reconstruction of the complex, it may be possible to get useful models of the components by comparative protein structure modeling (or homology modeling) (see the review by Topf and Sali, 2005). The number of models that can be constructed with useful accuracy, at least comparable to the resolution of the cryo-EM maps, is almost two orders of magnitude greater than the number of available experimentally determined structures, which indicates the huge potential for employing models in fitting EM maps (Topf and Sali, 2005).

Moreover, EM maps can be used to improve modeling in some cases (Topf and Sali, 2005). In cases where a structural homologue of the target component cannot be detected by sequence-based or threading search methods, it is possible to use the EM map (if the resolution is better than ~12 Å) for fold assignment of the constituting proteins: at ~12 Å resolution, it is usually possible to recognize boundaries between the individual components in the complex, while secondary structure features, such as long α-helices and large β-sheets, can begin to be identified at ~10 Å resolution, and short helices and individual strands at ~4 Å (Chiu et al., 2002). For example, Jiang et al. (2001) developed the Helixhunter program, which is capable of reliably identifying helix position, orientation, and length using a five-dimensional cross-correlation search of a three-dimensional density map followed by feature extraction; its results can in turn be used to probe a library of secondary structure elements derived from the structures in the PDB. This readily provides for the structure-based recognition of folds containing α-helices. They also developed the Foldhunter program, which uses a six-dimensional cross-correlation search that allows a probe structure to be fitted within a region or component of a target structure. These two methods have been successfully tested with simulated structures modeled from the PDB at resolutions of 6–12 Å. In cases where the fold of the protein component is known, the density maps can be useful in selecting the best template structures for comparative modeling, since a more accurate model fits the EM density map more tightly (Topf et al., 2005). Topf et al. (2005) developed a method for finding an optimal atomic model of a given assembly subunit and its position within an assembly by fitting alternative comparative models (created by MODELLER from different sequence alignments between the modeled protein and template structures) into a cryo EM map, using Foldhunter (Jiang et al., 2001) or Mod-EM (a density fitting module of MODELLER).

1.3.4 Large-Scale Modeling

In the era of many fully sequencing genomes and a focus on systemwide, integrated biological research from proteomics to metabolomics, the introduction of an "omics" for structural biology, that is, structural genomics, was a natural development, and one that reflected the maturity of structural biology as well as the need to obtain structures in order to annotate genomes fully and obtain explicit insight into the information implicit in genome sequences. The most often stated goal of structural genomics is to provide "structural coverage" of protein space by solving enough structures that all known proteins could be accurately modeled (Brenner, 2001; Vitkup et al., 2001). In the United States, the efforts of structural genomics groups also resulted in the launch of the NIH-funded Protein Structure Initiative, which currently supports four large structural genomics centers, as well as a larger number of smaller, technology-focused or specialized centers.

The success of structural genomics will, by definition, rely on both experimental structure determination and computational approaches. A question therefore raised is to ascertain to what extent have the high throughput and comprehensive aspects of genomics and the pipelines for structure determination reached efforts on computational structure prediction. Threading and comparative modeling methods have already been applied on a genomic scale. For example, ModPipe was developed for modeling known protein sequences using the comparative modeling program MODELLER on a larger scale; the models are deposited in a comprehensive database of comparative models, ModBase (Sanchez et al., 2000) (as of July 05, 2005, the database had 3,094,524 reliable models or fold assignments for domains in 1,094,750 proteins). The Web interface to the database allows flexible querying for the models, the fold assignments, sequence–structure alignments, and assessments of models of interest. Automation and large-scale modeling with *de novo* methods lag behind those of comparative modeling methods, because of the relatively poor quality of the models produced, and the relatively large amount of computer time required. Nevertheless, Rosetta initiated the successful use of large-scale modeling calculations done with ab initio methods (Baker and Sali, 2001).

1.3.4.1 Structure Modeling for Structural Genomics

It is clear that the eventual success of structural genomics will be brought about by the growing synergy between experimental structure determination and computational approaches, including the comparative modeling and *ab initio* fold prediction methods (see the review by Friedberg et al., 2004). The efficiency of using comparative modeling will be determined by the advances of distant homology detection and fold recognition algorithms, while the efficiency of using *ab initio* methods will be largely determined by the improvement of the quality of models and the reduction in computing time.

Predictions done through comparative modeling and *ab initio* methods can compensate each other and thus play a particularly important role for structural genomics; namely, target selection and modeling of the structure of proteins that are not selected for experimental determination (see the review by Baker and Sali, 2001). Of course, even with the worldwide initiatives in high-throughput structural determination, the structures for the vast majority of the proteins in nature will (at most) only be modeled and will never be determined by experiment.

Structural genomics as conducted to date generally omits several groups of proteins since they are considered to be very difficult targets; these largely excluded proteins include the membrane proteins (despite some focused attention on membrane proteins), and those with disordered structures that may fold only in the presence of appropriate interaction partners (Bracken et al., 2004). These "special" proteins, nevertheless, constitute a large portion of the whole proteome, for example membrane proteins constitute 20–30% of the proteome (Krogh et al., 2001). Achieving large-scale experimental and/or computational structural determination of these proteins would be as important as for any other proteins, and certainly as important as for those proteins already within the structural genomics scope.

1.3.4.2 Large-Scale Modeling of Human (Disease-Related) Proteins

Disease-related proteins are of great research interest for both experimental and computational scientists. Their high value for research in biomedicine and clinical medicine, and potentially in health care, stems from the fact that they provide a molecular picture of disease processes, which is a necessary prerequisite for rational drug development. Thousands of genes (proteins) have already been identified as associated with various diseases in humans. Computational modeling will play a more important role in predicting the structure of eukaryotic proteins than that of prokaryotic proteins, since eukaryotic proteins are more difficult to carry through a crystallography pipeline, and fewer, as a consequence, are likely ever to be determined experimentally. Several efforts have been carried out in order to model the structures of human proteins. For example, generated models are extensively used for studying the human disease proteins in association with SNP data. LS-SNP is a resource providing large-scale annotation of coding nonsynonymous SNPs (nsSNPs) based on multiple information sources (including structural models) in human proteins (Karchin et al., 2005). Yip et al. created the Swiss-Prot variant page and the ModSNP database for sequence and structure information on human protein variants (Yip et al., 2004). Ye et al. specifically created models of all human disease-related proteins collected in Swiss-Prot and studied the spatial distribution of disease-related nsSNPs on the models (Ye et al., 2006). These analyses provided some explanation for nsSNPs with known effects (harmful or neutral), and might in turn provide a basis for predicting the effects of nsSNPs.

1.3.4.3 Genome-Scale Modeling of Complexes

Proteins function via interactions with other macromolecules, and most cellular processes are carried out by multiprotein complexes. The identification and analysis of the components of these complexes provides insight into how the ensemble of expressed proteins (the proteome) is organized into functional units. Large-scale identifications of protein–protein interactions in many genomes are now possible due to the genome-scale discovery approaches for identifying interacting proteins; these methods include the yeast two-hybrid system and protein chips, which have been very widely employed. Using an approach with more potential for quantitative information, Gavin et al. (2002) used tandem-affinity purification (TAP) and mass spectrometry in a large-scale approach to characterize multiprotein complexes in *S. cerevisiae*.

Employing biophysical and computational methods for studying protein–protein interactions and complexes from a structural perspective would be similarly important. A significant step toward understanding how proteins assemble has been taken by Aloy et al. (2004). Starting from the large set of identified complexes of yeast by TAP (Gavin et al. 2002), they screened the complexes using low-resolution EM images. These images were used to assemble and validate models (see the "Integrating Homology Modeling and EM Density Map for Modeling Protein Assemblies" section). They also predicted links between complexes and provide a higher-order, structure-based network of connected molecular machines within the cell. The network they derived currently gives the most complete view available for complexes and their interrelationships.

1.4 Summary

From understanding single molecules, to a simple complex, to large assemblies to the biological networks, we are moving toward an understanding of life. Structure information (derived experimentally or computationally) helps us to understand the mechanisms by which the biochemical processes of cells occur and provides insight beyond chemical architecture—mechanism implications, for example, in suggesting features about evolution of function. In turn, structure prediction has made an increasing number of contributions to our understanding of biology [which has been described elsewhere both in detail and eloquently (Petrey and Honig, 2005)]. Advances have been achieved in computational predictions of structure at each level, and each advance brings new potential to impact our understanding of biology. Yet, challenges remain. We can expect that the computational challenges will be more daunting at a network level, characterizing the metabolic pathways, signal transduction cascades, and genetic circuits through which protein interactions determine cellular and organismic function; existing methods need improvement or new methods need to be developed that must deal with individual proteins, complexes, and sophisticated dynamic networks that connect them. The remainder of this book deals with contemporary efforts toward those advances. The structure-based network derived by Aloy et al. provides a useful initial framework for further studies. "Its beauty is that the whole is greater than the sum of its parts: Each new structure can help to understand multiple interactions. The complex predictions and the associated network will thus improve exponentially as the numbers of structures and interactions increase, providing an ever more complete molecular anatomy of the cell" (Aloy et al. 2004)

References

Alberts, I. L., Todorov, N. P. and Dean, P. M. 2005. Receptor flexibility in de novo ligand design and docking. *J. Med. Chem.* 48:6585–6596.

Alexandrov, N. N., Nussinov, R., and Zimmer, R. M. 1996. Fast protein fold recognition via sequence to structure alignment and contact capacity potentials. *Pac. Symp. Biocomput.* pp. 53–72.

Alexandrov, N., and Shindyalov, I. 2003. PDP: Protein domain parser. *Bioinformatics* 19:429–430.

Aloy, P., Bottcher, B., Ceulemans, H., Leutwein, C., Mellwig, C., Fischer, S., Gavin, A.-C., Bork, P., Superti-Furga, G., Serrano, L., and Russell, R. B. 2004. Structure-based assembly of protein complexes in yeast. *Science* 303:2026–2029.

Anfinsen, C. B. 1973. Principles that govern the folding of protein chains. *Science* 181:223–230.

Anfinsen, C. B., Haber, E., Sela, M., and White, F. H. J. 1961. The kinetics of formation of native ribonuclease during oxidation of the reduced polypeptide chain. *Proc. Natl. Acad. Sci., USA* 47:1309–1314.

Anfinsen, C. B., Redfield, R. R., Choate, W. I., Page, J., and Carroll, W. R. 1954. Studies on the gross structure, cross-linkages, and terminal sequences in ribonuclease. *J. Biol. Chem.* 207:201–210.

Baaden, M., Meier, C., and Sansom, M. S. P. 2003. A molecular dynamics investigation of mono and dimeric states of the outer membrane enzyme OMPLA. *J. Mol. Biol.* 331:177–189.

Baker, D., and Sali, A. 2001. Protein structure prediction and structural genomics. *Science* 294:93–96.

Baker, T. S., and Johnson, J. E. 1996. Low resolution meets high: Towards a resolution continuum from cells to atoms. *Curr. Opin. Struct. Biol.* 6:585–594.

Berven, F. S., Flikka, K., Jensen, H. B., and Eidhammer, I. 2004. BOMP: A program to predict integral β-barrel outer membrane proteins encoded within genomes of Gram-negative bacteria. *Nucleic Acids Res.* 32(Web Server Issue):W394–W399.

Bond, P. J., Faraldo-Gomez, J. D., and Sansom, M. S. P. 2002. OmpA: A pore or not a pore? Simulation and modeling studies. *Biophys. J.* 83:763–775.

Bowie, J. U., Luthy, R., and Eisenberg, D. 1991. A method to identify protein sequences that fold into a known three-dimensional structure. *Science* 253:164–170.

Bracken, C., Iakoucheva, L. M., Romero, P. R., and Dunker, A. K. 2004. Combining prediction, computation and experiment for the characterization of protein disorder. *Curr. Opin. Struct. Biol.* 14:570–576.

Brenner, S. E. 2001. A tour of structural genomics. *Nature Rev. Genet.* 2:801–809.

Brooks, B. R., Bruccoleri, R. E., Olafson, B. D., States, D. J., Swaminathan, S., and Karplus, M. 1983. CHARMM: A program for macromolecular energy, minimization, and dynamics calculations. *J. Comp. Chem.* 4:187–217.

Browne, W. J., North, A. C., Phillips, D. C., Brew, K., Vanaman, T. C., and Hill, R. L. 1969. A possible three-dimensional structure of bovine alpha-lactalbumin based on that of hen's egg-white lysozyme. *J. Mol. Biol.* 42:65–86.

Bryant, S. H., and Lawrence, C. E. 1993. An empirical energy function for threading protein sequence through the folding motif. *Proteins* 16:92–112.

Bujnicki, J. M., Elofsson, A., Fischer, D., and Rychlewski, L. 2001. Structure prediction meta server. *Bioinformatics* 17:750–751.

Busetta, B., and Barrans, Y. 1984. The prediction of protein domains. *Biochim. Biophys. Acta Protein Struct. Mol. Enzymol.* 790:117–124.

Bystroff, C., and Baker, D. 1998. Prediction of local structure in proteins using a library of sequence-structure motifs. *J. Mol. Biol.* 281:565–577.

Candler, A., Featherstone, M., Ali, R., Maloney, L., Watts, A., and Fischer, W. B. 2005. Computational analysis of mutations in the transmembrane region of Vpu from HIV-1. *Biochim. Biophys. Acta Biomembranes* 1716:1–10.

Canutescu, A. A., Shelenkov, A. A., and Dunbrack, R. L., Jr. 2003. A graph-theory algorithm for rapid protein side-chain prediction. *Protein Sci.* 12:2001–2014.

Casari, G., and Sippl, M. J. 1992. Structure-derived hydrophobic potential: Hydrophobic potential derived from X-ray structures of globular proteins is able to identify native folds. *J. Mol. Biol.* 224:725–732.

Ceulemans, H., and Russell, R. B. 2004. Fast fitting of atomic structures to low-resolution electron density maps by surface overlap maximization. *J. Mol. Biol.* 338:783–793.

Chen, C. P., and Rost, B. 2002. State-of-the-art in membrane protein prediction. *Appl. Bioinformatics* 1:21–35.

Chen, R., Li, L., and Weng, Z. 2003. ZDOCK: An initial-stage protein-docking algorithm. *Proteins* 52:80–87.

Chiu, W., Baker, M. L., Jiang, W., Dougherty, M., and Schmid, M. F. 2005. Electron cryomicroscopy of biological machines at subnanometer resolution. *Structure* 13:363–372.

Chiu, W., Baker, M. L., Jiang, W., and Zhou, Z. H. 2002. Deriving folds of macromolecular complexes through electron cryomicroscopy and bioinformatics approaches. *Curr. Opin. Struct. Biol.* 12:263–269.

Clore, G. M. 2000. Accurate and rapid docking of protein–protein complexes on the basis of intermolecular nuclear Overhauser enhancement data and dipolar couplings by rigid body minimization. *Proc. Natl. Acad. Sci. USA* 97:9021–9025.

Contreras-Moreira, B., and Bates, P. A. 2002. Domain Fishing: A first step in protein comparative modelling. *Bioinformatics* 18:1141–1142.

Cserzo, M., Wallin, E., Simon, I., von Heijne, G., and Elofsson, A. 1997. Prediction of transmembrane alpha-helices in prokaryotic membrane proteins: The dense alignment surface method. *Protein Eng.* 10:673–676.

Daley, D. O., Rapp, M., Granseth, E., Melen, K., Drew, D., and von Heijne, G. 2005. Global topology analysis of the *Escherichia coli* inner membrane proteome. *Science* 308:1321–1323.

Daniel, F. 2003. 3D-SHOTGUN: A novel, cooperative, fold-recognition meta-predictor. *Proteins Struct. Funct. Genet.* 51:434–441.

Deane, C. M., and Blundell, T. L. 2001. CODA: A combined algorithm for predicting the structurally variable regions of protein models. *Protein Sci.* 10:599–612.

Deber, C. M., Wang, C., Liu, L.-P., Prior, A. S., Agrawal, S., Muskat, B. L., and Cuticchia, A. J. 2001. TM Finder: A prediction program for transmembrane protein segments using a combination of hydrophobicity and nonpolar phase helicity scales. *Protein Sci.* 10:212–219.

Desmet, J., Maeyer, M. D., Hazes, B., and Lasters, I. 1992. The dead-end elimination theorem and its use in protein side-chain positioning. *Nature* 356:539–542.

Dill, K. A., Fiebig, K. M., and Chan, H. S. 1993. Cooperativity in protein-folding kinetics. *Proc. Natl. Acad. Sci. USA* 90:1942–1946.

Dominguez, C., Boelens, R., and Bonvin, A. M. J. J. 2003. HADDOCK: A protein–protein docking approach based on biochemical or biophysical information. *J. Am. Chem. Soc.* 125:1731–1737.

Donate, L. E., Rufino, S. D., Canard, L. H., and Blundell, T. L. 1996. Conformational analysis and clustering of short and medium size loops connecting regular secondary structures: A database for modeling and prediction. *Protein Sci.* 5:2600–2616.

Duan, Y., and Kollman, P. A. 1998. Pathways to a protein folding intermediate observed in a 1-microsecond simulation in aqueous solution. *Science* 282:740–744.

Dumontier, M., Yao, R., Feldman, H. J., and Hogue, C. W. V. 2005. Armadillo: Domain boundary prediction by amino acid composition. *J. Mol. Biol.* 350:1061–1073.

Dunbrack, J. R. L., and Karplus, M. 1993. Backbone-dependent rotamer library for proteins application to side-chain prediction. *J. Mol. Biol.* 230:543–574.

Edgar, R. C., and Sjolander, K. 2004. COACH: Profile–profile alignment of protein families using hidden Markov models. *Bioinformatics* 20:1309–1318.

Fahmy, A., and Wagner, G. 2002. TreeDock: A tool for protein docking based on minimizing van der Waals energies. *J. Am. Chem. Soc.* 124:1241–1250.

Fanelli, F., and DeBenedetti, P. G. 2005. Computational modeling approaches to structure–function analysis of G protein-coupled receptors. *Chem. Rev.* 105:3297–3351.

Fischer, D., Lin, S. L., Wolfson, H. L., and Nussinov, R. 1995. A geometry-based suite of molecular docking processes. *J. Mol. Biol.* 248:459–477.

Friedberg, I., Jaroszewski, L., Ye, Y., and Godzik, A. 2004. The interplay of fold recognition and experimental structure determination in structural genomics. *Curr. Opin. Struct. Biol.* 14:307–312.

Gabb, H. A., Jackson, R. M., and Sternberg, M. J. E. 1997. Modelling protein docking using shape complementarity, electrostatics and biochemical information. *J. Mol. Biol.* 272:106–120.

Galzitskaya, O. V., and Melnik, B. S. 2003. Prediction of protein domain boundaries from sequence alone. *Protein Sci.* 12:696–701.

Gardiner, E. J., Willett, P., and Artymiuk, P. J. (2001). Protein docking using a genetic algorithm. *Proteins Struct. Funct. Genet.* 44:44–56.

Gavin, A.-C., Bosche, M., Krause, R., Grandi, P., Marzioch, M., Bauer, A., Schultz, J., Rick, J. M., Michon, A.-M., Cruciat, C.-M., Remor, M., Hofert, C., Schelder, M., Brajenovic, M., Ruffner, H., Merino, A., Klein, K., Hudak, M., Dickson, D., Rudi, T., Gnau, V., Bauch, A., Bastuck, S., Huhse, B., Leutwein, C., Heurtier, M.-A., Copley, R. R., Edelmann, A., Querfurth, E., Rybin, V., Drewes, G., Raida, M., Bouwmeester, T., Bork, P., Seraphin, B., Kuster, B., Neubauer, G. and Superti-Furga, G. 2002. Functional organization of the yeast proteome by systematic analysis of protein complexes. *Nature* 415:141–147.

George, R. A., and Heringa, J. 2002a. Protein domain identification and improved sequence similarity searching using PSI-BLAST. *Proteins* 48:672–681.

George, R. A., and Heringa, J. 2002b. SnapDRAGON: A method to delineate protein structural domains from sequence data. *J. Mol. Biol.* 316:839–851.

Gibson, K. D., and Scheraga, H. A. 1967a. Minimization of polypeptide energy, I. Preliminary structures of bovine pancreatic ribonuclease S-peptide. *Proc. Natl. Acad. Sci. USA* 58:420–427.

Gibson, K. D., and Scheraga, H. A. 1967b. Minimization of polypeptide energy. II. Preliminary structures of oxytocin, vasopressin, and an octapeptide from ribonuclease. *Proc. Natl. Acad. Sci. USA* 58:1317–1323.

Ginalski, K., Elofsson, A., Fischer, D., and Rychlewski, L. 2003. 3D-Jury: A simple approach to improve protein structure predictions. *Bioinformatics* 19:1015–1018.

Giorgetti, A., and Carloni, P. 2003. Molecular modeling of ion channels: Structural predictions. *Curr. Opin. Chem. Biol.* 7:150–156.

Gray, J. J., Moughon, S., Wang, C., Schueler-Furman, O., Kuhlman, B., Rohl, C. A., and Baker, D. 2003. Protein–protein docking with simultaneous optimization of rigid-body displacement and side-chain conformations. *J. Mol. Biol.* 331:281–299.

Greer, J. 1981. Comparative model-building of the mammalian serine proteases. *J. Mol. Biol.* 153:1027–1042.

Greer, J., and Bush, B. L. 1978. Macromolecular shape and surface maps by solvent exclusion. *Proc. Natl. Acad. Sci. USA* 75:303–307.

Guan, X., and Du, L. 1998. Domain identification by clustering sequence alignments. *Bioinformatics* 14:783–788.

Guo, J.-T., Ellrott, K., Chung, W. J., Xu, D., Passovets, S., and Xu, Y. 2004. PROSPECT-PSPP: An automatic computational pipeline for protein structure prediction. *Nucleic Acids Res.* 32(Suppl. 2):W522–525.

Guo, J. T., Xu, D., Kim, D., and Xu, Y. 2003. Improving the performance of DomainParser for structural domain partition using neural network. *Nucleic Acids Res.* 31:944–952.

Heijne, V. 1986. The distribution of positively charged residues in bacterial inner membrane proteins correlates with the trans-membrane topology. *EMBO J.* 5:3021–3027.

Henderson, R., Baldwin, J. M., Ceska, T. A., Zemlin, F., Beckmann, E., and Downing, K. H. 1990. Model for the structure of bacteriorhodopsin based on high-resolution electron cryo-microscopy. *J. Mol. Biol.* 213:899–929.

Hirokawa, T., Boon-Chieng, S., and Mitaku, S. 1998. SOSUI: Classification and secondary structure prediction system for membrane proteins. *Bioinformatics* 14:378–379.

Holm, L., and Sander, C. 1994. Parser for protein folding units. *Proteins* 19:256–268.

Huo, S., Massova, I., and Kollman, P. A. 2002. Computational alanine scanning of the 1:1 human growth hormone–receptor complex. *J. Comp. Chem.* 23:15–27.

Inbar, Y., Benyamini, H., Nussinov, R., and Wolfson, H. J. 2003. Protein structure prediction via combinatorial assembly of sub-structural units. *Bioinformatics* 19(Suppl. 1):i158–i168.

Jiang, W., Baker, M. L., Ludtke, S. J., and Chiu, W. 2001. Bridging the information gap: Computational tools for intermediate resolution structure interpretation. *J. Mol. Biol.* 308:1033–1044.

Jones, D. T. 1999. GenTHREADER: An efficient and reliable protein fold recognition method for genomic sequences. *J. Mol. Biol.* 287:797–815.

Jones, D. T., Bryson, K., Coleman, A., McGuffin, L. J., Sadowski, M. I., Sodhi, J. S., and Ward, J. J. 2005. Prediction of novel and analogous folds using fragment assembly and fold recognition. *Proteins* 61(Suppl. 7):143–151.

Jones, G., Willett, P., Glen, R. C., Leach, A. R., and Taylor, R. 1997. Development and validation of a genetic algorithm for flexible docking. *J. Mol. Biol.* 267:727–748.

Karchin, R., Diekhans, M., Kelly, L., Thomas, D. J., Pieper, U., Eswar, N., Haussler, D., and Sali, A. 2005. LS-SNP: Large-scale annotation of coding non-synonymous SNPs based on multiple information sources. *Bioinformatics* 21:2814–2820.

Karplus, K., Barrett, C., and Hughey, R. 1998. Hidden Markov models for detecting remote protein homologies. *Bioinformatics* 14:846–856.

Katchalski-Katzir, E., Shariv, I., Eisenstein, M., Friesem, A. A., Aflalo, C., and Vakser, I. A. 1992. Molecular surface recognition: Determination of geometric fit between proteins and their ligands by correlation techniques. *Proc. Natl. Acad. Sci. USA* 89:2195–2199.

Kelley, L. A., MacCallum, R. M., and Sternberg, M. J. E. 2000. Enhanced genome annotation using structural profiles in the program 3D-PSSM. *J. Mol. Biol.* 299:501–522.

Kihara, D., Lu, H., Kolinski, A., and Skolnick, J. 2001. TOUCHSTONE: An ab initio protein structure prediction method that uses threading-based tertiary restraints. *Proc. Natl. Acad. Sci. USA* 98:10125–10130.

Kim, D. E., Chivian, D., and Baker, D. 2004. Protein structure prediction and analysis using the Robetta server. *Nucleic Acids Res.* 32(Suppl. 2):W526–531.

Klein, P., Kanehisa, M., and DeLisi, C. 1985. The detection and classification of membrane-spanning proteins. *Biochim. Biophys. Acta Prot. Struct. Mol. Enzymol.* 815:468–476.

Koehl, P., and Delarue, M. 1995. A self consistent mean field approach to simultaneous gap closure and side-chain positioning in homology modelling. *Nat. Struct. Biol.* 2:163–170.

Koh, I. Y. Y., Eyrich, V. A., Marti-Renom, M. A., Przybylski, D., Madhusudhan, M. S., Eswar, N., Grana, O., Pazos, F., Valencia, A., Sali, A., and Rost, B. 2003. EVA: Evaluation of protein structure prediction servers. *Nucleic Acids Res.* 31:3311–3315.

Kolinski, A., and Skolnick, J. 1994a. Monte Carlo simulation of protein folding. II. Application to protein A, ROP, and crambin. *Proteins* 18:353–366.

Kolinski, A., and Skolnick, J. 1994b. Monte Carlo simulations of protein folding. I. Lattice model and interaction scheme. *Proteins* 18:338–352.

Kolinski, A., and Skolnick, J. 1998. Assembly of protein structure from sparse experimental data: An efficient Monte Carlo model. *Proteins* 32:475–494.

Kolinski, A., and Skolnick, J. 2004. Reduced models of proteins and their applications. *Polymer* 45:511–524.

Kosinski, J., Cymerman, I. A., Feder, M., Kurowski, M. A., Sasin, J. M., and Bujnicki, J. M. 2003. A "FRankenstein's monster" approach to comparative modeling: Merging the finest fragments of Fold-Recognition models and iterative model refinement aided by 3D structure evaluation. *Proteins* 53(S6):369–379.

Kriventseva, E. V., Koch, I., Apweiler, R., Vingron, M., Bork, P., Gelfand, M. S., and Sunyaev, S. 2003. Increase of functional diversity by alternative splicing. *Trends Genet.* 19:124–128.

Krogh, A., Brown, M., Mian, I. S., Sjolander, K., and Haussler, D. 1994. Hidden Markov models in computational biology: Applications to protein modeling. *J. Mol. Biol.* 235:1501–1531.

Krogh, A., Larsson, B., von Heijne, G., and Sonnhammer, E. L. L. 2001. Predicting transmembrane protein topology with a hidden Markov model: Application to complete genomes. *J. Mol. Biol.* 305:567–580.

Kurowski, M. A., and Bujnicki, J. M. 2003. GeneSilico protein structure prediction meta-server. *Nucleic Acids Res.* 31:3305–3307.

Kyte, J., and Doolittle, R. F. 1982. A simple method for displaying the hydropathic character of a protein. *J. Mol. Biol.* 157:105–132.

Lau, K. F., and Dill, K. A. 1989. A lattice statistical mechanics model of the conformational and sequence spaces of proteins. *Macromolecules* 22:3986–3997.

Lee, C. 1994. Predicting protein mutant energetics by self-consistent ensemble optimization. *J. Mol. Biol.* 236:918–939.

Lee, C. 1995. Testing homology modeling on mutant proteins: Predicting structural and thermodynamic effects in the Ala98→Val mutants of T4 lysozyme. *Fold Des.* 1:1–12.

Lee, J., Kim, S.-Y., and Lee, J. 2005. Protein structure prediction based on fragment assembly and parameter optimization. *Biophys. Chem.* 115:209–214.

Levinthal, C. 1968. Are there pathways for protein folding? *J. Chem. Phys.* 65: 44–45.

Levitt, M., and Lifson, S. 1969. Refinement of protein conformations using a macromolecular energy minimization procedure. *J. Mol. Biol.* 46:269–279.

Levitt, M., and Warshel, A. 1975. Computer simulation of protein folding. *Nature* 253:694–698.

Liwo, A., Lee, J., Ripoll, D. R., Pillardy, J., and Scheraga, H. A. 1999. Protein structure prediction by global optimization of a potential energy function. *Proc. Natl. Acad. Sci. USA* 96:5482–5485.

Lundstrom, J., Rychlewski, L., Bujnicki, J., and Elofsson, A. 2001. Pcons: A neural-network-based consensus predictor that improves fold recognition. *Protein Sci.* 10:2354–2362.

Luthy, R., Bowie, J. U., and Eisenberg, D. 1992. Assessment of protein models with three-dimensional profiles. *Nature* 356:83–85.

Madej, T., Gibrati, J.F., and S.H. Bryant 1995 'Threading a database of protein cores.' *Proteins* 32:289–306.

Marsden, R. L., McGuffin, L. J., and Jones, D. T. 2002. Rapid protein domain assignment from amino acid sequence using predicted secondary structure. *Protein Sci.* 11:2814–2824.

Marti-Renom, M. A., Madhusudhan, M. S., and Sali, A. 2004. Alignment of protein sequences by their profiles. *Protein Sci.* 13:1071–1087.

Melen, K., Krogh, A., and von Heijne, G. 2003. Reliability measures for membrane protein topology prediction algorithms. *J. Mol. Biol.* 327:735–744.

Mintseris, J., Wiehe, K., Pierce, B., Anderson, R., Chen, R., Janin, J., and Weng, Z. 2005. Protein–protein docking benchmark 2.0: An update. *Proteins* 60:214–216.

Misura, K. M. S., and Baker, D. 2005. Progress and challenges in high-resolution refinement of protein structure models. *Proteins* 59:15–29.

Moult, J. 2005. A decade of CASP: Progress, bottlenecks and prognosis in protein structure prediction. *Curr. Opin. Struct. Biol.* 15:285–289.

Moult, J., Fidelis, K., Tramontano, A., Rost, B., and Hubbard, T. 2005. Critical assessment of methods of protein structure prediction (CASP)—Round VI. *Proteins* 61(S7):3–7.

Moult, J., Hubbard, T., Fidelis, K., and Pedersen, J. T. 1999. Critical assessment of methods of protein structure prediction (CASP): Round III. *Proteins*(Suppl. 3):2–6.

Moult, J., and James, M. N. G. 1986. An algorithm for determining the conformation of polypeptide segments in proteins by systematic search. *Proteins* 1:146–163.

Murzin, A. G., Brenner, S. E., Hubbard, T., and Chothia, C. 1995. SCOP: A structural classification of proteins database for the investigation of sequences and structures. *J. Mol. Biol.* 247:536–540.

Natt, N. K., Kaur, H., and Raghava, G. P. 2004. Prediction of transmembrane regions of β-barrel proteins using ANN- and SVM-based methods. *Proteins* 56:11–18.

Neuwald, A. F., Liu, J. S., and Lawrence, C. E. 1995. Gibbs motif sampling: Detection of bacterial outer membrane protein repeats. *Protein Sci.* 4:1618–1632.

Oldziej, S., Czaplewski, C., Liwo, A., Chinchio, M., Nanias, M., Vila, J. A., Khalili, M., Arnautova, Y. A., Jagielska, A., Makowski, M., Schafroth, H. D., Kazmierkiewicz, R., Ripoll, D. R., Pillardy, J., Saunders, J. A., Kang, Y. K., Gibson, K. D., and Scheraga, H. A. 2005. Physics-based protein-structure prediction using a hierarchical protocol based on the UNRES force field: Assessment in two blind tests. *Proc. Natl. Acad. Sci. USA* 102:7547–7552.

Oliva, B., Bates, P. A., Querol, E., Aviles, F. X., and Sternberg, M. J. 1997. An automated classification of the structure of protein loops. *J. Mol. Biol.* 266:814–830.

Orengo, C. A., Michie, A. D., Jones, S., Jones, D. T., Swindells, M. B., and Thornton, J. M. 1997. CATH—a hierarchic classification of protein domain structures. *Structure* 5:1093–1108.

Patthy, L. 1999. *Protein Evolution*. Malden, MA, Blackwell Science.

Pearlman, D. A., Case, D. A., Caldwell, J. W., Ross, W. R., Cheatham, T. W., DeBolt, S., Ferguson, D., Seibel, G., and Kollman, P. 1995. AMBER, a computer program for applying molecular mechanics, normal mode analysis, molecular dynamics and free energy calculations to elucidate the structures and energies of molecules. *Comput. Phys. Commun.* 91:1–41.

Pedersen, J., and Moult, J. 1995. Ab initio structure prediction for small polypeptides and protein fragments using genetic algorithms. *Proteins* 23:454–460.

Peitsch, M. C. 1996. ProMod and Swiss-Model: Internet-based tools for automated comparative protein modelling. *Biochem. Soc. Trans.* 24:274–279.

Peitsch, M. C., and Jongeneel, V. 1993. A 3-dimensional model for the CD40 ligand predicts that it is a compact trimer similar to the tumor necrosis factors. *Int. Immunol.* 5:233–238.

Petrey, D., and Honig, B. 2005. Protein structure prediction: Inroads to biology. *Mol. Cell* 20:811–819.

Petrey, D., Xiang, X., Tang, C. L., Xie, L., Gimpelev, M., Mitors, T., Soto, C. S., Goldsmith-Fischman, S., Kernytsky, A., Schlessinger, A., Koh, I. Y. Y., Alexov, E., and Honig, B. 2003. Using multiple structure alignments, fast model building, and energetic analysis in fold recognition and homology modeling. *Proteins Struct. Funct. Genet.* 53:430–435.

Ponder, J. W., and Richards, F. M. 1987. Tertiary templates for proteins. Use of packing criteria in the enumeration of allowed sequences for different structural classes. *J. Mol. Biol.* 193:775–791.

Qian, B., Ortiz, A. R., and Baker, D. 2004. Improvement of comparative model accuracy by free-energy optimization along principal components of natural structural variation. *Proc. Natl. Acad. Sci. USA* 101(43):15346–15351.

Rigden, D. J. 2002. Use of covariance analysis for the prediction of structural domain boundaries from multiple protein sequence alignments. *Protein Eng.* 15:65–77.

Rohl, C. A., Strauss, C., Chivian, D., and Baker, D. 2004. Modeling structurally variable regions in homologous proteins with Rosetta. *Proteins* 55:656–677.

Rose, G. D. 1978. Prediction of chain turns in globular proteins on a hydrophobic basis. *Nature* 272:586–590.

Roseman, A. M. 2000. Docking structures of domains into maps from cryo-electron microscopy using local correlation. *Acta Crystallogr. Sect. D Biol. Crystallogr.* 56 (Pt 10):1332–1340.

Rossman, M. G., and Liljas, A. 1974. Recognition of structural domains in globular proteins. *J. Mol. Biol.* 85:177–181.

Rost, B., Casadio, R., Fariselli, P., and Sander, C. 1995. Transmembrane helices predicted at 95% accuracy. *Protein Sci.* 4:521–533.

Rufino, S. D., Donate, L. E., Canard, L. H. J., and Blundell, T. L. 1997. Predicting the conformational class of short and medium size loops connecting regular secondary structures: Application to comparative modelling. *J. Mol. Biol.* 267:352–367.

Rychlewski, L., Jaroszewski, L., Li, W., and Godzik, A. 2000. Comparison of sequence profiles. Strategies for structural predictions using sequence information. *Protein Sci.* 9:232–241.

Sadreyev, R. I., Baker, D., and Grishin, N. V. 2003. Profile–profile comparisons by COMPASS predict intricate homologies between protein families. *Protein Sci.* 12:2262–2272.

Sali, A., and Blundell, T. L. 1993. Comparative protein modelling by satisfaction of spatial restraints. *J. Mol. Biol.* 234:779–815.

Samudrala, R., Xia, Y., Huang, E., and Levitt, M. 1999. Ab initio protein structure prediction using a combined hierarchical approach. *Proteins* 37(S3):194–198.

Sanchez, R., Pieper, U., Mirkovi, N., de Bakker, P. I. W., Wittenstein, E., and Ali, A. (2000). MODBASE, a database of annotated comparative protein structure models. *Nucleic Acids Res.* 28:250–253.

Sanger, F., Thompson, E. O., and Kitai, R. 1955. The amide groups of insulin. *Biochem. J.* 59:509–518.

Schroder, R. R., Manstein, D. J., Jahn, W., Holden, H., Rayment, I., Holmes, K. C., and Spudich, J. A. 1993. Three-dimensional atomic model of F-actin decorated with Dictyostelium myosin S1. *Nature* 364:171–174.

Schueler-Furman, O., Wang, C., and Baker, D. 2005a. Progress in protein–protein docking: Atomic resolution predictions in the CAPRI experiment using RosettaDock with an improved treatment of side-chain flexibility. *Proteins* 60:187–194.

Schueler-Furman, O., Wang, C., Bradley, P., Misura, K., and Baker, D. 2005b. Progress in modeling of protein structures and interactions. *Science* 310:638–642.

Scott, R. A., Vanderkooi, G., Tuttle, R. W., Shames, P. M., and Scheraga, H. A. 1967. Minimization of polypeptide energy, III. Application of a rapid energy minimization technique to the calculation of preliminary structures of gramicidins. *Proc. Natl. Acad. Sci. USA* 58:2204–2211.

Shi, J., Blundell, T. L., and Mizuguchi, K. 2001. FUGUE: Sequence–structure homology recognition using environment-specific substitution tables and structure-dependent gap penalties. *J. Mol. Biol.* 310:243–257.

Siddiqui, A. S., and Barton, G. J. 1995. Continuous and discontinuous domains: An algorithm for the automatic generation of reliable protein domain definitions. *Protein Sci.* 4:872–884.

Simons, K. T., Bonneau, R., Ruczinski, I., and Baker, D. 1999a. Ab initio protein structure prediction of CASP III targets using ROSETTA. *Proteins* 37(S3):171–176.

Simons, K. T., Kooperberg, C., Huang, E., and Baker, D. 1997. Assembly of protein tertiary structures from fragments with similar local sequences using simulated annealing and Bayesian scoring functions. *J. Mol. Biol.* 268:209–225.

Simons, K. T., Ruczinski, I., Kooperberg, C., Fox, B. A., Bystroff, C., and Baker, D. 1999b. Improved recognition of native-like protein structures using a combination of sequence-dependent and sequence-independent features of proteins. *Proteins* 34:82–95.

Simons, K. T., Strauss, C., and Baker, D. 2001. Prospects for ab initio protein structural genomics. *J. Mol. Biol.* 306:1191–1199.

Sippl, M. J. 1990. Calculation of conformational ensembles from potentials of mean force: An approach to the knowledge-based prediction of local structures in globular proteins. *J. Mol. Biol.* 213:859–883.

Skolnick, J., Kolinski, A., Brooks, C. L., III, Godzik, A., and Rey, A. 1993. A method for predicting protein structure from sequence. *Curr. Biol.* 3:414–423.

Sucha, S., Dubose, R. F., March, C. J., and Subhashini, S. 1995. Modeling protein loops using a {phi}(i+1), {psi}(i) dimer database. *Protein Sci.* 4:1412–1420.

Sutcliffe, M. J., Haneef, I., Carney, D., and Blundell, T. L. 1987. Knowledge based modelling of homologous proteins, Part I: Three-dimensional frameworks derived from the simultaneous superposition of multiple structures. *Protein Eng.* 1:377–384.

Swindells, M. B. 1995. A procedure for detecting structural domains in proteins. *Protein Sci.* 4:103–112.

Tieleman, D. P., and Berendsen, H. J. 1998. A molecular dynamics study of the pores formed by Escherichia coli OmpF porin in a fully hydrated palmitoyloleoylphosphatidylcholine bilayer. *Biophys. J.* 74:2786–2801.

Topf, M., Baker, M. L., John, B., Chiu, W., and Sali, A. 2005. Structural characterization of components of protein assemblies by comparative modeling and electron cryo-microscopy. *J. Struct. Biol.* 149:191–203.

Topf, M., and Sali, A. 2005. Combining electron microscopy and comparative protein structure modeling. *Curr. Opin. Struct. Biol.* 15:578–585.

Tusnady, G. E., and Simon, I. 1998. Principles governing amino acid composition of integral membrane proteins: Application to topology prediction. *J. Mol. Biol.* 283:489–506.

Ubbink, M., Ejdeback, M., Karlsson, B. G., and Bendall, D. S. 1998. The structure of the complex of plastocyanin and cytochrome f, determined by paramagnetic NMR and restrained rigid-body molecular dynamics. *Structure* 6:323–335.

Vakser, I. A. 1995. Protein docking for low-resolution structures. *Protein Eng.* 8:371–377.

van Dijk, A. D. J., Boelens, R., and Bonvin, A. M. J. J. 2005. Data-driven docking for the study of biomolecular complexes. *FEBS J.* 272:293–312.

van Gunsteren, W. F., and Berendsen, H. J. C. 1990. Computer simulation of molecular dynamics: Methodology, applications and perspectives in chemistry. *Angew. Chem. Int. Ed. Engl.* 29:992–1023.

van Vlijmen, H. W. T., and Karplus, M. 1997. PDB-based protein loop prediction: parameters for selection and methods for optimization. *J. Mol. Biol.* 267:975–1001.

Vasquez, M. 1996. Modeling side-chain conformation. *Curr. Opin. Struct. Biol.* 6:217–221.

Vitkup, D., Melamud, E., Moult, J., and Sander, C. 2001. Completeness in structural genomics. *Nat. Struct. Biol.* 8:559–566.

von Heijne, G. 1992. Membrane protein structure prediction. Hydrophobicity analysis and the positive-inside rule. *J. Mol. Biol.* 225:487–494.

Vonderviszt, F., and Simon, I. 1986. A possible way for prediction of domain boundaries in globular proteins from amino acid sequence. *Biochem. Biophys. Res. Commun.* 139:11–17.

Warshel, A. 1976. Bicycle-pedal model for the first step in the vision process. *Nature* 260:679–683.

Warshel, A. 2002. Molecular dynamics simulations of biological reactions. *Acc. Chem. Res.* 35:385–395.

Wetlaufer, D. B. 1973. Nucleation, rapid folding, and globular intrachain regions in proteins. *Proc. Natl. Acad. Sci. USA* 70:697–701.

Wheelan, S. J., Marchler-Bauer, A., and Bryant, S. H. 2000. Domain size distributions can predict domain boundaries. *Bioinformatics* 16:613–618.

White, S. H. 2004. The progress of membrane protein structure determination. *Protein Sci.* 13:1948–1949.

Wiehe, K., Pierce, B., Mintseris, J., Tong, W. W., Anderson, R., Chen, R., and Weng, Z. 2005. ZDOCK and RDOCK performance in CAPRI rounds 3, 4, and 5. *Proteins* 60:207–213.

Wodak, S. J., and Janin, J. 1981. Location of structural domains in protein. *Biochemistry* 20:6544–6552.

Wriggers, W., Milligan, R. A., and McCammon, J. A. 1999. Situs: A package for docking crystal structures into low-resolution maps from electron microscopy. *J. Struct. Biol.* 125:185–195.

Wriggers, W., Milligan, R. A., Schulten, K., and McCammon, J. A. 1998. Self-organizing neural networks bridge the biomolecular resolution gap. *J. Mol. Biol.* 284:1247–1254.

Xu, D., Baburaj, K., Peterson, C. B., and Xu, Y. 2001. Model for the three-dimensional structure of vitronectin: Predictions for the multi-domain protein from threading and docking. *Proteins* 44:312–320.

Xu, J., Li, M., Kim, D., and Xu, Y. 2003. RAPTOR: Optimal protein threading by linear programming. *J. Bioinform. Comput. Biol.* 1:95–117.

Xu, Y., and Xu, D. 2000. Protein threading using PROSPECT: Design and evaluation. *Proteins* 40:343–354.

Ye, Y., Li, Z., and Godzik, A. 2006. Modeling and analyzing three-dimensional structures of human disease proteins. *Pac. Symp. Biocomput.* (Maui).

Yip, Y. L., Scheib, H., Diemand, A. V., Gattiker, A., Famiglietti, L. M., Gasteiger, E., and Bairoch, A. 2004. The Swiss-Prot variant page and the ModSNP database: A resource for sequence and structure information on human protein variants. *Hum. Mutat.* 23:464–470.

Yona, G., and Levitt, M. 2002. Within the twilight zone: A sensitive profile–profile comparison tool based on information theory. *J. Mol. Biol.* 315:1257–1275.

Yuan, Z., Mattick, J. S., and Teasdale, R. D. 2004. SVMtm: Support vector machines to predict transmembrane segments. *J. Comp. Chem.* 25:632–636.

Zhai, Y., and Saier, M. H. J. R. 2002. The β-barrel finder (BBF) program, allowing identification of outer membrane β-barrel proteins encoded within prokaryotic genomes. *Protein Sci.* 11:2196–2207.

Zhang, Y., and Skolnick, J. 2004. Automated structure prediction of weakly homologous proteins on a genomic scale. *Proc. Natl. Acad. Sci. USA* 101:7594–7599.

Zheng, Q., and Kyle, D. J. 1996. Accuracy and reliability of the scaling-relaxation method for loop closure: An evaluation based on extensive and multiple copy conformational samplings. *Proteins* 24:209–217.

Zhou, H., and Zhou, Y. 2005. Fold recognition by combining sequence profiles derived from evolution and from depth-dependent structural alignment of fragments. *Proteins* 58:321–328.

2 Empirical Force Fields

Alexander D. MacKerell, Jr.

Protein structure and dynamics and, therefore, their biological functions are dictated by a collection of forces that vary from those associated with covalent linkages, such as bonds, to long-range through space forces, such as electrostatic or coulombic interactions. Accordingly, to be able to apply theoretical approaches to understand the behavior of proteins, it is necessary to be able to accurately predict the change in energy of a protein as a function of the change in conformation. Importantly, such predictions must include contributions from the environment in which the protein is immersed. While quantum-mechanical (QM) methods are attractive in their ability to model complex chemical phenomena at the level of electronic structure, such methods are typically inappropriate for proteins due to the large size of these macromolecules as well as the need to treat their environment in an explicit fashion. Rather, molecular mechanics (MM), which rely on potential energy functions or empirical force fields, afford the computational speed to allow for calculations on proteins along with their environment.

2.1 Potential Energy Functions

The computational speed associated with molecular mechanics is based on the simplicity of the mathematical models used in the potential energy function to relate the structure of the system to its energy. This simplicity is based on the smallest particles in the model typically being atoms, which are treated as point masses centered on the nucleus of each atom in the molecules comprising the system under study. The potential energy function therefore describes the interactions between the atoms in the system.

An example of the potential energy function used in the additive CHARMM force fields (Brooks et al., 1983; MacKerell et al., 1998b) is shown in Eq. (2.1); similar energy functions are used in the common macromolecular force fields for proteins including OPLS/AA (Jorgensen and Tirado-Rives, 1988), AMBER (Cornell

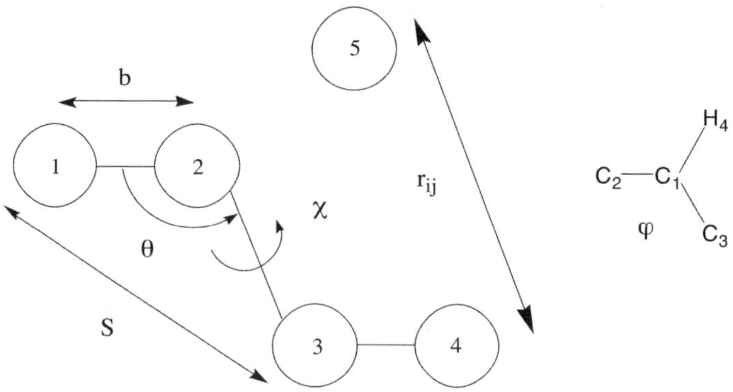

Fig. 2.1 Schematic diagram of the terms used to describe the energy as a function of the conformation in the potential energy function. The bond length between two covalently attached atoms 1 and 2 is b, θ is the valence angle between atoms 1, 2, and 3, χ is the dihedral angle involving atoms 1, 2, 3, and 4, S is the Urey-Bradley distance between atoms 1 and 3, and r_{ij} is the through space distance between atoms 1 and 4. The inset shows an example of an improper dihedral, φ, which is defined as the dihedral C1–C2–C3–H4.

et al., 1995), and GROMOS (van Gunsteren, 1987).

$$\begin{aligned} U(\vec{R}) = & \sum_{\text{bounds}} K_b(b-b_0)^2 + \sum_{\text{angle}} K_\theta(\theta-\theta_0)^2 + \sum_{\text{UB}} K_{\text{UB}}(S-S_0)^2 \\ & + \sum_{\text{dihedrals}} K_\chi(1+\cos(n\chi-\delta)) + \sum_{\text{impropers}} K_{\text{imp}}(\varphi-\varphi_0)^2 \\ & + \sum_{nonbound} \varepsilon_{ij}\left[\left(\frac{R_{\min,ij}}{r_{ij}}\right)^{12} - 2\left(\frac{R_{\min,ij}}{r_{ij}}\right)^{6}\right] + \frac{q_i q_j}{4\pi e r_{ij}} \end{aligned} \tag{2.1}$$

Equation (2.1), where the potential energy, U, is calculated as a function of the atomic positions, $\vec{R}$, includes terms for the internal (i.e., bonded) and external (i.e., interaction or nonbond) contributions. Internal terms include the bonds, valence angles, Urey-Bradley, dihedral or torsion angles, and improper dihedral terms while the external terms include the van der Waals (vdW) interations, treated via the Lennard-Jones (LJ) 6–12 term and the electrostatic interactions. In Eq. (2.1), terms describing the geometry of the molecule include the bond length, b, the valence angle, θ, the distance between atoms separated by two covalent bonds (Urey-Bradley term, 1,3 distance), S, the dihedral or torsion angle, χ, the improper angle, φ, and the distance between atoms i and j, r_{ij}. The schematic diagram in Fig. 2.1 illustrates the terms included in Eq. (2.1).

In order for the potential energy function to represent different types of, for example, bonds (e.g., C–C single versus double bonds) or atom types, parameters are used for each type of bond, angle, atom type, and so on in the molecules in the system. These parameters include the bond force constant and equilibrium distance, K_b and

b_0, respectively; the valence angle force constant and equilibrium angle, K_θ, and θ_0; the Urey-Bradley force constant and equilibrium distance, K_{UB} and S; the dihedral angle force constant, multiplicity, and phase angle, K_χ, n, and δ; and the improper force constant and equilibrium improper angle, K_ϕ and ϕ_0. External parameters that describe the interactions between atoms i and j include the partial atomic charges, q_i, and the LJ well-depth, ε_{ij}, and minimum interaction radius, $R_{\min,ij}$, used to treat the vdW interactions. Typically, ε_i and $R_{\min,i}$ are obtained for individual atom types and then combined to yield ε_{ij} and $R_{\min,ij}$ for the interacting atoms via combining rules. In CHARMM, ε_{ij} values are obtained via the geometric mean ($\varepsilon_{ij} = \mathrm{sqrt}(\varepsilon_i * \varepsilon_j)$ and $R_{\min,ij}$ via the arithmetic mean, $R_{\min,ij} = (R_{\min,i} + R_{\min,j})/2$. The dielectric constant, e, is set to one in all calculations where solvent is considered explicitly (see below), corresponding to the permittivity of vacuum.

Essential for the modeling of proteins, as well as all biomolecules, is the proper treatment of hydrogen bonding. Earlier force fields included explicit terms for hydrogen bonds (Weiner and Kollman, 1981); however, it has been shown that the combination of the Lennard-Jones and coulombic interactions produces an accurate representation of both the distance and angle dependencies of hydrogen bonds (Reiher, 1985). This success has allowed for the omission of explicit terms to treat hydrogen bonding from the majority of empirical force fields. It should be noted that the LJ and electrostatic parameters are highly correlated, such that LJ parameters determined for a set of partial atomic charges will not be applicable to another set of charges. Moreover, the internal parameters are dependent on the external parameters. For example, the barrier to rotation about the C–O bond in ethanol includes contributions from the electrostatic and vdW interactions between the hydroxyl hydrogen and the rest of the molecule as well as contributions from the bond, angle, and dihedral terms. Thus, if the LJ parameters or charges are changed, the internal parameters will have to be reoptimized to produce the correct energy barrier. Finally, condensed phase properties obtained from empirical force field calculations contain contributions for the conformations of the molecules being studied as well as external interactions between those molecules, emphasizing the importance of accurate treatment of both internal and external portions of the force field for accurate condensed phase simulations.

Beyond the terms included in Eq. (2.1), additional terms may be included in a potential energy function; such extended energy functions are typically referred to as Class II force fields. Class II force fields can include higher order corrections for the bond and valence angle terms and/or cross terms between, for example, bonds and valence angles or valence angles and dihedrals (Lii and Allinger, 1991; Derreumaux and Vergoten, 1995; Halgren, 1996a; Sun, 1998; Ewig et al., 2001; Palmo et al., 2003). Other alternative terms include the use of a Morse function for bonds (Burkert and Allinger, 1982). This function allows for bond breaking to be included in an empirical force field. Another alternative is the use of a cosine-based valence angle term that is well behaved for near-linear valence angles (Mayo et al., 1990; Rappé et al., 1992). For the dihedral term a recent improvement that avoids singularities associated with derivatives of torsion angle cosines and allows for application of any value of the phase has been presented (Blondel and Karplus, 1996) and, more recently, the introduction of a two-dimensional (2D) grid-based dihedral energy correction map (CMAP) (MacKerell et al., 2004a,b) that allows for any 2D dihedral surface (e.g., a QM ϕ, ψ surface of the alanine dipeptide) to be reproduced nearly exactly by the force field (see below). These two terms are included in the recent version of the CHARMM force field for proteins. Typically inclusion of these terms in an energy function increases its accuracy in treating conformational energies, especially at geometries far from the minimum-energy or equilibrium values as well as yield improved treatment of vibrational spectra. However, it should be emphasized that Class I force fields [i.e., those based on Eq. (2.1)] can yield accuracies similar to the Class II force fields when the parameters are properly optimized. In general, Class I force fields, when applied to biomolecular simulations in the vicinity of room temperature, adequately treat the intramolecular distortions, including relative conformational energies associated with large structural changes.

The external portion of a potential energy function may also be extended beyond that in Eq. (2.1), including alternate forms of both the vdW interactions and the electrostatics. The three primary alternatives to the LJ 6–12 term included in Eq. (2.1) are designed to "soften" the repulsive wall associated with Pauli exclusion. For example, the Buckingham potential (Buckingham and Fowler, 1985) uses an exponential term to treat repulsion while a buffered 14–7 term is used in the MMFF force field (Halgren, 1996b). A simple alternative is to replace the r^{12} repulsion with an r^9 term. All of these forms more accurately treat the repulsive wall as judged by high-level QM calculations (Halgren, 1992). However, as with the harmonic internal terms in Class I force fields, the LJ term appears to be adequate for biomolecular simulations at or near room temperature.

2.2 Implementation of Potential Energy Functions

As stated above, once an empirical force field is available, it may be used, in combination with the necessary software, to calculate the change of energy of a system as a function of coordinates. However, more useful is the combination of an empirical force field with numerical approaches allowing for sampling of relevant conformations via, e.g., a molecular dynamics (MD) simulation to be performed (Tuckerman and Martyna, 2000). Such approaches can be used to predict a variety of structural and thermodynamic properties, including free energies, via statistical mechanics (McQuarrie, 1976). Importantly, such approaches allow for comparisons with experimental thermodynamic data and the atomic details of interactions between molecules that dictate the thermodynamic properties can be obtained. Such atomic details are often difficult to access via experimental approaches, motivating the application of computational approaches.

Proper application of an empirical force field when performing MD simulations or other calculations on proteins is an essential consideration. Due to the central role of the external interactions in the energy function, it is important that all nonbond interactions between all atom–atom pairs be considered. The Ewald method can be used to treat the long-range electrostatic interactions for periodic systems (Ewald, 1921). Recent variations of the Ewald method that are computationally more tractable include the particle Mesh Ewald approach (Darden, 2001). Alternatively, reaction field methods can be used to simulate finite (e.g., spherical) systems (Beglov and Roux, 1994; Bishop et al., 1997; Im et al., 2001). Concerning the vdW or LJ interactions, the long-range contributions to this term beyond the atom–atom truncation distance (i.e., those beyond a distance where the atom–atom interactions are calculated explicitly) can be corrected for by assuming those contributions are homogeneous in nature (Allen and Tildesley, 1989; Lague et al., 2004).

The integrators that generate proper ensembles in MD simulations are another important consideration, as attaining the proper ensemble in an MD simulation is essential for direct comparison with experimental data (Tuckerman et al., 1992; Martyna et al., 1994; Feller et al., 1995; Barth and Schlick, 1998; Tuckerman and Martyna, 2000). Extensions of MD simulations have been developed that significantly increase the sampling of conformational space including locally enhanced sampling (Elber and Karplus, 1990; Hansmann, 1997; Simmerling et al., 1998) and replica-exchange or parallel tempering (Hansmann, 1997; Sugita and Okamoto, 1999; Nymeyer et al., 2004). It should be noted that the deterministic nature of MD simulations is typically lost when using such approaches, although the replica-exchange method can produce results that correspond to a proper ensemble. As always, the appropriate use of these different methods greatly facilitates investigations of molecular interactions via condensed phase simulations.

2.3 Treatment of Aqueous Solvation

Protein structure and function is greatly influenced by the condensed phase (i.e., aqueous) environment in which they exist. Accordingly, an empirical force field for proteins must treat the condensed phase environment in an accurate manner. Treatment of the protein environment may be performed using either explicit or implicit models. Explicit models, where the water, ions, and so on, are included explicitly in the simulation, are more microscopically accurate while implicit or continuum models can produce savings in computer time over explicit models and have the advantage of directly yielding free energies of solvation.

A number of explicit water models have been used in protein simulations including the TIP3P, TIP4P (Jorgensen et al., 1983), SPC, extended SPC/E (Berendsen et al., 1987) and F3C (Levitt et al., 1997) models. TIP3P is the most commonly used water model. Its three-point design (i.e., one oxygen and two hydrogen atoms) makes it computationally tractable and it yields the correct thermodynamic properties of water. Structurally the model treats the first and second solvation shells with reasonable accuracy. However, the second or tetrahedral peak in the O–O radial distribution is underestimated and the diffusion constant of the model is significantly larger than the corresponding experimental value (Feller et al., 1996). Another widely used three-point model is SPC. This model uses an internal tetrahedral geometry (i.e., H–O–H angle = 109.47°) leading to increased structure over TIP3P, as evidenced by a more-defined tetrahedral peak in the O–O radial distribution function. A variant of SPC, the SPC/E model, includes a correction for the polarization self-energy that yields improved structure and diffusion properties. However, this correction leads to an overestimation of the water potential energy in the bulk phase, which may perturb the energetic balance of solvent–solvent, solute–solvent, and solute–solute interactions. This problem must be considered when using this model in biomolecular simulations. TIP4P is a four-point water model that includes an additional particle along the H–O–H bisector. The additional particle overcomes many of the limitations listed above, although the computer demands of the model are higher. Recently, new water models have been presented (Mahoney and Jorgensen, 2000; Glättli et al., 2003), although they have not seen wide use in biomolecular simulations.

Selection of the proper water model is important for a successful simulation. The most important consideration is the compatibility of the model with the biomolecular force field being used. Such compatibility is important due to most force fields being developed in conjunction with a specific water model (e.g., AMBER, OPLS and CHARMM with TIP3P, OPLS also with TIP4P, GROMOS with SPC, ENCAD with F3C), such that it is best to use a force field with its prescribed water model unless special solvent requirements are important.

Implicit solvation models treat the protein environment as a continuum, for example, by treating regions not "inside" the protein with the dielectric constant of water (Davis and McCammon, 1990; Honig, 1993). Such models offer significant computational savings while yielding reasonably accurate treatment of solvation. Accordingly, implicit models are useful when extensive sampling of conformational space is required, as in protein folding. However, these models can fail when highly specific water–biomolecule interactions have an important structural or energetic role. The most widely used implicit solvation models are Poisson-Boltzmann (PB) and generalized Born (GB) models. In the PB model, contributions from solvent polarization along with the asymmetric shapes of biological molecules are taken into account (Gilson and Honig, 1988), from which free energies of solvation may be determined. Advances in this approach have included the optimization of atomic radii to reproduce experimental free energies of solvation of model compounds representative of proteins (Nina et al., 1997; Banavali and Roux, 2002). GB approaches (Still et al., 1990) are an alternative to PB that also yield free energies of solvation while being less computationally expensive, thereby facilitating their use in MD simulations. Several GB models have been developed that yield free energies of solvation at a level of accuracy similar to PB methods (Schaefer and Karplus, 1996; Jayaram et al., 1998; Onufriev et al., 2000; Zhang et al., 2001; Lee et al., 2003). Both the PB and GB methods can be combined with free energy solvent accessibility (SA) terms that account for the hydrophobic effect (Qui et al., 1997; Gallicchio et al., 2003), referred to as PB/SA or GB/SA approaches. Recent developments based on the GB method involve an improved treatment of vdW dispersion contributions beyond the typical solvent accessibility related terms (Gallicchio and Levy, 2004). Other implicit models that have been used in biomolecular simulations include the Langevin Dipoles Model (Florián and Warshel, 1997) and the EEF1 model (Lazaridis and Karplus, 1999). More information on implicit solvating models can be obtained from a recent review by Feig and Brooks (Feig and Brooks, 2004).

The PB/SA and GB/SA methods can be used for postprocessing of trajectories from MD simulations to obtain free energies of solvation. In this approach an MD simulation of the biomolecule(s) is performed using an explicit solvent representation followed by estimation of the free energy of solvation using the solute coordinates from the simulation (i.e., biomolecule only with the solvent omitted) (Kollman et al., 2000). This allows for determination of the free energy of solvation of a biomolecule averaged over the length of a simulation, using structures obtained with an explicit solvent representation. This approach is particularly attractive for the calculation of free energies of binding of macromolecular complexes (Jayaram et al., 2002; Gohlke et al., 2003; Habtemariam et al., 2005). This type of approach also has great utility for the estimation of ligand–protein binding (Ferrara et al., 2004), at a computationally reasonable cost as required for testing of large numbers of drug candidates.

2.4 Empirical Force Field Optimization

The ability of a simple potential energy function such as that in Eq. (2.1) to accurately model the energies as a function of protein conformation is based on proper optimization of the parameters used in the energy function. Indeed, until the parameters are available, one does not truly have an empirical force field. And the quality of that empirical force field is judged by its ability to accurately reproduce the experimental regimen.

Parameter optimization is based on reproducing a set of target data, including information on small model compounds representative of proteins as well as on proteins themselves. Target data are ideally obtained from experiments, though a majority of the data are often obtained from QM calculations. QM calculations are readily applicable to most small molecules; however, limitations in QM level of theory, especially with respect to the treatment of dispersion interactions (Chalasinski and Szczesniak, 1994; Chen et al., 2002), require the use of experimental data when available (MacKerell, 2004).

Details on the optimization of internal parameters have been presented previously by a number or workers (Halgren, 1996c; Ewig et al., 2001; MacKerell, 2001; Wang and Kollman, 2001). Briefly, the equilibrium bond, valence angle, and Urey-Bradley parameters along with the dihedral multiplicity and phase are optimized to reproduce internal geometries of the model compounds. The target data are often QM data, although it has been shown that condensed phase effects can influence the internal geometry of a molecule, such that survey data from structures in the Cambridge Structural Database (CSD, http://www.ccdc.cam.ac.uk/) (Allen et al., 1979) may be considered the ideal. The value of such data for treatment of the peptide bond has previously been discussed (MacKerell et al., 1998; MacKerell, 2004). Force constants for the bond, valence angle, Urey-Bradley, dihedral angle, and improper angles are optimized to reproduce vibrational spectra, including both the frequencies as well as the potential energy distribution (PED) (i.e., the contribution of internal degrees of freedom to the individual frequencies). Again the ideal data are obtained from condensed phase vibrational studies, although such data are typically limited making vibrational data from QM calculations the most commonly used. It should be emphasized that QM vibrational analysis allows for detailed assignment of the PED and, even when good experimental data are available, QM calculations are often advantageous to perform the assignments. When performing optimization of vibrational spectra, it should be noted that the low-frequency modes represent the largest structural distortions that occur in a molecule, such that their proper treatment is important for accurately treating the structural distortions that occur during MD simulations. Conformational energies from QM calculations, including barrier heights for rotations about dihedrals, are typically used for the final optimization of the dihedral angle parameters. In the CHARMM force fields the dihedral parameters are initially optimized based on vibrational data with only the parameters associated with dihedrals that involve all nonhydrogen atoms adjusted to reproduce potential energy surfaces. This final optimization is again important as the rotations about dihedrals represent the largest structural changes that occur in MD simulations of proteins. Recent work on lipids emphasizes the importance of proper treatment of the conformational energies (Klauda et al., 2005). In addition, empirical optimization of dihedral parameters to reproduce experimental distributions of conformers, such at the phi, psi angle distributions in proteins, have been shown to be important (MacKerell et al. 2004a,b). Those efforts have included optimization of the grid-based energy correction map discussed below.

Significant effort by a number of groups has gone into the determination of the electrostatic parameters; the partial atomic charges, q_i. Of the methods currently in use, the most common methods for proteins are the supramolecular and QM electrostatic potential (ESP) approaches. Other variations include bond charge increments (Bush et al., 1999; Jakalian et al., 2000) and electronegativity equilization methods (Gilson et al., 2003), although these methods are typically applied to small, drug-like molecules. An important consideration with the determination of partial atomic charges, related to the Coulombic treatment of electrostatics, is the omission of the explicit treatment of electronic polarizability. Due to this omission, it is necessary for static, partial atomic charges to reproduce the average polarization that occurs in the condensed phase environment. This is achieved by "enhancing" the charges of a molecule leading to an overestimation of the dipole moment as compared to the gas phase value. This is referred to as an implicitly polarized model. For example, many of the water models used in protein empirical force fields (e.g., TIP3P, TIP4P, SPC) have dipole moments in the vicinity of 2.2 debye (Jorgensen et al., 1983), versus the gas phase value of 1.85 debye. Inclusion of implicit polarizability allows for empirical force fields based on Eq. (2.1), which are often referred to as additive, to reproduce a variety of condensed phase properties (Rizzo and Jorgensen, 1999).

These additive models have been extensively used for simulations of proteins, as well as other biological molecules; however, they are limited in that they do not reproduce the change in electrostatic interactions due to inductive effects associated with changes in the polarity of the environment.

The supramolecular approach for the determination of partial atomic charges is used in the OPLS (Jorgensen and Tirado-Rives, 1988; Jorgensen et al., 1996) and CHARMM (MacKerell et al., 1998b; Foloppe and MacKerell, 2000; Feller et al., 2002) force fields. This approach involves optimization of the charges to reproduce QM-determined minimum interaction energies and geometries of a model compound with individual water molecules or for model compound dimers. Typically, the HF/6-31G* level of theory was used for the QM calculations, due to its overestimation of dipole moments (Cieplak et al., 1995), leading to the implicitly polarizable model discussed above. An additional advantage of the supramolecular approach is that in the QM calculation, local polarization effects due to the charge induction caused by the two interacting molecules are included, facilitating determination of charge distributions appropriate for the condensed phase.

It should be noted that although it has recently been shown that QM methods can accurately reproduce gas phase experimental interaction energies for a range of model compound dimers (Kim and Friesner, 1997; Huang and MacKerell, 2002), it is important to maintain the QM level of theory that was historically used for a particular force field when extending that force field to novel molecules. This assures that the balance of the nonbond interactions between different molecules in the system being studied is maintained. Finally, when considering the transferability of charges obtained from the supramolecular approach, it should be noted that the charges are typically obtained for functional groups such that they may be directly transferred between molecules.

The other commonly applied approach for charge determination in empirical force fields is ESP charge fitting. This methodology is based on the adjustment of charges to reproduce a QM-determined ESP mapped onto a grid surrounding the model compound. ESP methods are widely used and a number of charge fitting methods based on this approach have been developed (Singh and Kollman, 1984; Chirlian and Francl, 1987; Merz, 1992; Bayly et al., 1993; Henchman and Essex, 1999). Application of ESP fitting approaches is hampered by difficulties in unambiguously fitting charges to an ESP (Francl et al., 1996) and charges on "buried" atoms (e.g., a carbon to which three or four nonhydrogen atoms are covalently bound) tend to be underdetermined, requiring the use of restraints during fitting (Bayly et al., 1993). The latter method is referred to as Restrained ESP (RESP). In addition, the QM ESP is typically determined via gas phase calculations, which may yield charges that are not consistent with the condensed phase. Recent developments are addressing this limitation (Laio et al., 2002). Another problem is that multiple conformations of flexible molecules must also be taken into account (Cieplak et al., 1995), although it should be noted that the last two problems are also present to varying extents in the supramolecular approach. For ESP fitting, the QM level of theory has historically been HF/6-31G*, as used in the AMBER force fields (Cornell et al., 1995), although higher level QM calculations have been applied more recently in conjunction with the RESP approach (Duan et al., 2003). In summary, the supramolecular and ESP methods are both useful for the determination of partial atomic charges and, as with the water models, the method to use should be that which is consistent with the remainder of the force field.

The most difficult aspect of empirical force fields to optimize are the LJ terms, although proper treatment of these terms is essential for obtaining accurate condensed phase properties from empirical force fields. A big part of the difficulty in optimizing LJ parameters are limitations in the quality of QM calculations in treating dispersion interactions (Chalasinski and Szczesniak, 1994; Chen et al., 2002), requiring the use of experimental condensed phase data as the target data (Jorgensen, 1984, 1986). LJ parameters are generally optimized to reproduce experimentally measured values such as heats of vaporization, densities, isocompressibilities, and heat capacities of small model compound pure solvents. Alternative target data include heats or free energies of aqueous solvation, partial molar volumes or heats of sublimation and lattice geometries of crystals (Warshel and Lifson, 1970; MacKerell et al., 1995). While these approaches have acted as the basis for several protein force fields, it should be emphasized that LJ parameters optimized in this fashion are underdetermined due to the small number of experimental observables available for the optimization of a significantly larger number of LJ parameters. This leads to the parameter correlation problem where LJ parameters for different atoms in a molecule (e.g., H and C in ethane) can compensate for each other (MacKerell, 2001). The parameter correlation problem with respect to LJ parameters has been addressed via an approach that determines the absolute values of the LJ parameters based on experimental data, as above, while their relative values are optimized using high-level QM data as the target data (Yin and MacKerell, 1996; Chen et al., 2002). In general, determination of LJ parameters is quite time consuming; however, in many instances it is feasible to directly transfer the LJ parameters between functional groups in the context of different molecules.

2.5 Protein Force Fields

Current protein MD simulations are typically performed using additive all-atom protein models, including the OPLS/AA (Jorgensen and Tirado-Rives, 1988), CHARMM22 (MacKerell et al., 1998b), and AMBER (PARM99) (Cornell et al., 1995) force fields. An alternative is the use of extended or united atom models, where the nonpolar hydrogens are treated as part of the carbon to which they are covalently bound, with polar hydrogens important for hydrogen bonding included. United atom models offer computational savings over all-atom models, and are often used with implicit solvent models. Details of the approaches used for development of the commonly used all-atom models—CHARMM22, AMBER, and OPLS—are summarized in the following paragraphs, with a brief discussion of the extended atom models given below.

Internal parameters for AMBER and CHARMM22 were derived via reproduction of both experimental and QM data for small model compounds, including reproduction of geometries and vibrational spectra. The internal parameters for the OPLS force field (Jorgensen and Tirado-Rives, 1988) were initially taken from AMBER and have been subsequently optimized to reproduce conformational energies from QM calculations yielding OPLS/AA (Jorgensen et al., 1996). Additional optimization of selected torsions has been performed using higher level QM target data (Kaminski et al., 2001). Supporting the quality of these force fields in the treatment of proteins are MD simulations, showing the three force fields to reproduce selected experimental structures in a similar fashion (Price and Brooks, 2002).

Care was taken in the optimization of the external aspects of the three force fields. In OPLS/AA and CHARMM22, the partial atomic charges are based on HF/6-31G* supramolecular data while the standard AMBER release (PARM99) is based on RESP charges fit to the same level of theory. Condensed phase simulations were used as target data for the optimization of the LJ parameters in all three force fields. In CHARMM22 and AMBER, the charges were optimized to be consistent with the TIP3P water model, whereas OPLS was developed to work with the TIP3P, TIP4P, and SPC models. Based on water dimer interaction energies, it may be anticipated that the TIP4P and SPC models will also work well with CHARMM22 and AMBER, although rigorous tests have yet to be performed. Despite similarities in the optimization of the charges for the three force fields, differences in the local charge distributions have been noted (Ponder and Case, 2003). Such differences in charges may lead to differences in the balance of the local interactions (e.g., relative hydrogen bonding strength at the peptide bond NH versus CO), which may impact the atomic details of interactions obtained from the three force fields. Based on the presence of such differences, results from MD simulations using these force fields should be interpreted taking into account the applied optimization approach.

In all of the protein force fields, a significant and ongoing effort has been made with respect to the treatment of the Ramachandran map or ϕ, ψ energy surface (Ramachandran, et al., 1963). This is due to the conformational energy as a function of the ϕ, ψ dihedral angles dictating the region of conformational space being sampled in peptide and protein simulations. The quintessential model compound for optimizing the dihedral parameters related to ϕ, ψ is the alanine dipeptide (Fig. 2.2), along with related compounds such as the glycine dipeptide and the proline dipep-

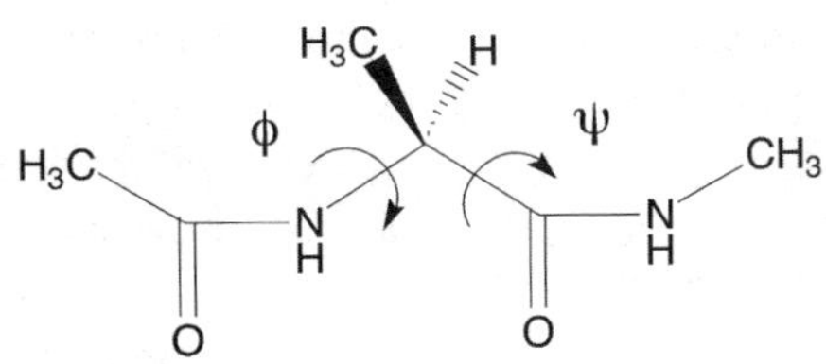

Fig. 2.2 Diagram of the alanine dipeptide including the ϕ, ψ dihedral angles used to define the Ramachandran diagram.

tide. The size of these compounds is such that they are accessible to high-level QM calculations (Head-Gordon et al., 1991; Beachy et al., 1997; Ono et al., 2002; Vargas et al., 2002; Duan et al., 2003; MacKerell et al., 2004b), the data from which can be used as target data for the parameter optimization. While targeting QM energetic data for the optimization is simple and well defined, studies have shown that directly reproducing gas phase QM data can lead to systematic problems in the conformational properties of the protein backbone (MacKerell et al., 1998b). This was shown in great detail using the CMAP approach for the treatment of the conformational energies of ϕ, ψ (MacKerell et al., 2004a,b). To overcome this limitation with the use of QM data, it has been shown that empirical adjustment of the dihedral or CMAP terms to better reproduce ϕ, ψ distributions from surveys of the protein databank (PDB) (Berman et al., 2002) leads to significant improvement in the treatment of the conformational properties of the backbone. Indeed, the recently published CHARMM22/CMAP all-atom protein force field represents a significant improvement over current force fields for proteins with respect to the treatment of both structural (Freedberg et al., 2004; MacKerell et al., 2004; Steinbach, 2004) and dynamic (Buck et al., 2006) properties.

To better understand the improvements in the treatment of the protein backbone conformational properties associated with the use of CMAP, Fig. 2.3 shows ϕ, ψ potentials of mean force (PMFs) and distributions for proteins (upper frames) and for the alanine dipeptide (lower frames). PMFs, or free energy surfaces, were obtained from MD simulations of proteins in their crystal environments using the CHARMM22 (MacKerell et al., 1998b) and CMAP modified CHARMM22 (MacKerell et al. 2004a,b) energy functions along with distributions from a survey of the PDB (Dunbrack and Cohen, 1997; Dunbrack, 2002). Comparison of the three surfaces shows the CMAP PMF to better reproduce the shape of the surface derived from the PDB. Notable are the improvements in the overall shape of the beta sheet ($\phi \sim -120°$, $\psi \sim 150°$) and alpha-helical ($\phi \sim -60°$, $\psi \sim -40°$) regions versus CHARMM22. Importantly, the improvements also occur at the model compound level, where the overall shape of the distribution of ϕ, ψ in MD simulations of the alanine dipeptide using CMAP is in better agreement with the distributions from a QM/MM simulation (SCCDFT) (Hu et al., 2003). The improved agreement at both the protein and model compound levels indicates that the overall force field is well balanced, such that the proper treatment of local contributions as evidenced by the alanine dipeptide MD results leads to the desired behavior at the macromolecular (i.e., protein) level by the force field.

Finally, it should be emphasized that the protein force fields are undergoing continual optimization in the context of the Class I energy function as well as via extension of the force field via, for example, the CMAP approach. Motivating such additional optimization is the ability to access additional target data by which to judge the quality of the force fields as well as improved algorithms and computational resources, allowing for more rigorous tests of the force fields. For example, the free energies of solvation of model compounds representative of protein side chains have recently been calculated for the AMBER, CHARMM, and OPLS/AA force

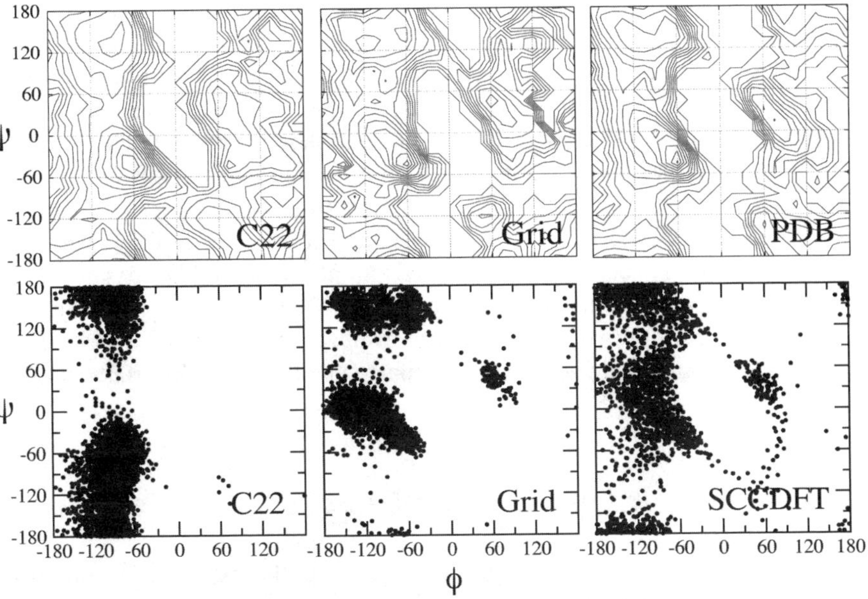

Fig. 2.3 ϕψ PMFs based on MD simulations using the CHARMM22 and CHARMM22 grid-corrected empirical force fields and from a survey of the PDB (upper frames) and ϕψ distributions from MD simulations of the alanine dipeptide (Ace-Ala-Nme, lower frames) in solution using the CHARMM22 (MacKerell et al., 1998) and CHARMM22 grid-corrected empirical force fields and previously published data from a QM/MM model (SCCDFT). PMF contours are in 0.5 kcal/mol increments up to 6 kcal/mol above the global minimum. PMFs were obtained from the respective probability distributions based on a Boltzmann distribution (McQuarrie, 1976). See reference (MacKerell et al., 2004a) for more details. Reproduced with permission from *J. Am. Chem. Soc.* 2004, 126:698–699. Copyright 2004 American Chemical Society.

fields and compared with experiment (Shirts et al., 2003); similar studies have been reported elsewhere (Villa and Mark, 2002; MacCallum and Tieleman, 2003; Deng and Roux, 2004). These studies show the force fields to yield reasonable free energies of solvation, although the need for improvements is evident.

Examples of recent adjustments of the protein force fields are numerous. Motivated by the free energy of solvation results discussed in the preceding paragraph, adjustments have been made to the tryptophan side chain parameters in CHARMM (Macias and MacKerell, 2005). Multiple adjustments have been made to the AMBER Cornell et al. force field (i.e., PARM94) (Cornell et al., 1995). In recent years, optimization of the ϕ, ψ related dihedral parameters was performed to improve agreement with QM data for both the alanine dipeptide and tetrapeptide (Beachy et al., 1997), yielding PARM99. More recently, adjustments have been made to deal with the tendency of PARM99 to favor α-helical conformations. Modifications include alteration of the ϕ, ψ dihedral parameters and changes in the charge distribution for the entire protein force field. The dihedral parameters were changed in two studies to alter the conformational space being sampled in peptide simulations

(Garcia and Sanbonmatsu, 2002; Okur et al., 2003). In another study, the partial atomic charges were redetermined via RESP fits to B3LYP/cc-pVTZ//HF/6-31G* QM data (Duan et al., 2003). This was followed by readjustment of the ϕ, ψ-related dihedral parameters to reproduce QM maps of the alanine and glycine dipeptide maps obtained at the MP2/cc-pVTZ//HF/6-31G* level with a dielectric constant of 4, where the dielectric was selected to mimic the environment on the protein interior. While the efforts at additional optimization of the force fields are admirable, care must be taken that the changes are done in an orderly, well-defined manner. For example, as discussed above, alteration of the charges will impact the remaining parameters in the force field, possibly requiring their readjustment. In addition, the development of a collection of variants of a force field may become problematic in that difficulties in comparing results from different studies may arise.

2.6 Extended or United Atom Protein Force Fields

An alternative to the all-atom protein force fields is the extended or united atom force fields. Such force fields typically omit nonpolar hydrogen atoms to save computer time; these models dominated the early protein force fields. Examples of extended atom models include CHARMM PARAM19 (Neria et al., 1996), OPLS/UA (Jorgensen and Tirado-Rives, 1988), the early AMBER force fields (Weiner and Kollman, 1981; Weiner et al., 1984, 1986), and GROMOS87 and 96 (van Gunsteren et al., 1996). The GROMOS force field is still widely used in MD simulations that include explicit solvent representations. GROMOS96 has been subjected to tests in the condensed phase (Daura et al., 1996) and improved LJ parameters have recently been reported (Schuler et al., 2001). The other united atom force fields are primarily used for simulations on long time scales via the use of implicit solvent models, with the majority of these studies being based on PARAM19. PARAM19 can be used with several continuum solvation models including EEF1 (Lazaridis and Karplus, 1999), ACE (Schaefer et al., 2001), several GB models (Lee et al., 2002; Im et al., 2003; Lee et al., 2003), and a buried surface area model by Caflisch and co-workers (Ferrara et al., 2002). A summary of recent applications of both united and all-atom protein force fields combined with implicit solvent models has been presented (Feig and Brooks, 2004).

Several other force fields have been used for protein simulations, although they have not been used extensively. These include ENCAD (Levitt, 1990; Levitt et al., 1995), CEDAR (Ferro et al., 1980; Hermans et al., 1984), MMFF (Halgren, 1999) and CVFF (Ewig et al., 2001), among others. A more comprehensive list has been presented by Ponder and Case (Ponder and Case 2003). While these and other force fields may be used for protein simulations, it should be emphasized that they should be subjected to tests to ensure that they are appropriate for the problem under study.

Recently, the first reports of MD simulations of force fields that include explicit treatment of electronic polarizability (Halgren and Damm, 2001; Rick and Stuart, 2002) have appeared in the literature. These include studies using a fluctuating charge model (Patel and Brooks, 2004; Patel et al., 2004) and a point-dipole model (Kaminski et al., 2002; Harder et al., 2005). In addition, an MD simulation of DNA using a classical Drude oscillator to treat electronic polarizability has been published (Anisimov et al., 2005; Lamoureux et al., 2006) and this model is expected to be extended to proteins in the near future (A.D. MacKerell, Jr. and B. Roux, work in progress). The main advantage of polarizable force fields is the ability to more accurately model environments of different polarities, such as the polar environment on the surfaces of a protein versus the more hydrophobic interior of the protein. This improvement is at the cost of computer time, as treatment of the polarizability (Rick et al., 1995; Tuckerman and Martyna, 2000) can significantly increase the computational costs. In addition, the currently available first-generation polarizable force fields will probably require improvements as they are more rigorously tested on a wide variety of proteins.

2.7 Summary

A wide variety of empirical force fields for proteins are available and have been tested on a large number of peptides and proteins as well as on small molecule model systems. Currently, the all-atom force fields based on an additive model (e.g., no explicit treatment of electronic polarizability) are the most commonly used and have been shown to reproduce many types of experimental data in a wide variety of systems. While many protein simulations are performed in the presence of explicit solvent, improvements in implicit or continuum solvent models allow for protein simulations to be performed at a considerable savings of computational resources. While implicit models have been particularly useful in protein folding studies, it should be emphasized that in situations where individual water molecules play an essential role in protein structure and function, they will typically fail. With respect to the future, force fields that include explicit treatment of electronic polarization offer the potential of more accurately treating the wide range of environments in and around proteins. However, the current additive models will often be the method of choice when extended sampling of conformational space or time ranges is required.

Acknowledgments

Financial support from the NIH (GM51501 and GM 072558) and the University of Maryland, School of Pharmacy, Computer-Aided Drug Design Center is acknowledged.

Appendix

Web sites associated with commonly used protein force fields and simulation packages

CHARMM	*www.charmm.org* *www.pharmacy.umaryland.edu/faculty/amackere/* CHARMM also allows for simulations using AMBER, MMFF, and OPLS/AA force fields.
AMBER	*amber.scripps.edu/*
GROMOS	*www.igc.ethz.ch/gromos/*
OPLS	*zarbi.chem.yale.edu/software.html* *www.cs.sandia.gov/projects/towhee/forcefields/oplsaa.html*
Tinker	*dasher.wustl.edu/tinker/* Includes the CHARMM, AMBER, and OPLS force fields among others.
CVFF	*www.accelrys.com*
CEDAR	*femto.med.unc.edu/Hermans/jhermans.html*

References

Allen, F. H., S. Bellard, M. D. Brice, B. A. Cartwright, A. Doubleday, H. Higgs, T. Hummelink, B. G. Hummelink-Peters, O. Kennard, W. D. S. Motherwell, J. R. Rodgers, and D. G. Watson. 1979. The Cambridge Crystallographic Data Centre: Computer-based search, retrieval, analysis and display of information. *Acta Crystallogr. Sect. B* 35:2331–2339.

Allen, M. P., and D. J. Tildesley. 1989. *Computer Simulation of Liquids*. New York, Oxford University Press.

Anisimov, V. M., G. Lamoureux, I. V. Vorobyov, N. Huang, B. Roux, and A. D. MacKerell, Jr. 2005. Determination of electrostatic parameters for a polarizable force field based on the classical drude oscillator. *J. Chem. Theory Comput.* 1:153–168.

Banavali, N. K., and B. Roux. 2002. Atomic radii for continuum electrostatic calculations on nucleic acids. *J. Phys. Chem. B* 106:11026–11035.

Barth, E., and T. Schlick. 1998. Extrapolation versus impulse in multiple-timestepping schemes. II. Linear analysis and applications to Newtonian and Langevin dynamics. *J. Chem. Phys.* 109:1633–1642.

Bayly, C. I., P. Cieplak, W. D. Cornell, and P. A. Kollman. 1993. A well-behaved electrostatic potential based method using charge restraints for deriving atomic charges: The RESP model. *J. Phys. Chem.* 97:10269–10280.

Beachy, M. D., D. Chasman, R. B. Murphy, T. A. Halgren, and R. A. Friesner. 1997. Accurate ab initio quantum chemical determination of the relative energetics of peptide conformations and assessment of empirical force fields. *J. Am. Chem. Soc.* 119:5908–5920.

Beglov, D., and B. Roux. 1994. Finite representation of an infinite bulk system: Solvent boundary potential for computer simulations. *J. Chem. Phys.* 100:9050–9063.

Berendsen, H. J. C., J. R. Grigera, and T. P. Straatsma. 1987. The missing term in effective pair potentials. *J. Phys. Chem.* 91:6269–6271.

Berman, H. M., T. Battistuz, T. N. Bhat, W. F. Bluhm, P. E. Bourne, K. Burkhardt, Z. Feng, G. L. Gilliland, L. Iype, S. Jain, P. Fagan, J. Marvin, D. Padilla, V. Ravichandran, B. Schneider, N. Thanki, H. Weissig, J. D. Westbrook, and C. Zardecki. 2002. The protein data bank. *Acta. Crystallogr. D. Biol. Crystallogr.* 58:899–907.

Bishop, T. C., R. D. Skeel, and K. Schulten. 1997. Difficulties with multiple time stepping and fast multipole algorithm in molecular dynamics. *J. Comput. Chem.* 18:1785–1791.

Blondel, A., and M. Karplus. 1996. New formulation of derivatives of torsion angles and improper torsion angles in molecular mechanics: Elimination of singularities. *J. Comput. Chem.* 17:1132–1141.

Brooks, B. R., R. E. Bruccoleri, B. D. Olafson, D. J. States, S. Swaminathan, and M. Karplus. 1983. CHARMM: A program for macromolecular energy, minimization, and dynamics calculations. *J. Comput. Chem.* 4:187–217.

Buck, M., S. Bonnet, R. W. Pastor, and A. D. MacKerell, Jr. 2006. Importance of the CMAP correction to the CHARMM22 Protein Force Field: Dynamics of hen lysozyme. *Biophys. J.* 90:L36–L38.

Buckingham, A. D., and P. W. Fowler. 1985. A model for the geometries of van der Waals complexes. *Can. J. Chem.* 63:2018.

Burkert, U., and N. L. Allinger. 1982. *Molecular Mechanics*. Washington, DC, American Chemical Society.

Bush, B. L., C. I. Bayly, and T. A. Halgren. 1999. Consensus bond-charge increments fitted to electrostatic potential or field of many compounds: Application of MMFF94 training set. *J. Comput. Chem.* 20:1495–1516.

Chalasinski, G., and M. M. Szczesniak. 1994. Origins of structure and energetics of van der Waals clusters from ab initio calculations. *Chem. Rev.* 94:1723–1765.

Chen, I.-J., D. Yin, and A. D. MacKerell, Jr. 2002. Combined ab initio/empirical optimization of Lennard-Jones parameters for polar neutral compounds. *J. Comput. Chem.* 23:199–213.

Chirlian, L. E., and M. M. Francl. 1987. Atomic charges derived from electrostatic potentials: A detailed study. *J. Comput. Chem.* 8:894–905.

Cieplak, P., W. D. Cornell, C. I. Bayly, and P. K. Kollman. 1995. Application of the multimolecule and multiconformational RESP methodlogy to biopolymers: Charge derivation for DNA, RNA, and proteins. *J. Comput. Chem.* 16:1357–1377.

Cornell, W. D., P. Cieplak, C. I. Bayly, I. R. Gould, K. M. Merz, D. M. Ferguson, D. C. Spellmeyer, T. Fox, J. W. Caldwell, and P. A. Kollman. 1995. A second generation force field for the simulation of proteins, nucleic acids, and organic molecules. *J. Am. Chem. Soc.* 117:5179–5197.

Darden, T. 2001. Treatment of long-range forces and potentials. In *Computational Biochemistry and Biophysics* (O. M. Becker, A. D. MacKerell, Jr., B. Roux, and M. Watanabe, Eds.). New York, Dekker, pp. 91–114.

Daura, X., P. H. Hünenberger, A. E. Mark, E. Querol, F. X. Avilés, and W. F. van Gunsteren. 1996. Free energies of transfer of Trp analogs from chloroform to water: Comparison of theory and experiment and the importance of adequate treatment of electrostatics and internal interactions. *J. Am. Chem. Soc.* 118:6285– 6294.

Davis, M. E., and J. A. McCammon. 1990. Electrostatics in biomolecular structure and dynamics. *Chem. Rev.* 90:509–521.

Deng, Y., and B. Roux. 2004. Hydration of amino acid side chains: Non-polar and electrostatic contributions calculated from staged molecular dynamics free energy simulations with explicit water molecules. *J. Phys. Chem. B* 108:16567–16576.

Derreumaux, P., and G. Vergoten. 1995. A new spectroscopic molecular mechanics force field. Parameters for proteins. *J. Chem. Phys.* 102:8586–8605.

Duan, Y., C. Wu, S. Chowdhury, M. C. Lee, G. Xiong, W. Zhang, R. Yang, P. Ceiplak, R. Luo, T. Lee, J. Caldwell, J. Wang, and P. Kollman. 2003. A point-charge force field for molecular mechanics simulations of proteins based on condensed-phase quantum mechanical calculations. *J. Comput. Chem.* 24:1999–2012.

Dunbrack, R. L., Jr. 2002. Culledpdb: Non-redundant set of protein sidechains from the PDB. Philadelphia.

Dunbrack, R. L., Jr., and F. E. Cohen. 1997. Bayesian statistical analysis of protein sidechain rotamer preferences. *Protein Sci.* 6:1661–1681.

Elber, R., and M. Karplus. 1990. Enhanced sampling in molecular dynamics: Use of the time-dependent hartree approximation for a simulation of carbon monoxide diffusion through myoglobin. *J. Am. Chem. Soc.* 112:9161–9175.

Ewald, P. P. 1921. Die Berechnung optischer und elektrostatischer Gitterpotentiale. *Ann. Phys.* 64:253–287.

Ewig, C. S., R. Berry, U. Dinur, J.-R. Hill, M.-J. Hwang, H. Li, C. Liang, J. Maple, Z. Peng, T. P. Stockfisch, T. S. Thacher, L. Yan, X. Ni, and A. T. Hagler. 2001. Derivation of class II force fields. VIII. Derivation of a general quantum mechanical force field for organic compounds. *J. Comput. Chem.* 22:1782–1800.

Feig, M., and C. L. Brooks III. 2004. Recent advances in the development and application of implicit solvent models in biomolecular simulations. *Curr. Opin. Struct. Biol.* 14:217–224.

Feller, S. E., K. Gawrisch, and A. D. MacKerell, Jr. 2002. Polyunsaturated fatty acids in lipid bilayers: Intrinsic and environmental contributions to their unique physical properties. *J. Am. Chem. Soc.* 124:318–326.

Feller, S. E., R. W. Pastor, A. Rojnuckarin, S. Bogusz, and B. R. Brooks. 1996. Effect of electrostatic force truncation on interfacial and transport properties of water. *J. Phys. Chem.* 100:17011–17020.

Feller, S. E., Y. Zhang, R. W. Pastor, and R. W. Brooks. 1995. Constant pressure molecular dynamics simulation: The Langevin piston method. *J. Chem. Phys.* 103:4613–4621.

Ferrara, P., J. Apostolakis, and A. Caflisch. 2002. Evaluation of a fast implicit solvent model for molecular dynamics simulations. *Proteins* 46:24–33.

Ferrara, P., H. Gohlke, D. J. Price, G. Klebe, and C. L. I. Brooks. 2004. Assessing scoring functions for protein–ligand interactions. *J. Med. Chem.* 47:3032–3047.

Ferro, D. R., J. E. McQueen, J. T. McCown, and J. Hermans. 1980. Energy minimization of rubredoxin. *J. Mol. Biol.* 136:1–18.

Florián, J., and A. Warshel. 1997. Langevin dipoles model for ab initio calculations of chemical processes in solution: Parameterization and application to hydration free energies of neutral and ionic solutes and conformational analysis in aqueous sdolution. *J. Phys. Chem. B* 101:5583–5595.

Foloppe, N., and A. D. MacKerell, Jr. 2000. All-atom empirical force field for nucleic acids: 1. Parameter optimization based on small molecule and condensed phase macromolecular target data. *J. Comput. Chem.* 21:86–104.

Francl, M. M., C. Carey, L. E. Chirlian, and D. M. Gange. 1996. Charge fit to electrostatic potentials. II. Can atomic charges be unambiguously fit to electrostatic potentials? *J. Comput. Chem.* 17:367–383.

Freedberg, D. I., R. M. Venable, A. Rossi, T. E. Bull, and R. W. Pastor. 2004. Discriminating the helical forms of peptides by NMR and molecular dynamics simulations. *J. Am. Chem. Soc.* 126:10478–10484.

Gallicchio, E., and R. M. Levy. 2004. AGBNP: An analytic implicit solvent model suitable for molecular dynamics simulations and high-resolution modeling. *J. Comput. Chem.* 25:479–499.

Gallicchio, E., L. Y. Zhang, and R. M. Levy. 2003. The SGB/NP hydration free energy model based on the surface generalized Born solvent reaction field and novel nonpolar hydration free energy estimators. *J. Comput. Chem.* 23:517–529.

Garcia, A. E., and K. Y. Sanbonmatsu. 2002. α-Helical stabilization by side chain sheilding of backbone hydrogen bonds. *Proc. Natl. Acad. Sci. USA* 99:2782–2787.

Gilson, M. K., H. S. Gilson, and M. J. Potter. 2003. Fast assignment of accurate partial atomic charges: An electronegativity equilization method that accounts for alternate resonance forms. *J. Chem. Inf. Comput. Sci.* 43:1982–1997.

Gilson, M. K., and B. Honig. 1988. Calculation of the total electrostatic energy of a macromolecular system: Solvation energies, binding energies, and conformational analysis. *Proteins* 4:7–18.

Glättli, A., X. Daura, and W. F. van Gunsteren. 2003. A novel approach for designing simple point charge models for liquid water with three interaction sites. *J. Comput. Chem.* 24:1087–1096.

Gohlke, H., C. Kiel, and D. Case. 2003. Protein–protein docking with simultaneous optimization of rigid-body displacement and side-chain conformations. *J. Mol. Biol* 330:891–913.

Habtemariam, B., V. M. Anisimov, and A. D. MacKerell, Jr. 2005. Cooperative binding of DNA and CBFβ to the runt domain of the CBFα studied via MD simulations. *Nucleic Acids Res.* 33:4212–4222.

Halgren, T. A. 1992. Representation of van der Waals (vdW) interactions in molecular mechanics force fields: Potential form, combination rules, and vdW parameters. *J. Am. Chem. Soc.* 114:7827–7843.

Halgren, T. A. 1996a. Merck molecular force field. I. Basis, form, scope, parameterization, and performance of MMFF94. *J. Comput. Chem.* 17: 490–519.

Halgren, T. A. 1996b. Merck molecular force field. II. MMFF94 van der Waals and electrostatic parameters for intermolecular interactions. *J. Comput. Chem.* 17:520–552.

Halgren, T. A. 1996c. Merck molecular force field. III. Molecular geometries and vibrational frequencies for MMFF94. *J. Comput. Chem.* 17:553–586.

Halgren, T. A. 1999. MMFF VII. Characterization of MMFF94, MMFF94s, and other widely available force fields for conformational energies and for intermolecular-interaction energies and geometries. *J. Comput. Chem.* 20:730–748.

Halgren, T. A., and W. Damm. 2001. Polarizable force fields. *Curr. Opin. Struct. Biol.* 11:236–242.

Hansmann, U. H. E. 1997. Parallel tempering algorithm for conformational studies of biological molecules. *Chem. Phys. Lett.* 281:140–150.

Harder, E., B. Kim, R. A. Friesner, and B. J. Berne. 2005. Efficient simulation method for polarizable protein force fields: Application to the simulation of BPTI in liquid water. *J. Chem. Theory Comput.* 1:169–180.

Head-Gordon, T., M. Head-Gordon, M. J. Frisch, C. L. Brooks, and J. A. Pople. 1991. Theoretical study of blocked glycine and alanine peptide analogues. *J. Am. Chem. Soc.* 113:5989–5997.

Henchman, R. H., and J. W. Essex. 1999. Generation of OPLS-like charges from molecular electrostatic potential using restraints. *J. Comput. Chem.* 20:483–498.

Hermans, J., H. J. C. Berendsen, W. F. van Gunsteren, and J. P. M. Postma. 1984. A consistent empirical potential for water–protein interactions. *Biopolymers* 23:1513–1518.

Honig, B. 1993. Macroscopic models of aqueous solutions: Biological and chemical applications. *J. Phys. Chem.* 97:1101.

Hu, H., M. Elstner, and J. Hermans. 2003. Comparison of a QM/MM force field and molecular mechanics force fields in simulations of alanine and glycine "dipeptides" (Ace-Ala-Nme and Ace-Gly-Nme) in water in relation to the problem of modeling the unfolded peptide backbone in solution. *Proteins Struct Funct. Genet.* 50:451–463.

Huang, N., and A. D. MacKerell, Jr. 2002. An *ab initio* quantum mechanical study of hydrogen-bonded complexes of biological interest. *J. Phys. Chem. B* 106:7820–7827.

Im, W., S. Bernéche, and B. Roux. 2001. Generalized solvent boundary potential for computer simulations. *J. Chem. Phys.* 114:2924–2937.

Im, W., M. S. Lee, and C. L. Brooks III. 2003. Generalized Born model with a simple smoothing function. *J. Comput. Chem.* 24:1691–702.

Jakalian, A., B. L. Bush, D. B. Jack, and C. I. Bayly. 2000. Fast, efficient generation of high-quality atomic charges. AM1-BCC model: I. Method. *J. Comput. Chem.* 21:132–146.

Jayaram, B., K. J. McConnell, S. B. Dixit, and D. L. Beveridge. 2002. Free-energy component analysis of 40 protein–DNA complexes: A consensus view on the thermodynamics of binding at the macromolecular level. *J. Comput. Chem.* 23:1–14.

Jayaram, B., D. Sprous, and D. L. Beveridge. 1998. Solvation free energy of biomacromolecules: Parameters for a modified generalized Born model consistent with the AMBER force field. *J. Phys. Chem. B* 102: 9571–9576.

Jorgensen, W. L. 1984. Optimized intermolecular potential functions for liquid hydrocarbons. *J. Am. Chem. Soc.* 106:6638–6646.

Jorgensen, W. L. 1986. Optimized intermolecular potential functions for liquid alcohols. *J. Phys. Chem.* 90:1276–1284.

Jorgensen, W. L., J. Chandrasekhar, J. D. Madura, R. W. Impey, and M. L. Klein. 1983. Comparison of simple potential functions for simulating liquid water. *J. Chem. Phys.* 79:926–935.

Jorgensen, W. L., D. S. Maxwell, and J. Tirado-Rives. 1996. Development and testing of the OPLS all-atom force field on conformational energetics and properties of organic liquids. *J. Am. Chem. Soc.* 118:11225–11236.

Jorgensen, W. L., and J. Tirado-Rives. 1988. The OPLS potential function for proteins. Energy minimizations for crystals of cyclic peptides and crambin. *J. Am. Chem. Soc.* 110:1657–1666.

Kaminski, G., R. A. Friesner, J. Tirado-Rives, and W. L. Jorgensen. 2001. Evaluation and reparametrization of the OPLS-AA force field for proteins via comparison with accurate quantum chemical calculations on peptides. *J. Phys. Chem. B* 105:6474–6487.

Kaminski, G. A., H. A. Stern, B. J. Berne, R. A. Friesner, Y. X. Cao, R. B. Murphy, R. Zhou, and T. A. Halgren. 2002. Development of a polarizable force field for proteins via ab initio quantum chemistry: First generation model and gas phase tests. *J. Comput. Chem.* 23: 1515–1531.

Kim, K., and R. A. Friesner. 1997. Hydrogen bonding between amino acid backbone and side chain analogues: A high-level ab initio study. *J. Am. Chem. Soc.* 119:12952–12961.

Klauda, J. B., B. R. Brooks, A. D. MacKerell, Jr., R. M. Venable, and R. W. Pastor. 2005. An ab initio study on the torsional surface of alkanes and its effect on molecular simulations of alkanes and a DPPC bilayer. *J. Phys. Chem. B* 109:5300–5311.

Kollman, P. A., I. Massova, C. Reyes, B. Kuhn, S. Huo, L. Chong, M. Lee, T. Lee, Y. Duan, W. Wang, O. Donini, P. Cieplak, J. Srinivasan, D. A. Case, and T. E. Cheatham III. 2000. Calculating structures and free energies of complex molecules: Combining molecular mechanics and continuum models. *Acc. Chem. Res.* 33:889–897.

Lague, P., R. W. Pastor, and B. R. Brooks. 2004. A pressure-based long-range correction for Lennard Jones interactions in molecular dynamics simulations: Application to alkanes and interfaces. *J. Phys. Chem. B* 108:363–368.

Laio, A., J. VandeVondele, and U. Rothlisberger. 2002. D-RESP: Dynamically generated electrostatic potential derived charges from quantum mechanics/molecular mechanics simulations. *J. Phys. Chem. B* 106:7300–7307.

Lamoureux, G., E. Harder, I. V. Vorobyov, B. Roux, and A. D. MacKerell, Jr. 2005. A polarizable model of water for molecular dynamics simulations of biomolecules. *Chem. Phys. Lett.* 418:241–245.

Lazaridis, T., and M. Karplus. 1999. Effective energy function for proteins in solution. *Proteins* 35:133–152.

Lee, M. S., M. Feig, F. R. Salsbury, Jr., and C. L. Brooks III. 2003. New analytical approximation of the standard molecular volume definition and its application to generalized Born calculations. *J. Comput. Chem.* 24:1348–1356.

Lee, M. S., F. R. Salsbury, Jr., and C. L. Brooks III. 2002. Novel generalized Born methods. *J. Chem. Phys.* 116:10606–10614.

Levitt, M. 1990. ENCAD—Energy Calculations and Dynamics. Stanford, CA and Rehovot, Israel, Molecular Applications Group.

Levitt, M., M. Hirshberg, R. Sharon, and V. Daggett. 1995. Potential energy function and parameters for simulations of the molecular dynamics of proteins and nucleic acids in solution. *Comput. Phys. Commun.* 91:215–231.

Levitt, M., M. Hirshberg, R. Sharon, K. E. Laidig, and V. Daggett. 1997. Calibration and testing of a water model for simulation of the molecular dynamics of proteins and nucleic acids in solution. *J. Phys. Chem. B* 101:5051–5061.

Lii, J.-L., and N. L. Allinger. 1991. The MM3 force field for amides, polypeptides and proteins. *J. Comput. Chem.* 12:186–199.

MacCallum, J. L., and P. Tieleman. 2003. Calculation of the water–cyclohexane transfer free energies of amino acid side-chain analogs using the OPLS all-atom force field. *J. Comput. Chem.* 24:1930–1935.

Macias, A. T., and A. D. MacKerell, Jr. 2005. CH/pi interactions involving aromatic amino acids: Refinement of the CHARMM tryptophan force field. *J. Comput. Chem.* 26:1452–1463.

MacKerell, A. D., Jr. 2001. Atomistic models and force fields. In *Computational Biochemistry and Biophysics* (O. M. Becker, A. D. MacKerell, Jr., B. Roux, and M. Watanabe, Eds.). New York, Dekker, pp. 7–38.

MacKerell, A. D., Jr. 2004. Empirical force fields for biological macromolecules: Overview and issues. *J. Comput. Chem.* 25:1584–1604.

MacKerell, A. D., Jr., D. Bashford, M. Bellott, R. L. Dunbrack, Jr., J. Evanseck, M. J. Field, S. Fischer, J. Gao, H. Guo, S. Ha, D. Joseph, L. Kuchnir, K. Kuczera, F. T. K. Lau, C. Mattos, S. Michnick, T. Ngo, D. T. Nguyen, B. Prodhom, I. Reiher, W. E., B. Roux, M. Schlenkrich, J. Smith, R. Stote, J. Straub, M. Watanabe, J. Wiorkiewicz-Kuczera, D. Yin, and M. Karplus. 1998a. All-atom empirical potential for molecular modeling and dynamics studies of proteins. *J. Phys. Chem. B* 102:3586–3616.

MacKerell, A. D., Jr., B. Brooks, C. L. Brooks III, L. Nilsson, B. Roux, Y. Won, and M. Karplus. 1998b. CHARMM: The energy function and its paramerization with an overview of the program. In *Encyclopedia of Computational Chemistry* (P. v. R. Schleyer et al., Eds.) Chichester, John Wiley & Sons. pp. 271–277.

MacKerell, A. D., Jr., M. Feig, and C. L. Brooks III. 2004a. Accurate treatment of protein backbone conformational energetics in empirical force fields. *J. Am. Chem. Soc.* 126:698–699.

MacKerell, A. D., Jr., M. Feig, and C. L. Brooks III. 2004b. Extending the treatment of backbone energetics in protein force fields: Limitations of gas-phase quantum mechanics in reproducing protein conformational distributions in molecular dynamics simulations. *J. Comput. Chem.* 25:1400–1415.

MacKerell, A. D., Jr., J. Wiórkiewicz-Kuczera, and M. Karplus. 1995. An all-atom empirical energy function for the simulation of nucleic acids. *J. Am. Chem. Soc.* 117:11946–11975.

Mahoney, M. W., and W. L. Jorgensen. 2000. A five-site model for liquid water and the reproduction of the density anomaly by rigid, nonpolarizable potential functions. *J. Chem. Phys.* 112:8910–8922.

Martyna, G. J., D. J. Tobias, and M. L. Klein. 1994. Constant pressure molecular dynamics algorithms. *J. Chem. Phys.* 101:4177–4189.

Mayo, S. L., B. D. Olafson, and W. A Goddard III. 1990. 'DREIDING: A generic force field for molecular simulations.' *J. Phys. Chem.* 94:8897–8909.

McQuarrie, D. A. 1976. *Statistical Mechanics*. New York, Harper & Row.

Merz, K. M. 1992. Analysis of a large data base of electrostatic potential derived atomic charges. *J. Comput. Chem.* 13:749–767.

Neria, E., S. Fischer, and M. Karplus. 1996. Simulation of activation free energies in molecular systems. *J. Chem. Phys.* 105:1902–1919.

Nina, M., D. Beglov, and B. Roux. 1997. Atomic radii for continuum electrostatics calculation based on molecular dynamics free energy simulations. *J. Phys. Chem. B* 101:5239–5248.

Nymeyer, H., S. Gnanakaran, and A. E. Garcia. 2004. Atomic simulations of protein folding, using the replica exchange algorithm. *Methods Enzymol.* 383:119–149.

Okur, A., B. Strockbine, V. Hornak, and C. Simmerling. 2003. Using PC clusters to evaluate the transferability of molecular mechanics force fields for proteins. *J. Comput. Chem.* 24:21–31.

Ono, S., M. Kuroda, J. Higo, N. Nakajima, and H. Nakamura. 2002. Calibration of force-field dependency in free energy landscapes of peptide conformations by quantum chemical calculations. *J. Comput. Chem.* 23:470–476.

Onufriev, A., D. Bashford, and D. A. Case. 2000. Modification of the generalized born model suitable for macromolecules. *J. Phys. Chem. B* 104:3712–3720.

Palmo, K., B. Mannfors, N. G. Mirkin, and S. Krimm. 2003. Potential energy functions: From consistent force fields to spectroscopically determined polarizable force fields. *Biopolymers* 68:383–394.

Patel, S., and C. L. Brooks III. 2004. CHARMM fluctuating charge force field for proteins: I Parameterization and application to bulk organic liquid simulations. *J. Comput. Chem.* 25:1–15.

Patel, S., A. D. MacKerell, Jr., and C. L. Brooks III. 2004. CHARMM fluctuating charge force field for proteins: II Protein/solvent properties from molecular dynamics simulations using a nonadditive electrostatic model. *J. Comput. Chem.* 25:1504–1514.

Ponder, J. W., and D. A. Case. 2003. Force fields for protein simulations. *Adv. Protein Chem.* 66:27–85.

Price, D. J., and C. L. Brooks III. 2002. Modern protein force fields behave comparably in molecular dynamics simulations. *J. Comput. Chem.* 23:1045–1057.

Qui, D., P. S. Shenkin, F. P. Hollinger, and W. C. Still. 1997. The GB/SA continuum model for solvation. A fast analytical method for the calculation of approximate Born radii. *J. Phys. Chem. A* 101:3005–3014.

Ramachandran, G. N., C. Ramakrishnan, and V. Sasisekharan. 1963. Stereochemistry of polypeptide chain configurations. *J. Mol. Biol.* 7:95–99.

Rappé, A. K., C. J. Colwell, W. A. Goddard III, and W. M. Skiff. 1992. UFF, a full periodic table force field for molecular mechanics and molecular dynamics simulations. *J. Am. Chem. Soc.* 114:10024–10035.

Reiher, W. E. 1985. Theoretical Studies of Hydrogen Bonding Ph.D. thesis, Harvard University.

Rick, S. W., and S. J. Stuart. 2002. Potentials and algorithms for incorporating polarizability in computer simulations. *Rev. Comput. Chem.* 18:89–146.

Rick, S. W., S. J. Stuart, J. S. Bader, and B. J. Berne. 1995. Fluctuating charge force fields for aqueous solutions. *J. Mol. Liq.* 65/66:31–40.

Rizzo, R. C., and W. L. Jorgensen. 1999. OPLS all-atom model for amines: Resolution of the amine hydration problem. *J. Am. Chem. Soc.* 121:4827–4836.

Schaefer, M., C. Bartels, F. LeClerc, and M. Karplus. 2001. Effective atom volumes for implicit solvent models: Comparison between Voronoi volumes and minimum fluctuation volumes. *J. Comput. Chem.* 22:1857–1879.

Schaefer, M., and M. Karplus. 1996. A comprehensive analytical treatment of continuum electrostatics. *J. Phys. Chem.* 100:1578–1599.

Schuler, L. D., X. Daura, and W. F. van Gunsteren. 2001. An improved GROMOS96 force field for aliphatic hydrocarbons in the condensed phase. *J. Comput. Chem.* 22:1205–1218.

Shirts, M. R., J. W. Pitera, W. C. Swope, and V. S. Pande. 2003. Extremely precise free energy calculations of amino acid side chain analogs: Comparison of common molecular mechanics force fields for proteins. *J. Chem. Phys.* 119:5740–5761.

Simmerling, C., T. Fox, and P. A. Kollman. 1998. Use of locally enhanced sampling in free energy calculations: Testing and application to the $\alpha \rightarrow \beta$ anomerization of glucose. *J. Am. Chem. Soc.* 120:5771–5782.

Singh, U. C., and P. A. Kollman. 1984. An approach to computing electrostatic charges for molecules. *J. Comput. Chem.* 5:129–145.

Steinbach, P. J. 2004. Exploring peptide energy landscapes: A test of force fields and implicit solvent models. *Proteins* 57:665–677.

Still, W. C., A. Tempczyk, R. C. Hawley, and T. Hendrickson. 1990. Semianalytical treatment of solvation for molecular mechanics and dynamics. *J. Am. Chem. Soc.* 112:6127–6129.

Sugita, Y., and Y. Okamoto. 1999. Replica-exchange molecular dynamics methods for protein folding. *Chem. Phys. Lett.* 314:141–151.

Sun, H. 1998. COMPASS: An ab inito force-field optimized for condensed-phase applications-overview with details on alkane and benzene compounds. *J. Phys. Chem. B* 102:7338–7364.

Tuckerman, M., B. J. Berne, and G. J. Martyna. 1992. Reversible multiple time scale molecular dynamics. *J. Chem. Phys.* 97:1990–2001.

Tuckerman, M. E., and G. J. Martyna. 2000. Understanding modern molecular dynamics: Techniques and applications. *J. Phys. Chem. B* 104:159–178.

van Gunsteren, W. F. 1987. GROMOS. Groningen Molecular Simulation Program Package. Groningen, University of Groningen.

van Gunsteren, W. F., S. R. Billeter, A. A. Eising, P. H. Hünenberger, P. Krüger, A. E. Mark, W. R. P. Scott, and I. G. Tironi. 1996. *Biomolecular Simulation: The GROMOS96 Manual and User Guide*. Zürich, BIOMOS b.v.

Vargas, R., J. Garza, B. P. Hay, and D. A. Dixon. 2002. Conformational study of the alanine dipeptide at the MP2 and DFT levels. *J. Phys. Chem. A* 106:3213–3218.

Villa, A., and A. E. Mark. 2002. Calculation of the free energy of solvation for neutral analogs of amino acid side chains. *J. Comput. Chem.* 23:548–553.

Wang, J., and P. A. Kollman. 2001. Automatic parameterization of force field by systematic search and genetic algorithms. *J. Comput. Chem.* 22:1219–1228.

Warshel, A., and S. Lifson. 1970. Consistent force field calculations. II. Crystal structures, sublimation energies, molecular and lattice vibrations, molecular conformations, and enthalpy of alkanes. *J. Chem. Phys.* 53:582–594.

Weiner, P. K., and P. A. Kollman. 1981. AMBER: Assisted Model Building with Energy Refinement. A general program for modeling molecules and their interactions. *J. Comput. Chem.* 2:287–303.

Weiner, S. J., P. A. Kollman, D. A. Case, U. C. Singh, C. Ghio, G. Alagona, S. Profeta, and P. Weiner. 1984. A new force field for molecular mechanical simulation of nucleic acids and proteins. *J. Am. Chem. Soc.* 106:765–784.

Weiner, S. J., P. A. Kollman, D. T. Nguyen, and D. A. Case. 1986. An all atom force field for simulations of proteins and nucleic acids. *J. Comput. Chem.* 7:230–252.

Yin, D., and A. D. MacKerell, Jr. 1996. Ab initio calculations on the use of helium and neon as probes of the van der Waals surfaces of molecules. *J. Phys. Chem.* 100:2588–2596.

Yin, D., and A. D. MacKerell, Jr. 1998. Combined ab initio/empirical approach for the optimization of Lennard-Jones parameters. *J. Comput. Chem.* 19:334–348.

Zhang, L. Y., E. Gallicchio, R. A. Friesner, and R. M. Levy. 2001. Solvent models for protein–ligand binding: Comparison of implicit solvent Poisson and surface generalized Born models with explicit solvent simulations. *J. Comput. Chem.* 22:591–607.

3 Knowledge-Based Energy Functions for Computational Studies of Proteins

Xiang Li and Jie Liang

3.1 Introduction

This chapter discusses theoretical framework and methods for developing knowledge-based potential functions essential for protein structure prediction, protein–protein interaction, and protein sequence design. We discuss in some detail the Miyazawa–Jernigan contact statistical potential, distance-dependent statistical potentials, as well as geometric statistical potentials. We also describe a geometric model for developing both linear and nonlinear potential functions by optimization. Applications of knowledge-based potential functions in protein-decoy discrimination, in protein–protein interactions, and in protein design are then described. Several issues of knowledge-based potential functions are finally discussed.

In the experimental work that led to the recognition of the 1972 Nobel prize in chemistry, Christian Anfinsen showed that a completely unfolded protein ribonuclease could refold spontaneously to its biologically active conformation. This observation indicates that the sequence of amino acids of a protein contains all of the information needed to specify its three-dimensional structure (Anfinsen et al., 1961; Anfinsen, 1973). The automatic *in vitro* refolding of denatured proteins was further confirmed in many other protein systems (Janicke, 1987). Anfinsen's experiments led to the thermodynamic hypothesis of protein folding, which postulates that a native protein folds into a three-dimensional structure in equilibrium, in which the state of the whole protein–solvent system corresponds to the global minimum of free energy under physiological conditions.

Based on this thermodynamic hypothesis, computational studies of proteins, including structure prediction, folding simulation, and protein design, all depend on the use of a potential function for calculating the effective energy of the molecule. In protein structure prediction, the potential function is used either to guide the conformational search process, or to select a structure from a set of possible sampled candidate structures. Potential functions have been developed through an inductive approach (Sippl, 1993), where the parameters are derived by matching the results from quantum-mechanical calculations on small molecules to experimentally measured thermodynamic properties of simple molecular systems. These potential functions are then generalized to the macromolecular level based on the assumption that the complex phenomena of macromolecular systems result from the combination of a large number of interactions as found in the most basic molecular systems. This type of potential function is often referred to as the "physics-based," "physical," or "semiempirical" effective potential function, or a force field (Levitt and Warshel, 1975; Wolynes et al., 1995; Momany et al., 1975; Karplus and Petsko, 1990). The physics-based potential functions have been extensively studied over the last three decades, and have found wide uses in protein folding studies (Duan and Kollman, 1998; Lazaridis and Karplus, 2000). Nevertheless, it is difficult to use physics-based potential functions for protein structure prediction, because they are based on full atomic models and therefore require high computational cost. In addition, a physical model may not fully capture all of the important physical interactions. Readers are referred to Chapter 2 for more discussion of physics-based potential functions.

Another type of potential function is developed through a deductive approach by extracting the parameters of the potential functions from a database of known protein structures (Sippl, 1993). Because this approach implicitly incorporates many physical interactions (electrostatic, van der Walls, cation-π interactions) and the extracted potentials do not necessarily reflect true energies, it is often referred to as the "knowledge-based" or "statistical" effective potential function. In the recent past, this approach quickly gained momentum due to the rapidly growing database of experimentally determined three-dimensional protein structures. Impressive successes in protein folding, protein–protein docking, and protein design have been achieved recently using knowledge-based potential functions (Russ and Ranganathan, 2002; Venclovas et al., 2003; Méndez et al., 2005). In this chapter, we focus our discussion on this type of potential functions.

3.2 General Framework

Several approaches have been proposed to extract knowledge-based potential functions from protein structures. They can be categorized roughly into two groups. One prominent group of knowledge-based potentials are those derived from statistical analysis of a database of protein structures (Tanaka and Scheraga, 1976; Miyazawa and Jernigan, 1985; Sippl, 1990). In this class of potentials, the interacting potential between a pair of residues is estimated from its relative frequency in the database when compared with that in a reference state or a null model (Miyazawa and Jernigan, 1996; Wodak and Rooman, 1993; Sippl, 1995; Lemer et al., 1995; Jernigan and Bahar, 1996). A different class of knowledge-based potentials is based on optimization. In this case, the set of parameters for the potential functions are optimized by some criterion, e.g., by maximizing the energy gap between known native conformation and a set of alternative (or decoy) conformations (Goldstein et al., 1992; Maiorov and Crippen, 1992; Thomas and Dill, 1996a; Tobi et al., 2000; Vendruscolo and Domanyi, 1998; Vendruscolo et al., 2000; Bastolla et al., 2001; Dima et al., 2000; Micheletti et al., 2001; Dobbs et al., 2002; Hu et al., 2004).

There are three main ingredients for developing a knowledge-based potential function. We first need *protein descriptors* to describe the sequence and the shape of the native protein structure in a format that is suitable for computation. We then need to decide on a *functional form* of the potential function. Finally, we need a *method to derive the values of the parameters* for the potential function.

3.2.1 Protein Representation and Descriptors

To describe the geometric shape of a protein and its sequence of amino acid residues, a protein is frequently represented by a d-dimensional descriptor $\boldsymbol{c} \in \mathbb{R}^d$. For example, a method that is widely used is to count nonbonded contacts of 210 types of amino acid residue pairs in a protein structure. In this case, the count vector $\boldsymbol{c} \in \mathbb{R}^d$, $d = 210$, is used as the protein descriptor. Once the structural conformation of a protein $\boldsymbol{s}$ and its amino acid sequence $\boldsymbol{a}$ are given, the protein descriptions $f : (\boldsymbol{s}, \boldsymbol{a}) \mapsto \mathbb{R}^d$ will fully determine the d-dimensional vector $\boldsymbol{c}$. In the case of contact descriptor, f corresponds to the mapping provided by specific contact definition, e.g., two residues are in contact if their distance is below a cutoff threshold distance. At the residue level, the coordinates of C_α, C_β, or side-chain center can be used to represent the location of a residue. At the atom level, the coordinates of atoms are directly used, and contact may be defined by the spatial proximity of atoms. In addition, other features of protein structures can be used as protein descriptors as well, including distances between residue or atom pairs, solvent-accessible surface areas, dihedral angles of backbones and sidechains, and packing densities.

3.2.2 Functional Form

The form of the potential function $H : \mathbb{R}^d \mapsto \mathbb{R}$ determines the mapping of a d-dimensional descriptor $\boldsymbol{c}$ to a real energy value. A widely used functional form for protein potential function H is the weighted linear sum of pairwise contacts (Tanaka and Scheraga, 1976; Miyazawa and Jernigan, 1985; Tobi et al., 2000; Vendruscolo and Domanyi, 1998; Samudrala and Moult, 1998; Lu and Skolnick, 2001). The linear sum H is

$$H(f(\boldsymbol{s}, \boldsymbol{a})) = H(\boldsymbol{c}) = \boldsymbol{w} \cdot \boldsymbol{c} = \sum_i w_i c_i, \tag{3.1}$$

where "·" denotes inner product of vectors; c_i is the number of occurrence of the i-th type of descriptor. As soon as the weight vector $\boldsymbol{w}$ is specified, the potential function is fully defined. In Section 3.4.3, we will discuss a nonlinear form potential function.

3.2.3 Deriving Parameters of Potential Functions

For statistical knowledge-based potential functions which are generally linear, the weight vector $\boldsymbol{w}$ is derived by characterization of the frequency distributions of structural descriptors from a database of experimentally determined protein structures.

For optimized knowledge-based linear potential functions, $\boldsymbol{w}$ is obtained through optimization. We describe the details of these two approaches below.

3.3 Statistical Method

3.3.1 Background

In statistical methods, the observed statistical frequencies of various protein structural features are converted into effective free energies, based on the assumption that frequently observed structural features correspond to low-energy states (Tanaka and Scheraga, 1976; Miyazawa and Jernigan, 1985; Sippl, 1990). This is the Boltzmann assumption, an idea first proposed by Tanaka and Scheraga (1976) to estimate potentials for pairwise interaction between amino acids (Tanaka and Scheraga, 1976). Miyazawa and Jernigan (1985) significantly extended this idea and derived a widely used statistical potential function, where solvent terms are explicitly considered and the interactions between amino acids are modeled by contact potentials. Sippl (1990) and others (Samudrala and Moult, 1998; Lu and Skolnick, 2001; Zhou and Zhou, 2002) derived distance-dependent energy functions to incorporate both short-range and long-range pairwise interactions. The pairwise terms were further augmented by incorporating dihedral angles (Nishikawa and Matsuo, 1993; Kocher et al., 1994), solvent accessibility and hydrogen-bonding (Nishikawa and Matsuo, 1993). Singh and Tropsha (1996) derived potentials for higher-order interactions (Singh et al., 1996). More recently, Ben-Naim (1997) presented three theoretical examples to demonstrate the nonadditivity of three-body interactions (Ben-Naim, 1997). Li and Liang (2005a) identified three-body interactions in native proteins based on an accurate geometric model, and quantified systematically the nonadditivities of three-body interactions (Li and Liang, 2005b).

3.3.2 Theoretical Model

At the equilibrium state, an individual molecule may adopt many different conformations or microscopic states with different probabilities. The distribution of protein molecules among the microscopic states follows the Boltzmann distribution, which connects the potential function $H(\boldsymbol{c})$ for a microstate $\boldsymbol{c}$ to its *probability of occupancy* $\pi(\boldsymbol{c})$. This probability $\pi(\boldsymbol{c})$ or the Boltzmann factor is

$$\pi(\boldsymbol{c}) = \exp[-H(\boldsymbol{c})/kT]/Z(\boldsymbol{a}), \tag{3.2}$$

where k and T are the Boltzmann constant and the absolute temperature measured in Kelvin, respectively. The partition function $Z(\boldsymbol{a})$ is defined as

$$Z(\boldsymbol{a}) \equiv \sum_{\boldsymbol{c}} \exp[-H(\boldsymbol{c})/kT]. \tag{3.3}$$

It is a constant under the true energy function once the sequence $\boldsymbol{a}$ of a protein is specified, and is independent of the representation $f(\boldsymbol{s}, \boldsymbol{a})$ and descriptor $\boldsymbol{c}$ of the protein. If we are able to measure the probability distribution $\pi(\boldsymbol{c})$ accurately, we can obtain the knowledge-based potential function $H(\boldsymbol{c})$ from the Boltzmann distribution:

$$H(\boldsymbol{c}) = -kT \ln \pi(\boldsymbol{c}) - kT \ln Z(\boldsymbol{a}). \tag{3.4}$$

The partition function $Z(\boldsymbol{a})$ cannot be obtained directly from experimental measurements. However, at a fixed temperature, $Z(\boldsymbol{a})$ is a constant and has no effect on the different probability of occupancy for different conformations.

In order to obtain a knowledge-based potential function that encodes the sequence–structure relationship of proteins, we have to remove background energetic interactions $H'(\boldsymbol{c})$ that are independent of the protein sequence and the protein structure. These generic energetic contributions are referred to collectively as the *reference state* (Sippl, 1990). An *effective potential energy* $\Delta H(\boldsymbol{c})$ is then obtained as

$$\Delta H(\boldsymbol{c}) = H(\boldsymbol{c}) - H'(\boldsymbol{c}) = -kT \ln\left[\frac{\pi(\boldsymbol{c})}{\pi'(\boldsymbol{c})}\right] - kT \ln\left[\frac{Z(\boldsymbol{a})}{Z'(\boldsymbol{a})}\right], \tag{3.5}$$

where $\pi'(\boldsymbol{c})$ is the probability of a sequence adopting a conformation specified by the vector $\boldsymbol{c}$ in the reference state. Since $Z(\boldsymbol{a})$ and $Z'(\boldsymbol{a})$ are both constants, $-kT \ln(Z(\boldsymbol{a})/Z'(\boldsymbol{a}))$ is also a constant that does not depend on the descriptor vector $\boldsymbol{c}$. If we assume that $Z(\boldsymbol{a}) \approx Z'(\boldsymbol{a})$ as in Sippl (1990), the effective potential energy can be calculated as

$$\Delta H(\boldsymbol{c}) = -kT \ln\left[\frac{\pi(\boldsymbol{c})}{\pi'(\boldsymbol{c})}\right]. \tag{3.6}$$

To calculate $\pi(\boldsymbol{c})/\pi'(\boldsymbol{c})$, one can further assume that the probability distribution of each descriptor is independent, and we have $\pi(\boldsymbol{c})/\pi'(\boldsymbol{c}) = \prod_i [\frac{\pi(c_i)}{\pi'(c_i)}]$. Furthermore, by assuming each occurrence of the i-th descriptor is independent, we have $\prod_i [\frac{\pi(c_i)}{\pi'(c_i)}] = \prod_i \prod_{c_i} [\frac{\pi_i}{\pi_i'}]$, where π_i and π_i' are the probability of the i-th type structural feature in native proteins and the reference state, respectively. In a linear potential function, the right-hand side of Eq. (3.6) can be calculated as

$$-kT \ln\left[\frac{\pi(\boldsymbol{c})}{\pi'(\boldsymbol{c})}\right] = -kT \sum_i c_i \ln\left[\frac{\pi_i}{\pi_i'}\right]. \tag{3.7}$$

Correspondingly, to calculate the effective potential energy $\Delta H(\boldsymbol{c})$ of the system, one often assumes that $\Delta H(\boldsymbol{c})$ can be decomposed into various basic energetic terms. For a linear potential function, $\Delta H(\boldsymbol{c})$ can be calculated as:

$$\Delta H(\boldsymbol{c}) = \sum_i \Delta H(c_i) = \sum_i c_i w_i. \tag{3.8}$$

If the distribution of each c_i is assumed to be linearly independent of the others in the native protein structures, we have

$$w_i = -kT \ln \left[\frac{\pi_i}{\pi_i'} \right]. \tag{3.9}$$

In other words, the probability of each structural feature in native protein structures follows the Boltzmann distribution. This is the *Boltzmann assumption* made in nearly all statistical potential functions. Finkelstein et al. (1995) summarized protein structural features which are observed to correlate with the Boltzmann distribution. These include the distribution of residues between the surface and interior of globules, the occurrence of various ϕ, ψ, χ angles, *cis* and *trans* prolines, ion pairs, and empty cavities in protein globules (Finkelstein et al., 1995).

The probability π_i can be estimated by counting frequency of the i-th structural feature after combining all structures in the database. Clearly, the probability π_i is determined once a database of crystal structures is given. The probability π_i' is calculated as the probability of the i-th structural feature in the reference state. Therefore, the choice of the reference state has large effects and is critical for developing knowledge-based statistical potential functions.

3.3.3 Miyazawa–Jernigan Contact Potential Function

Because of the historical significance of the Miyazawa–Jernigan model in developing statistical knowledge-based potential and its wide use, we will discuss the Miyazawa–Jernigan contact potential in detail. This also provides an exposure to different technical aspects of developing statistical knowledge-based statistical functions.

Residue representation and contact definition: In the Miyazawa–Jernigan model, the l-th residue is represented as a single ball located at its side-chain center z_l. If the l-th residue is a Gly residue, which lacks a side chain, the position of the C^{α} atom is taken as z_l. A pair of residues (l, m) are defined to be in contact if the distance between their side-chain centers is less than a threshold $\theta = 6.5$ Å. Neighboring residues l and m along amino acid sequences ($|l - m| = 1$) are excluded from statistical counting because they are likely to be in spatial contact that does not reflect the intrinsic preference for interresidue interactions. Thus, a contact between the l-th and m-th residues is defined using $\Delta_{(l,m)}$:

$$\Delta_{(l,m)} = \begin{cases} 1, & \text{if } |z_l - z_m| \leq \theta \text{ and } |l - m| > 1, \\ 0, & \text{otherwise,} \end{cases}$$

where $|z_l - z_m|$ is the Euclidean distance between the l-th and m-th residues. Hence, the total number count of (i, j) contacts of residue type i with residue type j in protein p is

$$n_{(i,j);\, p} = \sum_{\substack{l,m,\\ l<m}} \Delta_{(l,m)}, \quad \text{if } (\mathbb{I}(l), \mathbb{I}(m)) = (i, j) \text{ or } (j, i), \tag{3.10}$$

where $\mathbb{I}(l)$ is the residue type of the l-th amino acid residue. The total number count of (i, j) contacts in all proteins is then

$$n_{(i,j)} = \sum_{p} n_{(i,j);\, p}, \quad i, j = 1, 2, \ldots, 20. \tag{3.11}$$

Coordination and solvent assumption: The number of different types of pairwise residue–residue contacts $n_{(i,j)}$ can be counted directly from the structure of proteins following Eq. (3.11). We also need to count the number of residue–solvent contacts. Since solvent molecules are not consistently present in X-ray crystal structures, and therefore cannot be counted exactly, Miyazawa and Jernigan made an assumption based on the model of an effective solvent molecule, which has the volume of the average volume of the 20 types of residues. Physically, one effective solvent molecule may represent several real water molecules or other solvent molecules. The number of residue–solvent contacts $n_{(i,0)}$ can be estimated as

$$n_{(i,0)} = q_i n_i - \left(\sum_{\substack{j=1;\\ j\neq i}}^{20} n_{(i,j)} + 2n_{(i,i)} \right), \tag{3.12}$$

where the subscript 0 represents the effective solvent molecule; the other indices i and j represent the types of amino acids; $n_{(i)}$ is the number of residue type i in the set of proteins; q_i is the mean coordination number of buried residue i, calculated as the number of contacts formed by a buried residue of type i averaged over a structure database. Here the assumption is that residues make the same number of contacts on average, with either effective solvent molecules [first term in Eq. (3.12] or other residues [second term in Eq. (3.12)].

For convenience, we calculate the total numbers of residues $n_{(r)}$, of residue–residue contacts $n_{(r,r)}$, of residue–solvent contacts $n_{(r,0)}$, and of pairwise contacts of any type $n_{(\cdot,\cdot)}$ as follows:

$$n_{(r)} = \sum_{i=1}^{20} n_i; \quad n_{(i,r)} = n_{(r,i)} = \sum_{j=1}^{20} n_{(i,j)}; \quad n_{(r,r)} = \sum_{i=1}^{20} n_{(i,r)};$$

$$n_{(r,0)} = n_{(0,r)} = \sum_{i=1}^{20} n_{(i,0)}; \qquad n_{(\cdot,\cdot)} = n_{(r,r)} + n_{(r,0)} + n_{(0,0)}.$$

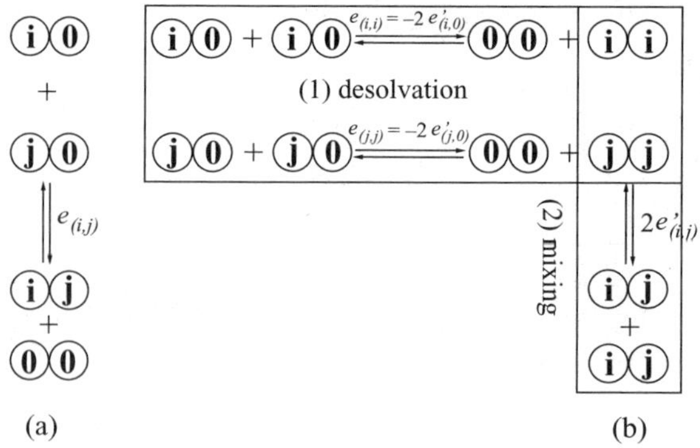

Fig. 3.1 The Miyazawa–Jernigan model of chemical reaction. Amino acid residues first go through the desolvation process, and then mix together to form pair contact interactions. The associated free energies of desolvation $e_{(i,i)}$ and mixing $e'_{(i,j)}$ can be obtained from the equilibrium constants of these two processes.

Chemical reaction model: Miyazawa and Jernigan (1985) developed a physical model based on hypothetical chemical reactions. In this model, residues of type i and j in solution need to be desolvated before they can form a contact. The overall reaction is the formation of (i, j) contacts, depicted in Fig. 3.1a. The total free energy change to form one pair of (i, j) contact from fully solvated residues of i and j is (Fig. 3.1a)

$$e_{(i,j)} = (E_{(i,j)} + E_{(0,0)}) - (E_{(i,0)} + E_{(j,0)}), \tag{3.13}$$

where $E_{(i,j)}$ is the absolute contact energy between the i-th and j-th types of residues, and $E_{(i,j)} = E_{(j,i)}$; $E_{(i,0)}$ are the absolute contact energy between the i-th residue and effective solvent, and $E_{(i,0)} = E_{(0,i)}$; likewise for $E_{(j,0)}$; $E_{(0,0)}$ are the absolute contact energies of solvent–solvent contacts (0, 0).

The overall reaction can be decomposed into two steps (Fig. 3.1b). In the first step, residues of type i and type j, initially fully solvated, are desolvated or "demixed from solvent" to form self-pairs (i, i) and (j, j). The free energy changes $e_{(i,i)}$ and $e_{(j,j)}$ upon this desolvation step can be easily seen from the desolvation process (horizontal box) in Fig. 3.1 as

$$\begin{aligned} e_{(i,i)} &= E_{(i,i)} + E_{(0,0)} - 2E_{(i,0)}, \\ e_{(j,j)} &= E_{(j,j)} + E_{(0,0)} - 2E_{(j,0)}, \end{aligned} \tag{3.14}$$

where $E_{(i,i)}$, $E_{(j,j)}$ are the absolute contact energies of self-pairs (i, i) and (j, j), respectively. In the second step, the contacts in (i, i) and (j, j) pairs are broken and residues of type i and residues of type j are mixed together to form two (i, j) pairs.

The free energy change upon this mixing step $2e'_{(i,j)}$ is (vertical box in Fig. 3.1)

$$2e'_{(i,j)} = 2E_{(i,j)} - (E_{(i,i)} + E_{(j,j)}). \tag{3.15}$$

Denote the free energy changes upon the mixing of residues of type i and solvent as $e'_{(i,0)}$, We have

$$-2e'_{(i,0)} = e_{(i,i)} \quad \text{and} \; -2e'_{(j,0)} = e_{(j,j)}, \tag{3.16}$$

which can be obtained from Eqs. (3.14) and (3.15) after substituting "j" with "0." Following the reaction model of Fig. 3.1b, the total free energy change to form one pair of (i, j) can be written as

$$2e_{(i,j)} = 2e'_{(i,j)} + e_{(i,i)} + e_{(j,j)} \tag{3.17a}$$

$$= 2e'_{(i,j)} - 2e'_{(i,0)} - 2e'_{(j,0)}. \tag{3.17b}$$

Contact energy model: The total energy of the system is due to the contacts between residue–residue, residue–solvent, solvent–solvent:

$$\begin{aligned} E_c &= \sum_{i=0}^{20} \sum_{\substack{j=0;\\ j\geq i}}^{20} E_{(i,j)} n_{(i,j)} \\ &= \sum_{i=1}^{20} \sum_{\substack{j=1;\\ j\geq i}}^{20} E_{(i,j)} n_{(i,j)} + \sum_{i=1}^{20} E_{(i,0)} n_{(i,0)} + E_{(0,0)} n_{(0,0)}. \end{aligned} \tag{3.18}$$

Because the absolute contact energy $E_{(i,j)}$ is difficult to measure and knowledge of this value is unnecessary for studying the dependence of energy on protein conformation, we can simplify Eq. (3.18) further. Our goal is to separate out terms that do not depend on contact interactions and hence do not depend on the conformation of the molecule. Equation (3.18) can be rewritten as

$$E_c = \sum_{i=0}^{20} (2E_{(i,0)} - E_{(0,0)}) q_i n_{(i)}/2 + \sum_{i=1}^{20} \sum_{\substack{j=1;\\ j\geq i}}^{20} e_{(i,j)} n_{(i,j)} \tag{3.19a}$$

$$= \sum_{i=0}^{20} E_{(i,i)} q_i n_{(i)}/2 + \sum_{i=0}^{20} \sum_{\substack{j=0;\\ j\geq i}}^{20} e'_{(i,j)} n_{(i,j)} \tag{3.19b}$$

by using Eqs. (3.12) and (3.13). Here only the second terms in Eqs. (3.19a) and (3.19b) are dependent on protein conformations. Therefore, only either $e_{(i,j)}$ or $e'_{(i,j)}$

needs to be estimated. Since the number of residue–residue contacts can be counted directly while the number of residue–solvent contacts is more difficult to obtain, Eq. (3.19a) is more convenient for calculating the total contact energy of protein conformations. Both $e_{(i,j)}$ and $e'_{(i,j)}$ are termed *effective contact energies* and their values were reported in Miyazawa and Jernigan (1996).

Estimating effective contact energies: quasi-chemical approximation: The effective contact energies $e_{(i,j)}$ in Eq. (3.19a) can be estimated in kT units by assuming that the solvent and solute molecules are in quasi-chemical equilibrium for the reaction depicted in Fig. 3.1a:

$$e_{(i,j)} = -\ln \frac{[m_{(i,j)}/m_{(\cdot,\cdot)}][m_{(0,0)}/m_{(\cdot,\cdot)}]}{[m_{(i,0)}/m_{(\cdot,\cdot)}][m_{(j,0)}/m_{(\cdot,\cdot)}]} = -\ln \frac{m_{(i,j)}m_{(0,0)}}{m_{(i,0)}m_{(j,0)}}, \tag{3.20}$$

where $m_{(i,j)}$, $m_{(i,0)}$, and $m_{(0,0)}$ are the contact numbers of pairs between residue type i and j, residue type i and solvent, and solvent and solvent, respectively. $m_{(\cdot,\cdot)}$ is the total number of contacts in the system and is canceled out. Similarly, $e'_{(i,j)}$ and $e'_{(i,0)}$ can be estimated from the model depicted in Fig. 3.1b:

$$2e'_{(i,j)} = -\ln \frac{[m_{(i,j)}]^2}{m_{(i,i)}m_{(j,j)}}, \tag{3.21a}$$

$$2e'_{(i,0)} = -\ln \frac{[m_{(i,0)}]^2}{m_{(i,i)}m_{(0,0)}}. \tag{3.21b}$$

Based on these models, two different techniques have been developed to obtain effective contact energy parameters. Following the hypothetical reaction in Fig. 3.1a, $e_{(i,j)}$ can be directly estimated from Eq. (3.20), as was done by Zhang and Kim (2000). Alternatively, one can follow the hypothetical two-step reaction in Fig. 3.1b and estimate each term in Eq. (3.17b) for $e_{(i,j)}$ by using Eq. (3.21). Because the second approach leads to additional insight about the desolvation effects ($e'_{(i,0)}$) and the mixing effects ($e'_{(i,j)}$) in contact interactions, we follow this approach in subsequent discussions. The first approach will become self-evident after our discussion.

Models of reference state: In reality, the true fraction $\frac{m_{(i,j)}}{m_{(\cdot,\cdot)}}$ of contacts of (i, j) type among all pairwise contacts $(\cdot, \cdot)$ is unknown. One can approximate this by calculating its mean value from sampled structures in the database. We have

$$\frac{m_{(i,j)}}{m_{(\cdot,\cdot)}} \approx \frac{\sum_p n_{(i,j);p}}{\sum_p n_{(\cdot,\cdot);p}}; \quad \frac{m_{(i,0)}}{m_{(\cdot,\cdot)}} \approx \frac{\sum_p n_{(i,0);p}}{\sum_p n_{(\cdot,\cdot);p}}; \quad \frac{m_{(0,0)}}{m_{(\cdot,\cdot)}} \approx \frac{\sum_p n_{(0,0);p}}{\sum_p n_{(\cdot,\cdot);p}},$$

where i and $j \neq 0$. However, this yields a biased estimation of $e'_{(i,j)}$ and $e_{(i,j)}$. When effective solvent molecules, residues of i-th type and residues of j-th type are randomly mixed, $e'_{(i,j)}$ will not be equal to 0 as should be because of differences in amino acid composition among proteins in the database. Therefore, a reference state must be used to remove this bias.

In the work of Miyazawa and Jernigan, the effective contact energies for mixing two types of residues $e'_{(i,j)}$ and for solvating a residue $e'_{(i,0)}$ are estimated based on two different random mixture reference states (Miyazawa and Jernigan, 1985). In both cases, the contacting pairs in a structure are randomly permuted, but the global conformation is retained. Hence, the total number of residue–residue, residue–solvent, solvent–solvent contacts remain unchanged.

The first random mixture reference state for desolvation contains the same set of residues of the protein p and a set of effective solvent molecules. We denote the overall number of (i, i), $(i, 0)$, $(0, 0)$ contacts in this random mixture state after summing over all proteins as $c'_{(i,i)}$, $c'_{(i,0)}$, and $c'_{(0,0)}$, respectively. $c'_{(i,i)}$ can be computed as

$$c'_{(i,i)} = \sum_p \left[\frac{q_i\, n_{i;p}}{\sum_k q_k\, n_{k;p}} \right]^2 \cdot n_{(\cdot,\cdot);p}, \tag{3.22}$$

where Miyazawa and Jernigan assumed that the average coordination number of residue i in all proteins is q_i. Therefore, a residue of type i makes $q_i n_{i;p}$ number of contacts in protein p. Similarly, the number of $(i, 0)$ contacts $c'_{(i,0)}$ can be computed as

$$c'_{(i,0)} = \sum_p \left[\frac{q_i\, n_{i;p}}{\sum_k q_k\, n_{k;p}} \right] n_{(\cdot,0);p}. \tag{3.23}$$

From the horizontal box in Fig. 3.1, the effective contact energy $e'_{(i,0)}$ can now be computed as

$$2e'_{(i,0)} = -\ln \left[\frac{n^2_{(i,0)}}{n_{(i,i)} n_{(0,0)}} \Big/ \frac{c'^{\,2}_{(i,0)}}{c'_{(i,i)} c'_{(0,0)}} \right] \qquad (i \neq 0). \tag{3.24}$$

The second random mixture reference state for mixing contains the exact same set of residues as the protein p, but all residues are randomly mixed. We denote the number of (i, j) contacts in this random mixture as $c_{(i,j);p}$. The overall number of (i, j) contacts in the full protein set $c_{(i,j)}$ is the sum of $c_{(i,j);p}$ over all proteins:

$$c_{(i,j)} = \sum_p \left[\frac{n_{(i,\cdot);p}}{n_{(\cdot,\cdot);p}} \right] \left[\frac{n_{(j,\cdot);p}}{n_{(\cdot,\cdot);p}} \right] \cdot n_{(\cdot,\cdot);p}. \tag{3.25}$$

From the vertical box in Fig. 3.1, the effective contact energy $e'_{(i,j)}$ can now be computed as

$$2e'_{(i,j)} = -\ln\left[\frac{n^2_{(i,j)}}{n_{(i,i)}n_{(j,j)}} \Big/ \frac{c^2_{(i,j)}}{c_{(i,i)}c_{(j,j)}}\right], \qquad i \text{ or } j \neq 0. \tag{3.26}$$

The compositional bias is removed by the denominator in Eq. (3.26), and $e'_{(i,j)}$ now equals 0.

Although $c'_{(0,0)}$ can be estimated from Eq. (3.21b) by assuming that $e'_{(i,0)} = 0$ in a reference state, Zhang and DeLisi (1997) simplified the Miyazawa–Jernigan process by further assuming that the number of solvent–solvent contacts in both reference states is the same as in the native state (Zhang et al., 1997):

$$c'_{(0,0)} = n_{(0,0)}. \tag{3.27}$$

Therefore, $c'_{(0,0)}$ and $n_{(0,0)}$ are canceled out in Eq. (3.24) and not needed for calculating $e'_{(i,0)}$. This treatment systematically subtracts a constant scaling energy from all effective energies $e_{(i,j)}$, and should produce exactly the same relative energy values for protein conformations as Miyazawa–Jernigan's original work, with the difference of a constant offset value. In fact, Miyazawa and Jernigan (1996) showed that this constant scaling energy is the effective contact energy $e_{\hat{r}\hat{r}}$ between the average residue $\hat{r}$ of the 20 residue types, and suggested that $e_{(i,j)} - e_{\hat{r}\hat{r}}$ be used to measure the stability of a protein structure (Miyazawa and Jernigan, 1996).

Hydrophobic nature of Miyazawa–Jernigan contact potential: In the relation of Eq. (3.17b), $e_{(i,j)} = e'_{(i,j)} - (e'_{(i,0)} + e'_{(j,0)})$, the Miyazawa–Jernigan effective contact energy $e_{(i,j)}$ is composed of two types of terms: the desolvation terms $e'_{(i,0)}$ and $e'_{(j,0)}$ and the mixing term $e'_{(i,j)}$. The desolvation term of residue type i, that is, $-e'_{(i,0)}$ or $e_{(i,i)}/2$ (Fig. 3.1), is the energy change due to the desolvation of residue i, the formation of the i–i self-pair, and the solvent–solvent pair. The value of this term $e_{(i,i)}/2$ should correlate well with the hydrophobicity of residue type i (Miyazawa and Jernigan, 1985; Li et al., 1997), although for charged amino acids this term also incorporates unfavorable electrostatic potentials of self-pairing. The mixing term $e'_{(i,j)}$ is the energy change accompanying the mixing of two different types of amino acids of i and j to form a contact pair i–j after breaking self-pairs i–i and j–j. Its value measures the tendency of different residues to mix together. For example, the mixing between two residues with opposite charges is more favorable than mixing between other types of residues, because of the favorable electrostatic interactions.

Important insights into the nature of residue–residue contact interactions can also be obtained by a quantitative analysis of the desolvation terms and the mixing terms. Among different types of contacts, the average difference of the desolvation terms is 9 times larger than that of the mixing terms [see Table 3.1 taken from

Table 3.1 Contact energies in kT units; $e_{(i,j)}$ for upper half and diagonal and $e'_{(i,j)}$ for lower half (from Miyazawa and Jernigan, 1996)

	Cys	Met	Phe	Ile	Leu	Val	Trp	Tyr	Ala	Gly	Thr	Ser	Asn	Gln	Asp	Glu	His	Arg	Lys	Pro
Cys	−5.44	−4.99	−5.80	−5.50	−5.83	−4.96	−4.95	−4.16	−3.57	−3.16	−3.11	−2.86	−2.59	−2.85	−2.41	−2.27	−3.60	−2.57	−1.95	−3.07
Met	0.46	−5.46	−6.56	−6.02	−6.41	−5.32	−5.55	−4.91	−3.94	−3.39	−3.51	−3.03	−2.95	−3.30	−2.57	−2.89	−3.98	−3.12	−2.48	−3.45
Phe	0.54	−0.20	−7.26	−6.84	−7.28	−6.29	−6.16	−5.66	−4.81	−4.13	−4.28	−4.02	−3.75	−4.10	−3.48	−3.56	−4.77	−3.98	−3.36	−4.25
Ile	0.49	−0.01	0.06	−6.54	−7.04	−6.05	−5.78	−5.25	−4.58	−3.78	−4.03	−3.52	−3.24	−3.67	−3.17	−3.27	−4.14	−3.63	−3.01	−3.76
Leu	0.57	0.01	0.03	−0.08	−7.37	−6.48	−6.14	−5.67	−4.91	−4.16	−4.34	−3.92	−3.74	−4.04	−3.40	−3.59	−4.54	−4.03	−3.37	−4.20
Val	0.52	0.18	0.10	−0.01	−0.04	−5.52	−5.18	−4.62	−4.04	−3.38	−3.46	−3.05	−2.83	−3.07	−2.48	−2.67	−3.58	−3.07	−2.49	−3.32
Trp	0.30	−0.29	0.00	0.02	0.08	0.11	−5.06	−4.66	−3.82	−3.42	−3.22	−2.99	−3.07	−3.11	−2.84	−2.99	−3.98	−3.41	−2.69	−3.73
Tyr	0.64	−0.10	0.05	0.11	0.10	0.23	−0.04	−4.17	−3.36	−3.01	−3.01	−2.78	−2.76	−2.97	−2.76	−2.79	−3.52	−3.16	−2.60	−3.19
Ala	0.51	0.15	0.17	0.05	0.13	0.08	0.07	0.09	−2.72	−2.31	−2.32	−2.01	−1.84	−1.89	−1.70	−1.51	−2.41	−1.83	−1.31	−2.03
Gly	0.68	0.46	0.62	0.62	0.65	0.51	0.24	0.20	0.18	−2.24	−2.08	−1.82	−1.74	−1.66	−1.59	−1.22	−2.15	−1.72	−1.15	−1.87
Thr	0.67	0.28	0.41	0.30	0.40	0.36	0.37	0.13	0.10	0.10	−2.12	−1.96	−1.88	−1.90	−1.80	−1.74	−2.42	−1.90	−1.31	−1.90
Ser	0.69	0.53	0.44	0.59	0.60	0.55	0.38	0.14	0.18	0.14	−0.06	−1.67	−1.58	−1.49	−1.63	−1.48	−2.11	−1.62	−1.05	−1.57
Asn	0.97	0.62	0.72	0.87	0.79	0.77	0.30	0.17	0.36	0.22	0.02	0.10	−1.68	−1.71	−1.68	−1.51	−2.08	−1.64	−1.21	−1.53
Gln	0.64	0.20	0.30	0.37	0.42	0.46	0.19	−0.12	0.24	0.24	−0.08	0.11	−0.10	−1.54	−1.46	−1.42	−1.98	−1.80	−1.29	−1.73
Asp	0.91	0.77	0.75	0.71	0.89	0.89	0.30	−0.07	0.26	0.13	−0.14	−0.19	−0.24	−0.09	−1.21	−1.02	−2.32	−2.29	−1.68	−1.33
Glu	0.91	0.30	0.52	0.46	0.55	0.55	0.00	−0.25	0.30	0.36	−0.22	−0.19	−0.21	−0.19	0.05	−0.91	−2.15	−2.27	−1.80	−1.26
His	0.65	0.28	0.39	0.66	0.67	0.70	0.08	0.09	0.47	0.50	0.16	0.26	0.29	0.31	−0.19	−0.16	−3.05	−2.16	−1.35	−2.25
Arg	0.93	0.38	0.42	0.41	0.43	0.47	−0.11	−0.30	0.30	0.18	−0.07	−0.01	−0.02	−0.26	−0.91	−1.04	0.14	−1.55	−0.59	−1.70
Lys	0.83	0.31	0.33	0.32	0.37	0.33	−0.10	−0.46	0.11	0.03	−0.19	−0.15	−0.30	−0.46	−1.01	−1.28	0.23	0.24	−0.12	−0.97
Pro	0.53	0.16	0.25	0.39	0.35	0.31	−0.33	−0.23	0.20	0.13	0.04	0.14	0.18	−0.08	0.14	0.07	0.15	−0.05	−0.04	−1.75

Miyazawa and Jernigan (1996)]. Thus, a comparison of the values of $(e_{(i,i)} + e_{jj})/2$ and $e'_{(i,j)}$ clearly shows that the desolvation term plays the dominant role in determining the energy difference among different conformations.

Similar conclusion can be drawn by an eigenvalue decomposition analysis of the Miyazawa–Jernigan matrix $\boldsymbol{M}$, which is made up of $e_{(i,j)}$ values alone, without the knowledge of the mixing terms $e'_{(i,j)}$ (Li et al., 1997). The $\boldsymbol{M}$ matrix is a 20×20 real symmetric matrix, and thus can be reconstructed based on the following spectral decomposition:

$$e_{(i,j)} = [\sum_{k=1}^{N} \lambda_k \boldsymbol{v}_k \boldsymbol{v}_k]_{ij} = \sum_{k=1}^{N} \lambda_k \boldsymbol{v}_k(i) \boldsymbol{v}_k(j), \tag{3.28}$$

where λ_k and $\boldsymbol{v}_k$ are the k-th largest eigenvalue and the corresponding eigenvector, respectively; $\boldsymbol{v}_k(i)$ is the i-th component of the k-th eigenvector. Li et al. (1997) found that there are two dominant eigenvalues λ_1 and λ_2, and the corresponding two eigenvectors are strongly correlated after a shift and a rescaling operation, i.e., $\boldsymbol{v}_2 = \alpha \boldsymbol{u} + \beta \boldsymbol{v}_1$. Here, $\boldsymbol{u}$ is the $\mathbf{1}$ vector with each component equal to 1 and α and β are scalars. Therefore, $\boldsymbol{M}$ can be well-approximated with only one eigenvector $\boldsymbol{v}_1$ corresponding to the largest eigenvalue λ_1. For each entry $e_{(i,j)}$ of the matrix $\boldsymbol{M}$, we have the following approximation:

$$e_{(i,j)} \approx \lambda_1 \boldsymbol{v}_1(i) \boldsymbol{v}_1(j) + \lambda_2 \boldsymbol{v}_2(i) \boldsymbol{v}_2(j) \approx c_0 + c_1(q_i + q_j) + c_2 q_i q_j, \tag{3.29}$$

where $q_i \equiv v_1(i)$, and c_0, c_1, and c_2 are constants. To better understand the underlying physical implications, Eq. (3.29) can be rewritten in the following form:

$$e_{(i,j)} \approx h_i + h_j - c_2(q_i - q_j)^2/2, \tag{3.30}$$

where

$$h_i = c_0/2 + c_1 q_i + (c_2/2) q_i^2.$$

Here $h_i + h_j$ is a single-body term and is interpreted as the desolvation term in Li et al. (1997); $-c_2(q_i - q_j)^2/2$ is a two-body term interpreted as the mixing term and the magnitude of the mixing term is significantly smaller than that of $h_i + h_j$. This result is not surprising and is consistent with the original model of Miyazawa–Jernigan contact matrix $\boldsymbol{M}$, where $e_{(i,j)} \equiv e'_{(i,j)} - (e'_{(i,0)} + e'_{(j,0)})$.

To summarize, the quantitative analysis of Miyazawa–Jernigan contact energies reveals that hydrophobic effect is the dominant driving force for protein folding. To a large extent, this conclusion justifies the HP model proposed by Chan and Dill (1990) where only hydrophobic interactions are included in studies of simple models of protein folding (Chan and Dill, 1990).

3.3.4 Distance-Dependent Potential Functions

In the Miyazawa–Jernigan potential function, interactions between amino acids are assumed to be short-ranged and a distance cutoff is used to define the occurrence of a contact. This type of statistical potential is referred to as the "contact potential." Another type of statistical potential allows modeling of residue interactions that are distance-dependent. The distance of interactions is usually divided into a number of small intervals or bins, and the potential functions are derived by applying Eq. (3.9) for individual distance intervals.

Formulation of distance-dependent potential functions: In distance-dependent statistical potential functions, Eq. (3.9) can be written in several forms. To follow the conventional notations, we use (i, j) to represent the k-th protein descriptor c_k for pairwise interactions between residue type i and residue type j. From Eq. (3.9), we have

$$\begin{aligned} \Delta H(i, j; d) &= -\ln \frac{\pi(i, j; d)}{\pi'(i, j; d)} = -\ln \frac{n_{(i,j;d)}/n}{\pi'(i, j; d)} \\ &= -\ln \frac{n_{(i,j;d)}}{n'_{(i,j;d)}}, \end{aligned} \tag{3.31a}$$

where $(i, j; d)$ represents an interaction between a specific residue pair (i, j) at distance d, $\Delta H(i, j; d)$ is the contribution from the (i, j) type of residue pairs at distance d, $\pi(i, j; d)$ and $\pi'(i, j; d)$ are the observed and expected probabilities of this distance-dependent interaction, respectively, $n_{(i,j;d)}$ is the observed number of $(i, j; d)$ interactions, n is the observed total number of all pairwise interactions in a database, and $n'_{(i,j;d)}$ is the expected number of $(i, j; d)$ interactions when the total number of all pairwise interactions in the reference state is set to n.

Since the expected joint probability $\pi'(i, j; d)$ for the reference is not easy to estimate, Sippl (1990) replaces Eq. (3.9) with

$$\begin{aligned} \Delta H(i, j; d) &= -\ln \frac{\pi(i, j \mid d)}{\pi'(i, j \mid d)} = -\ln \frac{n_{(i,j;d)}/n_{(d)}}{\pi'(i, j \mid d)} \\ &= -\ln \frac{n_{(i,j;d)}}{n'_{(i,j;d)}}, \end{aligned} \tag{3.31b}$$

where $\pi(i, j \mid d)$ and $\pi'(i, j \mid d)$ are the observed and expected probability of interaction of residue pairs (i, j) given the distance interval d, respectively; $n_{(d)}$ is the observed total number of all pairwise interactions at the distance d; $n'_{(i,j;d)} = \pi'(i, j \mid d) \cdot n(d)$ is the expected number of (i, j) interactions at d when the total number of all pairwise interactions at this distance d in the reference state is set to $n_{(d)}$. There are several variations of potential function of this form, including the "Knowledge-Based Potential function" (KBP) by Lu and Skolnick (2001).

In the work of developing the "Residue-specific All-atom Probability Discriminatory Function" (RAPDF) (Samudrala and Moult, 1998), Samudrala and Moult alternatively replaced Eq. 3.9 with

$$\begin{aligned}\Delta H(i, j; d) &= -\ln \frac{\pi(d \mid i, j)}{\pi'(d \mid i, j)} = -\ln \frac{n_{(i,j;d)}/n_{(i,j)}}{\pi'(d \mid i, j)} \\ &= -\ln \frac{n_{(i,j;d)}}{n'_{(i,j;d)}},\end{aligned} \tag{3.31c}$$

where $\pi(d \mid i, j)$ and $\pi'(d \mid i, j)$ are the observed and expected probability of interaction at the distance d for a given pair of residues (i, j), respectively; $n_{(i,j)}$ is the observed total number of interactions for (i, j) pairs regardless of the distance. $n'_{(i,j;d)} = \pi'(d \mid i, j) \cdot n_{(i,j)}$ is the expected number of (i, j) interactions at distance d when the total number of (i, j) interactions in the reference state is set to $n_{(d)}$.

The knowledge-based potential functions of Eqs. (3.31a), (3.31b), and (3.31c) can all be written using the unifying formula based on the number counts of interactions:

$$\Delta H(i, j; d) = -\ln \left[\frac{n_{(i,j;d)}}{n'_{(i,j;d)}} \right]. \tag{3.32}$$

Clearly, the different ways of assigning $n'_{(i,j;d)}$ make the potential functions differ from each other substantially, since the method to calculate $n_{(i,j;d)}$ is essentially the same for many potential functions. In other words, the model of reference state used to compute $n'_{(i,j;d)}$ is critical for distance-dependent energy functions.

Different models of reference states: Sippl first proposed the "uniform density" model of reference state, where the probability density function for a pair of contacting residues (i, j) is uniformly distributed along the distance vector connecting them: $\pi'(i, j \mid d) = \pi'(i, j)$ (Sippl, 1990). Lu and Skolnick made use of this type of reference state to calculate the expected number of (i, j) interactions at distance d as (Lu and Skolnick, 2001)

$$n'_{(i,j;d)} = \pi'(i, j \mid d) \cdot n_{(d)} = \pi'(i, j) \cdot n_{(d)}.$$

The expected probability $\pi'(i, j)$ is estimated using the random mixture approximation as

$$\pi'(i, j) = \chi_i \chi_j,$$

where χ_i and χ_j are the mole fractions of residue type i and j, respectively.

Samudrala and Moult (1998) made use of another type of reference state, where the probability of the distance between a pair of residues (i, j) being d is independent of the contact types (i, j) (Samudrala and Moult, 1998):

$$\pi'(d \mid i, j) = \pi'(d).$$

The expected number of (i, j) interactions at distance d in Eq. (3.31c) becomes

$$n'_{(i,j;d)} = \pi'(d \mid i, j) \cdot n_{(i,j)} = \pi'(d) \cdot n_{(i,j)},$$

where $\pi'(r)$ is estimated from $\pi(r)$:

$$\pi'(d) = \pi(d) = n_{(d)}/n.$$

Ideal gas reference state: In the uniform density model of Sippl, the same density of a particular residue pair (i, j) along a line could result from very different volume distribution of (i, j) pairs in specific regions of the protein. For example, one spherical shell proximal to the molecular center could be sparsely populated with residues, and another distant shell could be densely populated, but all may have the same density of (i, j) pairs along the same radial vector. Zhou and Zhou (2002) developed a new reference state (called DFIRE for "Distance-scaled, Finite Ideal-gas REference state") where residues follow uniform distribution everywhere in the protein (Zhou and Zhou, 2002). Assuming that residues can be modeled as noninteracting points (i.e., as ideal gas molecules), the distribution of interacting pairs should follow the uniform distribution not only along any vector lines, but also in the whole volume of the protein.

When the distance between a pair of residues (i, j) is at a threshold distance $d_\theta = 14.5$ Å, the interaction energy between them can be considered to be 0. Therefore, residue type i and type j form pairs at the distance d_θ purely by random, and the observed number of (i, j) pairs at the distance d_θ can be considered the same as the expected number of (i, j) pairs at the distance d_θ in the reference state. Denote v_d as the volume of a spherical shell of width Δd at a distance d from the center. The expected number of interactions (i, j) at the distance d after volume correction is

$$n'_{(i,j;d)} = n_{(i,j;d_\theta)} \cdot \frac{v_d}{d_\theta} = n_{(i,j,d_\theta)} \cdot \left(\frac{d}{d_\theta}\right)^{\alpha} \frac{\Delta d}{\Delta d_\theta}.$$

For a protein molecule, $n'_{(i,j;d)}$ will not increase as r^2 because of its finite size. In addition, it is well-known that the volume of a protein molecule cannot be treated as a solid body, as there are numerous voids and pockets in the interior. This implies that the number density for a very large molecule will also not scale as d^2 (Liang and Dill, 2001). Zhou and Zhou (2002) assumed that $n'_{(i,j;d)}$ increase in d^α rather than d^2, where the exponent α needs to be determined. To estimate the α value, each protein p in the database is reshaped into a ball of radius $c_p R_{g;p}$, where $R_{g;p}$ is the radius of gyration of protein p, and residues are distributed uniformly in this reshaped ball. Here c_p takes the value so that in the reshaped molecule, the number of total interacting pairs at d_θ distance is about the same as that observed in the native protein p, namely:

$$\sum_{(i,j)} n'_{(i,j;\,d_\theta)} = \sum_{(i,j)} n_{(i,j;\,d_\theta)}$$

for protein p. Once the value of c_p is determined and hence the effective radius $c_p R_{g;\,p}$ for each native protein is known, the number of interacting pairs $n_{(d)}$ at distance d can be counted directly from the reshaped ball. Zhou and Zhou further defined a reduced distance-dependent function $f(d) = n_{(d)}/d^{\alpha}$ and the relative fluctuation δ of $f(d)$:

$$\delta = \left[\frac{1}{n_b} \sum_d (f(d) - \bar{f})^2 / (\bar{f}) \right]^{1/2},$$

where $\bar{f} = \sum_d f(d)/n_b$, and n_b is the total number of distance shells, all of which have the same thickness. α is then estimated by minimizing the relative fluctuation δ. The rationale is that since idealized residues are points and are uniformly distributed in the reshaped ball, δ should be 0. In their study, α was found to be 1.61 (Zhou and Zhou, 2002).

3.3.5 Geometric Potential Functions

The effectiveness of potential function also depends on the representation of protein structures. Another class of knowledge-based statistical potentials is based on the computation of various geometric constructs that reflect the shape of the protein molecules more accurately. These geometric constructs include the Voronoi diagram (McConkey et al., 2003), the Delaunay triangulation (Singh et al., 1996; Zheng et al., 1997; Carter et al., 2001; Krishnamoorthy and Tropsha, 2003), and the alpha shape (Li et al., 2003; Li and Liang, 2005a,b) of the protein molecules. Geometric potential functions have achieved significant successes in many fields. For example, the potential function developed by McConkey et al. is based on the Voronoi diagram of the atomic structures of proteins, and is among the best performing atom-level potential functions in decoy discrimination (McConkey et al., 2003). Because the alpha shape of the molecule contains rich topological, combinatorial, and metric information, and has a strong theoretical foundation, we discuss the geometric potential functions in more detail below as an example of this class of potential function.

Geometric model: In Miyazawa–Jernigan and other contact potential functions, pairwise contact interactions are declared if two residues or atoms are within a specific cutoff distance. Contacts by distance cutoff can potentially include many implausible noncontacting neighbors, which have no significant physical interaction

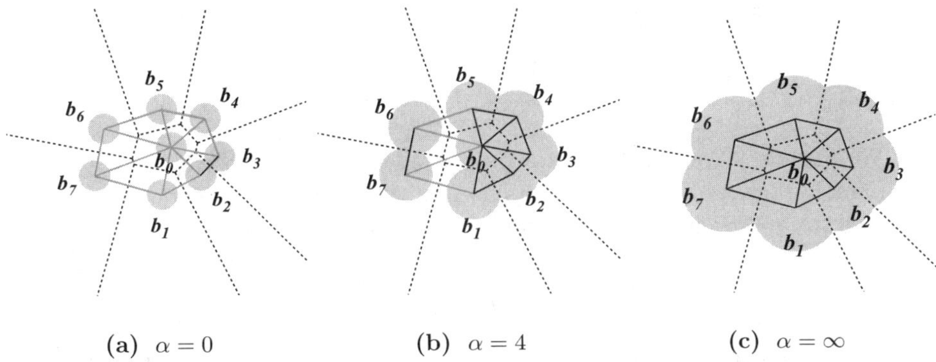

Fig. 3.2 Schematic drawing of the Delaunay complex and the alpha shape of a two-dimensional molecule. The Voronoi region of a ball is the set of points closest to it when measured in power distance. If two Voronoi regions share a boundary, i.e., if there is a Voronoi edge (dashed line), we draw a Delaunay edge (solid line in gray or black) between these two Voronoi vertices. A Delaunay edge is therefore the *dual* of a Voronoi edge. All Delaunay edges incident to ball residue b_i form the *1-star* for b_i, denoted as $St_1(b_i)$. When the balls are inflated by increasing the α value, more balls overlap, and more Voronoi edges intersect with the balls. Therefore, more *dual* Delaunay edges are included in the alpha shape (shown as black solid line segments). (a) When $\alpha = 0.0$, the balls are not inflated and there is only one alpha edge $\sigma_{2,3}$ between ball b_2 and ball b_3. (b) When $\alpha = 4.0$, the balls are inflated and their radii are $\sqrt{r^2 + 4.0}$. There are six alpha edges: $\sigma_{0,1}$, $\sigma_{0,2}$, $\sigma_{0,3}$, $\sigma_{0,4}$, $\sigma_{0,4}$, $\sigma_{0,5}$, and $\sigma_{6,7}$. For a ball b_i, the set of residue balls connected to it by alpha edges are called the near neighbors of the ball. The number of this set of residue balls is defined as the *degree of near neighbors* of the residue ball b_i, denoted as ρ_i. For example, $\rho_0 = 5$, and $\rho_7 = 1$. (c) When $\alpha = \infty$, all the Delaunay edges become alpha edges ($\alpha = 16.0$ is used for drawing). Hence, all long-range interactions not intervened by a third residue are included.

(Bienkowska et al., 1999). Whether or not a pair of residues can make physical contact depends not only on the distance between their center positions (such as C_α or C_β, or geometric centers of side chain), but also on the size and the orientations of side chains (Bienkowska et al., 1999). Furthermore, two atoms close to each other may in fact be shielded from contact by other atoms. By occupying the intervening space, other residues can block a pair of residues from directly interacting with each other. Inclusion of these fictitious contact interactions would be undesirable.

The geometric potential function solves this problem by identifying interacting residue pairs following the edges computed in the alpha shape. Details of alpha shape can be found in Chapter 6. When the parameter α is set to be 0, residue contact occurs if residues or atoms from nonbonded residues share a Voronoi edge, and this edge is at least partially contained in the body of the molecule. Fig. 3.2 illustrates the basic ideas.

Distance- and packing-dependent geometric potential function: For two nonbonded residue balls b_i of radius r_i with its center located at z_i and b_j of radius r_j at z_j, they form an alpha contact $(i, j \mid \alpha)$ if their Voronoi regions intersect and these residue balls also intersect after their radii are inflated to $r_i(\alpha) = (r_i^2 + \alpha)^{1/2}$ and

$r_j(\alpha) = (r_j^2 + \alpha)^{1/2}$, respectively. That is, the alpha contact $(i, j \mid \alpha)$ exists when

$$|z_i - z_j| < \left(r_i{}^2 + \alpha\right)^{1/2} + \left(r_j{}^2 + \alpha\right)^{1/2}, \quad \sigma_{i,j} \in \mathcal{K}_\alpha \text{ and } |i - j| > 1.$$

We further define the *1-star* for each residue ball b_i as: $St_1(b_i) = \{(b_i, b_j) \in \mathcal{K}_\alpha\}$, namely, the set of 1-simplices with b_i as a vertex. The *near neighbors* of b_i are derived from $St_1(b_i)$ and are defined as

$$\mathcal{N}_\alpha(b_i) \equiv \{b_j | \sigma_{i,j} \in \mathcal{K}_\alpha\}, \quad \alpha = 4.0,$$

and the *degree of near neighbors* ρ_i of residue b_i is defined as the size of this set of residues:

$$\rho_i \equiv |\mathcal{N}_\alpha(b_i)|, \quad \alpha = 4.0.$$

The degree of near neighbors ρ_i is a parameter related to the local packing density and hence indirectly the solvent accessibility around the residue ball b_i (Fig. 3.2b). A large ρ_i value indicates high local packing density and less solvent accessibility, and a small ρ_i value indicates low local packing density and high solvent accessibility. Similarly, the *degree of near neighbors* for a pair of residues is defined as

$$\rho_{(i,j)} \equiv |\mathcal{N}_\alpha(b_i, b_j)| = |\mathcal{N}_\alpha(b_i)| + |\mathcal{N}_\alpha(b_j)|, \quad \alpha = 4.0.$$

Reference state and collection of noninteracting pairs: We denote the shortest path length between residue b_i and residue b_j as $L_{(i,j)}$, which is the fewest number of alpha edges ($\alpha = 4$) that connects b_i and b_j. The reference state of the geometric potential is based on the collection of all non-interacting residue pairs (i, j):

$$\{(i, j) | L_{(i,j)} = 3\}.$$

Any (i, j) pair in this reference state is intercepted by two residues (Fig. 3.3). We assume that there is no attractive or repulsive interactions between them, because of the shielding effect by the two intervening residues. Namely, residue i and residue j form a pair only by random chance, and any properties associated with b_i, such as packing density, side-chain orientation, are independent of the same properties associated with b_j.

Statistical model: pairwise potential and desolvation potential: According to Eq. (3.9), the packing- and distance-dependent statistical potential of residue pair (k, l) at the packing environment $\rho_{(k,l)}$ and the distance specified by α is

$$H(k, l, \rho_{(k,l)} \mid \alpha) = -KT \ln \left(\frac{\pi_{(k,l,\rho_{(k,l)} \mid \alpha)}}{\pi'_{(k,l,\rho_{(k,l)})}} \right). \tag{3.33}$$

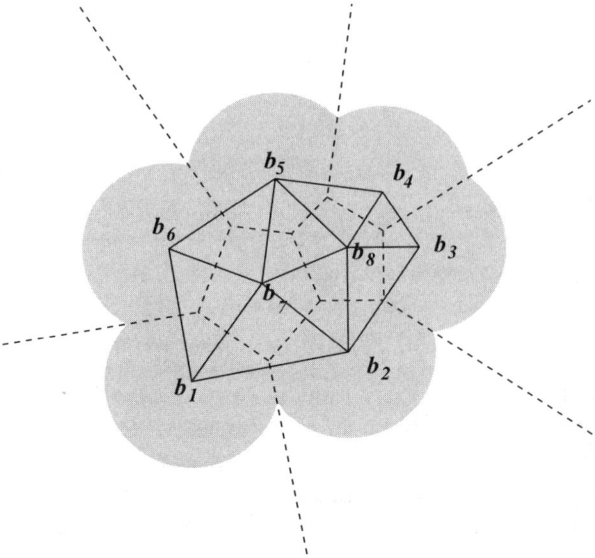

Fig. 3.3 Noninteracting pairs. (b_1, b_4) is considered as a noninteracting pair because the shortest length $L_{(1,4)}$ is equal to three, i.e., the interaction between b_1 and b_4 is blocked by two residues b_7 and b_8. Likewise, (b_3, b_6) is considered as a noninteracting pair as well.

Here, $\pi_{(k,l,\,\rho_{(k,l)}\,|\,\alpha)}$ is the observed probability:

$$\pi_{(k,l,\,\rho_{(k,l)}\,|\,\alpha)} = \frac{n_{(k,l,\,\rho_{(k,l)},\alpha)}}{n_{(\alpha)}}, \tag{3.34}$$

where $n_{(k,l,\,\rho_{(k,l)},\alpha)}$ is the number of residue pair (k, l) at the packing environment $\rho_{(k,l)}$ and the distance specified by α, and $n_{(\alpha)}$ is the total number of residue pairs at the distance specified by α. $\pi'_{(k,l,\,\rho_{(k,l)})}$ is the expected probability:

$$\pi'_{(k,l,\,\rho_{(k,l)})} = \frac{n'_{(k,l,\,\rho_{(k,l)})}}{n'}, \tag{3.35}$$

where $n'_{(k,l,z_{(k,l)})}$ is the number of residue pair (k, l) at the packing environment $z_{(k,l)}$ in the reference state, and n' is the total number of noninteracting residue pairs at the reference state.

The desolvation potential of residue type k to have ρ near neighbors $H(z \mid k)$ is estimated simply by following Eq. (3.9):

$$H(\rho \mid k) = \frac{\pi_{(\rho\,|\,k)}}{\pi'_{(\rho\,|\,k)}} = \frac{[n_{(k,\rho)}/n_{(k)}]}{[n_{(r,\rho)}/n_{(r)}]}, \tag{3.36}$$

where r represents all 20 residue types.

For a protein structure, the total internal energy is estimated by the summation of the desolvation energy and pairwise interaction energy in the particular desolvated environment:

$$\begin{aligned} H(s,a) = &\sum_{k,\rho} H(\rho \mid k) \cdot n_{(k,\rho)} \\ &+ \frac{1}{2} \sum_{k,l,\rho_{k,l},\alpha} H(k,l,\rho_{(k,l)} \mid \alpha) \cdot n_{(k,l,\rho_{(k,l)},\alpha)}. \end{aligned} \tag{3.37}$$

3.3.6 Sampling Weight of Proteins in Database

When developing statistical energy functions using a database consisting of many homologous sequences, undesirable sampling biases will be introduced. An easy way to avoid such sampling bias is to construct a database of structures in which no pair of proteins can have more than 25% sequence identity. By this criterion, a structure database may exclude a significant number of informative structures, which may be valuable for studying a specific type of proteins with very few known structures. An alternative method to avoid such sampling bias without neglecting these structures is to introduce weights that are properly adjusted for each structure, which may or may not be homologous to other structures in the database.

A similarity matrix $\boldsymbol{S}$ of all proteins in the database can be used to decide the weight for each protein structure (Miyazawa and Jernigan, 1996). The similarity between the k-th and l-th proteins is defined by Miyazawa and Jernigan based on the result of sequence alignment:

$$\begin{aligned} s_{kl} &\equiv \frac{2\theta_{kl}}{L_k + L_l}, \\ 0 \leq s_{kl} &= s_{lk} \leq 1, \\ s_{kk} &= 1, \end{aligned}$$

where θ_{kl} is the number of identical residues in the alignment, L_k and L_l are the lengths of sequences k and l, respectively. This similarity matrix $\boldsymbol{S}$ is symmetric and composed of real values. It has the spectral decomposition:

$$\boldsymbol{S} = \sum_i \lambda_i \boldsymbol{v}_i \boldsymbol{v}_i^T, \tag{3.38}$$

where λ_i and $\boldsymbol{v}_i$ are the i-th eigenvalue and eigenvector of $\boldsymbol{S}$, respectively. For a symmetric matrix, these eigenvectors form an orthonormal base. Because for the symmetric matrix $\boldsymbol{S}$, $\sum_i \lambda_i = \text{Trace}(\boldsymbol{S}) = n_{\text{prot}}$ and $\boldsymbol{S}$ is positive semidefinite, we have

$$0 \leq \lambda_i \leq n_{\text{prot}}, \tag{3.39}$$

where n_{prot} is the number of proteins included in the database. The value of λ_i reflects the weight of the corresponding orthogonal eigenvector $\boldsymbol{v}_i$ to the matrix $\boldsymbol{S}$. For the special case where there is one distinct sequence, which is completely dissimilar to any other $n_{\text{prot}} - 1$ sequences in the database, at least one eigenvalue will be exactly equal to 1 and the corresponding eigenvector represents this distinct sequence but contains no information about other sequences due to the orthogonality of the eigenvectors of matrix $\boldsymbol{S}$. In another case when there is one set of m sequences which are exactly the same within the group but are completely dissimilar to any other $n_{\text{prot}} - m$ sequences outside this set, at least one eigenvalue will be exactly equal to m and $m - 1$ eigenvalues will be equal to zero. The eigenvector corresponding to the nonzero eigenvalue represents the whole group of those m sequences but contains no information about other sequences.

On the basis of these characteristics, Miyazawa and Jernigan (1996) decreased all eigenvalues > 1 to 1 to reconstruct a new weight matrix $\boldsymbol{S}'$, so that redundant information from similar sequences is removed and the weight w_k for the k-th protein in the database is determined. In other words, we have before weighting:

$$w_k \equiv s_{kk} = \left[\sum_i \lambda_i \boldsymbol{v}_i \boldsymbol{v}_i^T \right]_{kk} = 1, \tag{3.40}$$

after weighting,

$$w_k \equiv s'_{kk} = \left[\sum_i \lambda'_i \boldsymbol{v}_i \boldsymbol{v}_i^T \right]_{kk}, \tag{3.41}$$

where

$$\lambda'_i = \begin{cases} \lambda_i, & \text{if } \lambda_i \le 1; \\ 1, & \text{if } \lambda_i > 1. \end{cases}$$

Therefore, if and only if a sequence is completely dissimilar to any other sequences ($\lambda'_i = \lambda_i = 1$), the sampling weight for that sequence will be 1. If all n_{prot} sequences in the database are identical, the sampling weights for these sequences will be $1/n_{\text{prot}}$. Generally, sampling weights take a value between one and $1/n_{\text{prot}}$, and are negatively proportional to the number of similar sequences.

3.4 Optimization Method

There are several drawbacks of knowledge-based potential functions derived from statistical analysis of a database. These include the neglect of chain connectivity in the reference state, and the problematic implicit assumption of Boltzmann distribution

(Thomas and Dill, 1996a,b; Ben-Naim, 1997). We defer a detailed discussion to Section 3.7.1.

An alternative method to develop potential functions for proteins is by optimization. For example, in protein design, we can use the thermodynamic hypothesis of Anfinsen to require that the native amino acid sequence $\boldsymbol{a}_N$ mounted on the native structure $\boldsymbol{s}_N$ has the best (lowest) fitness score compared to a set of alternative sequences (sequence decoys) taken from unrelated proteins known to fold into a different fold $\mathcal{D} = \{\boldsymbol{s}_N, \boldsymbol{a}_D\}$ when mounted on the same native protein structure $\boldsymbol{s}_N$:

$$H(f(\boldsymbol{s}_N, \boldsymbol{a}_N)) < H(f(\boldsymbol{s}_N, \boldsymbol{a}_D)) \quad \text{for all } (\boldsymbol{s}_N, \boldsymbol{a}_D) \in \mathcal{D}.$$

Equivalently, the native sequence will have the highest probability to fit into the specified native structure. This is the same principle described in Shakhnovich and Gutin (1993), Deutsch and Kurosky (1996), and Li et al. (1996). Sometimes we can further require that the score difference must be greater than a constant $b > 0$ (Shakhnovich, 1994):

$$H(f(\boldsymbol{s}_N, \boldsymbol{a}_N)) + b < H(f(\boldsymbol{s}_N, \boldsymbol{a}_D)) \quad \text{for all } (\boldsymbol{s}_N, \boldsymbol{a}_D) \in \mathcal{D}$$

Similarly, for protein structure prediction and protein folding, we require that the native amino acid sequence $\boldsymbol{a}_N$ mounted on the native structure $\boldsymbol{s}_N$ has the lowest energy compared to a set of alternative conformations (decoys) $\mathcal{D} = \{\boldsymbol{s}_D, \boldsymbol{a}_N\}$:

$$H(f(\boldsymbol{s}_N, \boldsymbol{a}_N)) < H(f(\boldsymbol{s}_D, \boldsymbol{a}_N)) \quad \text{for all } \boldsymbol{s}_D \in \mathcal{D}$$

and

$$H(f(\boldsymbol{s}_N, \boldsymbol{a}_N)) + b < H(f(\boldsymbol{s}_D, \boldsymbol{a}_S)) \quad \text{for all } (\boldsymbol{s}_D, \boldsymbol{a}_N) \in \mathcal{D}$$

when we insist on maintaining an energy gap between the native structure and decoy conformations. For linear potential function, we have

$$\boldsymbol{w} \cdot \boldsymbol{c}_N + b < \boldsymbol{w} \cdot \boldsymbol{c}_D \quad \text{for all } \boldsymbol{c}_D = f(\boldsymbol{s}_D, \boldsymbol{a}_N). \tag{3.42}$$

Our goal is to find a set of parameters through optimization for the potential function such that all these inequalities are satisfied.

As discussed earlier, there are three key steps in developing effective knowledge-based potential functions using optimization: (1) the functional form, (2) the generation of a large set of decoys for discrimination, and (3) the optimization techniques. The initial step of choosing an appropriate functional form is important. Knowledge-based pairwise potential functions are usually all in the form of weighted linear sum of interacting residue pairs. In this functional form, the weight coefficients are the parameters of the potential function, which are optimized for discrimination. This is the same functional form used in statistical potential, where the weight coefficients are derived from database statistics. The objectives of optimization are often maximization of the energy gap between native protein and the average of decoys, or the energy gap between native and decoys with the lowest score, or the z-score of the native protein (Goldstein et al., 1992; Maiorov and Crippen, 1992; Thomas and Dill, 1996a; Koretke et al., 1996, 1998; Hao and Scheraga, 1996; Mirny and Shakhnovich, 1996; Vendruscolo and Domanyi, 1998; Tobi et al., 2000; Vendruscolo et al., 2000; Dima et al., 2000; Micheletti et al., 2001; Bastolla et al., 2001).

3.4.1 Geometric Nature of Discrimination

There is a natural geometric view of the inequality requirement for weighted linear sum potential functions. A useful observation is that each of the inequalities divides the space of $\mathbb{R}^d$ into two halves separated by a hyperplane (Fig. 3.4a). The hyperplane for Eq. 3.42 is defined by the normal vector $(\boldsymbol{c}_N - \boldsymbol{c}_D)$ and its distance $b/||\boldsymbol{c}_N - \boldsymbol{c}_D||$ from the origin. The weight vector $\boldsymbol{w}$ must be located in the half-space opposite the direction of the normal vector $(\boldsymbol{c}_N - \boldsymbol{c}_D)$. This half-space can be written as $\boldsymbol{w} \cdot (\boldsymbol{c}_N - \boldsymbol{c}_D) + b < 0$. When there are many inequalities to be satisfied simultaneously, the intersection of the half-spaces forms a convex polyhedron (Edelsbrunner, 1987). If the weight vector is located in the polyhedron, all the inequalities are satisfied. Scoring functions with such weight vector $\boldsymbol{w}$ can discriminate the native protein sequence from the set of all decoys. This is illustrated in Fig. 3.4a for a two-dimensional toy example, where each straight line represents an inequality $\boldsymbol{w} \cdot (\boldsymbol{c}_N - \boldsymbol{c}_D) + b < 0$ that the potential function must satisfy.

For each native protein i, there is one convex polyhedron $\mathcal{P}_i$ formed by the set of inequalities associated with its decoys. If a potential function can discriminate simultaneously n native proteins from a union of sets of sequence decoys, the weight vector $\boldsymbol{w}$ must be located in a smaller convex polyhedron $\mathcal{P}$ that is the intersection of the n convex polyhedra:

$$\boldsymbol{w} \in \mathcal{P} = \bigcap_{i=1}^{n} \mathcal{P}_i.$$

There is yet another geometric view of the same inequality requirements. If we now regard $(\boldsymbol{c}_N - \boldsymbol{c}_D)$ as a point in $\mathbb{R}^d$, the relationship $\boldsymbol{w} \cdot (\boldsymbol{c}_N - \boldsymbol{c}_D) + b < 0$ for all sequence decoys and native proteins requires that all points $\{\boldsymbol{c}_N - \boldsymbol{c}_D\}$ are located on one side of a different hyperplane, which is defined by its normal vector $\boldsymbol{w}$ and its distance $b/||\boldsymbol{w}||$ to the origin (Fig. 3.4b). We can show that such a hyperplane exists if the origin is not contained within the convex hull of the set of points $\{\boldsymbol{c}_N - \boldsymbol{c}_D\}$ (see Appendix).

The second geometric view looks very different from the first view. However, the second view is dual and mathematically equivalent to the first geometric view. In the first view, a point $\boldsymbol{c}_N - \boldsymbol{c}_D$ determined by the structure–decoy pair $c_N = (\boldsymbol{s}_N, \boldsymbol{a}_N)$ and $c_D = (\boldsymbol{s}_N, \boldsymbol{a}_D)$ corresponds to a hyperplane representing an inequality,

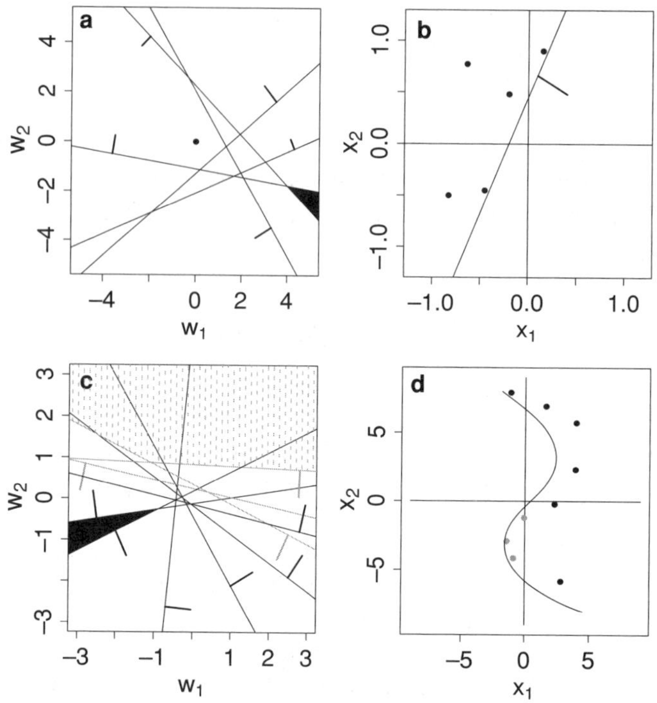

Fig. 3.4 Geometric views of the inequality requirement for protein scoring function. Here we use a two-dimensional toy example for illustration. (a). In the first geometric view, the space $\mathbb{R}^2$ of $\boldsymbol{w} = (w_1, w_2)$ is divided into two half-spaces by an inequality requirement, represented as a hyperplane $\boldsymbol{w} \cdot (\boldsymbol{c}_N - \boldsymbol{c}_D) + b < 0$. The hyperplane, which is a line in $\mathbb{R}^2$, is defined by the normal vector $(\boldsymbol{c}_N - \boldsymbol{c}_D)$ and its distance $b/||\boldsymbol{c}_N - \boldsymbol{c}_D||$ from the origin. In this figure, this distance is set to 1.0. The normal vector is represented by a short line segment whose direction points away from the straight line. A feasible weight vector $\boldsymbol{w}$ is located in the half-space opposite to the direction of the normal vector $(\boldsymbol{c}_N - \boldsymbol{c}_D)$. With the given set of inequalities represented by the lines, any weight vector $\boldsymbol{w}$ located in the shaped polygon can satisfy all inequality requirement and provides a linear scoring function that has perfect discrimination. (b) A second geometric view of the inequality requirement for linear protein scoring function. The space $\mathbb{R}^2$ of $\boldsymbol{x} = (x_1, x_2)$, where $\boldsymbol{x} \equiv (\boldsymbol{c}_N - \boldsymbol{c}_D)$, is divided into two half-spaces by the hyperplane $\boldsymbol{w} \cdot (\boldsymbol{c}_N - \boldsymbol{c}_D) + b < 0$. Here the hyperplane is defined by the normal vector $\boldsymbol{w}$ and its distance $b/||\boldsymbol{w}||$ from the origin. The origin corresponds to the native protein. All points $\{\boldsymbol{c}_N - \boldsymbol{c}_D\}$ are located on one side of the hyperplane away from the origin, therefore satisfying the inequality requirement. That is, a linear scoring function $\boldsymbol{w}$ such as the one represented by the straight line in this figure can have perfect discrimination. (c) In the second toy problem, a set of inequalities are represented by a set of straight lines according to the first geometric view. A subset of the inequalities require the weight vector $\boldsymbol{w}$ to be located in the shaded convex polygon on the left, but another subset of inequalities require $\boldsymbol{w}$ to be located in the dashed convex polygon on the top. Since these two polygons do not intersect, there is no weight vector $\boldsymbol{w}$ that can satisfy all inequality requirements. That is, no linear scoring function can classify these decoys from native protein. (d) According to the second geometric view, no hyperplane can separate all points $\{\boldsymbol{c}_N - \boldsymbol{c}_D\}$ from the origin. But a nonlinear curve formed by a mixture of Gaussian kernels can have perfect separation of all vectors $\{\boldsymbol{c}_N - \boldsymbol{c}_D\}$ from the origin: It has perfect discrimination.

and a solution weight vector $\boldsymbol{w}$ corresponds to a point located in the final convex polyhedron. In the second view, each structure–decoy pair is represented as a point $\boldsymbol{c}_N - \boldsymbol{c}_D$ in $\mathbb{R}^d$, and the solution weight vector $\boldsymbol{w}$ is represented by a hyperplane separating all the points $\mathcal{C} = \{\boldsymbol{c}_N - \boldsymbol{c}_D\}$ from the origin.

3.4.2 Optimized Linear Potential Functions

Several optimization methods have been applied to find the weight vector $\boldsymbol{w}$ of linear potential function. The Rosenblantt perceptron method works by iteratively updating an initial weight vector $\boldsymbol{w}_0$ (Vendruscolo and Domanyi, 1998; Micheletti et al., 2001). Starting with a random vector, e.g., $\boldsymbol{w}_0 = \boldsymbol{0}$, one tests each native protein and its decoy structure. Whenever the relationship $\boldsymbol{w} \cdot (\boldsymbol{c}_N - \boldsymbol{c}_D) + b < 0$ is violated, one updates $\boldsymbol{w}$ by adding to it a scaled violating vector $\eta \cdot (\boldsymbol{c}_N - \boldsymbol{c}_D)$. The final weight vector is therefore a linear combination of protein and decoy count vectors:

$$\boldsymbol{w} = \sum \eta(\boldsymbol{c}_N - \boldsymbol{c}_D) = \sum_{N \in \mathcal{N}} \alpha_N \boldsymbol{c}_N - \sum_{D \in \mathcal{D}} \alpha_D \boldsymbol{c}_D. \tag{3.43}$$

Here $\mathcal{N}$ is the set of native proteins, and $\mathcal{D}$ is the set of decoys. The set of coefficients $\{\alpha_N\} \cup \{\alpha_D\}$ gives a dual form representation of the weight vector $\boldsymbol{w}$, which is an expansion of the training examples including both native and decoy structures.

According to the first geometric view, if the final convex polyhedron $\mathcal{P}$ is nonempty, there can be an infinite number of choices of $\boldsymbol{w}$, all with perfect discrimination. But how do we find a weight vector $\boldsymbol{w}$ that is optimal? This depends on the criterion for optimality. For example, one can choose the weight vector $\boldsymbol{w}$ that minimizes the variance of score gaps between decoys and natives:

$$\arg_{\boldsymbol{w}} \min \frac{1}{|\mathcal{D}|} \sum (\boldsymbol{w} \cdot (c_N - c_D))^2 - \left[\frac{1}{|\mathcal{D}|} \sum_D (\boldsymbol{w} \cdot (\boldsymbol{c}_N - \boldsymbol{c}_D)) \right]^2$$

as used in Tobi et al. (2000), or minimizing the Z-score of a large set of native proteins, or minimizing the Z-score of the native protein and an ensemble of decoys (Chiu and Goldstein, 1998; Mirny and Shakhnovich, 1996), or maximizing the ratio R between the width of the distribution of the score and the average score difference between the native state and the unfolded ones (Goldstein et al., 1992; Hao and Scheraga, 1999). A series of important works using perceptron learning and other optimization techniques (Friedrichs and Wolynes, 1989; Goldstein et al., 1992; Tobi et al., 2000; Vendruscolo and Domanyi, 1998; Dima et al., 2000) showed that effective linear sum potential functions can be obtained.

There is another optimality criterion according to the second geometric view (Hu et al., 2004). We can choose the hyperplane $(\boldsymbol{w}, b)$ that separates the set of points $\{\boldsymbol{c}_N - \boldsymbol{c}_D\}$ with the largest distance to the origin. Intuitively, we want to characterize proteins with a region defined by the training set points $\{\boldsymbol{c}_N - \boldsymbol{c}_D\}$. It is desirable to define this region such that a new unseen point drawn from the same protein distribution as $\{\boldsymbol{c}_N - \boldsymbol{c}_D\}$ will have a high probability of falling within the defined region. Nonprotein points following a different distribution, which is assumed to be centered around the origin when no *a priori* information is available, will have a high probability of falling outside the defined region. In this case, we are more interested in modeling the region or support of the distribution of protein data, rather than estimating its density distribution function. For linear potential function, regions are half-spaces defined by hyperplanes, and the optimal hyperplane $(\boldsymbol{w}, b)$ is then the one with maximal distance to the origin. This is related to the novelty detection problem and single-class support vector machine studied in statistical learning theory (Vapnik and Chervonenkis, 1964, 1974; Schölkopf and Smola, 2002). In our case, any nonprotein points will need to be detected as outliers from the protein distribution characterized by $\{\boldsymbol{c}_N - \boldsymbol{c}_D\}$. Among all linear functions derived from the same set of native proteins and decoys, an optimal weight vector $\boldsymbol{w}$ is likely to have the least amount of mislabelings. The optimal weight vector $\boldsymbol{w}$ can be found by solving the following quadratic programming problem:

$$\text{Minimize} \qquad \tfrac{1}{2}||\boldsymbol{w}||^2 \tag{3.44}$$

$$\text{subject to } \boldsymbol{w}\cdot(\boldsymbol{c}_N - \boldsymbol{c}_D) + b < 0 \quad \text{for all } N \in \mathcal{N} \text{ and } D \in \mathcal{D}. \tag{3.45}$$

The solution maximizes the distance $b/||\boldsymbol{w}||$ of the plane $(\boldsymbol{w}, b)$ to the origin. We obtained the solution by solving the following support vector machine problem:

$$\begin{aligned} &\text{Minimize } \tfrac{1}{2}\|\boldsymbol{w}\|^2 \\ &\text{subject to } \boldsymbol{w}\cdot\boldsymbol{c}_N + d \le -1 \\ &\qquad\qquad\quad \boldsymbol{w}\cdot\boldsymbol{c}_D + d \ge 1, \end{aligned} \tag{3.46}$$

where $d > 0$. Note that a solution of Problem (3.46) satisfies the constraints in Inequalities (3.45), since subtracting the second inequality here from the first inequality in the constraint conditions of (3.46) will give us $\boldsymbol{w}\cdot(\boldsymbol{c}_N - \boldsymbol{c}_D) + 2 \le 0$.

3.4.3 Optimized Nonlinear Potential Functions

Optimized linear potential function can be obtained using the optimization strategy discussed above. However, it is possible that the weight vector $\boldsymbol{w}$ does not exist, i.e., the final convex polyhedron $\mathcal{P} = \bigcap_{i=1}^{n} \mathcal{P}_i$ may be an empty set. This occurs if a large number of native protein structures are to be simultaneously stabilized against a large number of decoy conformations, no such potential functions in the linear functional form can be found (Vendruscolo et al., 2000; Tobi et al., 2000).

According to our geometric pictures, there are two possible scenarios. First, for a specific native protein i, there may be severe restriction from some inequality constraints, which makes $\mathcal{P}_i$ an empty set. Some decoys are very difficult to discriminate due to perhaps deficiency in protein representation. In these cases, it is impossible to adjust the weight vector so the native protein has a lower score than the sequence decoy. Fig. 3.4c shows a set of inequalities represented by straight lines according to the first geometric view. In this case, there is no weight vector that can satisfy all these inequality requirements. That is, no linear potential function can classify all decoys from native protein. According to the second geometric view (Fig. 3.4d), no hyperplane can separate all points (black and green) $\{\boldsymbol{c}_N - \boldsymbol{c}_D\}$ from the origin, which corresponds to the native structures.

Second, even if a weight vector $\boldsymbol{w}$ can be found for each native protein, i.e., $\boldsymbol{w}$ is contained in a nonempty polyhedron, it is still possible that the intersection of n polyhedra is an empty set, i.e., no weight vector can be found that can discriminate all native proteins from the decoys simultaneously. Computationally, the question whether a solution weight vector $\boldsymbol{w}$ exists can be answered unambiguously in polynomial time (Karmarkar, 1984). If a large number (e.g., hundreds) of native protein structures are to be simultaneously stabilized against a large number of decoy conformations (e.g., tens of millions), no such potential functions can be found computationally (Vendruscolo et al., 2000; Tobi et al., 2000). A similar conclusion is drawn in a study on protein design, where it was found that no linear potential function can simultaneously discriminate a large number of native proteins from sequence decoys (Hu et al., 2004).

A fundamental reason for such failure is that the functional form of linear sum is too simplistic. It has been suggested that additional descriptors of protein structures such as higher order interactions (e.g., three-body or four-body contacts) should be incorporated in protein description (Betancourt and Thirumalai, 1999; Munson and Singh, 1997; Zheng et al., 1997). Functions with polynomial terms using up to 6 degrees of Chebyshev expansion have also been used to represent pairwise interactions in protein folding (Fain et al., 2002).

We now discuss an alternative approach. Let us still limit ourselves to pairwise contact interactions, although it can be naturally extended to include three- or four-body interactions (Li and Liang, 2005b). We can introduce a nonlinear potential function analogous to the dual form of the linear function in Eq. (3.43), which takes the following form:

$$H(f(\boldsymbol{s}, \boldsymbol{a})) = H(\boldsymbol{c}) = \sum_{D \in \mathcal{D}} \alpha_D K(\boldsymbol{c}, \boldsymbol{c}_D) - \sum_{N \in \mathcal{N}} \alpha_N K(\boldsymbol{c}, \boldsymbol{c}_N), \tag{3.47}$$

where $\alpha_D \geq 0$ and $\alpha_N \geq 0$ are parameters of the potential function to be determined, and $\boldsymbol{c}_D = f(\boldsymbol{s}_N, \boldsymbol{a}_D)$ from the set of decoys $\mathcal{D} = \{(\boldsymbol{s}_N, \boldsymbol{a}_D)\}$ is the contact vector of a sequence decoy D mounted on a native protein structure $\boldsymbol{s}_N$, and $\boldsymbol{c}_N = f(\boldsymbol{s}_N, \boldsymbol{a}_N)$ from the set of native training proteins $\mathcal{N} = \{(\boldsymbol{s}_N, \boldsymbol{a}_N)\}$ is the contact vector of a native sequence $\boldsymbol{a}_N$ mounted on its native structure $\boldsymbol{s}_N$. In this study, all decoy sequences $\{\boldsymbol{a}_D\}$ are taken from real proteins possessing different fold structures. The difference of this functional form from the linear function in Eq. (3.43) is that a kernel function $K(\boldsymbol{x}, \boldsymbol{y})$ replaces the linear term. A convenient kernel function K is

$$K(\boldsymbol{x}, \boldsymbol{y}) = e^{-||\boldsymbol{x}-\boldsymbol{y}||^2/2\sigma^2} \quad \text{for any vectors } \boldsymbol{x} \text{ and } \boldsymbol{y} \in \mathcal{N} \bigcup \mathcal{D},$$

where σ^2 is a constant. Intuitively, the surface of the potential function has smooth Gaussian hills of height α_D centered on the location $\boldsymbol{c}_D$ of decoy protein D, and has smooth Gaussian cones of depth α_N centered on the location $\boldsymbol{c}_N$ of native structures N. Ideally, the value of the potential function will be -1 for contact vectors $\boldsymbol{c}_N$ of native proteins, and will be $+1$ for contact vectors $\boldsymbol{c}_D$ of decoys.

3.4.4 Deriving Optimized Nonlinear Potential Functions

To obtain the nonlinear potential function, our goal is to find a set of parameters $\{\alpha_D, \alpha_N\}$ such that $H(f(\boldsymbol{s}_N, \boldsymbol{a}_N))$ has value close to -1 for native proteins, and the decoys have values close to $+1$. There are many different choices of $\{\alpha_D, \alpha_N\}$. We use an optimality criterion originally developed in statistical learning theory (Vapnik, 1995; Burges, 1998; Schölkopf and Smola, 2002). First, we note that we have implicitly mapped each structure and decoy from $\mathbb{R}^{210}$ through the kernel function of $K(\boldsymbol{x}, \boldsymbol{y}) = e^{-||\boldsymbol{x}-\boldsymbol{y}||^2/2\sigma^2}$ to another space with dimensions as high as tens of millions. Second, we then find the hyperplane of the largest margin distance separating proteins and decoys in the space transformed by the nonlinear kernel. That is, we search for a hyperplane with equal and maximal distance to the closest native proteins and the closest decoys in the transformed high dimensional space. Such a hyperplane can be found by obtaining the parameters $\{\alpha_D\}$ and $\{\alpha_N\}$ from solving the following Lagrange dual form of quadratic programming problem:

$$\begin{array}{ll}\text{Maximize} & \sum_{i\in\mathcal{N}\cup\mathcal{D},} \alpha_i - \frac{1}{2}\sum_{i,j\in\mathcal{N}\cup\mathcal{D}} y_i y_j \alpha_i \alpha_j e^{-||c_i - \boldsymbol{c}_j||^2/2\sigma^2} \\ \text{subject to} & 0 \leq \alpha_i \leq C,\end{array}$$

where C is a regularizing constant that limits the influence of each misclassified protein or decoy (Vapnik and Chervonenkis, 1964, 1974; Vapnik, 1995; Burges, 1998; Schölkopf and Smola, 2002), and $y_i = -1$ if i is a native protein, and $y_i = +1$ if i is a decoy. These parameters lead to optimal discrimination of an unseen test set (Vapnik and Chervonenkis, 1964, 1974; Vapnik, 1995; Burges, 1998; Schölkopf and Smola, 2002). When projected back to the space of $\mathbb{R}^{210}$, this hyperplane becomes a nonlinear surface. For the toy problem of Fig. 3.4, Fig. 3.4d shows that such a hyperplane becomes a nonlinear curve in $\mathbb{R}^2$ formed by a mixture of Gaussian kernels. It separates perfectly all vectors $\{\boldsymbol{c}_N - \boldsymbol{c}_D\}$ (black and green) from the origin. That is, a nonlinear potential function can have perfect discrimination.

3.4.5 Optimization Techniques

The techniques that have been used for optimizing potential function include perceptron learning, linear programming, gradient descent, statistical analysis, and support vector machine (Tobi et al., 2000; Vendruscolo et al., 2000; Xia and Levitt, 2000; Bastolla et al., 2000, 2001; Hu et al., 2004). These are standard techniques that can be found in optimization and machine learning literature. For example, there are excellent linear programming solvers based on simplex method, as implemented in CLP, GLPK, and LP_SOLVE (Berkelaar, 2004), and based on interior point method as implemented in the BPMD (Mészáros, 1996), the HOPDM, and the PCx packages (Czyzyk et al., 2004). We neglect the details of these techniques and point readers to the excellent treatises of Papadimitriou and Steiglitz (1998) and Vanderbei (1996).

3.5 Applications

Knowledge-based potential function has been widely used in the study of protein structure prediction, protein folding, and protein–protein interaction. In this section, we discuss briefly some of these applications. Additional details of applications of knowledge-based potential can be found in other chapters of this book.

3.5.1 Protein Structure Prediction

Protein structure prediction is an extraordinarily complex task that involves two major components: sampling the conformational space and recognizing the near native structures from the ensemble of sampled conformations.

In protein structure prediction, methods for conformational sampling will generate a huge number of candidate protein structures. These are often called *decoys*. Among these decoys, only a few are near native structures that are very similar to the native structure. A potential function must be used to discriminate the near-native structures from all other decoys for a successful structure prediction.

Several decoy sets have been developed as objective benchmarks to test if a knowledge-based potential function can successfully identify the native and near-native structures. For example, Park and Levitt (1996) constructed a 4-state-reduced decoy set. This decoy test set contains native and near-native conformations of seven sequences, along with about 650 misfolded structures for each sequence. The positions of C_α of these decoys were generated by exhaustively enumerating 10 selectively chosen residues in each protein using a 4-state off-lattice model. All other residues were assigned the phi/psi value based on the best fit of a 4-state model to the native chain (Park and Levitt, 1996).

A central depository of folding decoy conformations is the Decoys 'R'Us (Samudrala and Levitt, 2000). See Section 3.6 for the URL links to download several folding and docking decoy sets. A variety of knowledge-based potential functions have been developed and their performance in decoy discrimination has steadily improved (Zhou and Zhou, 2002; Lu and Skolnick, 2001; Li et al., 2003).

Figure 3.5 shows an example of decoy discrimination on the *4-state-reduced* decoy set. This result is based on the residue-level packing and distance-dependent geometric potential function discussed earlier. For all of the seven proteins in the 4-state-reduced set, the native structures have the lowest energy. In addition, all of the decoys with the lowest energy are within 2.5 Å RMSD to the native structure.

Table 3.2 lists the performance of the geometric potential function in folding and docking decoy discriminations. Several studies examine the comparative performance of different knowledge-based potential functions (Park and Levitt,

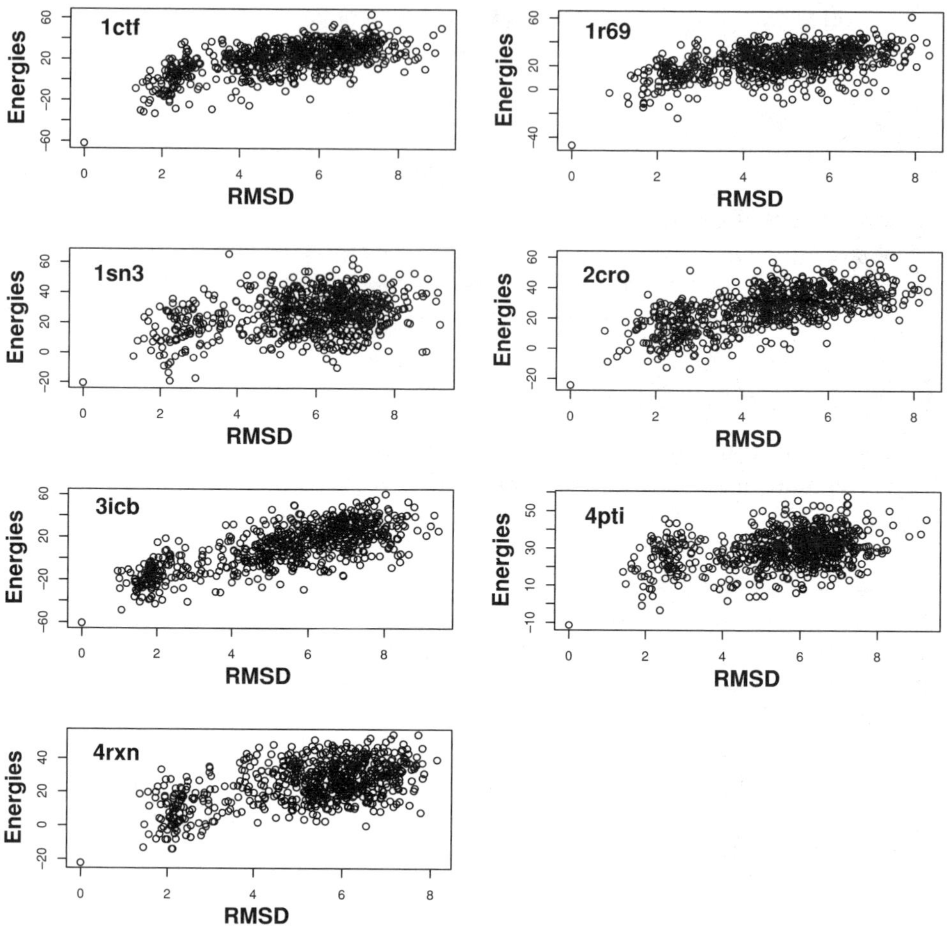

Fig. 3.5 Energies evaluated by packing- and distance-dependent residue contact potential plotted against the RMSD to native structures for conformations in Park & Levitt Decoy Set.

1996; Zhou and Zhou, 2002; Gilis, 2004). Such evaluations often are based on measuring the success in ranking native structure from a large set of decoy conformations and in obtaining a large z-score for the native protein structure. Because the development of potential function is a very active research field, the comparison of performances of different potential functions will be different as new models and techniques are developed and incorporated.

Not only can a knowledge-based potential function be applied at the end of the conformation sampling to recognize near-native structures, it can also be used during conformation generation to guide the efficient sampling of protein structures. Details of this application can be found in Jernigan and Bahar (1996) and Hao and Scheraga (1999). In addition, knowledge-based potential also plays an important role in protein threading studies. Chapter 12 provides further detailed discussion.

Table 3.2 Performance of geometric potential on folding and docking decoy discrimination

	4-state-reduced		lattice-ssfit		fisa-casp3		fisa		lmds	
Folding decoy sets	Native[a]	z^b	Native	z	Native	z	Native	z	Native	z
	7/7	4.46	8/8	7.70	3/3	5.23	3/4	5.42	7/10	1.45
	Rosetta-Bound-Perturb		Rosetta-Unbound-Perturb		Rosetta-Unbound-Global		Vakser's		Sternberg's	
Docking decoy sets	Native	z	Native	z	Native	z	Native	z	Native	z
	50/54	12.75	53/54	12.88	53/54	8.55	4/5	4.45	16/16	4.45
	RDOCK		29/42[c]							

[a] Number of native structures ranking first; e.g., 7/7 means seven out of seven native structures have the lowest energy among their corresponding decoy sets.
[b] $z = \overline{E} - E_{native}/\sigma$; $\overline{E}$ and σ are the mean and standard deviation of the energy values of conformations, respectively.
[c] Native complex is not included in these docking decoy sets. Thirty-two out of 42 decoy sets have at least one near-native structure (cRMSD < 2.5Å) in the top 10 structures.

3.5.2 Protein–Protein Docking Prediction

Knowledge-based potential functions can also be used to study protein–protein interactions. Here we give an example of predicting the binding surface of seven antibody or antibody-related-proteins (e.g., Fab fragment, T-cell receptor) (Li and Liang, 2005a). These protein–protein complexes are taken from the 21 CAPRI (Critical Assessment of PRedicted Interactions) target proteins. CAPRI is a communitywide competition designed to objectively assess the abilities in protein–protein docking prediction (Méndez et al., 2005). In CAPRI, a blind docking prediction starts from two known crystallographic or NMR structures of unbound proteins and ends with a comparison to a solved structure of the protein complex, to which the participants did not have access. Knowledge-based potential functions, together with geometric complementarity potential functions, can be used to recognize near-native docking complexes and to guide the generation of conformations for protein–protein docking.

When docking two proteins together, we say a *cargo* protein is docked to a fixed *seat* protein. To determine the binding surfaces on the cargo protein, we can examine all possible surface patches on the unbound structure of cargo protein as candidate binding interfaces. The alpha knowledge-based potential function is then used to identify native or near native binding surfaces. To evaluate the performance of the potential function, we assume the knowledge of the binding interface on the seat protein. We further assume the knowledge of the degree of near neighbors for interface residues.

We first partition the surface of the unbound cargo protein into candidate surface patches, each having the same size as the native binding surface of m residues. A candidate surface patch is generated by starting from a surface residue on the cargo protein, and following alpha edges on the boundary of the alpha shape by breadth-first search, until m residues are found (Fig. 3.6a). We construct n candidate surface patches by starting in turn from each of the n surface residues on the cargo protein. Because each surface residue is the center of one of the n candidate surface

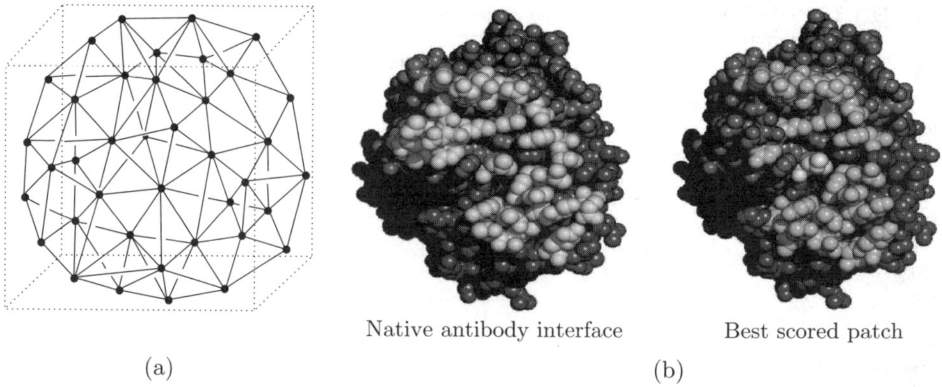

Fig. 3.6 Recognition of binding surface patch of CAPRI targets. (a) Boundary of alpha shape for a *cargo* protein. Each node represents a surface residue, and each edge represents the alpha edge between two surface residues. A candidate surface patch is generated by starting from a surface residue on the cargo protein, and following alpha edges on the boundary of the alpha shape by breadth-first search, until m residues are included. (b) Native interface and the surface patch with the best score on the antibody of the protein complex CAPRI Target T02. Only heavy chain (in red) and light chain (in blue) of the antibody are drawn. The antigen is omitted from this illustration for clarity. The best scored surface patch (in green) resembles the native interface (in yellow): 71% residues from this surface patch are indeed on the native binding interface. The residue in white is the starting residue used to generate this surface patch with the best score.

patches, the set of candidate surface patches covers exhaustively the whole protein binding interface.

Second, we assume that a candidate surface patch on the cargo protein has the same set of contacts as that of the native binding surface. The degree of near neighbors for each hypothetical contacting residue pair is also assumed to be the same. We replace the m residues of the native surface with the m residues from the candidate surface patch. There are $\frac{m!}{\prod_{i=1}^{20} m_i!}$ different ways to permute the m residues of the candidate surface patch, where m_i is the number of residue type i on the candidate surface patch. A typical candidate surface patch has about 20 residues, therefore the number of possible permutations is very large. For each candidate surface patch, we take a sample of 5000 random permutations. For a candidate surface patch SP_i, we assume that the residues can be organized so that they can interact with the binding partner at the lowest energy. Therefore, the binding energy $E(SP_i)$ is estimated as

$$E(SP_i) = \min_k E(SP_i)_k, \quad k = 1, \ldots, 5000.$$

Here $E(SP_i)_k$ is calculated based on the residue-level packing and distance-dependent potential for the k-th permutation. The value of $E(SP_i)$ is used to rank the candidate surface patches.

Table 3.3 Recognition of native binding surface of CAPRI targets

		Antibody[a]		Antigen	
Target	Complex	Rank^{b}_{native}	Overlap[c]	Rank_{native}	Overlap
T02	Rotavirus VP6-Fab	1/283[d]	0.71	1/639	0.68
T03	Flu hemagglutinin-Fab	1/297	0.56	1/834	0.71
T04	α-amylase-camelid Ab VH 1	56/89	0.60	102/261	0.03
T05	α-amylase-camelid Ab VH 2	23/90	0.57	57/263	0.25
T06	α-amylase-camelid Ab VH 3	1/88	0.70	1/263	0.62
T07	SpeA superantigen TCRβ	1/172	0.57	1/143	0.61
T13	SAG1-antibody complex	1/286	0.64	1/249	0.69

[a] "Antibody": Different surface patches on the antibody molecule are evaluated by the potential function, while the native binding surface on the antigen remains unchanged. "Antigen": similarly defined as "Antibody."
[b] Ranking of the native binding surface among all candidate surface patches.
[c] Fraction of residues from the best candidate surface patch that overlap with residues from the native binding surface patch.
[d] The first number is the rank of native binding surface and the second number is the number of total candidate surface patches.

We assess the statistical potential by taking the antibody/antigen protein in turn as the seat protein, and the antigen/antibody as the cargo protein. The native interface on the seat protein is fixed. We test if our statistical potential can discriminate native surface patch on the cargo protein from the set of candidate surface patches. We also test if the best scored patch resembles the native patch. The results are listed in Table 3.3 and the predicted antigen-binding interface of target T02 is shown in Fig. 3.6(b) as an example. For five of the seven protein complexes, we succeeded in discriminating the native patches on both the antibody and the antigen. Over 50% of the residues from the best scored patch overlap with the corresponding native patch. Our statistical potential does not work as well for targets T04 and T05, because the antibodies of these two complexes do not use their CDR domains to recognize the antigens as an antibody usually does, and such examples are not present in the data set of the 34 antibody–antigen complexes, based on which the geometric potential function was obtained.

3.5.3 Protein Design

Protein design aims to identify sequences compatible with a given protein fold but incompatible with any alternative folds (Koehl and Levitt, 1999a(,b). The goal is to design a novel protein that may not exist in nature but has enhanced or novel biological function. Several novel proteins have been successfully designed in recent years (Dahiyat and Mayo, 1997; Hill et al., 2000; Looger et al., 2003; Kuhlman et al., 2003). The problem of protein design is complex, because even a small protein of just 50 residues can have an astronomical number of sequences (10^{65}). This clearly precludes exhaustive search of the sequence space with any computational or experimental method. Instead, protein design methods rely on potential functions for biasing the search toward the feasible regions that encode protein sequences. To select the correct sequences and to guide the search process, a design potential function is critically important. Such a potential function should be able to characterize the global fitness landscape of many proteins simultaneously.

Here, we briefly describe the application of the optimized nonlinear design potential function discussed in Section 3.4.3 (Hu et al., 2004) in protein design. We aim to solve a simplified protein sequence design problem. Our goal is to distinguish each native sequence for a major portion of representative protein structures from a large number of alternative decoy sequences, each a fragment from proteins of different fold.

To train the nonlinear potential function, a list of 440 proteins was compiled from the 1998 release (WHATIF98) of the WHATIF database (Vendruscolo et al., 2000). Using gapless threading (Maiorov and Crippen, 1992), a set of 14,080,766 sequence decoys was obtained. The entries in the WHATIF99 database that are not present in WHATIF98 are used as a test set. After cleanup, the test set consists of 194 proteins and 3,096,019 sequence decoys.

To test the design potential functions for discriminating native proteins from sequence decoys, we take the sequence $\boldsymbol{a}$ from the conformation–sequence pair $(\boldsymbol{s}_N, \boldsymbol{a})$ for a protein with the lowest score as the predicted sequence. If it is not the native sequence $\boldsymbol{a}_N$, the discrimination failed and the design potential function does not work for this protein.

The nonlinear design potential function is capable of discriminating all of the 440 native sequences. In contrast, no linear potential function can succeed in this task. The nonlinear potential function also works well for the test set, where it succeeded in correctly identifying 93.3% (181 out of 194) of native sequences in the independent test set of 194 proteins. This compares favorably with results obtained using optimized linear folding potential function taken as reported in Tobi et al. (2000), which succeeded in identifying 80.9% (157 out of 194) of this test set. It also has better performance than optimized nonlinear potential function based on calculations using parameters reported in Bastolla et al. (2001), which succeeded in identifying 73.7% (143 out of 194) of proteins in the test set. The Miyazawa–Jernigan statistical potential succeeded in identifying 113 native proteins out of 194 (success rate 58.2%).

3.5.4 Protein Stability and Binding Affinity

Because the stability of protein in the native conformation is determined by the distribution of the full ensemble of conformations, namely, the partition function $Z(\boldsymbol{a})$ of the protein sequence $\boldsymbol{a}$, care must be taken when using statistical potentials to compare the stabilities of different protein sequences adopting the same given conformation as in protein design (Miyazawa and Jernigan, 1996; Sippl, 1990). This issue is discussed in some detail in Section 3.7.1.

Nevertheless, it is expected that statistical potentials should work well in estimating protein stability changes upon mutations, as the change in partition functions of the protein sequence is small. In most such studies and studies using physics-based empirical potential (see Chapter 2 in this book and Bordner and Abagyan (2004)), good correlation coefficient (0.6–0.8) between predicted and measured

Table 3.4 Database of folding and docking decoy sets

Decoys sets	Type	URL
Decoy 'R' Us[a]	folding	*http://dd.stanford.edu/*
Loop	folding	*http://francisco.compbio.ucsf.edu/~jacobson/decoy.htm*
CASP	folding	*http://predictioncenter.org/*
ZDOCK, RDOCK	docking	*http://zlab.bu.edu/~leely/RDOCK_decoy/*
Vakser decoy set	docking	*http://www.bioinformatics.ku.edu/files/vakser/decoys/*
Sternberg decoy set	docking	*http://www.sbg.bio.ic.ac.uk/docking/all_decoys.html*
Rosetta	docking, folding	*http://depts.washington.edu/bakerpg/*
CAPRI	docking	*http://capri.ebi.ac.uk/*

[a] The database of Decoys 'R' Us contains *multiple decoy sets*, *single decoy sets*, and *loop decoy sets*. *4-state-reduced* decoy set is included in the multiple decoy sets.

stability change can be achieved (Gilis and Rooman, 1996, 1997; Guerois et al., 2002; Bordner and Abagyan, 2004; Hoppe and Schomburg, 2005; Zhou and Zhou, 2002).

Several studies have shown that statistical potentials can also be used to predict quantitative binding free energy of protein–protein or protein–ligand interactions (DeWitte and Shakhnovich, 1996; Mitchell et al., 1999; Muegge and Martin, 1999; Liu et al., 2004; Zhang et al., 2005). In fact, Xu et al. showed that a simple number count of hydrophilic bridges across the binding interface is strongly correlated with binding free energies of protein–protein interaction (Xu et al., 1997). This study suggests that binding free energy may be predicted successfully by number counts of various types of interfacial contacts defined using some distance threshold. Such studies of number count provide an excellent benchmark to quantify the improvement in predicting binding free energy when using statistical potentials for different protein–protein and protein–ligand complexes. Similar to prediction of protein stability change upon mutation, knowledge based potential functions played an important role in a successful study of predicting binding free energy changes upon mutation (Kortemme and Baker, 2002; Kortemme et al., 2004).

3.6 Online Resources

A list of online sources of decoy data for folding and docking is provided in Table 3.4.

3.7 Discussion

3.7.1 Knowledge-Based Statistical Potential Functions

The statistical potential functions are often derived based on several assumptions: (1) protein energetics can be decomposed primarily into pairwise interactions; (2) interactions are independent from each other; (3) the partition function in native proteins Z and in reference states Z' are approximately equal; (4) the probability of occupancy of a state follows the Boltzmann distribution. These assumptions are often unrealistic and raise questions about the validity of the statistical potential functions: Can statistical potential functions provide energylike quantities such as the folding free energy of a protein, or the binding free energy of a protein–protein complex (Thomas and Dill, 1996b)? Can statistical potential functions correctly recognize the native structures from alternative conformations?

The assumptions of statistical knowledge-based potential functions: From Eq. (3.4), we can obtain the potential function $H(\boldsymbol{c})$ by estimating the probability $\pi(\boldsymbol{c})$. However, we need a number of assumptions for this approach to work. We need the independency assumption to have

$$\pi(\boldsymbol{c}) = \prod_i \pi(c_i) = \prod_i \prod_{c_i} \pi_i,$$

where c_i is the number of occurrences of the i-th structural feature, e.g., number of a specific residue pair contact; π_i is the probability of the i-th structural feature in the database. That is, we have to assume that the distribution of a specific structural feature is independent and not influenced by any other features, and is of no consequence for the distributions of other features as well. We also need to assume that $\boldsymbol{c}$ provides an adequate characterization of protein interactions, and the functional form of $\boldsymbol{w} \cdot \boldsymbol{c}$ provides the correct measurement of the energy of the interactions. We further need to assume that the energy for a protein–solvent system is decomposable, i.e., the overall energy can be partitioned into many basic energy terms, such as pairwise interactions, desolvation energies. Moreover, the partition functions Z' in a chosen reference state are approximately equal to the partition functions Z in native proteins. These above assumptions together lead to the Boltzmann assumption that the structural features contained in the protein database must be a population correctly sampled under the Boltzmann distribution. That is, for any protein descriptor, we have

$$\pi_i \propto \exp(-w_i).$$

To calculate π_i in practice, we have to rely on another assumption that all protein structures are crystallized at the same temperature. Therefore, the distribution π_i is reasonably similar for all proteins in the database, and hence the frequency counts of protein descriptors in different protein structures can be combined by simple summation with equal weight.

Clearly, none of these assumptions are strictly true. However, the successes of many applications of using the statistical knowledge-based potentials indicate that they do capture many important properties of proteins. The question for improving the statistical potential function is, how seriously each of these assumptions is violated and to what extent it affects the validity of the potential function. A few assumptions specific to a particular potential function (such as the coordination and solvation assumptions for the Miyazawa–Jernigan reaction model) have been described earlier. Here we discuss several assumptions in detail below.

Interactions are not independent: Using an HP (Hydrophobic-Polar) model on a two-dimensional lattice, Thomas and Dill (1996b) tested the accuracy of Miyazawa–Jernigan contact potentials and Sippl's distance-dependent potentials. In the HP model, a peptide chain contains only two types of monomer: H and P. The true energies are set as $H_{(H,H)} = -1$, $H_{(H,P)} = 0$, and $H_{(P,P)} = 0$. Monomers are in contact if they are nonbonded nearest neighbors on the lattice. The conformational space was exhaustively searched for all sequences with the chain length from 11 to 18. A sequence is considered to have a native structure if it has a unique ground energy state. All native structures were collected to build a structure database, from which the statistical potentials are extracted by following the Miyazawa–Jernigan or the Sippl method. The extracted energies are denoted as $e_{(H,H)}$, $e_{(H,P)}$, and $e_{(P,P)}$.

It was found that neither of these two methods can extract the correct energies. All extracted energies by these two methods depend on chain length, while the true energies do not. Using Miyazawa–Jernigan's method, the (H, H) contact is correctly determined as dominant and attractive. However, the estimated values for $e_{(H,P)}$ and $e_{(P,P)}$ are not equal to zero, whereas the true energies $H_{(H,P)}$ and $H_{(P,P)}$ are equal to zero. Using Sippl's method, the extracted potentials erroneously show a distance dependence, i.e., (H, H) interactions are favorable at short distances but unfavorable at long distances, and conversely for (P, P) interactions, whereas the true energies in the HP model only exist between a first-neighbor (H, H) contact, and become zero for all the interactions separated by two or more lattice units.

These systematic errors result from the assumption that the pairwise interactions are independent, and thus the volume exclusion in proteins can be neglected (Thomas and Dill, 1996b). However, (H, H) interactions indirectly affect the observed frequencies of (H, P) and (P, P) interactions. First, in both contact and distance-dependent potentials, because only a limited number of interresidue contacts can be made within the restricted volume at a given distance, the high density of (H, H) pairs at short distances is necessarily coupled with the low density (relative to reference state) of (H, P) and (P, P) pairs at the same distances, especially at the distance of one lattice unit. As a result, the extracted (H, P) and (P, P) energies are erroneously unfavorable at short distances. Second, for distance-dependent potentials, the energy of a specific type of pair interaction at a given distance is influenced by the same type of pair at different distances. For example, the high density of (H, H) pairs at short distances causes a compensating depletion (relative to the uniform density reference state) at certain longer distances, and conversely for (H, P) and (P, P) interactions. Admittedly this study was carried out using models of short chain lengths and a simple alphabet of residues where the foldable sequences may be very homologous, hence the observed artifacts are profound, and the deficiencies of the statistical potentials revealed in this study such as the excluded volume effect are likely to be significant in potential functions derived from real proteins.

Pairwise interactions are not additive: Interactions stabilizing proteins are often modeled by pairwise contacts at the atom or residue level. An assumption associated with this approach is the additivity of pairwise interactions, namely, the total energy or fitness score of a protein is the linear sum of all of its pairwise interactions.

However, the nonadditivity effects have been clearly demonstrated in cluster formation of hydrophobic methane molecules both in experiment (Ben-Naim, 1997) and in simulation (Rank and Baker, 1997; Shimizu and Chan, 2001, 2002; Czaplewski et al., 2000). Protein structure refinement will likely require higher order interactions (Betancourt and Thirumalai, 1999). Some three-body contacts have been introduced in several studies (Eastwood and Wolynes, 2001; Rossi et al., 2001; Godzik and Skolnick, 1992; Godzik et al., 1992), where physical models explicitly incorporating three-body interactions are developed. In addition, several studies of Delaunay four-body interactions clearly showed the importance of including higher order interactions in explaining the observed frequency distribution of residue contacts (Krishnamoorthy and Tropsha, 2003; Carter et al., 2001; Gan et al., 2001; Zheng et al., 1997; Singh et al., 1996; Munson and Singh, 1997).

Li and Liang (2005b) introduced a geometric model based on the Delaunay triangulation and alpha shape to collect three-body interactions in native proteins. A nonadditivity coefficient $\nu_{(i,j,k)}$ is introduced to compare the three-body potential energy $e_{(i,j,k)}$ with the summation of three pairwise interactions $e_{i,j}$, $e_{(i,k)}$, and $e_{(j,k)}$:

$$\nu_{(i,j,k)} = \exp[-e_{(i,j,k)}]/\exp[-(e_{(i,j)} + e_{(i,k)} + e_{(j,k)})].$$

There are three possibilities: (1) $\nu = 1$: interaction of a triplet type is additive in nature and can be well approximated by the sum of three pairwise interactions; (2) $\nu > 1$: three-body interactions are cooperative and their association is more favorable than three independent pairwise interactions; (3) $\nu < 1$: three-body interactions are anticooperative.

After systematically quantifying the nonadditive effects of all 1540 three-body contacts, it was found that hydrophobic interactions and hydrogen bonding interactions make nonadditive contributions to protein stability, but the nonadditive nature depends on whether such interactions are located in the protein interior or on the protein surface. When located in the interior, many hydrophobic interactions such as those involving alkyl residues are anticooperative, namely, $\nu < 1$. Salt-bridge and regular hydrogen-bonding interactions such as those involving ionizable residues and polar residues are cooperative in interior. When located on the protein surface, these salt-bridge and regular hydrogen-bonding interactions are anticooperative with $\nu < 1$, and hydrophobic interactions involving alkyl residues become cooperative (Li and Liang, 2005b).

Sequence dependency of the partition function Z(a): We can obtain the total effective energy $\Delta E(\boldsymbol{s}, \boldsymbol{a})$ given a structure conformation $\boldsymbol{s}$ and its amino acid sequence $\boldsymbol{a}$

from Eq. (3.5):

$$\begin{aligned}\Delta H(f(\boldsymbol{s}, \boldsymbol{a})) &= \Delta H(\boldsymbol{c}) = \sum_i \Delta H(c_i) \\ &= -kT \sum_{c_i} \ln\left(\frac{\pi(c_i)}{\pi'(c_i)}\right) - kT \ln\left(\frac{Z(\boldsymbol{a})}{Z'(\boldsymbol{a})}\right),\end{aligned} \tag{3.48}$$

where c_i is the total number count of the occurrence of the i-th descriptor, e.g., the total number of i-th type of pairwise contact. The summation involving $Z(\boldsymbol{a})$ and $Z'(\boldsymbol{a})$ is ignored during the evaluation of $\Delta H(c_i)$ by assuming $Z(\boldsymbol{a}) \approx Z'(\boldsymbol{a})$.

It is clear that both $Z(\boldsymbol{a})$ and $Z'(\boldsymbol{a})$ do not depend on the particular structural conformation $\boldsymbol{s}$. Therefore, the omission of the term of the partition functions $-kT\ln(\frac{Z(\boldsymbol{a})}{Z'(\boldsymbol{a})})$ will not affect the rank ordering of energy values of different conformations (i.e., decoys) for the same protein sequence. On the other hand, it is also clear that both $Z(\boldsymbol{a})$ and $Z'(\boldsymbol{a})$ depend on the specific sequence $\boldsymbol{a}$ of a protein. Therefore, there is no sound theoretical basis to compare the stabilities between different proteins using the same knowledge-based potential function, unless the ratio of $Z(\boldsymbol{a})/Z'(\boldsymbol{a})$ for each individual sequence is known and is included during the evaluation (Miyazawa and Jernigan, 1985; Samudrala and Moult, 1998; Sippl, 1990). Notably, DFIRE and other statistical energy functions have been successfully used to predict binding affinities across different protein–protein/peptide complexes. Nevertheless, the theoretical basis is not sound either, because the values of partition function $Z(\boldsymbol{a})$ for different protein complexes can be drastically different. It remains to be seen whether a similarly successful prediction of binding affinities can be achieved just by using the number of native interface contacts at some specific distance interval, i.e., the packing density along the native interface. This omission is probably benign for the problem of predicting the free energy change of a protein monomer or the binding free energy change of a protein–protein complex upon point mutations, because the distribution of the ensemble of protein conformations may not change significantly after one or several point mutations.

Evaluating potential function: The measure used for performance evaluation of potential functions is important. For example, the z-score of native protein among decoys is widely used as an important performance statistic. However, the z-score strongly depends on the properties of the decoy set. Imagine we have access to the true energy function. If a decoy set has a diverse distribution in true energy values, the z-score of the native structure will not be very large. However, this by no means suggests that a knowledge-based energy function that gives a larger z-score for native protein is better than the true energy function. Alternative measures may provide more accurate or useful performance evaluation. For example, the correlation r of energy value and CRMSD may be helpful in protein structure prediction. Since a researcher has no access to the native structure, he or she has to rely on the guidance of an energy function to search for better structures with lower CRMSD to the unknown native structure. For this purpose, a potential function with a large r will be very useful. Perhaps the performance of a potential function should be judged not by a single statistic but comprehensively by a number of measures.

3.7.2 Relationship of Knowledge-Based Energy Functions and Further Development

The Miyazawa–Jernigan contact potential is the first widely used knowledge-based potential function. Because it is limited by the simple spatial description of a cutoff distance, it cannot capture the finer spatial details. Several distance-dependent potentials have been developed to overcome this limitation, and in general have better performance (Lu and Skolnick, 2001; Samudrala and Moult, 1998; Zhou and Zhou, 2002). A major focus of works in this area is the development of models for the reference state. For example, the use of the ideal gas as reference state in the potential function DFIRE significantly improves the performance in folding and docking decoy discrimination (Zhang et al., 2004a).

Because protein surface, interior, and protein–protein interface are packed differently, the propensity of the same pairwise interaction can be different depending on whether the residues are solvent-exposed or are buried. The contact potential of Simons et al. considers two types of environment, i.e., buried and nonburied environments separately (Simons et al., 1999). The geometric potential function (Li and Liang, 2005a) described in Section 3.3.5 incorporates both dependencies on distance and fine-graded local packing, resulting in significant improvement in performance. Table 3.2 shows that this potential can be successfully used in both protein structure and docking prediction. Knowledge-based potential have also been developed to account for the loss of backbone, side-chain, and translational entropies in folding and binding (Amzel, 2000; Lee et al., 1994).

Another emphasis of recent development of potential functions is the orientational dependency of pairwise interaction (Kortemme et al., 2003; Buchete et al., 2003, 2004; Miyazawa and Jernigan, 2005). Kortemme et al. developed an orientation-dependent hydrogen bonding potential, which improved prediction of protein structure and specific protein–protein interactions (Kortemme et al., 2003). Miyazawa and Jernigan developed a fully anisotropic distance-dependent potential, with drastic improvements in decoy discrimination over the original Miyazawa–Jernigan contact potential (Miyazawa and Jernigan, 2005).

Computational efficiency: Given current computing power, all potential functions discussed above can be applied to large-scale discrimination of native or near-native structures from decoys. For example, the geometric potential requires complex computation of the Delaunay tetrahedrization and alpha shape of the molecule (see Chapter 6 for details). Nevertheless, the time complexity is only $\mathcal{O}(N \log N)$, where N is the number of residues for residual-level potentials or atoms for atom-level potentials. For comparison, a naive implementation of contact computing without the use of proper data structure such as a quad-tree or k-d tree is $\mathcal{O}(N^2)$.

In general, atom-level potentials have better accuracy in recognizing native structures than residue-level potentials, and are often preferred for the final refinement of predicted structures, but it is computationally too expensive to be applicable in every step of a folding or sampling computation.

Potential function for membrane protein: The potential functions we have discussed in Section 3 are based on the structures of soluble proteins. Membrane proteins are located in a very different physicochemical environment. They also have different amino acid composition, and they fold differently. Potential functions developed for soluble proteins are therefore not applicable to membrane proteins. For example, Cys–Cys has the strongest pairing propensity because of the formation of disulfide bond. However, Cys–Cys pairs rarely occur in membrane proteins. This and other differences in pairwise contact propensity between membrane and soluble proteins are discussed in Adamian and Liang (2001).

Nevertheless, the physical models underlying most potential functions developed for soluble proteins can be modified for membrane proteins (Adamian and Liang, 2001, 2002; Adamian et al. , 2003; Park et al., 2004; Jackups and Liang, 2005). For example, Sale et al. used the MHIP potential developed in Adamian and Liang (2001) to predict optimal bundling of TM helices. With the help of 27 additional sparse distance constraints from experiments reported in the literature, these authors succeeded in predicting the structure of dark-adapted rhodopsin to within 3.2 Å of the crystal structure (Sale et al., 2004). It is likely that statistical potentials can be similarly developed for protein–ligand and protein–nucleotide interactions using the same principle.

3.7.3 Optimized Potential Function

Knowledge-based potential functions derived by optimization have a number of characteristics that are distinct from statistical potentials. We discuss these in detail below.

Training set for optimized potential function: Unlike statistical potential functions where each native protein in the database contributes to the knowledge-based potential function, only a subset of native proteins contribute. In an optimized potential function, in addition, a small fraction of decoys also contribute to the potential function. In the study of Hu et al. (2004), about 50% of native proteins and $< 0.1\%$ of decoys from the original training data of 440 native proteins and 14 million sequence decoys contribute to the potential function.

As illustrated in the second geometric views, the discrimination of native proteins occurs at the boundary surface between the vector points and the origin. It does not help if the majority of the training data are vector points away from the boundary surface. This implies the need for optimized potentials to have appropriate training data. If no *a priori* information is known, it is likely many decoys (>millions) will be needed to accurately define the discrimination boundary surface, because of the usually large dimension of the descriptors for proteins. However, this imposes significant computational burden.

Various strategies have been developed to select only the most relevant vector points. Usually, one may only include the most difficult decoys during training, such as decoys with lower energy than native structures, decoys with lowest absolute energies, and decoys already contributing to the potential function in previous iteration (Micheletti et al., 2001; Tobi et al., 2000; Hu et al., 2004). In addition, an iterative training process is often necessary (Micheletti et al., 2001; Tobi et al., 2000; Hu et al., 2004).

Reduced nonlinear potential function: The use of nonlinear terms for potential function involves large data sets, because they are necessary *a priori* to define accurately the discrimination surface. This demands the solution of a huge optimization problem. Moreover, the representation of the boundary surface using a large basis set requires expensive computing time for the evaluation of a new unseen contact vector $\boldsymbol{c}$. To overcome these difficulties, the nonlinear potential function needs to be further simplified.

One simple approach is to use alternative optimal criterion, for example, by minimizing the distance expressed in 1-norm instead of the standard 2-norm Euclidean distance. The resulting potential function will automatically have reduced terms. Another promising approach is to use rectangle kernels (Hu, Dai, and Liang, manuscript).

Potential function by optimal regression: Currently, most optimized potential functions are derived based on decoy discrimination, which is a form of binary classification. Here we suggest a conceptual improvement that can significantly improve the development of optimized potential functions. If we can measure the thermodynamic stabilities of all major representative proteins under identical experimental conditions (e.g., temperature, pH, salt concentration, and osmolarity), we can attempt to develop potential functions with the objective of minimizing the regression errors of fitted energy values and measured energy values. The resulting energy surface will then provide quantitative information about protein stabilities. However, the success of this strategy will depend on coordinated experimental efforts in protein thermodynamic measurements. The scale of such efforts may need to be similar to that of genome sequencing projects and structural genomics projects.

3.7.4 Data Dependency of Knowledge-Based Potentials

There are many directions to improve knowledge-based potential functions. Often it is desirable to include additional descriptors in the energy functions to more accurately account for solvation, hydrogen bonding, backbone conformation (e.g., ϕ and ψ angles), and side chain entropies. Furthermore, potential functions with different descriptors and details may be needed for different tasks [e.g., backbone prediction versus structure refinement, (Rohl et al., 2004)].

An important issue in both statistical potentials and optimized potentials is their dependency on the amount of available training data and possible bias in such data. For example, whether knowledge-based potentials derived from a bias data set are applicable to a different class of proteins is the topic of several studies (Zhang et al., 2004b; Khatun et al., 2004). In addition, when the amount of data is limited, overfitting is a real problem if too many descriptors are introduced in either of the two types of potential functions. For statistical potentials, hierarchical hypothesis testing should help to decide whether additional terms are warranted. For optimized potentials, cross-validation will help to uncover possible overfitting (Hu et al., 2004).

3.8 Summary

In this chapter, we discussed the general framework of developing knowledge-based potential functions in terms of molecular descriptors, functional form, and parameter calculations. We also discussed the underlying thermodynamic hypothesis of protein folding. With the assumption that frequently observed protein features in a database of structures correspond to low energy state, frequency of observed interactions can be converted to energy terms. We then described in detail the models behind the Miyazawa–Jernigan contact potential, distance-dependent potentials, and geometric potentials. We also discussed how to weight sample structures of varying degree of sequence similarity in the structural database. In the section on optimization method, we described general geometric models for the problem of obtaining optimized knowledge-based potential functions, as well as methods for developing optimized linear and nonlinear potential functions. This was followed by a brief discussion of several applications of the knowledge-based potential functions. Finally, we pointed out general limitations and possible improvements for the statistical and optimized potential functions.

3.9 Further Reading

Anfinsen's thermodynamic hypothesis can be found in Anfinsen et al. (1961) and Anfinsen (1973). More technical details of the Miyazawa–Jernigan contact potential are described in Miyazawa and Jernigan (1985, 1996). The distance-dependent potential function was first proposed by Sippl (1990), with further development described in Lu and Skolnick (2001); Samudrala and Moult (1998). The development of geometric potentials can be found in Zheng et al. (1997), Carter et al. (2001), Li et al. (2003), Krishnamoorthy and Tropsha (2003), and McConkey et al. (2003). The gas-phase approximation of the reference state is discussed in Zhou and Zhou (2002). Thomas and Dill offered insightful comments about the deficiency of knowledge-based statistical potential functions (Thomas and Dill, 1996b). The development of optimized linear potential functions can be found in Vendruscolo et al. (2000), Micheletti et al. (2001) and Tobi et al. (2000). The geometric view for designing the optimized potential function and the nonlinear potential function are based on the results in Hu et al. (2004).

Acknowledgments

We thank Drs. Bob Jernigan, Hui Lu, Dong Xu, Hongyi Zhou, and Yaoqi Zhou for helpful discussions. This work is supported by grants from the National Science Foundation (CAREER DBI0133856), the National Institutes of Health (GM68958), the Office of Naval Research (N000140310329), and the Whitaker Foundation (TF-04-0023).

References

Adamian, L., Jackups, R., Binkowski, T.A., and Liang, J. 2003. Higher-order interhelical spatial interactions in membrane proteins. *J. Mol. Biol.* 327:251–272.

Adamian, L., and Liang, J. 2001. Helix–helix packing and interfacial pairwise interactions of residues in membrane proteins. *J. Mol. Biol.* 311:891–907.

Adamian, L., and Liang, J. 2002. Interhelical hydrogen bonds and spatial motifs in membrane proteins: Polar clamps and serine zippers. *Proteins* 47:209–218.

Amzel, L.M. 2000. Calculation of entropy changes in biological processes: Folding, binding, and oligomerization. *Methods Enzymol.* 323:167–177.

Anfinsen, C.B. 1973. Principles that govern the folding of protein chains. *Science* 181:223–230.

Anfinsen, C., Haber, E., Sela, M., and White, F. 1961. The kinetics of formation of native ribonuclease during oxidation of the reduced polypeptide chain. *Proc. Natl. Acad. Sci. USA* 47:1309–1314.

Bastolla, U., Farwer, J., Knapp, E.W., and Vendruscolo, M. 2001. How to guarantee optimal stability for most representative structures in the protein data bank. *Proteins*, 44:79–96.

Bastolla, U., Vendruscolo, M., and Knapp, E.W. 2000. A statistical mechanical method to optimize energy functions for protein folding. *Proc. Natl. Acad. Sci. USA* 97:3977–3981.

Ben-Naim, A. 1997. Statistical potentials extracted from protein structures: Are these meaningful potentials? *J. Chem. Phys.* 107:3698–3706.

Berkelaar, M. 2004. LP_Solve package. URL http://www.cs.sunysb.edu/algorith/implement/lpsolve/implement.shtml

Betancourt, M.R., and Thirumalai, D. 1999. Pair potentials for protein folding: Choice of reference states and sensitivity of predicted native states to variations in the interaction schemes. *Protein Sci.* 8:361–369.

Bienkowska, J.R., Rogers, R.G., and Smith, T.F. 1999. Filtered neighbors threading. *Proteins* 37:346–359.

Bordner, A.J., and Abagyan, R.A. 2004. Large-scale prediction of protein geometry and stability changes for arbitrary single point mutations. *Proteins* 57:400–413.

Buchete, N.V., Straub, J.E., and Thirumalai, D. 2003. Anisotropic coarse-grained statistical potentials improve the ability to identify nativelike protein structures. *J. Chem. Phys.* 118:7658–7671.

Buchete, N.V., Straub, J.E., and Thirumalai, D. 2004. Orientational potentials extracted from protein structures improve native fold recognition. *Protein Sci.* 13:862–874.

Burges, C.J.C. 1998. A tutorial on support vector machines for pattern recognition. Knowledge Discovery and Data Mining 2. URL /papers/Burges98.ps.gz

Carter, C., Jr., LeFebvre, B., Cammer, S., Tropsha, A., and Edgell, M. 2001. Four-body potentials reveal protein-specific correlations to stability changes caused by hydrophobic core mutations. *J. Mol. Biol.* 311:625–638.

Chan, H.S., and Dill, K.A. 1990. Origins of structure in globular proteins. *Proc. Natl. Acad. Sci. USA* 87:6388–6392.

Chiu, T.L., and Goldstein, R.A. 1998. Optimizing energy potentials for success in protein tertiary structure prediction. *Folding Des.* 3:223–228.

Czaplewski, C., Rodziewicz-Motowidlo, S., Liwo, A., Ripoll, D.R., Wawak, R.J., and Scheraga, H.A. 2000. Molecular simulation study of cooperativity in hydrophobic association. *Protein Sci.* 9:1235–1245.

Czyzyk, J., Mehrotra, S., Wagner, M., and Wright, S. 2004. PCx package. URL http://www-fp.mcs.anl.gov/otc/Tools/PCx/

Dahiyat, B.I., and Mayo, S.L. 1997. *De novo* protein design: Fully automated sequence selection. *Science* 278:82–87.

Deutsch, J.M., and Kurosky, T. 1996. New algorithm for protein design. *Phys. Rev. Lett.* 76:323–326.

DeWitte, R.S., and Shakhnovich, E.I. 1996. SMoG: de novo design method based on simple, fast and accurate free energy estimates. 1. Methodology and supporting evidence. *J. Am. Chem. Soc.* 118:11733–11744.

Dima, R.I., Banavar, J.R., and Maritan, A. 2000. Scoring functions in protein folding and design. *Protein Sci.* 9:812–819.

Dobbs, H., Orlandini, E., Bonaccini, R., and Seno, F. 2002. Optimal potentials for predicting inter-helical packing in transmembrane proteins. *Proteins* 49:342–349.

Duan, Y., and Kollman, P.A. 1998. Pathways to a protein folding intermediate observed in a 1-microsecond simulation in aqueous solution. *Science* 282:740–744.

Eastwood, M.P., and Wolynes, P.G. 2001. Role of explicitly cooperative interactions in protein folding funnels: A simulation study. *J. Chem. Phys.* 114:4702–4716.

Edelsbrunner, H. 1987. *Algorithms in Combinatorial Geometry*. Berlin, Springer-Verlag.

Fain, B., Xia, Y., and Levitt, M. 2002. Design of an optimal Chebyshev-expanded discrimination function for globular proteins. *Protein Sci.* 11:2010–2021.

Finkelstein, A.V., Badretdinov, A.Y., and Gutin, A.M. 1995. Why do protein architectures have boltzmann-like statistics? *Proteins* 23:142–150.

Friedrichs, M.S., and Wolynes, P.G. 1989. Toward protein tertiary structure recognition by means of associative memory hamiltonians. *Science* 246:371–373.

Gan, H., Tropsha, A., and Schlick, T. 2001. Lattice protein folding with two and four-body statistical potentials. *Proteins* 43:161–174.

Gilis, D. 2004. Protein decoy sets for evaluating energy functions. *J. Biomol. Struct. Dyn.* 21:725–736.

Gilis, D., and Rooman, M. 1996. Stability changes upon mutation of solvent-accessible residues in proteins evaluated by database-derived potentials. *J. Mol. Biol.* 257:1112–1126.

Gilis, D., and Rooman, M. 1997. Predicting protein stability changes upon mutation using database-derived potentials: Solvent accessibility determines the importance of local versus non-local interactions along the sequence. *J. Mol. Biol.* 272:276–290.

Godzik, A., Kolinski, A., and Skolnick, J. 1992. Topology fingerprint approach to the inverse protein folding problem. *J. Mol. Biol.* 227:227–238.

Godzik, A., and Skolnick, J. 1992. Sequence–structure matching in globular proteins: Application to supersecondary and tertiary structure determination. *Proc. Natl. Acad. Sci. USA* 89:12098–12102.

Goldstein, R., Luthey-Schulten, Z.A., and Wolynes, P.G. 1992. Protein tertiary structure recognition using optimized hamiltonians with local interactions. *Proc. Natl. Acad. Sci. USA* 89:9029–9033.

Guerois, R., Nielsen, J.E., and Serrano, L. 2002. Predicting changes in the stability of proteins and protein complexes: A study of more than 1000 mutations. *J. Mol. Biol.* 320:369–387.

Hao, M.H., and Scheraga, H.A. 1996. How optimization of potential functions affects protein folding. *Proc. Natl. Acad. Sci. USA* 93:4984–4989.

Hao, M.H., and Scheraga, H.A. 1999. Designing potential energy functions for protein folding. *Curr. Opin. Struct. Biol.* 9:184–188.

Hill, R.B., Raleigh, D.P., Lombardi, A., and DeGrado, W.F. 2000. *De novo* design of helical bundles as models for understanding protein folding and function. *Acc. Chem. Res.* 33:745–754.

Hoppe, C., and Schomburg, D. 2005. Prediction of protein thermostability with a direction- and distance-dependent knowledge-based potential. *Protein Sci.* 14:2682–2692.

Hu, C., Li, X., and Liang, J. 2004. Developing optimal non-linear scoring function for protein design. *Bioinformatics* 20:3080–3098.

Jackups, R., Jr. and Liang, J. 2005. Interstrand pairing patterns in β-barrel membrane proteins: The positive-outside rule, ar omatic rescue, and strand registration prediction. *J. Mol. Biol.* 354:979–993.

Janicke, R. 1987. Folding and association of proteins. *Prog. Biophys. Mol. Biol.* 49:117–237.

Jernigan, R.L., and Bahar, I. 1996 Structure-derived potentials and protein simulations. *Curr. Opin. Struct. Biol.* 6:195–209.

Karmarkar, N. 1984. A new polynomial-time algorithm for linear programming. *Combinatorica* 4:373–395.

Karplus, M., and Petsko, G.A. 1990. Molecular dynamics simulations in biology. *Nature* 347:631–639.

Khatun, J., Khare, S.D., and Dokholyan, N.V. 2004. Can contact potentials reliably predict stability of proteins? *J. Mol. Biol.* 336:1223–1238.

Kocher, J.A., Rooman, M.J., and Wodak, S.J. 1994. Factors influencing the ability of knowledge-based potentials to identify native sequence–structure matches. *J. Mol. Biol.* 235:1598–1613.

Koehl, P., and Levitt, M. 1999a. *De novo* protein design. I. In search of stability and specificity. *J. Mol. Biol.* 293:1161–1181.

Koehl, P., and Levitt, M. 1999b. *De novo* protein design. II. Plasticity of protein sequence. *J. Mol. Biol.* 293:1183–1193.

Koretke, K.K., Luthey-Schulten, Z., and Wolynes, P.G. 1996. Self-consistently optimized statistical mechanical energy functions for sequence structure alignment. *Protein Sci.* 5:1043–1059.

Koretke, K.K., Luthey-Schulten, Z., and Wolynes, P.G. 1998. Self-consistently optimized energy functions for protein structure prediction by molecular dynamics. *Proc. Natl. Acad. Sci. USA* 95:2932–2937.

Kortemme, T., and Baker, D. 2002. A simple physical model for binding energy hot spots in protein–protein complexes. *Proc. Natl. Acad. Sci. USA* 99:14116–14121.

Kortemme, T., Kim, D.E., and Baker, D. 2004. Computational alanine scanning of protein–protein interfaces. *Sci. STKE* 2004:pl2.

Kortemme, T., Morozov, A.V., and Baker, D. 2003. An orientation-dependent hydrogen bonding potential improves prediction of specificity and structure for proteins and protein–protein complexes. *J. Mol. Biol.* 326:1239–1259.

Krishnamoorthy, B., and Tropsha, A. 2003. Development of a four-body statistical pseudo-potential to discriminate native from non-native protein conformations. *Bioinformatics* 19:1540–1548.

Kuhlman, B., Dantas, G., Ireton, G.C., Varani, G., Stoddard, B.L., and Baker, D. 2003. Design of a novel globular protein fold with atomic-level accuracy. *Science* 302:1364–1368.

Lazaridis, T., and Karplus, M. 2000. Effective energy functions for protein structure prediction. *Curr. Opin. Struct. Biol.* 10:139–145.

Lee, K.H., Xie, D., Freire, E., and Amzel, L.M. 1994. Estimation of changes in side chain configurational entropy in binding and folding: General methods and application to helix formation. *Proteins* 20:68–84.

Lemer, C.M.R, Rooman, M.J., and Wodak, S.J. 1995. Protein-structure prediction by threading methods—Evaluation of current techniques. *Proteins* 23:337–355.

Levitt, M., and Warshel, A. 1975. Computer simulation of protein folding. *Nature* 253:694–698.

Li, H., Helling, R., Tang, C., and Wingreen, N. 1996. Emergence of preferred structures in a simple model of protein folding. *Science* 273:666–669.

Li, H. Tang, C., and Wingreen, N.S. 1997. Nature of driving force for protein folding: A result from analyzing the statistical potential. *Phys. Rev. Lett.* 79:765–768.

Li, X., Hu, C., and Liang, J. 2003. Simplicial edge representation of protein structures and alpha contact potential with confidence measure. *Proteins* 53:792–805.

Li, X., and Liang, J. 2005a. Computational design of combinatorial peptide library for modulating protein–protein interactions. *Pacific Symposium of Biocomputing.*

Li, X., and Liang, J. 2005b. Geometric cooperativity and anti-cooperativity of three-body interactions in native proteins. *Proteins* 60:46–65.

Liang, J., and Dill, K.A. 2001. Are proteins well-packed? *Biophys. J.* 81:751–766.

Liu, S., Zhang, C., Zhou, H., and Zhou, Y. 2004. A physical reference state unifies the structure-derived potential of mean force for protein folding and binding. *Proteins* 56:93–101.

Looger, L.L., Dwyer, M.A., Smith, J.J., and Hellinga, H.W. 2003. Computational design of receptor and sensor proteins with novel functions. *Nature* 423:185–190.

Lu, H., and Skolnick, J. 2001. A distance-dependent atomic knowledge-based potential for improved protein structure selection. *Proteins* 44:223–232.

Maiorov, V.N., and Crippen, G.M. 1992. Contact potential that recognizes the correct folding of globular proteins. *J. Mol. Biol.* 227:876–888.

McConkey, B.J., Sobolev, V., and Edelman, M. 2003. Discrimination of native protein structures using atom-atom contact scoring. *Proc. Natl. Acad. Sci. USA* 100:3215–3220.

Méndez, R., Leplae, R., Lensink, M.F., and Wodak, S.J. 2005. Assessment of capri predictions in rounds 3–5 shows progress in docking procedures. *Proteins* 60:150–169.

Mészáros, C.S. 1996. Fast Cholesky factorization for interior point methods of linear programming. *Comput. Math. Appl.* 31:49–51.

Micheletti, C., Seno, F., Banavar, J.R., and Maritan, A. 2001. Learning effective amino acid interactions through iterative stochastic techniques. *Proteins* 42: 422–431.

Mirny, L.A., and Shakhnovich, E.I. 1996. How to derive a protein folding potential? A new approach to an old problem. *J. Mol. Biol.* 264:1164–1179.

Mitchell, B.O., Laskowski, R.A., Alex, A., and Thornton, J.M. 1999. BLEEP: Potential of mean force describing protein–ligand interactions: II. Calculation of binding energies and comparison with experimental data. *J. Comput. Chem.* 20:1177–1185.

Miyazawa, S., and Jernigan, R.L. 1985. Estimation of effective interresidue contact energies from protein crystal structures: Quasi-chemical approximation. *Macromolecules* 18:534–552.

Miyazawa, S., and Jernigan, R.L. 1996. Residue–residue potentials with a favorable contact pair term and an unfavorable high packing density term. *J. Mol. Biol.* 256:623–644.

Miyazawa, S., and Jernigan, R.L. 2005. How effective for fold recognition is a potential of mean force that includes relative orientations between contacting residues in proteins? *J. Chem. Phys.* 122:024901.

Momany, F.A., McGuire, R.F., Burgess, A.W., and Scheraga, H.A. 1975. Energy parameters in polypeptides. VII. Geometric parameters, partial atomic charges, nonbonded interactions, hydrogen bond interactions, and intrinsic torsional potentials for the naturally occurring amino acids. *J. Phys. Chem.* 79:2361–2381.

Muegge, I., and Martin, Y.C. 1999. A general and fast scoring function for protein–ligand interactions: A simplified potential approach. *J. Med. Chem.* 42:791–804.

Munson, P.J., and Singh, R.K. 1997. Statistical significance of hierarchical multi-body potential based on Delaunay tessellation and their application in sequence–structure alignment. *Protein Sci.* 6:1467–1481.

Nishikawa, K., and Matsuo, Y. 1993. Development of pseudoenergy potentials for assessing protein 3-D–1-D compatibility and detecting weak homologies. *Protein Eng.* 6:811–820.

Papadimitriou, C., and Steiglitz, K. 1998. *Combinatorial Optimization: Algorithms and Complexity*. New York, Dover.

Park, B.H., and Levitt, M. 1996. Energy functions that discriminate X-ray and near-native folds from well-constructed decoys. *J. Mol. Biol.* 258:367–392.

Park, Y., Elsner, M., Staritzbichler, R., and Helms, V. 2004. Novel scoring function for modeling structures of oligomers of transmembrane alpha-helices. *Proteins* 57:577–585.

Rank, J.A., and Baker, D. 1997. A desolvation barrier to hydrophobic cluster formation may contribute to the rate-limiting step in protein folding. *Protein Sci.* 6:347–354.

Rohl, C.A., Strauss, C.E., Misura, K.M., and Baker, D. 2004. Protein structure prediction using Rosetta. *Methods Enzymol.* 383:66–93.

Rossi, A., Micheletti, C., Seno, F., and Maritan, A. 2001. A self-consistent knowledge-based approach to protein design. *Biophys. J.* 80:480–490.

Russ, W.P., and Ranganathan, R. 2002. Knowledge-based potential functions in protein design. *Curr. Opin. Struct. Biol.* 12:447–452.

Sale, K., Faulon, J., Gray, G., Schoeniger, J.S., and Young, M. 2004. Optimal bundling of transmembrane helices using sparse distance constraints. *Protein Sci.* 13:2613–2627.

Samudrala, R., and Levitt, M. 2000. Decoys 'R' Us: A database of incorrect conformations to improve protein structure prediction. *Protein Sci.* 9:1399–1401.

Samudrala, R., and Moult, J. 1998. An all-atom distance-dependent conditional probability discriminatory function for protein structure prediction. *J. Mol. Biol.* 275:895–916.

Schölkopf, B., and Smola, A.J. 2002. *Learning with Kernels: Support Vector Machines, Regularization, Optimization, and Beyond*. Cambridge, MA, MIT Press.

Shakhnovich, E.I. 1994. Proteins with selected sequences fold into unique native conformation. *Phys. Rev. Lett.* 72:3907–3910.

Shakhnovich, E.I., and Gutin, A.M. 1993. Engineering of stable and fast-folding sequences of model proteins. *Proc. Natl. Acad. Sci. USA* 90:7195–7199.

Shimizu, S., and Chan, H.S. 2001. Anti-cooperativity in hydrophobic interactions: A simulation study of spatial dependence of three-body effects and beyond. *J. Chem. Phys.* 115:1414–1421.

Shimizu, S., and Chan, H.S. 2002. Anti-cooperativity and cooperativity in hydrophobic interactions: Three-body free energy landscapes and comparison with implicit-solvent potential functions for proteins. *Proteins* 48:15–30.

Simons, K.T., Ruczinski, I., Kooperberg, C., Fox, B., Bystroff, C., and Baker, D. 1999. Improved recognition of native-like protein structures using a combination of sequence-dependent and sequence-independent features of proteins. *Proteins* 34:82–95.

Singh, R.K., Tropsha, A., and Vaisman, I.I. 1996. Delaunay tessellation of proteins: Four body nearest-neighbor propensities of amino acid residues. *J. Comput. Biol.* 3:213–221.

Sippl, M.J. 1990. Calculation of conformational ensembles from potentials of the main force. *J. Mol. Biol.* 213:167–180.

Sippl, M.J. 1993. Boltzmann's principle, knowledge-based mean fields and protein folding. An approach to the computational determination of protein structures. *J. Comput. Aided. Mol. Des.* 7:473–501.

Sippl, M.J. 1995. Knowledge-based potentials for proteins. *Curr. Opin. Struct. Biol.* 5:229–235.

Tanaka, S., and Scheraga, H.A. 1976. Medium- and long-range interaction parameters between amino acids for predicting three-dimensional structures of proteins. *Macromolecules* 9:945–950.

Thomas, P.D., and Dill, K.A. 1996a. An iterative method for extracting energy-like quantities from protein structures. *Proc. Natl. Acad. Sci. USA* 93:11628–11633.

Thomas, P.D., and Dill, K.A. 1996b. Statistical potentials extracted from protein structures: How accurate are they? *J. Mol. Biol.* 257:457–469.

Tobi, D., Shafran, G., Linial, N., and Elber, R. 2000. On the design and analysis of protein folding potentials. *Proteins* 40:71–85.

Vanderbei, R.J. 1996. *Linear Programming: Foundations and Extensions*. New York, Kluwer Academic Publishers.

Vapnik, V. 1995. *The Nature of Statistical Learning Theory*. New York, Springer. ISBN 0-387-94559-8.

Vapnik, V., and Chervonenkis, A. 1964. A note on one class of perceptrons. *Automation and Remote Control* 25.

Vapnik, V., and Chervonenkis, A. 1974. *Theory of Pattern Recognition* [in Russian]. Nauka, Moscow, (German Translation: W. Wapnik & A. Tscherwonenkis, *Theorie der Zeichenerkennung*, Akademie–Verlag, Berlin, 1979).

Venclovas, E., Zemla, A., Fidelis, K., and Moult, J. 2003. Comparison of performance in successive CASP experiments. *Proteins* 45:163–170.

Vendruscolo, M., and Domanyi, E. 1998. Pairwise contact potentials are unsuitable for protein folding. *J. Chem. Phys.* 109:11101–11108.

Vendruscolo, M., Najmanovich, R., and Domany, E. 2000. Can a pairwise contact potential stabilize native protein folds against decoys obtained by threading? *Proteins* 38:134–148.

Wodak, S.J., and Roomanm, M.J. 1993. Generating and testing protein folds. *Curr. Opin. Struct. Biol.* 3:247–259.

Wolynes, P.G., Onuchic, J.N., and Thirumalai, D. 1995. Navigating the folding routes. *Science* 267:1619–1620.

Xia, Y., and Levitt, M. 2000. Extracting knowledge-based energy functions from protein structures by error rate minimization: Comparison of methods using lattice model. *J. Chem. Phys.* 113:9318–9330.

Xu, D., Lin, S.L., and Nussinov, R. 1997. Protein binding versus protein folding: The role of hydrophilic bridges in protein associations. *J. Mol. Biol.* 2651:68–84.

Zhang, C., and Kim, S.H. 2000. Environment-dependent residue contact energies for proteins. *Proc. Natl. Acad. Sci. USA* 97:2550–2555.

Zhang, C., Liu, S., Zhou, H., and Zhou, Y. 2004a. An accurate, residue-level, pair potential of mean force for folding and binding based on the distance-scaled, ideal-gas reference state. *Protein Sci.* 13:400–411.

Zhang, C., Liu, S., Zhou, H., and Zhou, Y. 2004b. The dependence of all-atom statistical potentials on training structural database. *Biophys. J.* 86:3349–3358.

Zhang, C., Liu, S., Zhu, Q., and Zhou, Y. 2005. A knowledge-based energy function for protein–ligand, protein–protein, and protein–DNA complexes. *J. Med. Chem.* 48:2325–2335.

Zhang, C., Vasmatzis, G., Cornette, J.L., and DeLisi, C. 1997. Determination of atomic desolvation energies from the structures of crystallized proteins. *J. Mol. Biol.* 267:707–726.

Zheng, W., Cho, S.J., Vaisman, I.I., and Tropsha, A. 1997. A new approach to protein fold recognition based on Delaunay tessellation of protein structure. *Pac. Symp. Biocomput.* pp. 486–497.

Zhou, H., and Zhou, Y. 2002. Distance-scaled, finite ideal-gas reference state improves structure-derived potentials of mean force for structure selection and stability prediction. *Protein Sci.* 11:2714–2726.

4 Computational Methods for Domain Partitioning of Protein Structures

Stella Veretnik and Ilya Shindyalov

4.1 Introduction

Analysis of protein structures typically begins with decomposition of structure into more basic units, called "structural domains". The underlying goal is to reduce a complex protein structure to a set of simpler yet structurally meaningful units, each of which can be analyzed independently. Structural semi-independence of domains is their hallmark: domains often have compact structure and can fold or function independently. Domains can undergo so-called "domain shuffling" when they reappear in different combinations in different proteins thus implementing different biological functions (Doolittle, 1995). Proteins can then be conceived as being built of such basic blocks: some, especially small proteins, consist usually of just one domain, while other proteins possess a more complex architecture containing multiple domains. Therefore, the methods for partitioning a structure into domains are of critical importance: their outcome defines the set of basic units upon which structural classifications are built and evolutionary analysis is performed. This is especially true nowadays in the era of structural genomics. Today there are many methods that decompose the structure into domains: some of them are *manual* (i.e., based on human judgment), others are *semiautomatic*, and still others are completely *automatic* (based on algorithms implemented as software). Overall there is a high level of consistency and robustness in the process of partitioning a structure into domains (for ~80% of proteins); at least for structures where domain location is obvious. The picture is less bright when we consider proteins with more complex architectures—neither human experts nor computational methods can reach consistent partitioning in many such cases. This is a rather accurate reflection of biological phenomena in general since domains are formed by different mechanisms, hence it is nearly impossible to come up with a set of well-defined rules that captures all of the observed cases.

This chapter focuses on computational methods for domain partitioning: it begins with an analysis of basic premises about structural domains, reviews currently available methods, takes a detailed look into some partitioning algorithms, and finally discusses challenges and potential new approaches for the next generation of domain partitioning algorithms.

Note: terms "partitioning," "decomposition," "cutting," and "assignment" are used interchangeably here with no special meaning associated with a particular term.

4.2 Definitions of Structural Domains

There are many definitions of protein domains; while some are based on sequence information only (they are rather cursory), most others (see below) consider structure information in addition to sequence information. Evolutionary information, i.e., recurrence of domains in different proteins, is also employed (Murzin et al., 1995). The most common definition of domains is often referred to as structural domains. The analysis presented here is concerned with *structural domains*; however, often they will be referred to simply as "domains" throughout this chapter.

Structural domains are units of the structure that (1) are compact, (2) are stable, (3) contain a hydrophobic core, (4) can fold independently of the rest of the protein, (5) occur in combinations with different domains, and (6) perform a specific function. Right away we see that there are structural/thermodynamics (def. 1–4), evolutionary (def. 5), and functional (def. 6) aspects to structural domains. Methods for partitioning protein structure into domains use these definitions as their guides (for more detailed discussion see Veretnik et al., 2004). There are several manual methods for domain partitioning (SCOP, AUTHORS) (Islam et al., 1995; Murzin et al., 1995), semiautomatic method (CATH) (Orengo et al., 1997), and a score of automatic methods (Table 4.1). Human expertise is nearly always superior to algorithms in the area of domain decompositions. However, computational methods are critical in the current era of structural genomics, when the sheer number of solved structures overwhelms human experts. Moreover, for the theoretical/computational study of proteins it is essential to have a *consistent* domain partitioning, which only can be achieved with automatic methods. How do computational methods approach this problem and how well do they capture the principles of structural domains? Before addressing this question, let us point out that structure partitioning into domains is not always unequivocally agreed upon even by human experts. The reason for it lies in the underlying complexity of structural domains—they do not always match the above set of definitions: some domains do recur in different combinations, but they are small and are lacking a hydrophobic core, other domains that recur as a single unit are large and can clearly be decomposed into separate structural units. There are many cases in which domains are not globular, but they constitute parts of the compact protein–protein (or nucleic acid–protein) complex; finally there are many cases where the function is carried jointly by two or more domains. In all such cases domains will be defined differently depending on which aspect of the definition—structural, evolutionary, or functional—becomes the most important for a particular research. Among three existing manual or semiautomatic methods, SCOP focuses on evolutionary recurrent units of structures and CATH stresses structural integrity of the units, while the AUTHORS method considers the function as well as the contribution of the structural unit to a protein complex. The corresponding resources, based on these three methods, disagree in over 20% of the cases; the rate of disagreement is particularly high in proteins with complex architectures, indicating that multiple solutions may exist in such cases. This inherent ambiguity of structural

Table 4.1 Summary of domain decomposition methods.

Method	Year Generation	Strategy	Type of domains generated	Approaches/models used
Rossman and Liljas (Rossman and Liljas, 1974)	1974 first generation	top-down	contiguous	Distance plots of the structure against itself using Cα distances, search for strong interactions close to diagonal
Crippen (Crippen, 1978)	1978 first generation	bottom-up	contiguous and noncontiguous	Clustering of small structural units
Rose (Rose, 1979)	1979 first generation	top-down	contiguous	Cutting the projection of 3D structure onto 2D domain disclosing plain
Wodak and Janin (Wodak and Janin, 1981)	1981 first generation	top-down	contiguous	Finding minimum in the interface between two domains
PUU (Holm and Sander, 1994)	1994 second generation	top-down	contiguous and noncontiguous	Rendering of the contact matrix, constructed using rigid body oscillation
DETECTIVE (Swindells, 1995a)	1995 second generation	bottom-up	Contiguous and noncontiguous	Building of the hydrophobic core
Islam et al. (Islam et al., 1995)	1995 second generation	top-down	contiguous and noncontiguous	Finding minima in the interdomain contact density
DOMAK (Siddiqui and Barton, 1995)	1995 second generation	top-down	contiguous and noncontiguous	Splitting structure by maximizing intradomain/inter-domain contacts
Sowdhamini and Blundell (Sowdhamini and Blundell, 1995)	1995 second generation	bottom-up	contiguous and noncontiguous	Clustering of secondary structures
Taylor (Taylor, 1999)	1999 second generation	bottom-up	contiguous and noncontiguous	Clustering of residues in spatial proximity using Ising model
STRUDL (Wernisch et al., 1999)	1999 second generation	top-down	contiguous and noncontiguous	Finding minimum interdomain contacts using Kernighan–Lin graph heuristics
DomainParser (Xu et al., 2000)	2000 second generation	top-down	contiguous and noncontiguous	Finds minimum interdomain contacts using graph theoretical approach with maximum flow/minimum cut using Ford-Fulkerson algorithm
(Xuan et al., 2000)	2000 second generation	bottom-up	contiguous and noncontiguous	Assemble domains from rudimentary fragments using fuzzy clustering

(*cont.*)

Table 4.1 (*Continued*)

Method	Year Generation	Strategy	Type of domains generated	Approaches/models used
PDP .(Alexandrov and Shindyalov, 2003)	2003 second generation	top-down	contiguous and noncontiguous	Finding partitioning with minimal number of contacts between domains
HVdWD (Hierarchy of Van der Waals Domains) (Berezovsky, 2003)	2003 second generation	bottom-up	contiguous and noncontiguous	Clustering of short segments. Both initial segments and the clustering threshold are based primarily on van der Waals interactions among atoms
(Kundu et al., 2004)	2004 second generation	top-down	contiguous and noncontiguous	Decomposition of the structure using Gaussian Network Model; assumes semi-independent motion of domains

domain definition is one of the difficult issues for computational methods, as we will see below.

4.3 Computational Methods

As a general rule computational methods focus on structural integrity of the resulting domains—this is intuitively clear and practically achievable, while evolutionary and functional information is much trickier to capture, support, and use. Historically the most common principle of structural partitioning is based on the interpretation of the so-called “contact density,” i.e., on the simple fact that there are more residue-to-residue contacts within a structural unit than between structural units. The implementations of this principle can be very different in different methods, but the underlying idea always comes back to finding regions with a high density of interactions. The extensive atomic interactions within structural domain were first pointed out by Wetlaufer (1973) and implemented in the very first domain partitioning algorithm by Rossman and Liljas (1974), who use Cα–Cα distance maps to identify structural domains. New computational methods have been appearing ever since; overall approximately 20 different methods had been published in literature. Methodologically they can be separated into first generation (methods published from 1974 to 1993) and second-generation algorithms: from 1994 until now (Wernisch and Wodak, 2003). There were no attempts of formal training and validation of the early algorithms: parameters of the algorithms were tuned using all existing data (meaning overfitting algorithms to a particular data set); no validation on an independent set of data was performed. The lack of solved structures at that time might be the chief reason. The second-generation methods approach the problem differently. First, algorithms were trained to optimize parameters, and then in a separate step their performance was evaluated. While most of the first-generation methods partition structure only into contiguous domains (consisting of a single polypeptide fragment with single starting and ending points in a protein chain), second-generation methods universally allow for noncontiguous domains (consisting of several fragments separated by other residues). Domain partitioning can be divided into two fundamental approaches: *top-down* (starting from the entire structure and proceeding to partition it iteratively into smaller units) and *bottom-up* (defining very small structural units and assembling them into domains). Some methods use both approaches within their algorithm: first decomposition and then assembly or vice versa. Here we classify each method based on its chief or overall approach to domain identification. Generally, the process of domain decomposition is performed in two steps: (1) tentative domains are constructed (either by splitting the structure into domains or building domains from smaller units) and (2) tentatively defined domains are evaluated in a postprocessing step. Overall an amazing array of approaches has been put forward over the years to solve the domain decomposition problem. In the rest of this section, 16 different domain partitioning methods—from the earliest ones to the latest ones—are discussed in terms of their general approach and methodology used.

4.3.1 Rossman and Liljas (Rossman and Liljas, 1974)

Method: Cα–Cα distance plots of the structure against itself are generated and contours of interactions are examined. Domains are recognized as a series of shorter interactions closer to the diagonal. Similar pattern of contours indicates similar domain structure.

Results: Recurrence of different combinations of domains in various proteins is pointed out.

4.3.2 Crippen (Crippen, 1978)

Method: Method is based on the assumption that the stable conformation of a protein is largely due to the energetically favorable interactions of the residues which are frequently distant in sequence, i.e., long-range interactions. In the first step, protein chain is divided into segments—short stretches of polypeptide chain whose residues have no long-range interactions with any other residues. Long-range interactions between two residues are defined as follows: sequential distance is at least seven residues, distance between Cα–Cα atoms is <9 Å. The segments (which correspond frequently to isolated secondary structures or coiled regions) are then hierarchically clustered, using contact density criteria (contact density is normalized by the size of the individual segments). The process is repeated iteratively first joining segments in the most intimate contact, while last clusters have few contacts relative to the number of residues involved.

Results: The method allows one to view the crystal structure of a protein as a packing tree, with individual secondary structures at the bottom, structural domains close to the top, and the subdomain structure in order of increasing complexity in the intermediate levels of the tree. The author hypothesizes that the steps of the folding mechanism can possibly be discerned from such packing tree. Folding intermediates correspond to internal nodes of the packing tree; the order of their appearance can be predicted by following the packing tree from the bottom level upwards.

4.3.3 Rose (Rose, 1979)

Method: The protein is treated as a rigid body and a set of Cartesian axes are drawn through the center of its mass: in classical mechanics these are referred to as "principal axes" and they correspond to eigenvectors of inertia tensors. The three-dimensional structure of a protein is then reduced to two-dimensional space by projecting Cα atoms of each residue onto the so-called domain "disclosing plane." The plane is determined by two of the three principal axes: those with the larger moments of inertia. The Cα atoms are connected in sequential fashion. An arbitrary line is drawn through the plane to divide protein into at least two pieces. The best line is searched according to two criteria: first, the two parts of a protein must be contiguous fragments and separated as much as possible by the line. In the ideal situation the line will completely separate the two parts; deviation from this ideal case is measured by the length by which one part of the protein extends into the other (the length is defined as the sum of distances between Cα atoms on their projection onto the disclosing plane). Second, the length of two resulting domains is compared; in the ideal case the domains will be of identical length; deviation from this is measured with a nonlinear function. The entire process is then repeatedly applied to each of the two domains until termination criteria are met: the projection of the chain on the disclosing plane does not close upon itself, which indicates either very short domains or a lack of compactness within a domain.

Results: The author points out that the hierarchy of domains resulting from such a process, when placed in ascending order, is in accordance with the *local process* of the automata theory, in which current state can be derived from the previous state. The hierarchical domain structure might then reflect the folding process of the protein, which proceeds by hierarchical condensation. During hierarchical condensation the nearby hydrophobic elements coalesce to form folding primitives which in turn coalesce to form larger module until the entire hydrophobic core is complete.

4.3.4 Wodak and Janin (Wodak and Janin, 1981)

Method: The method evaluates size and properties of domain interfaces: a domain interface is defined as a surface area buried in contacts between two groups of residues. Minima in the interface area are found for the entire protein structure and then recursively for each of the partitioned units. The process stops when the significance of interface minima drops below a predefined threshold.

4.3.5 Holm and Sander (Holm and Sander, 1994)

Method: The method identifies folding units by finding groups of residues that have longest intergroup oscillation time. Oscillation time τ is proportional to the center of masses and inversely proportional to the interface strength, which is determined by nonbonded atomic interactions at the interface between two units (two atoms are interacting if their distance is ≤ 4.0 Å). Folding units are found by first creating a contact matrix and then the rendering of the contact matrix in such a way that strongly interacting residues are grouped together (using a reciprocal averaging) and rows/columns 1 through k belong to one unit while rows and columns of $k + 1$ through L(length of the protein) belong to the other unit. Bisection of the ordered contact matrix can continue recursively for each of the resulting folding units until some limit on unit size is reached (10–19 residues). At this point each autonomous folding unit becomes a domain. An additional hierarchical five-level filtering is applied during the process (once the higher level filter is met, the partition is accepted): (1) Domains should have 40 or more residues. Thus, units smaller than 80 residues are never cut. (2) Highly flexible units are always cut. (3) β-Sheets are never cut. (4) A cut is acceptable if both resulting units have a high globularity/compactness value. (5) A cut that produces a nonglobular domain with less than 40 residues is accepted on condition that the larger domain in the cut will be split into two domains upon recursive application of the filters.

4.3.6 Swindells (Swindells, 1995b)

Method: Hydrophobic cores of the protein structure are defined using a set of rules based on solvent exposure, minimal size of a core, fraction of the spatially adjacent residues, etc. The hydrophobic cores become the centers of the domains, which grow by iteratively including residues spatially proximal to the core until most of the residues are assigned. Isolated residues that generate contradictory results are removed from the assignment. In the final step the unassigned residues are assigned by extending domains to both ends of the structure and/or to the ends of the appropriate secondary structure.

4.3.7 Islam, Luo, and Sternberg (Islam et al., 1995)

Method: The protein chain is cut iteratively into domains by finding the minima in the interdomain contact density. The cutting is stopped when density of contacts reaches an empirically determined threshold F. The series of contiguous segments is evaluated in a postprocessing step: every two noncontiguous segments are tentatively combined, and if their contact density is over the threshold F, the segments are clustered together. This process is repeated iteratively until no more clustering of the segments occurs. If any of the remaining segments are less than 32 residues long, they are merged with most appropriate larger segment. At this point each segment becomes a domain.

4.3.8 Siddiqui and Barton (Siddiqui and Barton, 1995)

Method: Each position in the protein chain is considered as potential splitting point of the chain into two domains. The intraresidue contacts within each of two potential domains (called "split value") are calculated at each step; the chain is split at the position that has the highest *split value*. The process is continued iteratively on each of the resulting domains until one of the stopping criteria—minimum size of the domain or minimal split value (MSV)—is met. Two-segmented domains (in which one of the domains is surrounded by a noncontiguous second domain or both noncontiguous domains are arranged in an interdigitated manner) are derived in a similar process, but instead of one potential split position within a chain, two positions ($\mathbf{A}_1$ **B** $\mathbf{A}_2$ domain arrangement) or four positions ($\mathbf{A}_1$ $\mathbf{B}_1$ $\mathbf{A}_2$ $\mathbf{B}_2$ domain arrangement) within a chain are considered simultaneously for a potential split. The maximum split value is found by varying all two (or four) potential split positions. During postprocessing steps additional screening is performed to weed out unrealistic domains with: (a) a high number of segments in single-domain proteins, (b) short segments in multi-segmented domains, (c) small domains inserted into large domains.

4.3.9 Sowdhamini and Blundell (Sowdhamini and Blundell, 1995)

Method: The method clusters secondary structures into domains using a phylogenetic inference approach. The constructed dendrogram is based on a *proximity index*, which measures the extent of interaction between a pair of secondary structures. Proximity index is defined as an average of all possible distances between the α-carbon atoms of one secondary structure to the α-carbon atoms of the other. Domains are then defined by choosing the clusters in the dendrogram at the level in which *disjoint factor* >1. Disjoint factor measures the density of interactions between secondary structures within the domain relative to all interactions of secondary structures in the protein.

4.3.10 Taylor (Taylor, 1999)

Method: The method uses principles of the Ising model from statistical mechanics. Residues are assigned numerical labels; the label of each residue is iteratively increased or decreased based on the average value of labels in its neighborhood. The process continues until the label values of all the residues stabilize (or oscillate). Residues with the same label value constitute a domain. The neighborhood of residues is defined by the radius R: if the distance between α-carbons of residues i and j is less than R, then the residues are in each other's neighborhoods. Residues in small domains (less than 40 residues) are reassigned to the neighboring domains.

4.3.11 Wernisch, Hunting, and Wodak (Wernisch et al., 1999)

Method: The STRUDL (STRUctural Domain Limits) method uses Kernighan–Lin graph heuristics to iteratively partition all residues of a structure into two sets with minimum contacts between them, repeating the process until either defined termination criteria are met or no cut can be made any further based on compactness criteria. The search for substructures with minimum contact area begins by systematically subtracting residues (one at a time) from one subset (V) and moving them to its complement U ($V + U$ contain all residues in the structure) in such a way that contact between U and V increases minimally. After each move described above, a process is initiated in which all possible switches among residues in the two sets are tested. The arrangement which produces the lowest number of contacts between two sets is selected. The process is repeated by moving subsequent residues from V to U until $N/2$ residues are moved. From all partitions of residues between U and V the one with lowest contact number is chosen as the final prediction and the resulting domains go through a postprocessing step in which the validity of domains (based on various heuristics for compactness) is assessed. The entire process repeats for each domain.

The contact area between two sets of residues in U and V is based on the sum of contact areas of all residues in U and V, where the contact area between any two residues is based on the pairwise sum of all of its interacting atoms. Interaction between atoms is determined using Voronoi cell, which in turn is based on the van der Waals radii of the atoms.

4.3.12 Guo, Xu, and Xu (Guo et al., 2003; Xu et al., 2000)

Method: The DomainParser method uses a graph-theoretical approach to find the best partitioning of a given structure into two parts. A protein structure is modeled as a graph in which nodes represent residues and edges connecting nodes represent interactions between residues. The weight of connection is proportional to the strength of interaction between two residues. A minimum cut splitting the network into two is found by determining the maximum flow through the network, using the Ford–Fulkerson algorithm (a more detailed description is given below in Section 4.4).

The process is repeated iteratively for each of the resulting domains until stopping criteria are met. Multiple minima partitions are performed at each step; the best partition is determined during a post-processing step by checking several basic properties of the resulting domains. The properties include hydrophobic moment, the number of noncontiguous fragments, domain size, compactness, and relative motion of domains. The "acceptable" range for each property as well as a set of stopping criteria is determined in advance using a neural network trained on the set of known structures.

4.3.13 Xuan, Ling, and Chen (Xuan et al., 2000)

Method: The method (no name given) uses fuzzy clustering analysis for domain recognition. In the first step the chain is partitioned into elementary fragments which consist of residues adjacent in sequence and in similar "contact environment." The contact environment of a residue is defined in terms of all other residues which interact with that residue. In the next step the initial fragments are clustered with other fragments using fuzzy logic and lower threshold on the criterion of similarity of contact environment. At this point the rudimentary domains are formed. During the last step the final domains are assembled from the rudimentary domains using principles of minimum domain size, minimum fragment size, and integrity of secondary structure joins.

4.3.14 Alexandrov and Shindyalov (Alexandrov and Shindyalov, 2003)

Method: The PDP method partitions a polypeptide chain into two parts by introducing either a single cut through the chain which results in two contiguous domains or a double cut, which will produce a contiguous and a noncontiguous domain (the two cuts must be at least 35 residues apart). The optimal position of the cut(s) is chosen by minimizing the number of contacts between two resulting domains. Calculation of the number of contacts takes into account the sizes of resulting domains. The threshold for domain contacts is set to be one-half of the expected number of contacts. The expected number of contacts between domains is considered to be proportional to their surface area. The calculation of the surface of the domain assumes the shapes of the resulting domains to be close to spherical; thus, it is proportional to $n^{2/3}$, where n is the number of residues in a domain. The process of division is repeated iteratively for each of the resulting domains. A post-processing step involves rechecking each of the contacts among the resulting domains: filtering out very small domains (less than 30 residues) and combining domains together if they exhibit a high number of contacts.

4.3.15 Berezovsky (Berezovsky, 2003)

Method: Initially a protein chain is partitioned into segments based on the threshold (referred to as "potential barriers") between local maximum and minimum of van der Waals energies calculated using 6-12 Lennard-Jones potential. Next, the pairwise interactions between isolated segments are evaluated; segments are either joined together or granted the status of a domain. A segment becomes a domain if its van der Waals energy internal to the segment is at least 3 times higher than the sum of its interaction energies with all other segments (bottom up step). The chain segmentation step can be performed at different thresholds of "potential barrier": lowering the potential barrier increases the number of segments and decreases their size.

Results: Multiple solutions to the protein partitioning can be produced by applying a series of thresholds—a unique approach so far among existing methods. Multiplicity of solutions reflects observations that for some proteins there are several possible domain decompositions—all of which appear to be biologically meaningful. The authors also point out that the resulting domains are made of closed loops—contiguous subtrajectories of the folded chain with a short Cα–Cα distance between their ends. Such closed loops are proposed to be the elementary units of structural domains.

4.3.16 Kundu, Sorensen, and Phillips (Kundu et al., 2004)

Method: This approach to decomposition of the structure into domains assumes a semi-independent motion and local compactness of the domains. A concept of Gaussian Network Model (GNM) is employed to find structurally compact clusters which are connected internally, but motionally decoupled from the other parts of the structural units. The method employs connected graphs in which each residue is represented by a node and connection between any two residues is determined by Cα distances of the residues. The structure is recursively partitioned into domains by constructing Laplacian matrix, performing a single value decomposition to find the lowest eigenvalue which corresponds to the slowest motion within the structure. After each decomposition step, potential domains are evaluated using a layer of filters which include minimum size of the domain, minimum length of discontinuous fragments comprising the domain, and integrity of β-sheets.

4.4 In-depth Look into Algorithmic Domain Decomposition

To gain a better insight into how domain decomposition methods work, two methods are discussed here in detail—one is from the first generation of algorithms and the other is a recent method.

Crippen's method (Crippen, 1978) is one of the earliest methods published. It uses a bottom-up approach of clustering secondary structures and is actually unique among first-generation methods as it allows construction of noncontiguous domains. Twenty-five structures were analyzed. The same 25 proteins are employed for tuning several parameters used in the algorithm (described below). During the first step, protein structure is decomposed into so-called "basic units": each unit consists of consecutive residues that do not have long-range interactions with other residues in the unit. Long-range interaction between two residues is defined with the following condition: distance <9Å between Cα atoms of two residues that are at least seven residues apart in sequence. The definition of long-range interactions comes from fitting secondary structure assignments of myoglobin to the segments generated; it matches the definition of secondary structures by Levitt and Greer (1977) rather well. The assigned basic units are permitted to overlap during the assignment process; the final boundaries between the segments are drawn in the middle of units overlap. Once an entire structure is broken down into basic units, the process of clustering begins: units with the highest contact density are paired first to form a new larger unit. The contact density is determined by calculating all pairwise contacts between all possible pairs of residues in two units; this value is then normalized by the

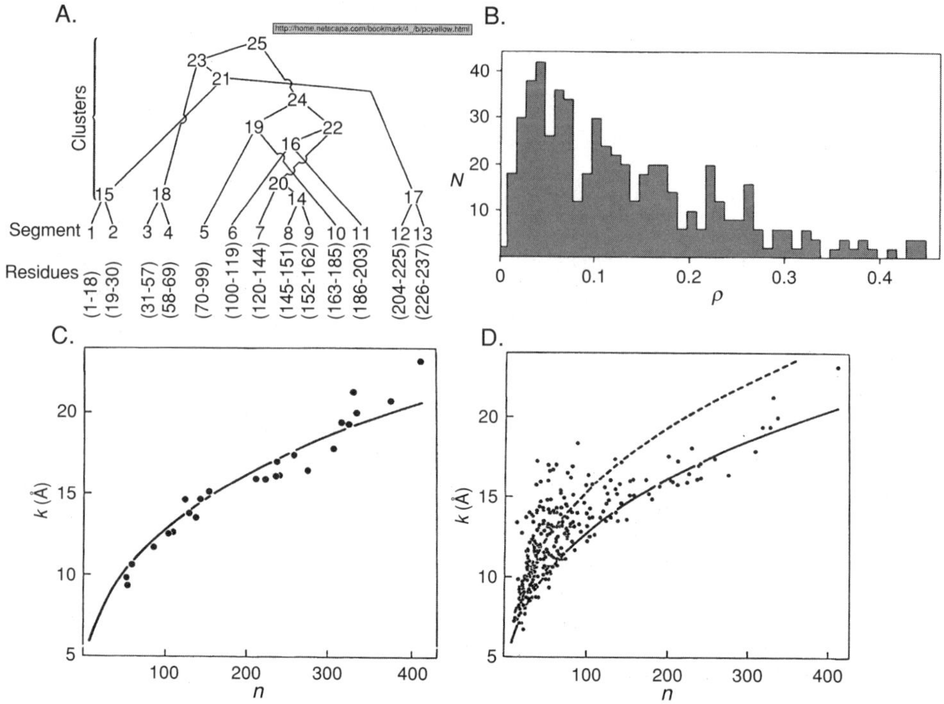

Fig. 4.1 Domain decomposition method by Crippen. (A) Binary tree representation of hierarchical clustering of basic and intermediate unit for concanavalin A. (B) Histogram of number of occurrences N of clusters formed with contact density ρ (data for 25 proteins). (C) Radius of gyration as a function of number of residues for 25 whole proteins. (D) Radius of gyration as a function of number of residues for each cluster for 25 proteins. Solid curves is the same as in (C), dotted curve is 1.2 times the solid curve.

size of the two units. The newly constructed cluster becomes a new unit and the process of pairing continues until only one single unit remains. Such clustering can be formalized using a binary tree, with basic units as leaves, intermediate nodes as clusters formed by two units at the level immediately below (Fig. 4.1A).

The root of the tree is the cluster that includes all of the residues. Intermediate clusters can be formed by joining groups of residues that are contiguous in sequence or alternatively by bringing two noncontiguous stretches together into a single cluster. A *break fraction* for a tree is defined as follows: zero indicates that all clusters in the tree consist of adjacent segments, while a *break fraction* of one indicates that all clusters in the tree are built from sequentially nonadjacent segments. In the real proteins *break fraction* is somewhere between 0 and 1, in myoglobin it is 0.262. Theoretically domains can be assigned at any level of clustering; the decision whether a given cluster in a tree constitutes a domain is defined by two parameters: contact density ρ and radius of gyration k. Both parameters are determined by inspecting properties of the 25 protein structures. From a histogram of contact density/residue^2

constructed for all 25 proteins, contact density between first and last assembled clusters varies significantly; however, the range of contact density is similar for all proteins. The threshold for contact density is chosen to be $\rho < 0.1$ contact/residue^2, which includes top one-third of strongest interacting clusters of the entire 300 clusters of 25 proteins (Fig. 4.1B). Similarly, the radius of gyration was calculated for each of the 25 complete proteins as well as for 300 clusters produced and plotted against the number of residues in the unit. Two curves—one formed by radius of gyration of complete proteins (Fig. 4.1C) and the other (produced by multiplying values of the first curve by 1.2)—bracket the range of gyration radii that are acceptable for domains (Fig. 4.1D). The domains are delineated by starting at the level of small clusters (close to the bottom level of the tree) and moving upwards. The status of each cluster is checked using parameters ρ gnd k: domain status is achieved if both parameters are within an acceptable range. The resulting domain clusters correspond well to the commonly accepted concept of spatially compact structures.

The DomainParser method (Guo et al., 2003; Xu et al., 2000) is among the most recent methods published; it uses a top-down graph-theoretical approach for domain decomposition and an extensive postprocessing step. During the training stage of the algorithm multiple parameters are tuned; the method is then validated on a SCOP data set. Domain decomposition is addressed by modeling the protein structure as a network consisting of nodes (residues) and edges (connections between residues). A connection between any two residues is drawn when they are adjacent in the sequence or alternatively are in physical proximity in the structure. The strength of the interaction between two residues is expressed as the capacity of the edge to connect the two nodes. This edge capacity is a function of (a) the number of atom–atom contacts between residues, (b) the number of backbone contacts between residues, (c) the existence of backbone interactions across a β-sheet, and (d) whether both residues belong to the same β-strand. The values for all parameters involved in edge capacity are optimized during the training stage of the algorithm.

The partitioning of the network into two parts is then equivalent to decomposing a given structure into two domains. Ideally, partitioning should be done using the edges with least capacity, which will result in partitioning structure along least dense interactions among residues. The problem of partitioning the network is solved using *maximum flow/minimum cut* theorem by Ford-Fulkerson and implemented by Edmond and Karp. The gist of the approach is as follows: artificial source and sink node are added to the network (Fig. 4.2A). A "bottleneck"—a set of critical edges in the network flow—is found by gradually increasing the flow of all edges in a network. Removing the set of critical edges from the network prevents flow from the source to the sink. At this point nodes that are connected to the source represent one interconnected part of the network, while nodes connected to the sink are the second interconnected part of the network. Since the node capacity is increased gradually it is expected that nodes with least capacity (least residue–residue contacts) will be the ones contributing to the bottleneck. The process of subdividing the network into two parts is repeated multiple times by connecting the source and sink to a different part of the network; a set of minimal cuts is collected and evaluated during a postprocessing

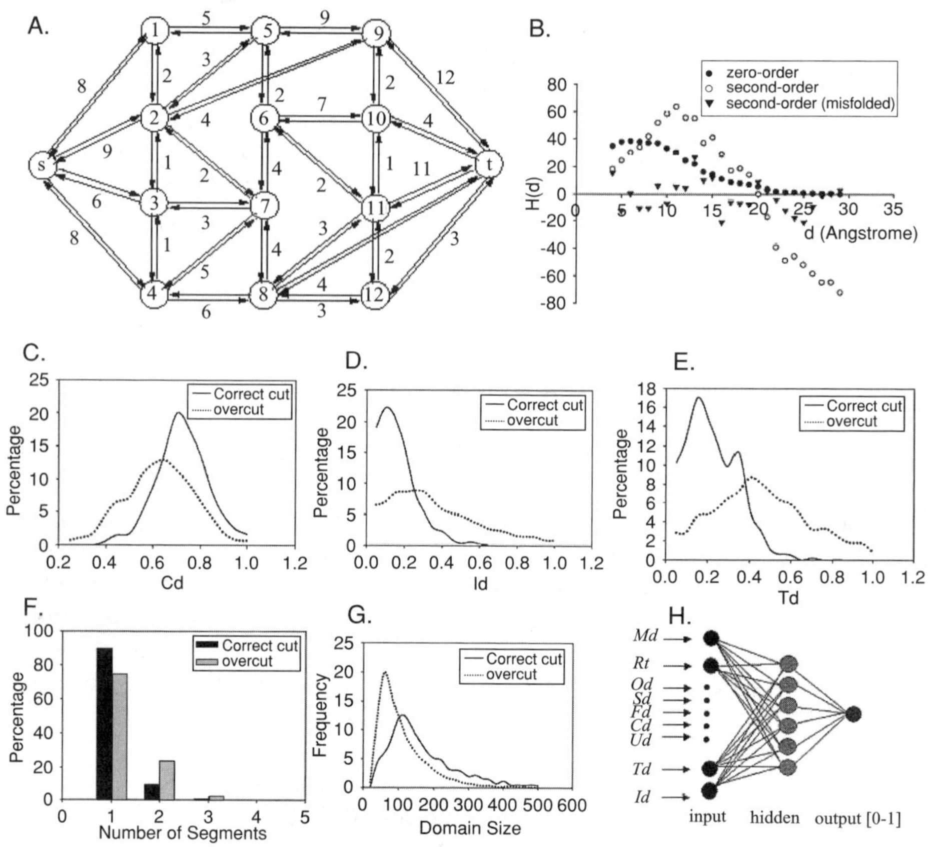

Fig. 4.2 Domain decomposition using DomainParser algorithm. (A) Schematic representation of protein structure a directed flow network. Value on each edge represents the edge's capacity. Artificial nodes "start" and "sink" are denoted by s and t, respectively. (B–G) Examples of parameters used by the algorithms: (B) zero- and second-order spherical moment profile of structure 2ilb; (C) compactness of structure; (D) size of domain interface relative to domain's volume; (E) measure of relative motion between domains; (F) distribution of the number of segments per domain; (G) distribution of domain sizes; (H) neural network architecture for evaluation of decomposed individual domains.

step. The entire procedure is then repeated in each of the resulting domains, until either domain's size drops below 80 residues or the partitioning produces domains that do not meet necessary conditions of domain definitions.

The stopping criteria are multifaceted and defined by (1) domain size: no less than 35 residues, (2) β-sheets kept intact, (3) compactness of domain above threshold g_m, (4) size of the domain–domain interface below threshold f_m, (5) ratio of number of residues and number of segments in domain is above threshold l_s. The values for g_m, f_m, l_s, and minimum domain size are determined during the training stage of the algorithm.

A suite of additional parameters exists; these are involved in a post-processing step of the algorithm, in which an assessment is made whether the substructure meets the additional criteria of a structural domain. These parameters are (1) hydrophobic moment (Fig. 4.2B), (2) the number of segments in the partitioned domain (Fig. 4.2F), (3) compactness (Fig. 4.2C), (4) the size of domain interface relative to domain's volume (Fig. 4.2D), (5) relative motion between compact domains (Fig. 4.2E). Distribution for each of the parameters for true versus false domains is collected during the training stage using 633 correctly partitioned domains and 928 incorrectly partitioned domains. Multiple neural networks are then investigated; the best one has 9 input nodes, 6 nodes in the hidden layer, and 1 output node (Fig. 4.2F). Performance of the DomainParser method is then evaluated using set of 1317 protein chains in which domains are defined by SCOP.

4.5 Evaluating Automatic Methods with Manual Consensus Benchmark

Why are there so many different automatic methods for domain decomposition? A chief reason is the complexity of the problem itself: it is nearly impossible to capture succinctly the principles of domain decomposition and apply them successfully to the entire universe of protein structures. Thus, every new method strives to reach a bit further beyond existing methods in its ability to decompose complex structures. With so many different methods available, it is essential to be able to compare the performance of the algorithms and to determine what fraction of known structures any given method predicts correctly. It is equally important to know what are the strengths and weaknesses of each algorithm, in particular what types of structures are difficult for a given method. Evaluation of automatic methods is an essential part of the algorithm development process; in fact, the performance of each method is typically reported along with the algorithm. However, each method uses its own data set for evaluation of the algorithm, thus it is nearly impossible to compare the performance of the algorithms to each other. An exception to this is a set of 55 chains (Jones et al., 1998) which is frequently used to test the performance. However, this is a very small data set, thus it is likely that its resolution will be insufficient to detect differences between methods. This situation was partially rectified recently, with construction of a large comprehensive benchmark data set developed specifically for evaluation of domain decomposition algorithms (Veretnik et al., 2004). The data set is assembled using a principle of consensus approach among expert methods: it includes proteins for which three expert methods (CATH, SCOP, and AUTHORS) produce similar domain decomposition. There are a total of 374 proteins in this benchmark, which was used to evaluate three recent automatic methods: PUU, DomainParser, and PDP (Fig. 4.3A).

The evaluation includes information about the success rate of each algorithm, analysis of errors in terms of predicting fewer domains (undercut) or too many domains (overcut). The analysis further looks into the tendencies of the methods

A.

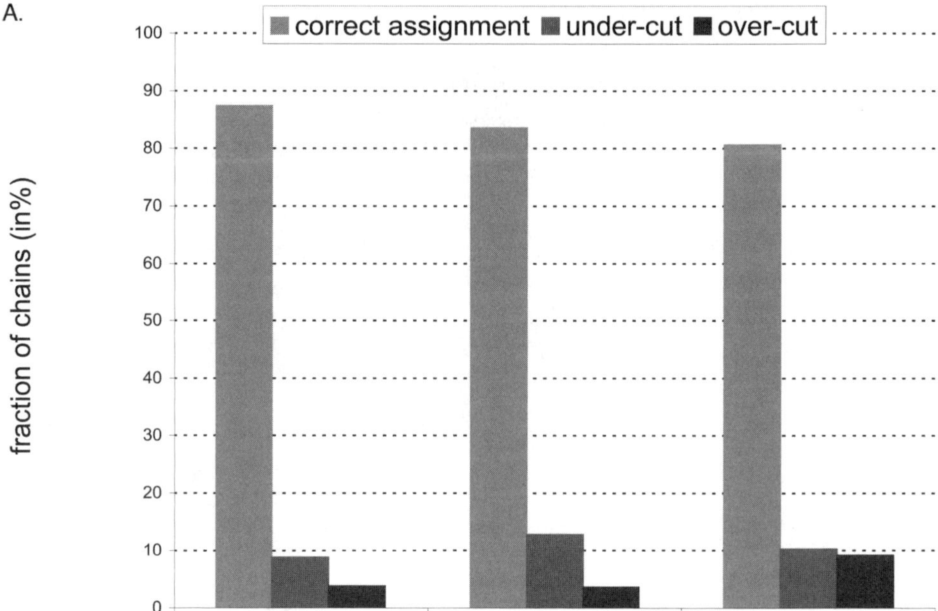

	PDP	DomainParser	PUU
correct assignment	87.4	83.6	80.6
under-cut	8.8	12.8	10.2
over-cut	3.8	3.63	9.2

method

B.

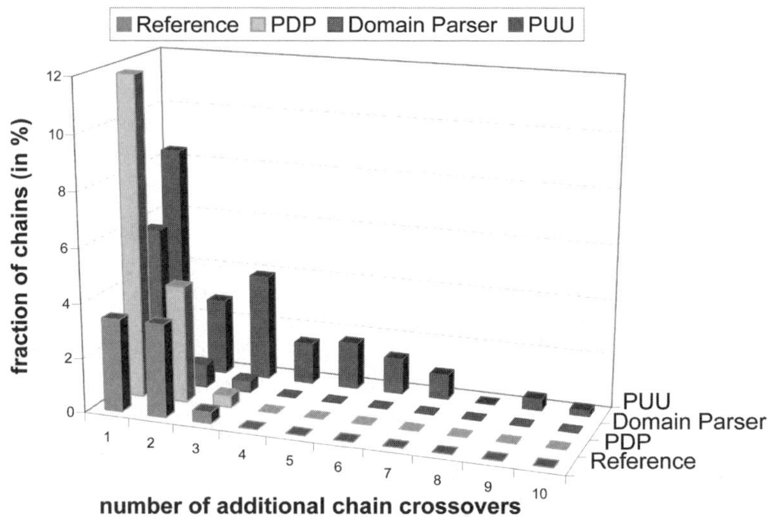

	1	2	3	4	5	6	7	8	9	10
Reference	3.4	3.4	0.4	0	0	0	0	0	0	0
PDP	11.8	4.3	0.43	0	0	0	0	0	0	0
Domain Parser	5.8	0.85	0.43	0	0	0	0	0	0	0
PUU	8.4	2.8	3.9	1.5	1.7	1.3	0.9	0	0.4	0.2

Fig. 4.3 Benchmarking of automatic domain assignment methods. (A) Performance of DomainParser, PDP, and PUU on consensus-based benchmark of 374 structures. (B) Evaluating tendency to partion domains into noncontiguous fragments.

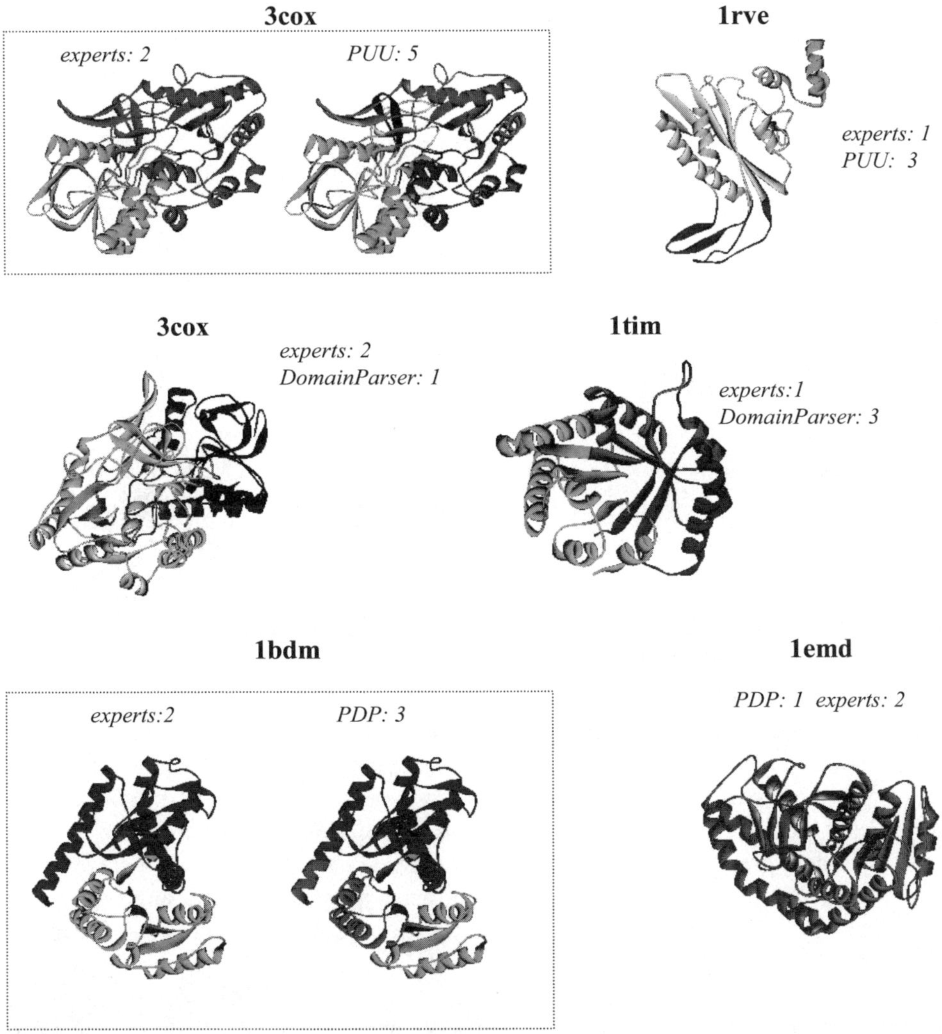

Fig. 4.4 Examples of incorrectly assigned structures by DomainParser, PDP, and PUU.

to fragment domains into noncontiguous stretches of polypeptide chain (Fig. 4.3B). A very important part of this evaluation process is the systematic analysis of what types of structures are difficult for each automatic method and what types of errors are typical for each algorithm. This analysis reveals that the PUU method tends to produce very compact domains which consist of many discontinuous fragments. Two methods, PUU and PDP, tend to overcut protein structures by continuing to partition actual domains; PUU exhibits more frequent and severe cases of overcuts (Fig. 4.4). The DomainParser method, on the other hand, tends to undercut structures—it produces fewer domains than human experts. This appears to be related to splitting structures with β-sheets close to the domain interface—a tricky

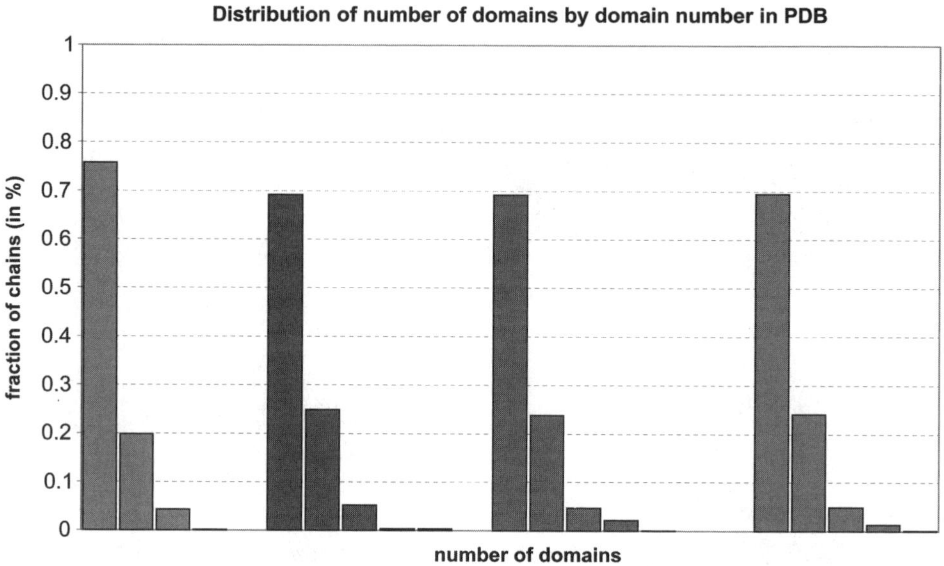

Fig. 4.5 Distribution of single- and multi-domain proteins in PDB. Domains are defined using CATH method. The distribution by domain number is given separately for archaea, bacteria, and eukaryotes as well as to all structures in PDB.

issue (there are several parameters dealing with β-structure in the DomainParser algorithm). A similar problem also exists for the PUU method, which frequently fails to cut β-structures, even though it tends to cut structures too much in other cases (PUU too has a parameter dealing with β-structure partitioning).

Another serious issue that hurts development of new methods is a severe bias of proteins in PDB toward one-domain proteins: nearly 70% of structures are single domains (Fig. 4.5). This overrepresentation of simple structures in PDB is due to the difficulties associated with obtaining the crystal structures of complex multidomain chains. This poverty of complex structures cripples the ability of computational methods to infer principles of decomposition of multidomain proteins. A new more comprehensive benchmarking data set, which covers many more architectures and topologies and includes a larger fraction of multidomain structures, has been recently published (Holland et al., 2006).

4.6 Future Goals

While benchmarking data sets are invaluable in cross-comparing methods as well as aiding in understanding the weaknesses of current methods, the very principles underlying benchmark design requires consensus among human experts. Thus, the most difficult and contentious cases of architectures are not even addressed by the above benchmark. The very existence of the architectures for which multiple plausible domain decompositions exist refutes our simpleminded tendency to fit one approach for partitioning to all structures. As more protein structures are solved, the fraction of such "controversial" proteins is likely to increase. The best way to address this inherent complexity of the protein structures might be to accept the possibility of alternative domain decompositions and implement this feature in new algorithms. One of the latest algorithms has such a capacity already (Berezovsky, 2003), which simply uses a series of thresholds instead of a single threshold during structure decomposition. In general it appears that the main difficulty the algorithms have is proceeding with partitioning too far or not far enough, rather than making partitions in incorrect regions of the structure. This situation can be possibly remedied by performing domain decomposition under multiple thresholds. Allowing multiple thresholds is likely to produce single solutions for simple structures and multiple solutions in the cases of complex architectures. The introduction of multiple thresholds into existing algorithms should be relatively simple. The future success of algorithms for domain decomposition may require a shift in our thinking about what constitutes a good solution for this complex problem; this is likely to involve considering alternative decomposition scenarios as an essential part of the solution.

4.7 Summary

Domain decomposition of 3D structures is an important and not completely solved problem. An astonishingly wide array of approaches had been implemented in an attempt to automate this process. Currently, the best algorithms resolve more than 80% of the structures for which human experts reach consensus for domain partitioning. As more complex structures are solved, we do not expect this success rate to increase dramatically. It would be timely to reevaluate our approach to the problem of domain decomposition and our expectation of reaching a single solution in every case. Rather it would be constructive to accept the fact that there are multiple legitimate decomposition schemes for complex architectures and adapt future algorithms to deal with such a possibility.

4.8 Suggested Further Reading

For early fundamental works on protein domain definitions there are seminal papers by Wetlaufer and Ristow (1973) and Richardson (1985). For a recent comprehensive review on structural domains the paper by Ponting and Russell (2002) is recommended. Evolution of domains is discussed well in Ponting et al. (2000) and Todd et al. (2001). The thorny topic of correspondence between protein domains and exons in genes is extensively discussed; a good start is the book *Protein Evolution* (Patthy, 1999). Finally, discussion of the protein universe through the lens of structural domains can be found in Holm and Sander (1996) and Orengo et al. (1994).

References

Alexandrov, N., and Shindyalov, I. 2003. PDP: Protein domain parser. *Bioinformatics*. 19:429–430.

Berezovsky, I. N. 2003. Discrete structure of van der Waals domains in globular proteins. *Protein Eng*. 16:161–167.

Crippen, G. M. 1978. The tree structural organization of proteins. *J. Mol. Biol.* 126:315–332.

Doolittle, R. F. 1995. The multiplicity of domains in proteins. *Annu. Rev. Biochem.* 64:287–314.

Guo, J. T., Xu, D., Kim, D., and Xu, Y. 2003. Improving the performance of DomainParser for structural domain partition using neural network. *Nucleic Acids Res.* 31:944–952.

Holland, T. A., Veretnik, S., Shindyalov, I. N., and Bourne, P. E. 2006. Partitioning protein structure into domains: Why is it so difficult? *J. Mol. Biol.* 361:562–590.

Holm, L., and Sander, C. 1994. Parser for protein folding units. *Proteins* 19, 256–268.

Holm, L., and Sander, C. 1996. Mapping the protein universe. *Science* 273:595–603.

Islam, S. A., Luo, J., and Sternberg, M. J. 1995. Identification and analysis of domains in proteins. *Protein Eng.* 8:513–525.

Jones, S., Stewart, M., Michie, A., Swindells, M. B., Orengo, C., and Thornton, J. M. 1998. Domain assignment for protein structures using a consensus approach: characterization and analysis. *Protein Sci.* 7:233–242.

Kundu, S., Sorensen, D. C., and Phillips, G. N., Jr. 2004. Automatic domain decomposition of proteins by a Gaussian network model. *Proteins* 57:725–733.

Levitt, M., and Greer, J. 1977. Automatic identification of secondary structure in globular proteins. *J. Mol. Biol.* 114:181–239.

Murzin, A. G., Brenner, S. E., Hubbard, T., and Chothia, C. 1995. SCOP: A structural classification of proteins database for the investigation of sequences and structures. *J. Mol. Biol.* 247:536–540.

Orengo, C. A., Jones, D. T., and Thornton, J. M. 1994. Protein superfamilies and domain superfolds. *Nature* 372:631–634.

Orengo, C. A., Michie, A. D., Jones, S., Jones, D. T., Swindells, M. B., and Thornton, J. M. 1997. CATH—A hierarchic classification of protein domain structures. *Structure* 5:1093–1108.

Patthy, L. 1999. *Protein Evolution*. Oxford, Blackwell Science.

Ponting, C. P., and Russell, R. R. 2002. The natural history of protein domains. *Annu. Rev. Biophys. Biomol. Struct.* 31:45–71.

Ponting, C. P., Schultz, J., Copley, R. R., Andrade, M. A., and Bork, P. 2000. Evolution of domain families. *Adv. Protein. Chem.* 54:185–244.

Richardson, J. S. 1985. Describing patterns of protein tertiary structure. *Methods Enzymol.* 115:341–358.

Rose, G. D. 1979. Hierarchic organization of domains in globular proteins. *J. Mol. Biol.* 134:447–470.

Rossman, M. G., and Liljas, A. 1974. Letter: Recognition of structural domains in globular proteins. *J. Mol. Biol.* 85:177–181.

Siddiqui, A. S., and Barton, G. J. 1995. Continuous and discontinuous domains: An algorithm for the automatic generation of reliable protein domain definitions. *Protein Sci.* 4:872–884.

Sowdhamini, R., and Blundell, T. L. 1995. An automatic method involving cluster analysis of secondary structures for the identification of domains in proteins. *Protein Sci.* 4:506–520.

Swindells, M. B. 1995a. A procedure for detecting structural domains in proteins. *Protein Sci.* 4:103–112.

Swindells, M. B. 1995b. A procedure for the automatic determination of hydrophobic cores in protein structures. *Protein Sci.* 4:93–102.

Taylor, W. R. 1999. Protein structural domain identification. *Protein Eng.* 12:203–216.

Todd, A. E., Orengo, C. A., and Thornton, J. M. 2001. Evolution of function in protein superfamilies, from a structural perspective. *J. Mol. Biol.* 307:1113–1143.

Veretnik, S., Bourne, P. E., Alexandrov, N. N., and Shindyalov, I. N. 2004. Toward consistent assignment of structural domains in proteins. *J. Mol. Biol.* 339:647–678.

Wernisch, L., Hunting, M., and Wodak, S. J. 1999. Identification of structural domains in proteins by a graph heuristic. *Proteins* 35:338–352.

Wernisch, L., and Wodak, S. J. 2003. Identifying structural domains in proteins. *Methods Biochem. Anal.* 44:365–385.

Wetlaufer, D. B. 1973. Nucleation, rapid folding, and globular intrachain regions in proteins. *Proc. Natl. Acad. Sci. USA* 70:697–701.

Wetlaufer, D. B., and Ristow, S. 1973. Acquisition of three-dimensional structure of proteins. *Annu. Rev. Biochem.* 42:135–158.

Wodak, S. J., and Janin, J. 1981. Location of structural domains in protein. *Biochemistry* 20:6544–6552.

Xu, Y., Xu, D., and Gabow, H. N. 2000. Protein domain decomposition using a graph-theoretic approach. *Bioinformatics* 16:1091–1104.

Xuan, Z. Y., Ling, L. J., and Chen, R. S. 2000. A new method for protein domain recognition. *Eur. Biophys. J.* 29:7–16.

5 Protein Structure Comparison and Classification

Orhan Çamoğlu and Ambuj K. Singh

5.1 Introduction

The success of genome projects has generated an enormous amount of sequence data. In order to realize the full value of the data, we need to understand its functional role and its evolutionary origin. Sequence comparison methods are incredibly valuable for this task. However, for sequences falling in the twilight zone (usually between 20 and 35% sequence similarity), we need to resort to structural alignment and comparison for a meaningful analysis. Such a structural approach can be used for classification of proteins, isolation of structural motifs, and discovery of drug targets.

The success of structural analysis rests on both the quality of the alignment of a query with a target protein and the speed with which relevant targets can be isolated in a large database. The multiple alignment of a set of protein structures denotes the extraction of their maximal common substructure. For such alignments to be computationally meaningful, the constraints of such approximate matching and a score function (or a distance measure) need to be defined.

Comparison of protein structures is the basic building block of structural analysis. Structure comparison is carried out by first aligning two structures in 3D space and then assessing the similarity between them. For this alignment, two structures are superimposed onto each other. Figure 5.1 shows two protein structures from the Protein Data Bank (PDB) (Berman et al., 2000). The similarity between these two structures becomes evident after superimposition, as in Fig. 5.2.

Protein structure alignment of two protein structures P and Q is defined as a one-to-one mapping between the residues of P and Q. Using this mapping, two structures can be superimposed and a similarity measure between them can be computed. Alignment of protein structures is useful in answering a number of questions:

- A protein with unknown function but known structure can be aligned to proteins with known functions and known structures. These alignments can provide clues regarding the function of the query protein.
- Structural alignments can be used to classify proteins. Since the structure of a protein is strongly linked to its function, the resulting classifications can capture the functional and evolutionary similarity of proteins.
- Multiple structure alignment of a set of proteins with a similar function can identify a structural motif pertaining to the function.

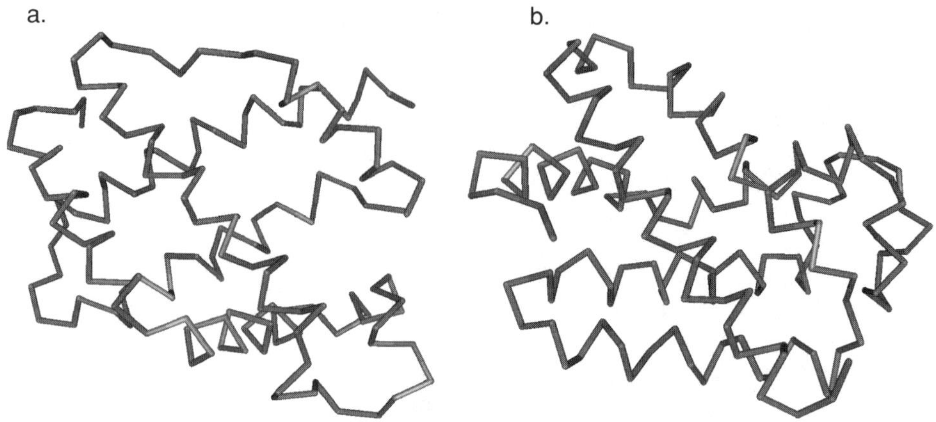

Fig. 5.1 Backbone trace of (a) 1lh1 (leghemoglobin from yellow lupin) (b) 1urv-A (cytoglobin from human).

- Remote homologies that are not apparent from sequence comparison can be discovered by structural alignment.

The exact alignment of protein structures is computationally expensive. This is due to the exponential number of correspondences between the point sets of two protein structures. [Once a correspondence has been found, the problem of aligning them in order to minimize the RMSD is quite fast—linear in the size of the proteins (Arun et al., 1987; Kabsch, 1978).] An early complexity result in this area is due to Lathrop (1994) who showed that the problem of protein threading (the alignment of a protein sequence to a protein structure) is NP-complete under variable-length gaps and nonlocal scoring functions.

Fig. 5.2 1urv-A is superimposed on 1lh1.

Goldman et al. (1999) have formulated the alignment problem using contact maps. A contact map of a protein structure is a graph in which nodes comprise the residues and edges are placed between two residues whose physical distances are below a given threshold. The alignment of two protein structures then amounts to finding the largest common subgraph of their contact maps. The authors show that even a simplified version of this problem (contact maps restricted to having a maximum degree of one) is hard to solve [NP-complete (Garey and Johnson, 1979)] and hard to approximate [MAXSNP-complete (Garey and Johnson, 1979)]. Approximate solutions within a factor of ϵ have also been investigated by Kolodny and Linial (2004). They show that if a scoring function similar to STRUCTAL (Levitt and Gerstein, 1998) is used, then an approximate solution can be found in time $O(n^{10}/\epsilon^6)$. Though exact approximation of the structural alignment problem is theoretically interesting, the currently available tools rely on heuristics. We focus on them in the remainder of the chapter.

The rest of the chapter is organized as follows. Section 5.2 discusses pairwise structure comparison and existing algorithms. Section 5.3 discusses multiple structure alignment and structural motifs. Section 5.4 presents techniques for querying protein structure databases. Section 5.5 presents classification of protein structures and automated classification techniques. Concluding remarks appear in Section 5.6. References and resources follow in Section 5.7. Some suggestions for further reading are included in Section 5.8.

5.2 Pairwise Alignment of Protein Structures

The goal of pairwise structure alignment is to find the best mapping between the residues of two given protein structures. The general methodology for achieving this can be summarized as follows.

1. **Structure representation:** Proteins have many characteristics including types of residues, positions of different atoms, types and properties of the various bonds. For the purposes of structural alignment, it is not feasible and usually not necessary to include all of these aspects in the representation of proteins. Many algorithms consider only 3D positions of a few atoms, and ignore their types. A further simplification can be obtained by considering only the positions of the backbone carbon atoms (C_α and C_β). These are used to represent a residue. This reduces the number of atoms considered from thousands to hundreds for a typical protein.
2. **Feature extraction:** Most features for structure alignment are based on secondary structure elements (SSEs) and interresidue (inter-C_α or C_β) distance matrices.
3. **Structure comparison and alignment optimization:** First, one finds similar features between two proteins (i.e., similar distance matrices for distance matrix-based methods or similar SSE layouts for geometric hashing-based methods). Sets of similar features define local alignments. These local alignments are then

merged iteratively into a global alignment. This global alignment is optimized further.

4. **Significance assessment:** The significance of the obtained alignment is computed, usually by estimating the likelihood of obtaining a similar alignment at random. This estimate is based on protein similarity, alignment length, size of the proteins, and gaps in the alignment.

5.2.1 Measuring the Quality of an Alignment

Given two proteins A and B, we are interested in identifying the largest common substructure of the proteins. A correspondence of size k identifies a subset of size k from each of the proteins and establishes an equivalence between the subsets. For example, if $A = a_1, a_2, \ldots, a_m$ and $B = b_1, b_2, \ldots, b_n$, then a correspondence of size five may be defined as $(a_1, a_2, a_3, a_5, a_6)$–$(b_3, b_4, b_5, b_6, b_8)$. A correspondence is order preserving if it maintains the backbone order of C_α atoms. For example, the above correspondence is order preserving whereas the following correspondence is not: $(a_1, a_2, a_3, a_5, a_6)$–$(b_3, b_4, b_7, b_8, b_5)$. In the order-preserving formulation, the problem of structure alignment simplifies to a 3D curve matching. The computational task becomes more difficult for non-order-preserving alignments. However, nonorder-preserving alignments are needed in order to discover non-sequential motifs such as molecular surface motifs and binding sites. They also allow a search of the database with partial structural information.

The similarity analysis of two protein structures is facilitated by rotations and translations so that the common substructures of the proteins are juxtaposed. Such rigid body transformations have been studied in detail in computer vision (Arun et al., 1987). Given a protein A, we denote its rigid body transformation using a mapping f as $f(A)$.

The root-mean-square distance (RMSD) between two proteins A and B under a correspondence R of size k and a transformation f is defined as

$$\mathrm{RMSD}(A,B,R,f) = \sqrt{\frac{\sum_{i=1}^{k} dist^2(a_i, f(R(a_i)))}{k}}$$

Given two proteins and a correspondence between them, it is computationally easy to find the transformation that minimizes the RMSD. RMSD between two proteins A and B under a correspondence R of size k is defined as $\mathrm{RMSD}(A,B,R) = \mathrm{argmin}_f\ \mathrm{RMSD}(A,B,R,f)$.

Another measure of distance between protein structures is defined using the interresidue distances directly (and without using any rigid body transformations). This distance, distance matrix error (DME), is defined as $\mathrm{DME}(A, B, R) = \frac{1}{N}\sqrt{\sum_{i=1}^{k}\sum_{j=1}^{k}(dist^2(a_i, a_j) - dist^2(R(a_i), R(a_j)))}$. The advantage of this representation is that two structures do not need to be superimposed and the measure can be computed directly using the distance matrices.

The interplay between the RMSD and the size of the identified common substructure is interesting. Since RMSD is an average measure, it decreases monotonically with the size of the common substructure. However, small alignments may not be meaningful. Irving et al. (2001) analyzed the variation of RMSD with the number of aligned residues. They found a linear dependence of RMSD on the number of residues for a small number of aligned residues followed by an exponential region.

Some researchers have defined distance measures so that the effect of distant pairs is reduced. For example, DALI (Holm and Sander, 1993) uses an elastic score defined as follows:

$$\phi^E(i,j) = \begin{cases} \left(\theta^E - \frac{|d_{ij}^A - d_{ij}^B|}{d_{ij}^*}\right) w(d_{ij}^*) & i \neq j \\ \theta^E & i = j \end{cases} \tag{5.1}$$

where d_{ij}^A and d_{ij}^B are the equivalenced elements in the distance matrices of proteins A and B, d_{ij}^* is the average of d_{ij}^A and d_{ij}^B, $\theta^E = 0.20$, and envelope function w is defined as $w(r) = \exp(-r^2/400)$. The envelope function gives lower weights to residues that are farther apart, thus reducing their relative contribution.

In a similar vein, Levitt and Gerstein (1998) defined a scoring scheme that depends more on the best-fitting residues. In this scheme, a scoring matrix is created based on an initial alignment of proteins. The score for each entry is defined as $S_{i,j} = \frac{M}{1+\left(\frac{d_{ij}}{d_0}\right)^2}$ where d_{ij} is the distance between residues i and j, $M = 20$, and $d_0 = 5$ Å.

Jia et al. (2004) proposed a new scoring scheme for CE. CE score is defined by $\frac{rmsd}{aligned\_length^{\alpha}}\left(1 + \frac{num\_gap}{aligned\_length^{\beta}}\right)$ where α and β are greater than 0. The significance of an observed score is computed by using the probability density function for the random scores.

Overall, the similarity of two protein structures is difficult to quantify using a number. Godzik (1996) investigated the alignment of a number of protein pairs and found that although different alignment methods produce similar results at the SSE level, there are significant differences at the residue level. The RMSD values and the length of aligned substructures varied widely. For example, the alignment of azurin (1azcA) and plastocyanin (1plc) produced RMSD in the range of 1.5–6.7 Å and the length of aligned substructures varied in the range of 13–108 residues. There is usually no unique answer to the structural alignment problem and additional input is required to characterize a good solution (e.g., in a given pair of proteins, one may want to focus more on the well-conserved core and less on the loop regions).

Once a set of similar candidates has been obtained by using a database search, the statistical distribution of similarity scores needs to be computed in order to assign a level of significance (p-value). It is difficult to distinguish between structure similarities that arise from evolutionary relationships and those resulting from physical constraints on protein folding. However, the question of significance can be answered in a purely mathematical manner by considering the space of possible configurations, the size of the alignment, and the distribution of the scores (Gibrat et al., 1996; Ye and Godzik, 2004; Holm and Sander, 1993; Shindyalov and Bourne, 1998).

5.2.2 Computational Approaches

Many methods have been proposed for pairwise protein structure comparison. [Excellent in-depth surveys can be found in Eidhammer et al. (2000) and Brown et al. (1996).] These methods propose different approximations and approaches to structure alignment. They use different types of representations (atoms, residues, secondary structure elements and groups of these) and algorithms (dynamic programming, geometric hashing, randomized algorithms). These methods are grouped based on their data representations and algorithms.

5.2.2.1 Dynamic Programming-Based Approaches

Dynamic programming (DP) algorithms have been used to find the optimal sequence alignment between pairs of sequences (Needleman and Wunsch, 1970). In structural alignment, however, the traditional DP algorithm cannot be used directly. The difference is that in structure alignment, the distance between two residues is dependent on the alignment of other residues. Sali and Blundell (1990) managed to overcome this difficulty by using rotation- and transformation-independent features on the scoring function. Employing simulated annealing, they first find possible equivalences between two proteins depending on a set of residue properties (sequence identity, hydrophobicity, charge, volume, torsional angles). Similarity of residue properties is then used to create a two-dimensional similarity matrix. Finally, the proteins are aligned by applying dynamic programming.

Orengo and Taylor (1996) proposed SSAP that uses two levels of dynamic programming (double dynamic programming). In the upper-level dynamic programming, score matrix entry S_{ij} represents the score of aligning the ith residue of the first protein to the jth residue of the second protein. The best path in this matrix finds the optimal alignment of the two proteins. The value S_{ij} is computed at the lower level of dynamic programming: the ith and jth residues are assumed to be aligned, and a score matrix is computed by using the difference of C_β–C_β vectors of residues. The resulting score S_{ij} represents how well the rest of the residues are aligned when the ith and jth residues are aligned.

An example of two levels of dynamic programming is given in Fig. 5.3. In the first lower level matrix, residue C of protein B is aligned to residue F of protein A. To obtain the matrix, C_β–C_β vectors from residue C to each residue in protein B are compared to C_β–C_β vectors from residue F to each residue in protein A. The best scoring path is identified in the score matrix. Scores in this path are copied to the main score matrix in the upper level. The second score matrix in the lower level is

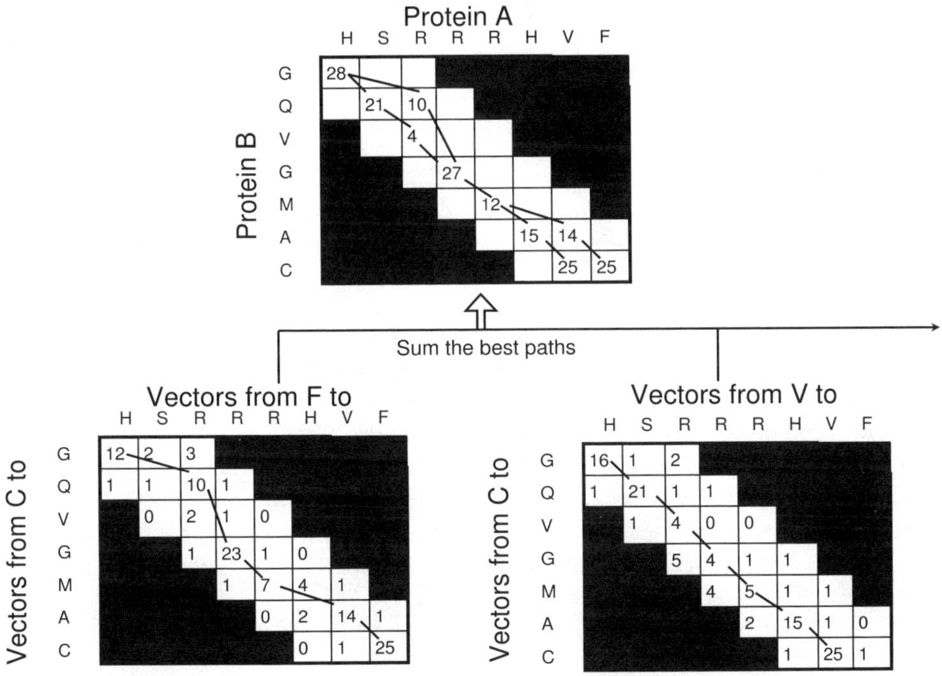

Fig. 5.3 SSAP algorithm uses two levels of dynamic programming. Lower level scoring matrices are calculated by aligning one residue from each structure. Then, C_β vectors relative to the aligned residues are compared to fill out lower level score matrix. The best path is computed and aggregated to the upper level (Orengo and Taylor, 1996).

computed by aligning residue *C* of protein B to residue *V* of protein A while using C_β–C_β vectors from residue *C* in protein B and from residue *V* in protein A. The best path is identified and copied to the main score matrix. After all lower matrices are processed, the best scoring path in the upper level scoring matrix is found by dynamic programming. This path defines the alignment between proteins.

Taylor (1999) proposed an extension to the SSAP algorithm by incorporating a stochastic component: suboptimal alignments are allowed and double dynamic programming is used to evaluate the effect of these suboptimal alignments iteratively.

Gerstein and Levitt (1996) proposed another approach to iterative dynamic programming. They start with a random alignment. Given an alignment, they optimize the RMSD by superimposing the proteins, and compute the interresidue distances. Dynamic programming is applied to the interresidue distance matrix to find the best residue correspondences. The current alignment is then modified using these correspondences. This procedure is repeated until convergence. The entire computation is carried out with different initial alignments and the best resulting alignment is reported.

One major drawback of dynamic programming-based approaches is that they preserve the sequence order of residues in the structural alignment. These sequential

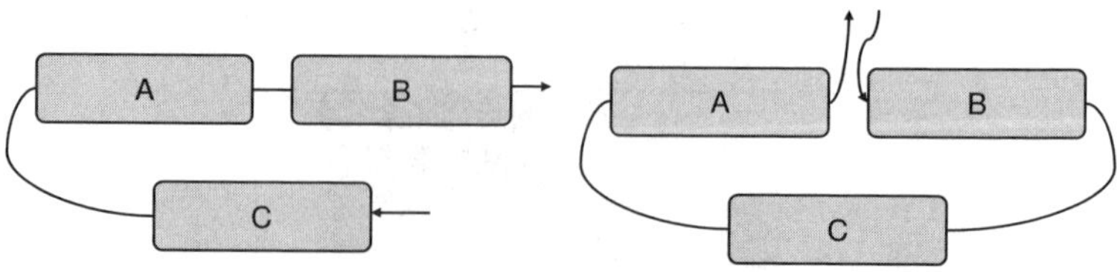

Fig. 5.4 Same layout of SSE structures can have different sequence orders due to circular mutations. The layout on the left has a CAB order on the SSEs while the layout on the right has a BCA order (Binkowski et al., 2004).

alignments cannot capture similarities between proteins where circular permutations have occurred. Two proteins with a similar structure layout of SSEs can have different sequence layouts of SSEs. In Fig. 5.4, SSEs A, B, and C have similar structural layouts in both of the proteins. However, their sequence orders are significantly different. The sequence order for the first protein is CAB whereas it is BCA for the second protein. To discover such structural similarities where sequence order is not preserved, a number of algorithms have been proposed. We elaborate on some of them.

5.2.2.2 Distance Matrices and Contact Maps

Algorithms in this category represent protein structures as two-dimensional distance matrices. For each residue in a protein, its distances to the remaining residues are computed to construct these matrices. DALI (Holm and Sander, 1993), one of the most popular algorithms, uses the distance matrices to find highly similar local structures. The intuition is that if two protein structures are similar, then their distance matrices should be similar too. In the first step of the algorithm, similar submatrices of size six in two proteins are found by comparing their distance matrices. These comparisons result in alignments of size six between two proteins. Then, compatible alignments are merged to obtain larger alignments called *seeds*. These seeds are randomly merged and extended by making optimal and suboptimal choices via a Monte Carlo algorithm. The best alignments are further improved by randomly removing parts of the alignment and realigning the proteins.

An example of merging of compatible alignments is shown in Fig. 5.5. Upon comparison of distance matrices of proteins A and B, matrix component (a, b) is aligned to (a', b'). Similarly, (b, c) is aligned to (b', c'). Since these two alignments are overlapping (i.e., they both align b to b'), they are checked for compatibility. If the nine matrix components for these two alignments are found similar to each other, alignments are merged to obtain a seed of $(a, b, c) - (a', b', c')$.

Two different measures of scoring are used in DALI: a rigid body similarity measure and an elastic scoring measure. This approach does not require the alignments to be order preserving. In recent versions of DALI, the computational performance has been improved by imposing SSE constraints on the alignments.

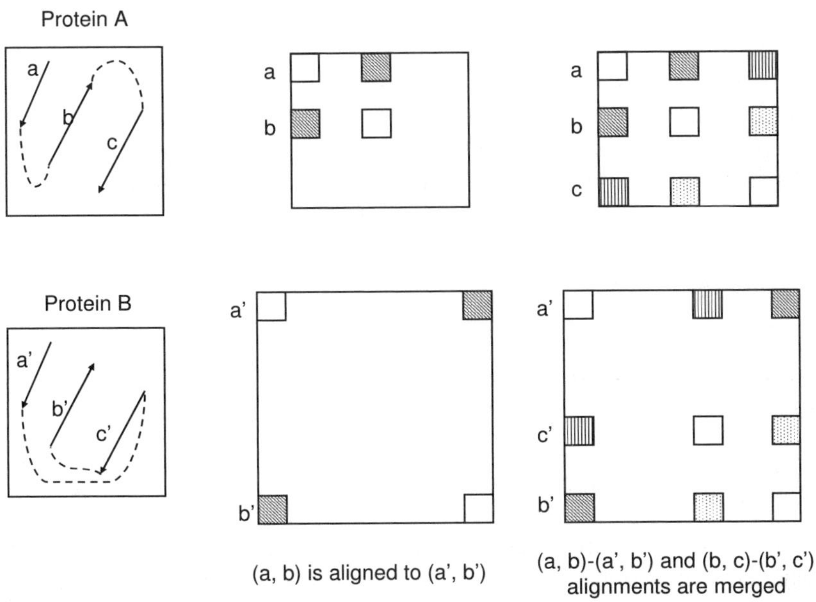

Fig. 5.5 DALI algorithm first aligns small parts of each protein. First (a, b) is aligned to (a', b') and (b, c) is aligned to (b', c'). These two alignments are merged to obtain a larger alignment $(a, b, c) - (a', b', c')$ (Holm and Sander, 1993).

CE (Shindyalov and Bourne, 1998) uses distance matrices to find small but highly similar fragments in a manner similar to DALI. The difference is that CE uses a combinatorial extension of these fragments. In the first stage, all combinations of 8 residue alignments between two proteins are tested and ones that meet a threshold are selected. Then starting from a fragment, the alignment is extended by adding new fragments. A new fragment is added if its addition will result in a better alignment score than some threshold. To improve the performance, the program does not allow gaps of length larger than 30 residues. After the best alignments of fragments are found, they are further improved by using a dynamic programming approach.

Chew et al. (1999) represent a protein structure by vectors defined by adjacent C_α atoms. These vectors represent the protein backbone. They place these vectors on a unit sphere and compare two proteins by their trace on this sphere.

5.2.2.3 Geometric Hashing

Geometric hashing allows rotation and translation invariant comparison of 3D objects. Nussinov and Wolfson (1991) adapt this idea for the comparison of protein structures. Their approach is to use a set of reference frames for each protein and map its residues into 3D grid cells for each reference frame. If two protein structures are similar, then there exists a pair of reference frames (one for each protein) such that a large number of residue pairs will be mapped to the same grid cell. A hash function can be associated with the grid cells, allowing efficient lookup. A reference frame can be defined using C_α, C_β, and N atoms of each residue (Pennec and Ayache, 1998), or using three or more residues at a time.

Holm and Sander (1995) use geometric hashing on the secondary structure elements. Reference frames are defined for pairs of SSEs. Based on each frame, the locations of the rest of the SSEs are hashed. A pair of frames that maps corresponding SSEs into similar locations is found. Such a matching frame forms an initial alignment that is refined by an iterative process.

Proteins are flexible molecules capable of undergoing structural conformations such as hinge-based motion (Rose and Stroud, 1998). Incorporation of such flexibility implies moving away from the common assumptions of proteins as rigid bodies. Some tools have been developed recently (Shatsky, 2004; Ye and Godzik, 2003) that incorporate flexible alignments. Verbitsky et al. (1999) use geometric hashing to align structures allowing hinge bending.

5.2.2.4 Hierarchical Algorithms

Hierarchical algorithms are based on rapidly identifying mappings between *similar* SSE fragments of two proteins. The similarity of two fragments is defined using length and angle constraints. Fragment pairs that align well form the seed for expensive residue-level alignments.

The VAST program (Madej et al., 1995) carries out a hierarchical alignment using SSEs. It begins with a bipartite graph: vertices on one side consist of pairs of SSEs from the query protein and vertices on the other side consist of pairs of SSEs from the target protein. An edge is inserted between two pairs of SSEs if they can be aligned well. A maximal clique is found in this bipartite graph; this defines the initial SSE alignment. This initial alignment is extended to C_α atoms by Gibbs sampling. A nice feature of the VAST program is its ability to report on the unexpectedness of the match through a *p-value*. This is computed by considering the size of the match, the size of the proteins, and the quality of the alignment.

A number of other algorithms also use hierarchical alignment (Alexandrov and Fischer, 1996; Singh and Brutlag, 1997; Holm and Sander, 1995). LOCK (Singh and Brutlag, 1997) represents SSEs by vectors and matches pairs of SSEs based on similarity of angles and distances between them. The alignment of SSEs is extended using iterative dynamic programming. Proteins are aligned and the nearest neighbor of each residue on the other structure is computed. The residue pairs that identify each other as the nearest neighbor define the new alignment. This process is repeated until convergence.

Novotny et al. (2004) evaluated various pairwise structure comparison tools for identification of similar folds. The CATH (Orengo et al., 1997) structure classification database was used as a reference. None of the tools were able to achieve a 100% success rate, but CE, DALI, and VAST showed best performance. Sierk and Pearson (2004) performed a similar analysis to evaluate the performance of comparison tools in detecting homologues. They showed that DALI was more selective than the other tools.

5.3 Multiple Structure Comparison and Structural Motif Search

Protein motifs, specifically active sites and binding sites, play important roles in biochemical reactions. Motifs are defined as substructures that are common to a set of proteins that possess functional or evolutionary relationships. Querying proteins for specific motifs and discovering motifs in a set of proteins are crucial for the functional classification and understanding of proteins. Since active sites of proteins are determined by structure of the participating amino acids rather than their sequence order, structural motifs can be more useful than sequence motifs. However, structural motif detection is more challenging. First, the size of the search space is much larger than in the sequence domain. Each residue can be one of 20 amino acids, so a motif of length 5 amounts to 20^5 possibilities in the sequence domain. This number is much larger for structural motifs if one considers the number of structural conformations of 5 residues without the sequence constraint. Also, for sequences, the type of each residue is known with almost certainty, but for structures the data has limited resolution. So even if two proteins have the same structural motif, their alignment may not produce the perfect score.

Multiple structure alignment of a set of related proteins results in a consensus structure which has the minimum RMSD sum to the protein structures in the set. [This is similar to finding the Steiner string for a set of sequences (Gusfield, 1997).]

5.3.1 Motif Detection

An important characteristic of motif finding programs is the granularity of the discovered motifs as shown in Fig. 5.6. Some of these algorithms consider only C_α atoms, while others consider all the atoms in the protein structures (Pennec and

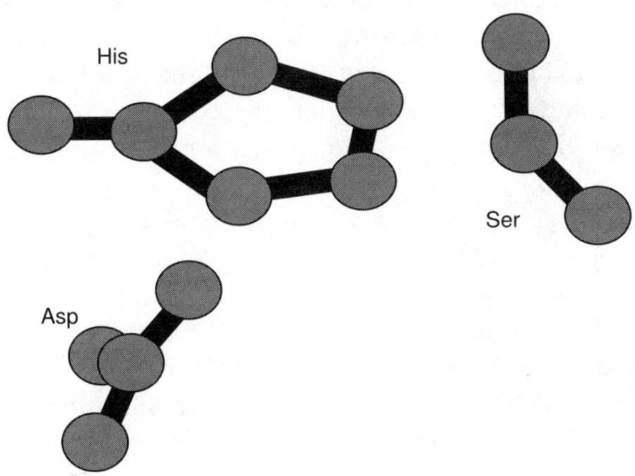

Fig. 5.6 Motif definition for TESS program: relative conformation of the catalytic triads from chymotrypsin 1cho.

Ayache, 1998; Singh and Saha, 2003). Another set of methods defines motifs not by atoms but by secondary structure elements (Kato and Takahashi, 2001). Methods have also been proposed for the detection of small active sites while preserving atom types in motifs (Wallace et al., 1997). There are also methods that use both sequence and structure information simultaneously (Bradley et al., 2002). A majority of these methods adopt RMSD as an indication of quality.

Singh and Saha (2003) proposed a framework for structural motif queries. They incorporate the label information as well as the coordinates of the atoms. Starting from an initial alignment, the nearest neighbors of points in two structures are found and this information is used to modify the alignment. This process is carried out until convergence. The distance function on the structures is modified to account for the label information.

TESS (Wallace et al., 1997) is based on geometric hashing and finds small (usually two or three residues long) active sites. The method considers all of the atoms in the protein, and defines reference frames for each residue by using a combination of C, O, N, S atoms, depending on the residue types. For each reference frame, the relative positions of the atoms that are in the 18-Å neighborhood are stored in the grid. The grid positions that are heavily populated are analyzed. The types of atoms as well as their grid locations have to match for the resulting motif.

TRILOGY (Bradley et al., 2002) finds sequence–structure patterns, which are as small as three residues, across diverse families. In this automated approach, three-residue patterns are extracted and a sequence feature of residue types and a structure feature of relative positions of the residues are computed for each pattern. Structure feature is based on the C_α–C_α distances between residues and, C_α–C_β vectors for each residue in the pattern. Sequence feature is based on the types of residues in the pattern and their distance from each other on the protein sequence. Residues are grouped into seven classes and these class types are used in the definition of sequence features to increase the flexibility of the representation. Patterns that have similar structure and sequence features are identified. These three-residue patterns are then extended by merging patterns that vary by a single residue. This process is carried out until the patterns fail to cover at least three SCOP superfamilies. At the final step, significance scores are assigned to these patterns based on the likelihood of obtaining them at random. Figure 5.7 depicts a pattern whose sequence feature is $Ax^{4-5}[FYW]x^{7-8}N$.

Chen and Bahar (2004) proposed an unsupervised approach to discover frequent patterns in protein families. In this approach, each residue is characterized by its dynamic features, and biochemical and geometric features of the neighboring residues. Dynamic features summarize the rigidness of the local structure around the residue based on the interactions between neighboring residues; biochemical features summarize the amino acid types and chemical properties of neighboring residues; and geometric features summarize the relative positions of the neighboring residues in a similar manner to Pennec and Ayache (1998). After the feature extraction step, the frequency for each feature is computed. Features that occur at a frequency higher than a predefined threshold are considered to be frequent patterns. Then, these frequent patterns are merged to obtain longer augmented frequent patterns.

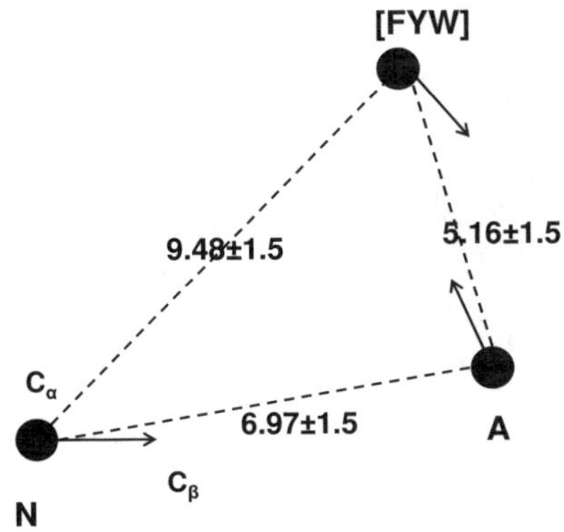

Fig. 5.7 Features of three-residue patterns used in TRILOGY program. Points represent $C\alpha$ atoms and arrows represent $C\alpha - C\beta$ vectors (Bradley et al., 2002).

Jonassen et al. (1999) find local packing motifs in protein structures. For each residue, residues in close proximity are identified and an ordered list is created based on these residue types. Residues that have similar ordered lists are considered as candidates that participate in a local packing motif. The similarity of these candidates is finally verified in the structure domain by superimposing local structures around them.

5.3.2 Multiple Structure Alignment

Although many algorithms have been proposed for the pairwise structure alignment problem, there are only a few algorithms available for multiple structure alignment. A popular technique is to compute pairwise alignments, and to construct a multiple alignment from these (Orengo and Taylor, 1996; Gerstein and Levitt, 1996; Guda et al., 2001). Star alignment-based approaches are usually adopted for this construction. First, all pairwise comparisons are performed and a *pivot* protein, whose RMSD sum to other proteins is minimum, is chosen. Then, the proteins are aligned to the pivot and a multiple structure alignment is constructed. Figure 5.8 depicts the multiple alignment of 15 proteins to protein 1arb. A shortcoming of this approach is that it may miss global patterns since only two proteins are considered at a time. This is reminiscent of the problems arising during multiple alignment of sequences (Gusfield, 1997). A set of recent algorithms capture the global relationship by first extracting common substructures in protein structures and then by constructing multiple structure alignment using the common substructures (Leibowitz et al., 2001; Shatsky et al., 2002; Dror et al., 2003).

Gerstein and Levitt (1996) extend iterative dynamic programming to obtain a multiple structure alignment. Upon performing all pairwise alignments, the pivot that has the smallest average distance to other structures is found. Then, all structures are aligned to this structure. So, if position i in the median structure is aligned to position j in the first structure, and to position k in the second structure, then positions j and k are aligned to each other.

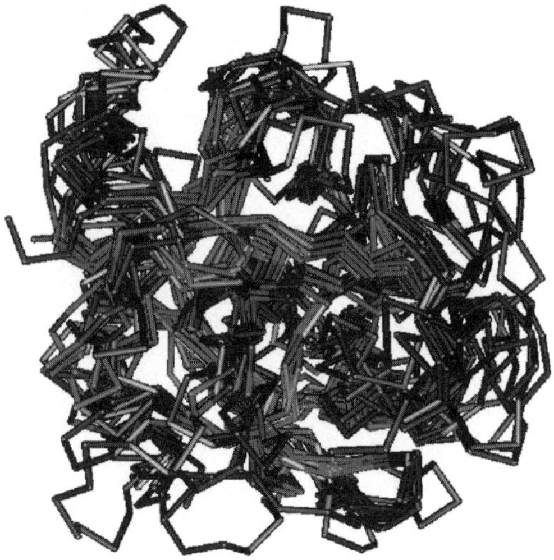

Fig. 5.8 Multiple structure alignment of 10 proteins using 1arb as the pivot by using VAST. Molecular representation is made by Cn3D (http://www.ncbi.nlm.nih.gov/Structure/CN3D/cn3d.shtml).

Guda et al. (2001) extend CE (Shindyalov and Bourne, 1998) to perform multiple structure alignment. They first compute pairwise alignments of proteins, and choose a pivot structure. After an initial alignment is found using the pivot structure, it is refined by a Monte Carlo algorithm.

Leibowitz et al. (2001) use geometric hashing to find common substructures. These common substructures are used as a seed for the multiple alignment. The seeds are merged with each other to obtain larger seeds and the seed that produces the highest scoring alignment is reported. Shatsky et al. (2002) developed a technique that is faster and does not require all proteins to participate in the alignment. Each protein is used as pivot in turn and the one with the best multiple alignment is chosen. Dror et al. (2003) further improved these algorithms by computing the common core of the query proteins using secondary structure elements. The performance is better since SSEs are used instead of residues. In this approach, each SSE is represented as a line segment. A fingerprint for each SSE pair is computed based on the distance and angle between SSEs. These fingerprints are mapped to a 5D grid. Pairs in the same and adjacent buckets in this grid are used as the bases for multiple alignments. For each base, rigid body transformations of proteins are computed. The bases which have similar rigid body transformations of proteins are merged to obtain larger bases. These large bases are used as the common core of the multiple alignment.

5.4 Structure Search in Protein Databases

As the sizes of experimentally determined (Berman et al., 2000) and theoretically estimated (Pieper et al., 1999) protein structures grow, there is a need for scalable searching techniques. Current pairwise alignment techniques produce high-quality pairwise alignments. However, such tools incur excessive running times for database queries, i.e., when a query protein is searched against a large target set. Current tools alleviate this problem by building online databases that contain precomputed results for known protein structures [MMDB (Wang et al., 2002) for VAST, FSSP (Holm and Sander, 1996) for DALI, web database for CE]. In general, scalable techniques are needed for:

1. Searching a protein against a large set of proteins,
2. Clustering a large set of protein structures,
3. Comparing two sets of proteins, and
4. Multiple structure alignment based on pairwise comparisons.

To solve the above problems, some index-based approaches have been proposed (Aung and Tan, 2004; Camoglu et al., 2004). ProtDex2 (Aung and Tan, 2004) uses an inverted file index to index features based on SSEs. It extracts features on SSEs by using their relative distances. The overall similarity of proteins is estimated and the promising candidates are compared using pairwise alignment techniques. Next, the approach developed in Camoglu et al. (2004) is discussed in detail. First, we discuss principles of index-based approaches, then we present an index-based structure comparison technique. The query algorithm is explained in Section 5.4.3, and experimental results are presented in Section 5.4.4.

5.4.1 Index Based Approaches

In numerous applications from databases to image analysis, index-based solutions have been applied to demanding search problems. The main advantage of using index structures is that they reduce the computational complexity of searching. A sequential search compares all of the n elements in the database with the query and reports the best results with a complexity of $O(n)$. This complexity can be reduced to $O(\log(n))$ using an index structure.

The goal of using an index structure is to reduce the number of pairwise comparisons by quickly selecting the promising candidates. In order to build an index structure, some features are extracted from proteins. Then, an index structure is built on the extracted features. Given a query protein, its features are extracted and the database index is queried with these features. Proteins whose features are similar to the query protein's features are aligned using a more extensive algorithm, such as the ones described in Section 5.5.2.

5.4.2 PSI: An Index-Based Structure Comparison Program

PSI (Protein Structure Index) (Camoglu et al., 2004) is an SSE-based approach. It finds high-quality seeds by aligning the SSEs of a database protein with those of a given query protein. For a query protein (or a set of query proteins), this technique can be used to find similar proteins in a target dataset efficiently.

PSI extracts feature vectors corresponding to triplets of neighboring SSEs. These triplets are regarded as the building blocks of the larger protein alignment. Later, these features are embedded in Euclidean space and stored in an index structure.

The index construction consists of the following four steps:

1. **SSE approximation:** The structural features of SSEs are simplified by representing them as line segments in 3D. This is achieved by first splitting up an SSE in two halves at the middle residue. Centers of mass are computed for each half, and a line segment is defined between them. This line segment is extended in both directions to cover the length of the SSE.
2. **Triplet construction:** Triplets of SSEs are used as primitive elements for building an alignment. A set of matching triplets between two protein structures can be used to align the given proteins using rigid body transformations (Arun et al., 1987). Since residues that are spatially close are more valuable for structure similarity (Eidhammer and Jonassen, 2001), neighborhood constraints are imposed while building the SSE triplets. A sphere of radius 50 Å is centered at each SSE and up to four nearest SSEs are chosen. All possible pairs of these SSEs are combined with the center SSE to define the relevant triplets. An example of triplet construction is shown in Fig. 5.9. Empirically, the average number of triplets per SSE turns out to be 3.8 for the proteins in PDB.
3. **Feature vector extraction:** After the triplets of protein structures are determined, feature vectors are constructed. Let $<s_i, s_j, s_k>$ be a triplet. The line segment approximation of each SSE is split into three equisized, non-overlapping intervals. The middle interval of each SSE is used to represent the SSE. The pair of SSEs, $<s_i, s_j>$, contributes three values to the feature vector:
   (a) min_{ij} = minimum distance between the midregions of s_i and s_j.
   (b) max_{ij} = maximum distance between the midregions of s_i and s_j.
   (c) θ_{ij} = the angle between the line segment approximations of s_i and s_j.

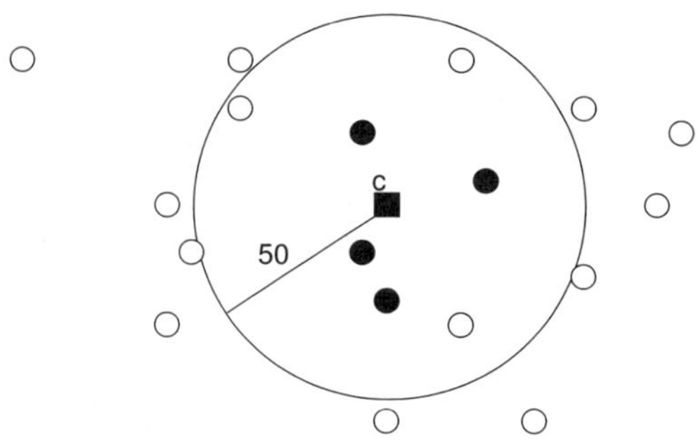

Fig. 5.9 The local neighborhood set of the SSE, c, of a protein. The black square corresponds to the midpoint of c. The circles represent the midpoints of the remaining SSEs of the same protein. 50 Å is the threshold distance for local neighborhood. Four best neighboring SSEs (shown with filled circles) are chosen.

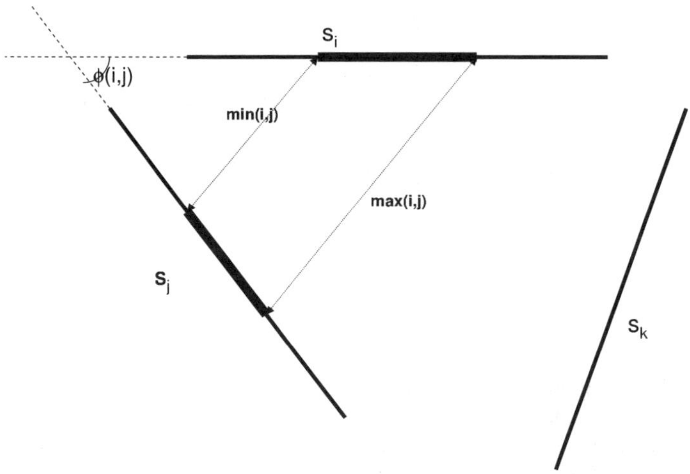

Fig. 5.10 SSE pair $<s_i, s_j>$ contributes three values to the feature vector for triplet $<s_i, s_j, s_k>$: $\min(i, j)$ and $\max(i, j)$, minimum and maximum distance between the mid-portions of s_i and s_j, and $\theta(i, j)$, angle between s_i and s_j.

Figure 5.10 shows the extraction of these values for a pair of SSEs. This set of values is extracted for each pair in the triplet, resulting in a feature vector of size nine for each triplet.

4. **Multidimensional index structure construction:** The feature vector for an SSE triplet has three range values (minimum and maximum distance values) and three angle values. Each feature vector defines an extent in a six-dimensional Euclidean space. The feature vectors are embedded and indexed using an R*-tree (Beckmann et al., 1990). (R*-trees are dynamic index structures that provide efficient range search queries for multidimensional data.)

5.4.3 Query Algorithm

For a given query protein, the search technique (Camoglu et al., 2003) runs in four steps:

Step 1: Similar triplets between database proteins and query protein are computed and stored.

Step 2: A triplet pair graph (TPG) is constructed on similar triplet pairs. An example of such a graph is shown in Fig. 5.11.

Step 3: A bipartite graph is constructed using the TPG. The largest matching in this graph defines the initial alignment seed at SSE level. A *p-value* is computed for each seed. An example bipartite graph is shown in Figure 5.12.

Step 4: The proteins that have large *p-values* are removed without further consideration. The C_α alignment of the remaining proteins are determined using a pairwise structural alignment algorithm.

- (1): $(q_1, q_2, q_3) \Rightarrow (p_1, p_2, p_3)$, s_1
- (2): $(q_1, q_2, q_4) \Rightarrow (p_1, p_2, p_4)$, s_2
- (3): $(q_2, q_3, q_5) \Rightarrow (p_2, p_3, p_5)$, s_3
- (4): $(q_1, q_6, q_7) \Rightarrow (p_1, p_6, p_7)$, s_4
- (5): $(q_6, q_7, q_8) \Rightarrow (p_6, p_7, p_8)$, s_5
- (6): $(q_6, q_8, q_{10}) \Rightarrow (p_6, p_9, p_{10})$, s_6
- (7): $(q_1, q_3, q_5) \Rightarrow (p_{11}, p_3, p_5)$, s_7

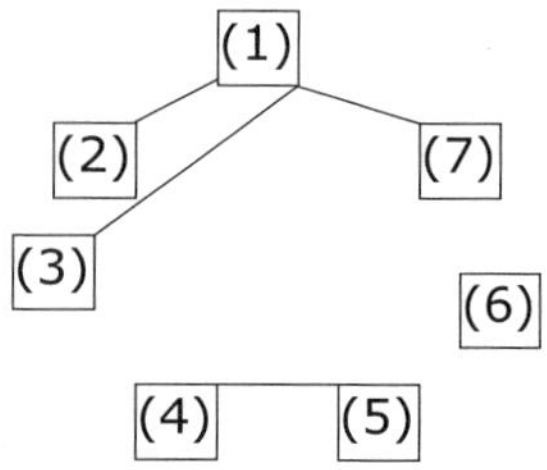

Fig. 5.11 The triplet pairs between the two proteins and their scores are shown on the left. q_i is a secondary structure element of query protein and p_i is a secondary structure of target protein. The corresponding triplet pair graph (TPG) is shown on the right. The TPG has three connected components: one with four triplet pairs, the others with two and one.

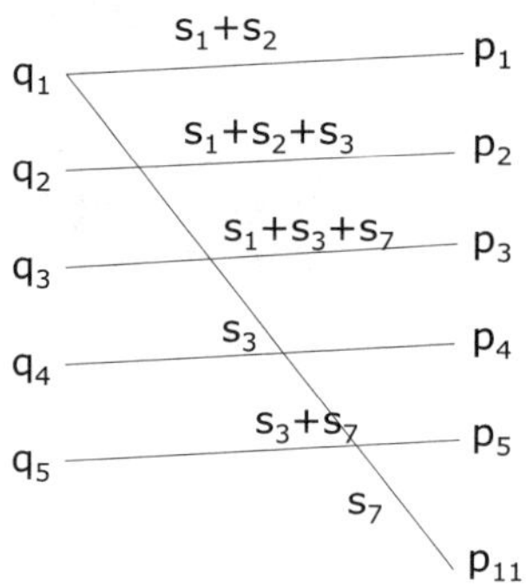

Fig. 5.12 The bipartite graph of the largest weight component in Fig. 5.11 (assuming $s_1 + s_2 + s_3 + s_4 + s_7 > \max(s_6, s_4 + s_5)$). There is a conflict in the alignment of SSE q_1: it has an edge to both p_1 and p_{11}. The edge with the largest weight is chosen to resolve the conflict. Assuming $s_1 + s_2 > s_7$, the initial alignment will be $\{q_1 - p_1,\ q_2 - p_2,\ q_3 - p_3,\ q_4 - p_4,\ q_5 - p_5\}$.

5.4.4 Experimental Evaluation of PSI

The results of PSI are verified with SCOP classification and pairwise comparison tools. The first set of experiments analyzed how well the SCOP classification of a query protein matched with the SCOP classifications of the top-ranking proteins returned by PSI. On the average, 2.5 of the top 3 proteins and 5.4 of the top 10 proteins share the same superfamily as the query protein. This shows that PSI captures strong structure similarities. The same experiment was carried out for fold level classifications and similar results were obtained, showing the ability of PSI to capture remote homologies. Detailed explanation of these experiments can be found in Camoglu et al. (2004).

PSI was also compared with the pairwise comparison tools VAST, CE, and DALI. More than 98% of its results concurred with those of VAST. PSI ran 3 to 3.5

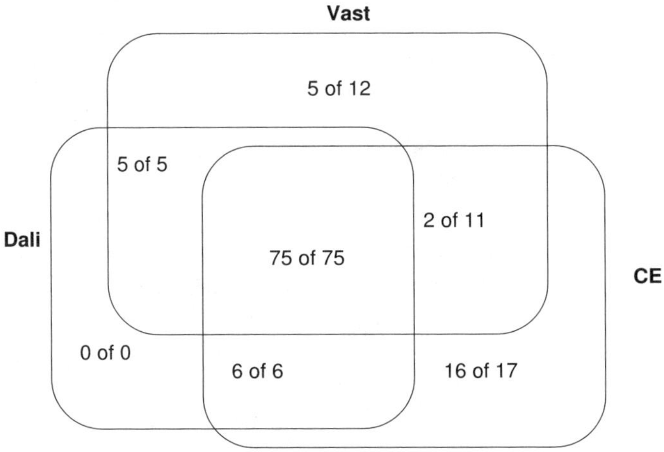

Fig. 5.13 The number of similar proteins of high significance for query protein 1d3g-A found by VAST, DALI, and CE, and the number of such proteins returned by PSI. For example, there are five proteins DALI and Vast found as similar but CE found dissimilar; all five of these are returned by PSI. Note that PSI returned all proteins for which VAST, DALI, and CE agree.

times faster than VAST's pruning step. Similar results were also obtained for CE and DALI.

PSI can be incorporated with other tools to accelerate them by discarding dissimilar proteins. PSI-DALI (for DALI) and PSI-CE (for CE) were implemented in order to demonstrate this. PSI-DALI ran 2 times faster than DALI with 98% recall. PSI-CE ran 2.7 times faster than CE with 83–92% recall.

PSI has a different recall when compared to the results of VAST, CE, and DALI. An in-depth analysis revealed that these three comparison tools produce different results themselves. In Fig. 5.13, the results of these tools for query protein 1d3g-A, dihydroorotate dehydrogenase from human, are displayed. As can be seen, all tools agree over a large portion (75 proteins) of the result; PSI returns these proteins as well. On the whole, PSI is able to find the consensus set of results for each query.

5.5 Protein Classification

Although the study of a single structure or the alignment of a small group of structures can reveal a great deal of information, a global comprehensive view of the protein space is essential to understanding the fold similarities and the evolutionary process. Such a hierarchical classification of proteins based on the analysis of structure has been pursued by a number of researchers (Murzin et al., 1995; Orengo et al., 1997; Holm and Sander, 1996). The resulting organization of protein structures brings a semblance of order to a dynamic field and also provides a valuable resource to other biologists for benchmarking studies. In one such study, Chothia et al. (2003) study the evolution of protein domains across pathways and species. They also study the distribution of domain combinations. They infer a number of relationships such as power law behavior and preferential order of domain combinations based on the study. Gerstein (1997) analyzed representative proteomes from the three kingdoms of life for the SSE content, small motifs of SSEs (alpha-alpha-alpha or beta-alpha-beta), and folds. Though the genomes had similar SSE content, there were marked differences in the observed motifs.

In this section, we introduce various structure classification databases and an automated classification scheme for protein structures. Then, we describe the use of pairwise comparison tools as component classifiers. In Section 5.5.3, we discuss the use of decision trees in automated classification. Some experimental results are presented in Section 5.5.4.

5.5.1 Structure Classification Databases

Protein structure classification databases employ different heuristics and similarity metrics, and can be fully automated (Holm and Sander, 1996), semiautomated (Orengo et al., 1997), or manual (Murzin et al., 1995).

SCOP (Structural Classification Of Proteins) (Murzin et al., 1995) is one of the most popular classification databases. It hierarchically classifies proteins into four categories:

- **Class:** This is the highest level of the hierarchy. The main similarity criteria at this level are the type and the general layout of SSEs in the proteins. There are four main classes: (1) all α, all the SSEs are helices, (2) all β, all the SSEs are strands, (3) α/β, helices and strands are in close proximity, (4) $\alpha + \beta$, helices and strands are apart. In addition to these main classes, there are some other less populated classes (membrane, multiple domain, small protein, etc.).
- **Fold:** This is the second level of the SCOP hierarchy. Members of the same fold have similar SSE arrangements (however, this similarity is not clearly defined). Disagreements between automated methods and SCOP appear mostly on this level. In general, fold-level relationships are observed between proteins that have possible remote evolutionary relationships.
- **Superfamily:** Proteins in the same superfamily possess functional and evolutionary relationships. They have the same active sites, or participate in similar reactions and pathways. They have significant structural similarity and some sequence similarity. This level is one of the most interesting classification levels, because functional relations between proteins may be inferred using similarities at this level.
- **Family:** This is the lowest level of classification. Members of a family possess strong evolutionary and functional relationships. Identification of family members is a fairly simple procedure since the main measure at this level is the sequence similarity. Proteins that have more than 30% sequence identity are classified into the same family. Proteins that have lower sequence similarity, but very strong structural similarity and functional relationships are also classified into the same family.

As can be seen, there are gray areas in the SCOP database and scientists managing SCOP use their discretion in deciding these classifications.

CATH (Orengo et al., 1997) is created using the SSAP (Orengo and Taylor, 1996) program, and also tuned manually. Like SCOP, it is hierarchical and has four levels: class, architecture, topology, and homologous topology. Class is similar to SCOP's class: proteins are categorized based on the types of SSEs they have. The architecture level classification of a protein is assigned manually based on the orientation of SSEs in 3D without considering their connectivity. At the topology level, SSE orientation and the connection between them are taken into account. At the lowest level, proteins are grouped into homologous topologies if there is sufficient evidence that they have an evolutionary relationship.

FSSP (Fold classification based on Structure-Structure alignment of Proteins) (Holm and Sander, 1996) is constructed using the DALI program (Holm and Sander, 1993). A representative set of all available protein structures is selected and all proteins in this representative set are aligned with each other. The resulting Z-scores are used to build a fold tree using an average linkage clustering algorithm. FSSP is fully automated.

Getz et al. (2002) assign SCOP and CATH classification to query proteins using FSSP; this assignment is based on the scores of the proteins in FSSP's answer set for the query protein. They also found strong correlations between these classifications. The CATH topology of a protein can be identified from its SCOP fold with 93% accuracy, and the SCOP fold of a protein can be identified from its CATH topology with 82% accuracy.

5.5.2 Automated Classification

With an exponential growth in the number of newly discovered protein structures, manual databases have become harder to manage. Automated classification schemes that can produce classifications with a similar quality to the manual classifications are needed. Automated techniques are needed to find proteins that have similar structural features in a database that contains thousands of structures, especially with various distance metrics. Additionally, for recently discovered structures, the identification of the appropriate categories for their classification is crucial. It is not reasonable to expect every researcher to examine thousands of proteins to decide the classification of the new protein. Automated techniques are needed in this regard.

A number of schemes toward automated classification (Getz et al., 2002; Lindahl and Eloffson, 2000), fold recognition (Lundstrom et al., 2001; Portugaly and Linial, 2000), and structure prediction servers (Fischer, 2003; Kim et al., 2004) from protein sequences have been proposed. But, these schemes consider only the sequence information and ignore other sources of information that are available such as structure. An approach developed in Can et al. (2004) uses sequence and structure information simultaneously for the automated classification of proteins. This is discussed in detail next.

In the protein classification problem, there is an existing classification scheme such as SCOP, and there is a set of new proteins that has to be classified based on the implicit rules and conventions of the existing classification. A classification algorithm simplifies the update of databases such as SCOP that have to accommodate many new proteins periodically. The main questions that need to be answered during classification are:

- Does the query protein belong to an existing category (family/superfamily/fold), or does it need a new category to be defined?
- If the query protein belongs to an existing category, what is its classification (label)?

The hierarchical nature of classification databases simplifies the classification problem. For example, if a protein belongs to a family, then its superfamily and fold are known. In the case of SCOP, the easiest classification level is family and the hardest one is fold. A hierarchical classifier can take advantage of this. The first attempt is to assign the protein to a family. If this is successful, then the other levels are already known. In case the protein does not belong to an existing family, a superfamily level classification is attempted. If the protein does not belong to an existing superfamily, a fold level classification is attempted.

5.5.3 Building a Classifier from a Comparison Tool

Given a protein comparison tool (sequence or structure), a classifier can be designed by comparing a given query with all proteins of known classification. Similarity scores obtained thus provide a measure of proximity of the query protein to the categories defined in the classification. After sorting these scores, the first classification question (does the query belong to an existing category?) can be answered as follows. If none of the categories show a similarity greater than some predefined cutoff, the query protein is classified as not belonging to an existing category. This cutoff can be determined by investigating the score distribution of each tool.

A 1-Nearest-Neighbor (1NN) classification can be used for the second classification problem (what is the category of the query protein?). The query protein is assigned to the category of the protein that is found to be most similar, i.e., gets the highest score, using the comparison tool.

In Can et al. (2004), five component classifiers were defined using two sequence and three structure comparison tools. The first sequence tool uses the Hidden Markov Model (HMM) library from the SUPERFAMILY database (Gough, 2002). This library is manually curated to classify proteins at the SCOP *superfamily* level. The models in the SUPERFAMILY database can be searched using HMM-based search tools such as HMMER (Eddy, 1998) and SAM (Hughey and Krogh, 1995). These tools assign a similarity score to a protein sequence according to its match with a model. This classifier is referred to as *HMMER*.

The second sequence comparison tool is PSI-Blast (Altschul and Koonin, 1998). PSI-Blast is an improved version of Blast that works in iterations. In the

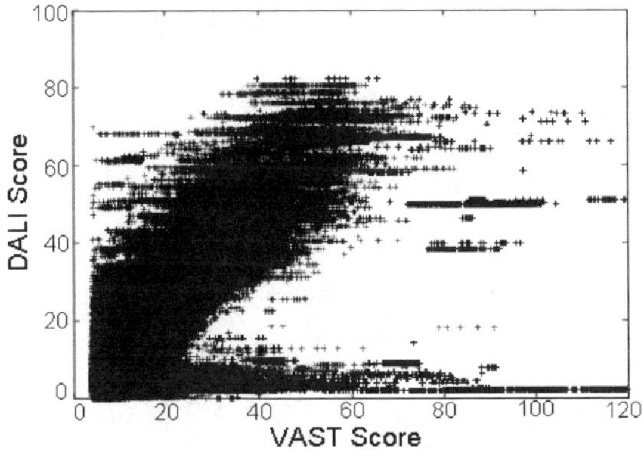

Fig. 5.14 Comparison of VAST and DALI scores for a set of proteins.

first iteration, Blast is run and a new scoring scheme is created based on the set of close neighbors. This process of searching and redefinition of score matrix can be repeated.

The structure comparison tools used were CE (Shindyalov and Bourne, 1998), VAST (Madej et al., 1995), and DALI (Holm and Sander, 1993). Each of these tools performs comparisons with a different technique. As a result, they provide a different view of structural relationships between proteins. For example, even though both VAST and DALI compare structures, they assign different scores to the same pair of proteins, as can be seen in Fig. 5.14. [In a similar study, Shindyalov and Bourne (2000) compared CE and DALI scores and showed that there were many proteins that were found similar by CE and dissimilar by DALI, and vice versa.] By exploiting these differences, one can achieve better performance with a combination of the tools.

Each sequence- and structure-comparison method described above assigns a score for a pair of proteins that indicates the statistical significance of the similarity between them. In particular, the z-scores reported by CE and DALI, p-values reported by VAST, and e-values ($-\log(e\text{-value})$) reported by HMMER and PSI-Blast are used as the similarity scores.

5.5.3.1 Performance of Component Classifiers

The individual performance (accuracy) of the tools when they are used as component classifiers was tested with a number of experiments. It was assumed that the classifications of all proteins in SCOP v1.59 (DS159) are known, and the goal of the component classifiers was to classify the new proteins introduced in SCOP v1.61 (QS161) into families, superfamilies, and folds. A hierarchical classification scheme is used (Camoglu et al., 2005). At the family level, all new proteins were queried. At the superfamily level, only proteins that do not have family-level similarities are queried. At the fold level, only proteins that do not have family/superfamily-level similarities are queried.

At the family level, the sequence tools outperform the structure tools by achieving 94.5 and 92.6% accuracy. The highest success rate of the structure tools is only 89% by VAST. For 83% of the query proteins, all five tools make correct decisions (as to whether the query protein belongs to an existing category), for 4.1% of the queries four tools, for 1.9% of the queries three tools, for 6.9% of the queries two tools, and for 2.7% of the queries only one tool makes the correct decision. An interesting point here is that for 98.2% of the proteins, at least one tool is successful. So, it is theoretically possible to classify up to 98.2% of these proteins correctly by combining the results of the individual tools.

At the superfamily level, the performance of the structure tools improves relative to the sequence tools, as expected. However, the overall performance of the tools drops significantly compared to the family level. This is expected since classification at the superfamily level is harder. HMMER has one of the best performances with a 79.1% accuracy; this is no surprise considering that HMMER is manually tuned for superfamily classifications. Among the structure tools, VAST has the best performance with a 78.6% success rate. PSI-Blast performs poorly with a success rate of only 66.1%. Only 44.7% of the queries can be classified correctly by all five tools. For 4% of the queries, none of them is successful. This again raises the possibility of achieving better accuracy through a combination of the tools. These results are depicted in Fig. 5.15.

Structure tools outperform sequence tools at the fold level. VAST has the best performance with an 85% success rate. PSI-Blast has the worst performance with

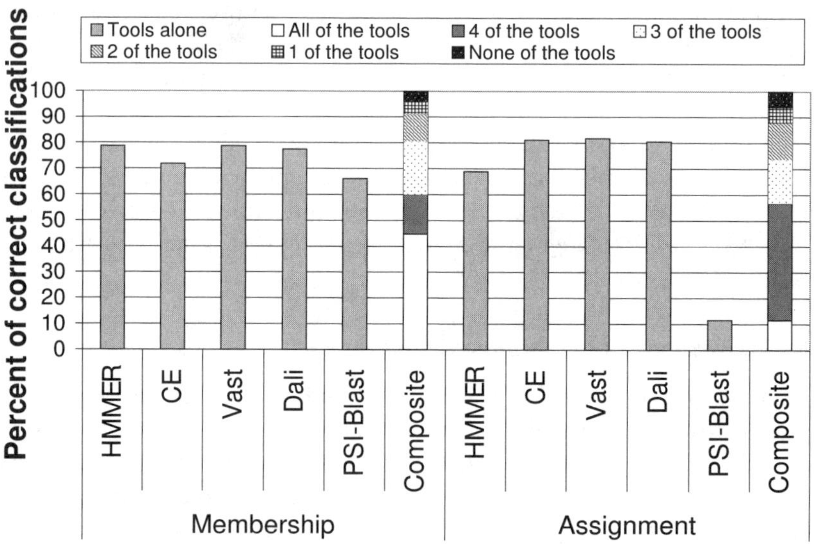

Fig. 5.15 Performance of individual classifiers on recognizing the members of existing superfamilies, and assigning categories to them for the new proteins in SCOP v.1.61. The first set of bars represents the performance for recognition of existing members and the second set represents the performance for the assignment of categories.

a 60.7% success rate. For only 30.9% of the queries, all five tools are successful (Camoglu et al., 2005).

Once a tool has marked a new protein as a member of an existing category, the classification of the query protein is complete, i.e., the query protein is assigned to the same category as its nearest neighbor. The next question is to judge the accuracy of this assignment, i.e., whether the assigned category is the correct one. The accuracies of the tools are high at the family level. All except DALI have success rates above 90%. HMMER has the best performance with 94.8% accuracy and is followed by PSI-Blast with 92.3% accuracy. For 76.5% of the queries, all five tools are able to assign the correct family label. For only 2.1% of the queries, none of them is successful. The superfamily results can be seen in Fig. 5.15. At the fold level, all tools seem to perform poorly. At the fold level, PSI-Blast is not able to make even one correct fold assignment, whereas VAST assigns correct folds to 54% of the queries. For 35.1% of the queries, none of the tools is able to assign the correct fold label.

5.5.4 Automated Classification Using Ensemble Classifier

As evident from the earlier experiments, an ensemble classifier can potentially obtain higher classification accuracy than any single component classifier. There are many studies in the area of machine learning and pattern recognition that address the intelligent design of ensemble classifiers (Duda et al., 2001; Meir and Ratsch, 2003; Schapire and Singer, 1999). These include both competitive models (e.g., bagging and binning) and collaborative models (e.g., boosting).

Camoglu et al. (2005) employ a hierarchical decision tree to answer the question whether the query protein belongs to an existing category. To combine different tools, their results need to be normalized to a consistent scale. This is achieved by dividing the scores into bins. A bin is a *tool-neutral* accuracy extent, e.g., 90–100%, 80–100%, instead of a *tool-specific* similarity score. Bins for each tool are manually crafted to achieve maximum performance. All proteins' 1NN scores are obtained and sorted. Bin boundaries are placed on this sorted list and accuracy is computed for each bin. For example, if bin i is labeled with an accuracy of x%, then the predicted accuracy of all proteins whose scores fall in the bin is x%.

After the bins for each tool are constructed, a decision tree is created. At each level of this tree, a combination of tools is run and depending on their decisions, query proteins are classified as a member of an existing category or new. The decision tree for the family level is presented in Fig. 5.16. As can be seen, at the first level of the decision tree, PSI-Blast and HMMER are run to classify the query protein. Each tool reports a confidence for their classification decision and these decisions and confidences are merged (Can et al., 2004). If the confidence of the consensus decision is greater than 95%, the query protein is classified as a member of an existing family. If the confidence is lower than 60%, it is classified as a member of a new family. If the confidence is between 60 and 95%, then tools at this level cannot make a confident decision, and the next level of the tree is used. At the next level, all of the structure tools are run for a consensus decision. There are again two thresholds,

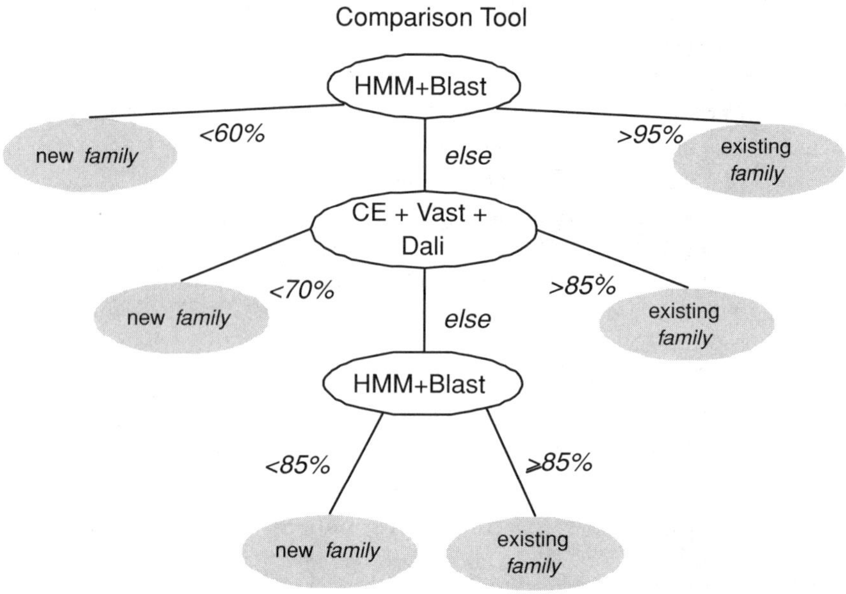

Fig. 5.16 The decision tree for recognizing if a query protein belongs to an existing family.

85 and 70%. At the last level, there is only one threshold, 80%. If the confidence is greater than this threshold, the query protein is classified as a member of an existing family, else it is classified as a member of a new family.

There are two components of a decision tree: (1) the combination of tools at each level and (2) the thresholds at each level. The first issue is addressed by using the domain knowledge of the classification schemes and the classification results for component classifiers (e.g., Fig. 5.15). For example, at the family level, it is known that sequence tools have priority, and at the superfamily level and fold levels, structure tools are more important. The problem of finding the right thresholds requires more care. The thresholds can be automatically determined by examining different choices and finding values that maximize accuracy for the training data. But, this approach tends to overfit. Thus, the distributions are manually analyzed and thresholds set after this analysis. An example of such an analysis is depicted in Fig. 5.17. Although placing the cutoff at point A_2 is more suitable for the training data, it is clear that there is a natural separation at point M_2. If A_2 is used as the threshold, it overfits the data and the eventual performance suffers. The determined thresholds for superfamily and fold level classifications are shown in Table 5.1.

5.5.5 Experimental Analysis

To validate that the consensus classifier indeed improves the classification performance, the standard validation technique in pattern recognition is applied (Duda et al., 2001). Two data sets are used: a training set and a test set. Ensemble classifier

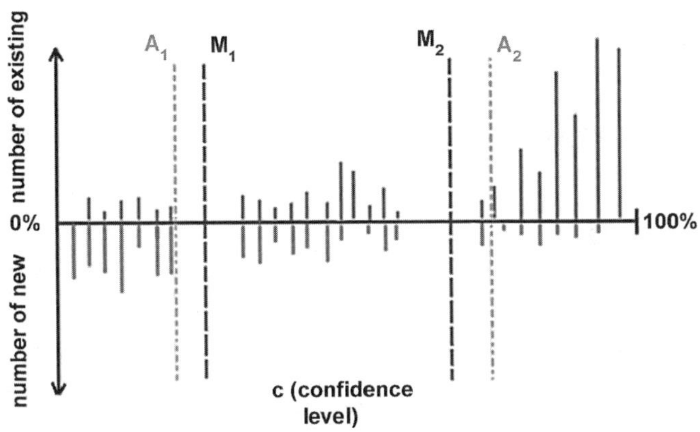

Fig. 5.17 An example histogram of the confidence levels for the training data. A_1 and A_2 are the thresholds found by the automated greedy approach. M_1 and M_2 are the manual thresholds. It is seen that A_1 and A_2 overfit the data, whereas M_1 and M_2 do not.

Table 5.1 Heuristic decision tree rules for recognition of members of existing superfamilies and folds. At each level, a combination of tools is run and the probability of being a member of an existing category is assigned to each protein. The proteins that have probabilities higher than the indicated range are assigned to the predicted category, the ones within the range are passed to the next step, and those below the range are deemed new. For the last level, only a single threshold exists.

	Level 1	Range	Level 2	Range	Level 3	Threshold
Superfamily	VAST	45%:93%	HMMER	40%:75%	CE+DALI	55%
Fold	VAST	50%:85%	CE	80%:90%	DALI	60%

is trained with proteins introduced in SCOP 1.61 using the classifications of proteins in SCOP 1.59, and it is validated on proteins introduced in SCOP 1.63 using the classifications of proteins in SCOP 1.61. Automated ensemble classifier performed well on assigning proteins to their correct classifications by achieving 98% success for family assignments, 83% success for superfamily assignments, and 61% success for fold assignments. The performance of each tool and the ensemble classifier for superfamily classification (training and evaluation phases) is shown in Table 5.2. The benefits of an ensemble classifier are obvious. Complete results appear in Camoglu et al. (2005).

5.6 Concluding Remarks

A number of algorithms for protein comparison were presented in this chapter. Besides a survey of the current state of the art of techniques for problems in structure analysis such as pairwise alignment, multiple alignment, and motif discovery,

Table 5.2 Performance of individual tools and ensemble classifier on training and test data sets. The performance is measured for two cases: Existing (ability to recognize the proteins from the existing classifications) and Assignment (ability to assign correct classifications)

		Existing		Assignment	
		Training	Evaluation	Training	Evaluation
Superfamily	HMMER	78.63	79.29	68.9	39.18
	CE	71.77	38.03	81.1	43.81
	VAST	78.63	66.67	81.71	55.67
	DALI	77.42	75.24	80.49	74.74
	PSI-Blast	66.13	31.39	11.59	24.23
	Ensemble	80.65	83.82	93.29	82.99

techniques that scale to large data sets were discussed in depth. In this vein, an approach based on index structures was presented for pairwise protein structure comparison. In this technique, SSE-based features were extracted from database proteins and inserted into an index structure. Given a query protein, its features were also extracted and compared using the index. A graph-based approach was used to extend and evaluate the matches. Embedding this algorithm into well-known pairwise structure analysis algorithms led to significant speed-ups. As data sets such as the PDB grow in size, scalability of structure comparison algorithms will be an important factor.

Classification of proteins was also presented. A number of existing classification techniques based on protein structure were discussed. A technique for automated classification of proteins was investigated in depth. The approach used a number of sequence and comparison tools and combined them in a decision tree to increase the classification accuracy. The use of multiple data sources is an increasingly useful criterion in biological analysis. Some aspects of such an approach were presented in the use of multiple sequence and structure comparison tools for protein classification.

Comparing protein structures quantitatively poses a number of challenges: defining the appropriate notion of similarity, developing new algorithms based on these notions of similarity that can scale to an exponentially increasing number of structures, and understanding the significance of a score. The substantial amount of research in this area is a reflection of the current challenges and activity in this area. Other opportunities for future research include: flexible models for alignment, simultaneous use of sequence and structure in pairwise and multiple alignment, and evolutionary characterization of structures and protein–protein interactions.

5.7 References and Resources

5.7.1 Definitions

- **Protein structure alignment:** one-to-one mapping between the residues of two protein structures.

- **Root-mean-square distance:** the root-mean-square distance (RMSD) between two proteins A and B under a correspondence R of size k and a transformation f is defined as

$$\mathrm{RMSD}(A,B,R,f) = \sqrt{\frac{\sum_{i=1}^{k} dist^2(a_i, f(R(a_i)))}{k}}$$

- **Distance matrix:** a two-dimensional matrix where each entry M_{ij} stores the Euclidean distance between the ith and jth residues of a protein.
- **Structural motif:** a substructure that is common in the structures of a set of proteins.
- **Multiple structure alignment:** the alignment of a set of related proteins that results in a consensus structure which has the minimum RMSD sum to the protein structures in the set.
- **Feature vector:** a vector composed of numerical values that summarizes the properties of an object.
- **Index structure:** a data structure that organizes a set of objects and supports efficient retrieval.

5.7.2 Resources

Name	Description	Link
CATH	Classification	*http://cathwww.biochem.ucl.ac.uk/latest/*
CE	Pairwise & multiple alignment	*http://cl.sdsc.edu/*
DALI	Pairwise alignment	*http://www.ebi.ac.uk/dali/*
FSSP	Classification	*http://ekhidna.biocenter.helsinki.fi/dali/start*
MASS	Multiple alignment	*http://bioinfo3d.cs.tau.ac.il MASS/*
MultiProt	Multiple alignment	*http://bioinfo3d.cs.tau.ac.il/MultiProt/*
PDB	Structure repository	*http://www.rcsb.org/pdb/Welcome.do*
ProtDex	Database search	*http://xena1.ddns.comp.nus.edu.sg/g̃enesis/PD2.htm*
PSI	Database search	*http://bioserver.cs.ucsb.edu/proteinstructuresimilarity.php*
SCOP	Classification	*http://scop.mrc-lmb.cam.ac.uk/scop/*
SSAP	Pairwise alignment	*http://www.cathdb.info/cgi-bin/cath/GetSsapRasmol.pl*
Trilogy	Motif finding	*http://theory.lcs.mit.edu/trilogy/*
URMS	Pairwise alignment	*http://cbsusrv01.tc.cornell.edu/urms/*
VAST	Pairwise alignment	*http://www.ncbi.nih.gov/Structure/VAST/vastsearch.html*

5.8 Further Reading

For readers interested in pairwise protein structure comparison, we recommend “Protein structure comparison by alignment of distance matrices” by Holm and Sander (1993), “Approximate protein structural alignment in polynomial time” by

Kolodny and Linial (2004), and "Sensitivity and selectivity in protein structure comparison" by Sierk and Pearson (2004). For more information on motif detection and multiple structure alignment, we recommend "Discovery of sequence–structure patterns across diverse proteins" by Bradley et al. (2002) and "MASS: multiple structural alignment by secondary structures" by Dror et al. (2003). More information about database searches can be found in "Index-based similarity search for protein structure databases" by Camoglu et al. (2004) and "Rapid 3d protein structure database searching using information retrieval techniques" by Aung and Tan (2004). For further discussion about the comparison of classification databases and automated classification techniques, we recommend "Automated assignment of SCOP and CATH protein structure classifications from FSSP scores" by Getz et al. (2002) and "Decision tree based information integration for automated protein classification" by Camoglu et al. (2005).

Acknowledgment

This work was supported in part by NSF grants DBI-0213903 and EF-0331697.

References

Alexandrov, N., and D. Fischer. 1996. Analysis of topological and nontopological structural similarities in the PDB: New examples from old structures. *Proteins* 25:354–365.

Altschul, S. F., and E. V. Koonin. 1998. Iterated profile searches with PSI-BLAST—a tool for discovery in protein databases. *Trends Biochem Sci.* 23:444–447.

Arun, K., T. Huang, and S. Blostein. 1987. Least-squares fitting of two 3-D point sets. *IEEE Trans. Pattern Anal. Mach. Intell.* 9:698–700.

Aung, Z., and K.-L. Tan. 2004. Rapid 3d protein structure database searching using information retrieval techniques. *Bioinformatics* 20:1045–1052.

Beckmann, N., H.-P. Kriegel, R. Schneider, and B. Seeger. 1990. The R*-tree: An efficient and robust access method for points and rectangles. In *SIGMOD*, pp. 322–331, Atlantic City, NJ.

Berman, H. M., J. Westbrook, Z. Feng, G. Gilliland, T. N. Bhat, H. Weissig, I. N. Shindyalov, and P. E. Bourne. 2000. The Protein Data Bank. *Nucleic Acids Res.* 28:235–242.

Binkowski, T. A., B. DasGupta, and J. Liang. 2004. Order independent structural alignment of circularly permuted proteins. In *IEEE EMBS*, July.

Bradley, P., P. S. Kim, and B. Berger. 2002. TRILOGY: Discovery of sequence–structure patterns across diverse proteins. *Proc. Natl. Acad. Sci. USA* 99:8500–8503.

Brown, N., C. Orengo, and W. Taylor. 1996. A protein structure comparison methodology. *Comput. Chem.* 20:359–380.

Camoglu, O., T. Can, A. K. Singh, and Y.-F. Wang. 2005. Decision tree based information integration for automated protein classification. *J. Bioinform. Comput. Biol.* 3(3):717–742.

Camoglu, O., T. Kahveci, and A. K. Singh. 2004. Index-based similarity search for protein structure databases. *J. Bioinform. Comput. Biol.* 2:99–126.

Camoglu, O., T. Kahveci, and A. K. Singh. 2003. Towards index-based similarity search for protein structure databases. In *CSB*, pp. 148–158.

Can, T., O. Camoglu, A. K. Singh, and Y.-F. Wang. 2004. Automated protein classification using consensus decision. In *CSB*, pp. 224–235.

Chen, S.-C., and I. Bahar. 2004. Mining frequent patterns in protein structures: A study of protease families. *Bioinformatics* 20:77–85.

Chew, L., D. Huttenlocher, K. Kedem, and J. Kleinberg. 1999. Fast detection of common geometric substructure in proteins. *J. Comput. Biol.* 6:313–325.

Chothia, C., J. Gough, C. Vogel, and S. A. Teichmann. 2003. Evolution of the protein repertoire. *Science* 300:1701–1703. URL *http://www.sciencemag.org/cgi/content/abstract/300/5626/1701*.

Dror, O., H. Benyamini, R. Nussinov, and H. Wolfson. 2003. MASS: Multiple structural alignment by secondary structures. *Bioinformatics* 19:i95–i104.

Duda, R. O., P. E. Hart, and D. G. Stork. 2001. *Pattern Classification*, 2nd edition. New York, Wiley–Interscience.

Eddy. S. R. 1998. Profile hidden Markov models. *Bioinformatics* 14:755–763.

Eidhammer, I., and I. Jonassen. 2001. Protein structure comparison and structure patterns—An algorithmic approach. ISMB tutorial.

Eidhammer, I., I. Jonassen, and W. Taylor. 2000. Structure comparison and structure patterns. *J. Comput Biol.* 7:685–716.

Fischer, D. 2003. 3D-SHOTGUN: A novel, cooperative, fold-recognition metapredictor. *Proteins Struct. Funct. Genet.* 51:434–441.

Garey, M., and D. Johnson. 1979. *Computers and Intractability: A Guide to the Theory of NP-Completeness*. San Francisco, Freeman.

Gerstein, M. 1997. A structural census of genomes: Comparing bacterial, eukaryotic, and archaeal genomes in terms of protein structure. *J. Mol. Biol.* 274:562–576.

Gerstein, M., and M. Levitt. 1996. Using iterative dynamic programming to obtain pairwise and multiple alignments of protein structures. In *ISMB*, pp. 59–66. PMID: 8877505.

Getz, G., M. Vendruscolo, D. Sachs, and E. Domany. 2002. Automated assignment of SCOP and CATH protein structure classifications from FSSP scores. *Proteins* 46:405–415.

Gibrat, J.-F., T. Madej, and S. Bryant. 1996. Surprising similarites in structure comparison. *Curr. Opin. Struct. Biol.* 6:377–385.

Godzik, A. 1996. The structural alignment between two proteins: Is there a unique answer? *Protein Sci.* 5:1325–1338.

Goldman, D., C. H. Papadimitriou, and S. Istrail. 1999. Algorithmic aspects of protein structure similarity. In *FOCS '99: Proceedings of the 40th Annual Symposium on Foundations of Computer Science*, p. 512, Washington, DC. IEEE Computer Society. ISBN 0-7695-0409-4.

Gough, J. 2002. The SUPERFAMILY database in structural genomics. *Acta Crystallogr.* D58:1897–1900.

Guda, C., E. D. Scheeff, P. E. Bourne, and N. Shindyalov. 2001. A new algoritm for the alignment of multiple protein structures using Monte Carlo optimization. In *PSB*.

Gusfield, D. 1997. *Algorithms on Strings, Trees, and Sequences: Computer Science and Computational Biology*. London, Cambridge University Press. ISBN 0-521-58519-8 (hardcover).

Holm, L., and C. Sander. 1993. Protein structure comparison by alignment of distance matrices. *J. Mol. Biol.* 233:123–138.

Holm, L., and C. Sander. 1995. 3-D lookup: Fast protein structure database searches at 90% reliability. In *ISMB*, pp. 179–187.

Holm, L., and C. Sander. 1996. Mapping the protein universe. *Science* 273:595–602.

Hughey, R., and A. Krogh. 1995. SAM: Sequence alignment and modeling software system. Technical Report, University of California at Santa Cruz.

Irving, J. A., J. C. Whisstock, and A. M. Lesk. 2001. Protein structural alignments and functional genomics. *Proteins* 42:378–382.

Jia, Y., T. G. Dewey, I. N. Shindyalov, and P. E. Bourne. 2004. A new scoring function and associated statistical significance for structure alignment by CE. *J. Comput. Biol.* 11:787–799.

Jonassen, I., I. Eidhammer, and W. R. Taylor. 1999. Discovery of local packing motifs in protein structures. *Proteins* 34:206–219.

Kabsch, W. 1978. A discussion of the solution for the best rotation to relate two sets of vectors. *Acta Crystallogr.* A34:827–828.

Kato, H., and Y. Takahashi. 2001. Automated identification of three-dimensional common structural features of proteins. *J. Chem. Software* 7:161–170.

Kim, D. E., D. Chivian, and D. Baker. 2004. Protein structure prediction and analysis using the Robetta server. *Nucleic Acids Res.* 32:526–531.

Kolodny, R., and N. Linial. 2004. From The Cover: Approximate protein structural alignment in polynomial time. *Proc. Natl. Acad. Sci. USA* 101:12201–12206. URL *http://www.pnas.org/cgi/content/abstract/101/33/12201*.

Lathrop, R. H. 1994. The protein threading problem with sequence amino acid interaction preferences is NP-complete. *Protein Eng.* 7:1059–1068.

Leibowitz, N., Z. Y. Fligelman, R. Nussinov, and H. J. Wolfson. 2001. Automated structure alignment and detection of a common substructural motif. *Proteins* 2001:235–245.

Levitt, M., and M. Gerstein. 1998. A unified statistical framework for sequence comparison and structure comparison. *Proc. Natl. Acad. Sci. USA* 95:5913–5920, URL http://www.pnas.org/cgi/content/abstract/95/11/5913.

Lindahl, E., and A. Eloffson. 2000. Identification of related proteins on family, superfamily and fold level. *J. Mol. Biol.* 295:613–625.

Lundstrom, J., L. Rychlewski, J. Bujnicki, and A. Elofsson. 2001. Pcons: A neural-network-based consensus predictor that improves fold recognition. *Protein Sci.* 10:2354–2362.

Madej, T., J.-F. Gibrat, and S. H. Bryant. 1995. Threading a database of protein cores. *Proteins* 23:356–369.

Meir, R., and G. Ratsch. 2003. An introduction to boosting and leveraging. In *Advanced Lectures on Machine Learning*. S. Mendelson and A. Smola (Eds.). Berlin, Springer-Verlag, pp. 119–184.

Murzin, A. G., S. E. Brenner, T. Hubbard, and C. Chothia. 1995. SCOP: A structural classification of proteins database for the investigation of sequences and structures. *J. Mol. Biol.* 247:536–540.

Needleman, S., and C. Wunsch. 1970. A general method applicable to the search for similarities in the amino acid sequence of two proteins. *J. Mol. Biol.* 48:443–53.

Novotny, M., D. Madsen, and G. J. Kleywegt. 2004. Evaluation of protein fold comparison servers. *Proteins Struct. Funct. Bioinform.* 54:260–270.

Nussinov, R., and H. Wolfson. 1991. Efficient detection of three-dimensional structural motifs in biological macromolecules by computer vision techniques. *Proc. Nat. Acad. Sci. USA* 88:10495–10499.

Orengo, C., and W. Taylor. 1996. SSAP: Sequential structure alignment program for protein structure comparison. *Methods Enzymol.* 266:617–635.

Orengo, C. A., A. D. Michie, S. Jones, D. T. Jones, M. B. Swindells, and J. M. Thorton. 1997. CATH–A hierarchic classification of protein domain structures. *Structure* 5:1093–1108.

Pennec, X., and N. Ayache. 1998. A geometric algorithm to find small but highly similar 3D substructures in proteins. *Bioinformatics* 14:516–522.

Pieper, U., N. Eswar, A. C. Stuart, V. A. Ilyin, and A. Sali. 1999. MODBASE, a database of annotated comparative protein structure models. *Bioinformatics* 15:1060–1061.

Portugaly, E., and M. Linial. 2000. Estimating the probability for a protein to have a new fold: A statistical computational model. *Proc. Natl. Acad. Sci. USA* 97:5161–5166.

Rose, R. B., and R. M. Stroud. 1998. Domain flexibility in retroviral proteases: Structural implications for drug resistant mutations. *Biochemistry* 37:2607–2621.

Sali, A., and T. Blundell. 1990. Definition of general topological equivalence in protein structures: A procedure involving comparison of properties and relationships through simulated annealing and dynamic programming. *J. Mol. Biol.* 212:403–428.

Schapire, R. E., and Y. Singer. 1999. Improved boosting algorithms using confidence-rated predictions. *Machine Learning* 37:297–336.

Shatsky, M. 2004. Flexprot: Alignment of flexible protein structures without a predefinition of hinge regions. *J. Comput. Biol.* 11:83–106.

Shatsky, M., R. Nussinov, and H. Wolfson. 2002. Flexible protein alignment and hinge detection. *Proteins* 48:242–256.

Shindyalov, I. N., and P. E. Bourne. 1998. Protein structure alignment by incremental combinatorial extension (CE) of the optimal path. *Protein Eng.* 11:739–747.

Shindyalov, I. N., and P. E. Bourne. 2000. An alternative view of the protein fold space. *Proteins* 38:247–260.

Sierk, M. L., and W. R. Pearson. 2004. Sensitivity and selectivity in protein structure comparison. *Protein Sci.* 13:773–785. URL http://www.proteinscience.org/cgi/content/abstract/13/3/773.

Singh, A., and D. Brutlag. 1997. Hierarchical protein structure superposition using both secondary structure and atomic representations. In *ISMB*, pp. 284–293. ISBN 1-57735-022-7.

Singh, R., and M. Saha. 2003. Identifying structural motifs in proteins. In *Pac. Symp. Biocomput.*

Taylor, W. R. 1999. Protein structure comparison using iterated double dynamic programming. *Protein Sci.* 8:654–665.

Verbitsky, G., R. Nussinov, and H. Wolfson. 1999. Flexible structural comparison allowing hinge-bending, swiveling motions. *Proteins* 34:232–254.

Wallace, A. C., N. Borkakoti, and J. M. Thorton. 1997. TESS: A geometric hashing algorithm for deriving 3D coordinate templates for searching structural databases. application to enzyme active sites. *Protein Sci.* 6:2308–2323.

Wang, Y., J. B. Anderson, J. Chen, L. Y. Geer, S. He, D. I. Hurwitz, C. A. Liebert, T. Madej, G. H. Marchler, A. Marchler-Bauer, A. R. Panchenko, B. A. Shoemaker, J. S. Song, P. A. Thiessen, R. A. Yamashita, and S. H. Bryant. 2002. MMDB: Entrez's 3D-structure database. *Nucleic Acids Res.* 30:249–252.

Ye, Y., and A. Godzik. 2003. Flexible structure alignment by chaining aligned fragment pairs allowing twists. *Bioinformatics* 19:ii246–255.

Ye, Y., and A. Godzik. 2004. Database searching by flexible protein structure alignment. *Protein Sci.* 13:1841–1850. URL *http://www.proteinscience.org/cgi/content/abstract/13/7/1841.*

6 Computation of Protein Geometry and Its Applications: Packing and Function Prediction

Jie Liang

6.1 Introduction

Three-dimensional atomic structures of protein molecules provide rich information for understanding how these working molecules of a cell carry out their biological functions. With the amount of solved protein structures rapidly accumulating, computation of geometric properties of protein structure becomes an indispensable component in studies of modern biochemistry and molecular biology. Before we discuss methods for computing the geometry of protein molecules, we first briefly describe how protein structures are obtained experimentally.

There are primarily three experimental techniques for obtaining protein structures: X-ray crystallography, solution nuclear magnetic resonance (NMR), and freeze-sample electron microscopy (cryo-EM). In X-ray crystallography, the diffraction patterns of X-ray irradiation of a high-quality crystal of the protein molecule are measured. Since the diffraction is due to scattering of the X-ray by the electrons of the molecules in the crystal, the position, the intensity, and the phase of each recorded diffraction spot provide information for the reconstruction of an *electron density map* of atoms in the protein molecule. Based on independent information of the amino acid sequence, a model of the protein conformation is then derived by fitting model conformations of residues to the electron density map. An iterative process called *refinement* is then applied to improve the quality of the fit of the electron density map. The final model of the protein conformation consists of the coordinates of each of the non-hydrogen atoms (Rhodes, 1999).

The solution NMR technique for solving protein structure is based on measuring the tumbling and vibrating motions of the molecule in solution. By assessing the chemical shifts of atomic nuclei with spins due to interactions with other atoms in the vicinity, a set of estimated distances between specific pairs of atoms can be derived from NOSEY spectra. When a large number of such distances are obtained, one can derive a set of conformations of the protein molecule, each consistent with all of the distance constraints (Crippen and Havel, 1988). Although determining conformations from either X-ray diffraction patterns or NMR spectra is equivalent to solving an ill-posed inverse problem, techniques such as Bayesian Markov chain Monte Carlo with parallel tempering have been shown to be effective in obtaining protein structures from NMR spectra (Rieping et al., 2005). The cryo-EM technique for obtaining protein structure is described in more detail in Chapter 11.

6.2 Theory and Model

6.2.1 The Idealized Ball Model

The shape of a protein molecule is complex. The chemical properties of atoms in a molecule are determined by their electron charge distribution. It is this distribution that generates the scattering patterns of the X-ray diffraction. Chemical bonds between atoms lead to transfer of electronic charges from one atom to another, and the resulting isosurfaces of the electron density distribution depend not only on the location of individual nuclei but also on interactions between atoms. This results in an overall complicated isosurface of electron density (Bader, 1994).

The geometric model of a macromolecule amenable to convenient computation is an idealized model, where the shapes of atoms are approximated by three-dimensional balls. The shape of a protein or a DNA molecule consisting of many atoms is then the space-filling shape taken by a set of atom balls. This model is often called the *interlocking hard-sphere model*, the *fused ball model*, the *space filling model* (Lee and Richards, 1971; Richards, 1974a, 1985; Richmond, 1984), or the *union of ball model* (Edelsbrunner, 1995). In this model, details on the distribution of electron density, e.g., the differences between regions of covalent bonds and noncovalent bonds, are ignored. This idealization is quite reasonable, as it reflects the fact that the electron density reaches a maximum at a nucleus, and its magnitude decays almost spherically away from the point of the nucleus. Despite possible inaccuracy, this idealized model has found wide acceptance, because it enables quantitative measurement of important geometric properties (such as area and volume) of molecules. Insights gained from these measurements correlate well with experimental observations (Lee and Richards, 1971; Richards, 1977, 1985; Connolly, 1983; Richards and Lim, 1994; Gerstein and Richards, 1999).

With this idealization, the shape of each atom is that of a ball, and its size parameter is the ball radius. There are many possible choices for the parameter set of atomic radii (Richards, 1974b; Tsai et al., 1999). Frequently, atomic radii are assigned the values of their van der Waals radii (Bondi, 1964). Among all these atoms, hydrogen atom has the smallest mass, and has a much smaller radius than those of other atoms. For simplification, the *united atom* model is often employed to approximate the union of a heavy atom and the hydrogen atoms connected by a covalent bond. In this case, the radius of the heavy atom is increased to approximate the size of the union of the two atoms. This practice significantly reduces the total number of atom balls in the molecule. However, this approach has been questioned for possible inadequacy (Word et al., 1999).

The mathematical model of this idealized model is that of the union of balls (Edelsbrunner, 1995). For a molecule M of n atoms, the ith atom is modeled as a ball b_i, whose center is located at $z_i \in \mathbb{R}^3$, and the radius of this ball is $r_i \in \mathbb{R}$, namely, we have $b_i \equiv \{\boldsymbol{x}|\boldsymbol{x} \in \mathbb{R}^3, ||\boldsymbol{x} - \boldsymbol{z}_i|| \leq r_i\}$ parameterized by $(\boldsymbol{z}_i, r_i)$. The molecule M is formed by the union of a finite number n of such balls defining the set $\mathcal{B}$:

$$M = \bigcup \mathcal{B} = \bigcup_{i=1}^{n} \{b_i\}.$$

It creates a space-filling body corresponding to the union of the excluded volumes $\mathrm{vol}(\bigcup_{i=1}^{n} b_i)$ (Edelsbrunner, 1995). When the atoms are assigned the van der Waals radii, the boundary surface $\partial \bigcup \mathcal{B}$ of the union of balls is called the *van der Waals surface*.

6.2.2 Surface Models: Lee–Richards and Connolly's Surfaces

Protein folds into native three-dimensional shape to carry out its biological functional roles. The interactions of a protein molecule with other molecules (such as ligand, substrate, or other protein) determine its functional roles. Such interactions occur physically on the surfaces of the protein molecule.

The importance of the protein surface was recognized very early on. Lee and Richards developed the widely used *solvent-accessible* (SA) *surface model*, which is also often called the *Lee–Richards surface model* (Lee and Richards, 1971). Intuitively, this surface is obtained by rolling a ball of radius r_s everywhere along the van der Waals surface of the molecule. The center of the solvent ball will then sweep out the solvent-accessible surface. Equivalently, the solvent-accessible surface can be viewed as the boundary surface $\partial \bigcup \mathcal{B}_{r_s}$ of the union of a set of inflated balls $\mathcal{B}_{r_s}$, where each ball takes the position of an atom, but with an inflated radius $r_i + r_s$ (Fig. 6.1a).

The solvent-accessible surface in general has many sharp crevices and sharp corners. In the hope of obtaining a smoother surface, one can take the surface swept out by the front instead of the center of the solvent ball. This surface is

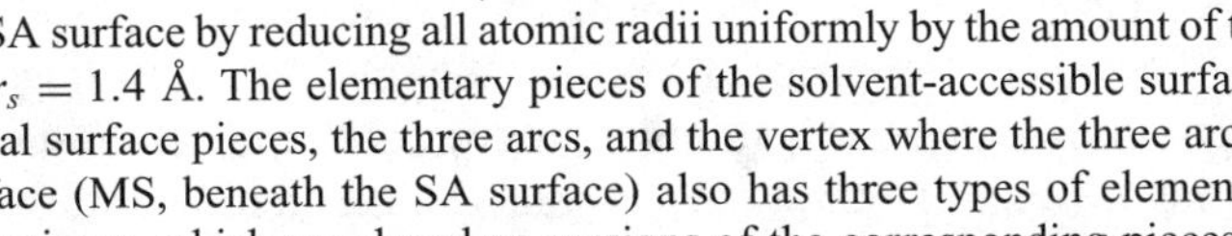

Fig. 6.1 Geometric models of protein surfaces. (a) The solvent-accessible (SA) surface is shown in the front. The van der Waals surface (beneath the SA surface) can be regarded as a shrunken version of the SA surface by reducing all atomic radii uniformly by the amount of the radius of the solvent probe $r_s = 1.4$ Å. The elementary pieces of the solvent-accessible surface are the three convex spherical surface pieces, the three arcs, and the vertex where the three arcs meet. (b) The molecular surface (MS, beneath the SA surface) also has three types of elementary pieces: the convex spheric pieces, which are shrunken versions of the corresponding pieces in the solvent-accessible surface, the concave toroidal pieces, and concave spheric surface. The latter two are also called the reentrant surface. (c) The toroidal surface pieces in the molecular surface correspond to the arcs in the solvent-accessible surface, and the concave spheric surface to the vertex. The set of elements in one surface can be continuously deformed to the set of elements in the other surface.

the *molecular surface* (MS model), which is often called the *Connolly's surface* after Michael Connolly who developed the first algorithm for computing molecular surface (Connolly, 1983). Both solvent-accessible surface and molecular surface are formed by elementary pieces of simpler shape.

Elementary pieces: For the solvent-accessible surface model, the boundary surface of a molecule consists of three types of elements: the convex spherical surface pieces, arcs or curved line segments (possibly a full circle) formed by two intersecting spheres, and a vertex that is the intersection point of three atom spheres. The whole boundary surface of the molecules can be thought of as a surface formed by stitching these elements together.

Similarly, the molecular surface swept out by the front of the solvent ball can also be thought of as being formed by elementary surface pieces. In this case, they are the convex spherical surface pieces, the toroidal surface pieces, and the concave or inverse spherical surface pieces (Fig. 6.1b) . The latter two types of surface pieces are often called the "reentrant surfaces" (Connolly, 1983; Richards, 1985).

The surface elements of the solvent-accessible surface and the molecular surface are closely related. Imagine a process where atom balls are shrunk. The vertices in the solvent-accessible surface become the concave spherical surface pieces, the arcs become the toroidal surfaces, and the convex surface pieces become smaller convex surface pieces (Fig. 6.1c). Because of this mapping, these two types of surfaces are combinatorially equivalent and have similar topological properties, i.e., they are homotopy equivalent.

However, the SA surface and the MS surface differ in their metric measurement. In concave regions of a molecule, often the front of the solvent ball can sweep out a larger volume than the center of the solvent ball. A void of size close to zero in the solvent-accessible surface model will correspond to a void of the size of a solvent ball ($4\pi r_s^3/3$). It is therefore important to distinguish these two types of measurement when interpreting the results of volume calculations of protein molecules. The intrinsic structures of these fundamental elementary pieces are closely related to several geometric constructs we describe below.

6.2.3 Geometric Constructs

Voronoi diagram: A Voronoi diagram (Fig. 6.2a), also known as Voronoi tessellation, is a geometric construct that has been used for analyzing protein packing since the early days of protein crystallography (Richards, 1974a; Finney, 1975; Gellatly and Finney, 1982). For a two-dimensional Voronoi diagram, we consider the following analogy. Imagine a vast forest containing a number of fire observation towers. Each fire ranger is responsible for putting out any fire closer to his/her tower than to any other tower. The set of all trees for which a ranger is responsible constitutes the Voronoi cell associated with his/her tower, and the map of ranger responsibilities, with towers and boundaries marked, constitutes the Voronoi diagram.

We formalize this for three-dimensional space. Consider the point set S of atom centers in three-dimensional space $\mathbb{R}^3$. The *Voronoi region* or *Voronoi cell* V_i of an

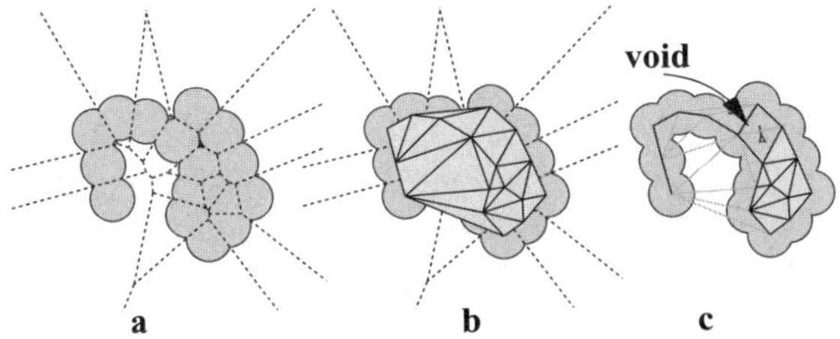

Fig. 6.2 Geometry of a simplified two-dimensional model molecule, to illustrate the geometric constructs and the procedure mapping the Voronoi diagram to the Delaunay triangulation. (a) The molecule formed by the union of atom disks of uniform size. Voronoi diagram is in dashed lines. (b) The shape enclosed by the boundary polygon is the *convex hull*. It is tessellated by the *Delaunay triangulation*. (c) The alpha shape of the molecule is formed by removing those Delaunay edges and triangles whose corresponding Voronoi edges and Voronoi vertices do not intersect with the body of the molecule. A molecular void is represented in the alpha shape by two empty triangles.

atom b_i with atom center $z_i \in \mathbb{R}^3$ is the set of all points that are at least as close to z_i as to any other atom centers in $\mathcal{S}$:

$$V_i = \{x \in \mathbb{R}^3 | ||x - z_i|| \leq ||x - z_j||, z_j \in \mathcal{S}\}. \quad (6.1)$$

We can have an alternative view of the Voronoi cell of an atom b_i. Consider the distance relationship of atom center z_i with atom center z_k of another atom b_k. The plane bisecting the line segment connecting points z_i and z_k divides the full $\mathbb{R}^3$ space into two half spaces, where points in one half space are closer to z_i than to z_k, and points in the other half space are closer to z_k than to z_i. If we repeat this process and take z_k in turn from the set of all atom centers other than z_i, we will have a number of half spaces where points are closer to z_i than to each atom center z_k. The Voronoi region V_i is then the common intersections of these half spaces, which is convex. When we consider atoms of different radii, we replace the Euclidean distance $||x - z_i||$ with the *power distance* defined as: $\pi_i(x) \equiv ||x - z_i||^2 - r_i^2$.

Delaunay tetrahedrization: Delaunay triangulation in $\mathbb{R}^2$ or Delaunay tetrahedrization in $\mathbb{R}^3$ is a geometric construct that is closely related to the Voronoi diagram (Fig. 6.2b). In general, it uniquely tessellates the space of the *convex hull* of the atom centers in $\mathbb{R}^3$ with tetrahedra. Convex hull for a point set is the smallest convex body that contains the point set.[1] The Delaunay tetrahedrization of a molecule can be obtained from the Voronoi diagram. Consider that the Delaunay tetrahedrization

[1] For a two-dimensional toy molecule, we can imagine that we put nails at the locations of the atom centers, and tightly wrap a rubber band around these nails. The rubber band will trace out a polygon. This polygon and the region enclosed within is the convex hull of the set of points corresponding to the atom centers. Similarly, imagine that if we can tightly wrap tinfoil around a set of points in three-dimensional space, the resulting convex body formed by the tinfoil and space enclosed within is the convex hull of this set of points in $\mathbb{R}^3$.

is formed by gluing four types of primitive elements together: vertices, edges, triangles, and tetrahedra. Here vertices are just the atom centers. We obtain a Delaunay edge by connecting atom centers z_i and z_j if and only if the Voronoi regions V_i and V_j have a common intersection, which is a planar piece that may be either bounded or extend to infinity. We obtain a Delaunay triangle connecting atom centers z_i, z_j, and z_k if the common intersection of Voronoi regions V_i, V_j, and V_k exists, which is either a line segment, or a half-line, or a line in the Voronoi diagram. We obtain Delaunay tetrahedra connecting atom centers z_i, z_j, z_k, and z_l if and only if the Voronoi regions V_i, V_j, V_k, and V_l intersect at a point.

6.2.4 Topological Structures

Delaunay complex: The structures in both Voronoi diagram and Delaunay tetrahedrization are better described with concepts from algebraic topology. We focus on the intersection relationship in the Voronoi diagram and introduce concepts formalizing the primitive elements. In $\mathbb{R}^3$, between two and four Voronoi regions may have common intersections. We use *simplices* of various dimensions to record these intersection or overlap relationships. In Delaunay tetrahedrization, we have vertices σ_0 as 0-simplices, edges σ_1 as 1-simplices, triangles σ_2 as 2-simplices, and tetrahedra σ_3 as 3-simplices. Each of the Voronoi plane, Voronoi edge, and Voronoi vertices corresponds to a 1-simplex (Delaunay edge), 2-simplex (Delaunay triangle), and 3-simplex (Delaunay tetrahedron), respectively. If we use 0-simplices to represent the Voronoi cells, and add them to the simplices induced by the intersection relationship, we can think of the Delaunay tetrahedrization as the structure obtained by "gluing" these simplices properly together. Formally, these simplices form a *simplicial complex* $\mathcal{K}$:

$$\mathcal{K} = \{\sigma_{|I|-1} | \bigcap_{i \in I} V_i \neq \emptyset\}, \tag{6.2}$$

where I is an index set for the vertices representing atoms whose Voronoi cells overlap, and $|I| - 1$ is the dimension of the simplex.

Alpha shape and protein surfaces: Imagine we can turn a knob to increase or decrease the size of all atoms simultaneously. We can then have a model of growing balls and obtain further information from the Delaunay complex about the shape of a protein structure. Formally, we use a parameter $\alpha \in \mathbb{R}$ to control the size of the atom balls. For an atom ball b_i of radius r_i, we modified its radius r_i at a particular α value to $r_i(\alpha) = (r_i^2 + \alpha)^{1/2}$. When $-r_i < \alpha < 0$, the size of an atom is shrunk. The atom could even disappear if $\alpha < 0$ and $|\alpha| > r_i$. We start to collect the simplices at different α value as we increase α from $-\infty$ to $+\infty$ (see Fig. 6.3 for a two-dimensional example). At the beginning, we only have vertices. When α is increased such that two atoms are close enough to intersect, we collect the corresponding Delaunay edge that connects these two atom centers. When three atoms intersect, we collect the corresponding Delaunay triangle spanning these three atom centers. When four atoms intersect, we collect the corresponding Delaunay tetrahedron. At

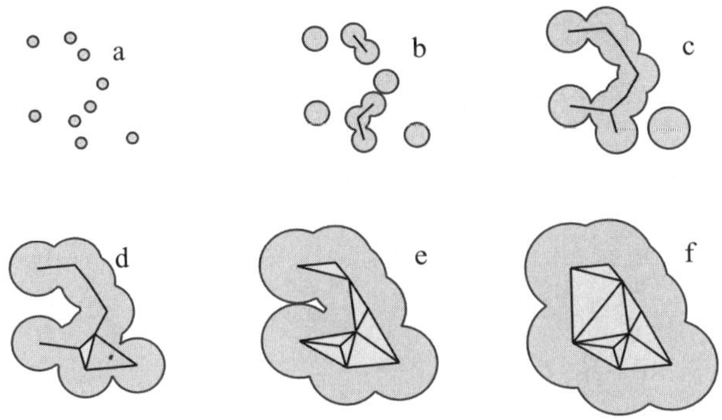

Fig. 6.3 The family of alpha shapes or dual simplicial complexes for a two-dimensional toy molecule. (a) We collect simplices from the Delaunay triangulation as atoms grow by increasing the α value. At the beginning as α grows from $-\infty$, atoms are in isolation and we only have vertices in the alpha shape. (b, c) When α is increased such that some atom pairs start to intersect, we collect the corresponding Delaunay edges. (d) When three atoms intersect as α increases, we collect the corresponding Delaunay triangles. When $\alpha = 0$, the collection of vertices, edges, and triangles form the dual simplicial complex $\mathcal{K}_0$, reflecting the topological structure of the protein molecule. (e) More edges and triangles from the Delaunay triangulation are now collected as atoms continue to grow. (d) Finally, all vertices, edges, and triangles are now collected as atoms are grown to large enough size. We get back the full original Delaunay complex.

any specific α value, we have a *dual simplicial complex* or *alpha complex* $\mathcal{K}_\alpha$ formed by the collected simplices. If all atoms take the incremented radius of $r_i + r_s$ and $\alpha = 0$, we have the dual simplicial complex $\mathcal{K}_0$ of the protein molecule. When α is sufficiently large, we have collected all simplices and we get the full Delaunay complex. This series of simplicial complexes at different α value form a family of shapes (Fig. 6.3), called *alpha shapes*, each faithfully representing the geometric and topological property of the protein molecule at a particular resolution parametrized by the α value.

An equivalent way to obtain the alpha shape at $\alpha = 0$ is to take a subset of the simplices, with the requirement that the corresponding intersections of Voronoi cells must overlap with the body of the union of the balls. We obtain the dual complex or alpha shape $\mathcal{K}_0$ of the molecule at $\alpha = 0$ (Fig. 6.2c):

$$\mathcal{K}_0 = \{\sigma_{|I|-1} | \bigcap_{i \in I} V_i \cap \bigcup \mathcal{B} \neq \emptyset\}. \tag{6.3}$$

Alpha shape provides a guide map for computing geometric properties of the structures of biomolecules. Take the molecular surface as an example; the reentrant surfaces are formed by the concave spherical patch and the toroidal surface. These can be mapped from the boundary triangles and boundary edges of the alpha shape, respectively (Edelsbrunner et al., 1995). Recall that a triangle in the Delaunay tetrahedrization corresponds to the intersection of three Voronoi regions, i.e., a Voronoi edge. For a triangle on the boundary of the alpha shape, the corresponding Voronoi edge intersects with the body of the union of balls by definition. In this case, it intersects with the solvent-accessible surface at the common intersecting vertex when the three atoms overlap. This vertex corresponds to a concave spherical surface patch in the molecular surface. For an edge on the boundary of the alpha shape, the corresponding Voronoi plane coincides with the intersecting plane when two atoms meet, which intersect with the surface of the union of balls on an arc. This line segment corresponds to a toroidal surface patch. The remaining parts of the surface are convex pieces, which correspond to the vertices, namely, the atoms on the boundary of the alpha shape.

The numbers of toroidal pieces and concave spherical pieces are exactly the numbers of boundary edges and boundary triangles in the alpha shape, respectively. Because of the restriction of bond length and the excluded volume effects, the number of edges and triangles in molecules are roughly on the order of $\mathcal{O}(n)$ (Liang et al., 1998a).

6.2.5 Metric Measurement

We have described the relationship between the simplices and the surface elements of the molecule. Based on this relationship, we can compute efficiently size properties of the molecule. We take the problem of volume computation as an example.

Consider a grossly incorrect way to compute the volume of a protein molecule using the solvent-accessible surface model. We could define that the volume of the molecule is the summation of the volumes of individual atoms, whose radii are inflated to account for the solvent probe. By doing so we would have significantly exaggerated the value of the true volume, because we neglected to consider volume overlaps. We can explicitly correct this by following the inclusion–exclusion formula: when two atoms overlap, we subtract the overlap; when three atoms overlap, we first subtract the pair overlaps, we then add back the triple overlap, etc. This continues when there are four, five, or more atoms intersecting. At the combinatorial level, the principle of inclusion–exclusion is related to the Gauss–Bonnet theorem used by Connolly (Connolly, 1983). The corrected volume $V(\mathcal{B})$ for a set of atom balls $\mathcal{B}$ can then be written as

$$V(\mathcal{B}) = \sum_{\substack{\mathrm{vol}(\bigcap T)>0 \\ T\subset\mathcal{B}}} (-1)^{\dim(T)-1} \mathrm{vol}\left(\bigcap T\right), \tag{6.4}$$

where $\mathrm{vol}(\bigcap T)$ represents volume overlap of various degree, and $T \subset \mathcal{B}$ is a subset of the balls with nonzero volume overlap: $\mathrm{vol}(\bigcap T) > 0$.

However, the straightforward application of this inclusion–exclusion formula does not work well. The degree of overlap can be very high: theoretical and simulation studies showed that the volume overlap can be up to 7–8 degrees (Kratky, 1981; Petitjean, 1994). It is difficult to keep track of these high degree of volume overlaps

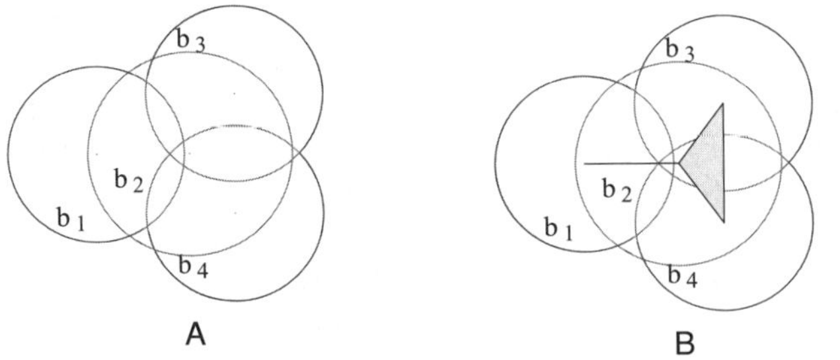

Fig. 6.4 An example of analytical area calculation. (A) Area can be computed using the direct inclusion-exclusion. (B) The formula is simplified without any redundant terms when using alpha shape.

correctly during computation, and it is also difficult to compute the volume of these overlaps because there are many different combinatorial situations, i.e., to quantify how large is the k-volume overlap of which one of the $\binom{7}{k}$ or $\binom{8}{k}$ overlapping atoms for all of $k = 2, \ldots, 7$ (Petitjean, 1994). It turns out that for three-dimensional molecules, overlaps of five or more atoms at a time can always be reduced to a "+" or a "−" signed combination of overlaps of four or fewer atom balls (Edelsbrunner, 1995). This requires that the 2-body, 3-body, and 4-body terms in Eq. (6.4) enter the formula if and only if the corresponding edge σ_{ij} connecting the two balls (1-simplex), triangles σ_{ijk} spanning the three balls (2-simplex), and tetrahedron σ_{ijkl} cornered on the four balls (3-simplex) all exist in the dual simplicial complex $\mathcal{K}_0$ of the molecule (Edelsbrunner, 1995; Liang et al., 1998a). Atoms corresponding to these simplices will all have volume overlaps. In this case, we have the simplified exact expansion:

$$
\begin{aligned}
V(\mathcal{B}) = & \sum_{\sigma_i \in \mathcal{K}} \text{vol}(b_i) - \sum_{\sigma_{ij} \in \mathcal{K}} \text{vol}(b_i \cap b_j) \\
& + \sum_{\sigma_{ijk} \in \mathcal{K}} \text{vol}(b_i \cap b_j \cap b_k) - \sum_{\sigma_{ijkl} \in \mathcal{K}} \text{vol}(b_i \cap b_j \cap b_k \cap b_l).
\end{aligned}
$$

The same idea is applicable for the calculation of surface area of molecules.

An example: An example of area computation by alpha shape is shown in Fig. 6.4. Let b_1, b_2, b_3, b_4 be the four disks. To simplify the notation we write A_i for the area of b_i, A_{ij} for the area of $b_i \cap b_j$, and A_{ijk} for the area of $b_i \cap b_j \cap b_k$. The total area of the union, $b_1 \cup b_2 \cup b_3 \cup b_4$, is

$$
\begin{aligned}
A_{\text{total}} = & (A_1 + A_2 + A_3 + A_4) \\
& - (A_{12} + A_{23} + A_{24} + A_{34}) \\
& + A_{234}.
\end{aligned}
$$

We add the area of b_i if the corresponding vertex belongs to the alpha complex (Fig. 6.4), we subtract the area of $b_i \cap b_j$ if the corresponding edge belongs to the alpha complex, and we add the area of $b_i \cap b_j \cap b_k$ if the corresponding triangle belongs to the alpha complex. Note without the guidance of the alpha complex, the inclusion–exclusion formula may be written as

$$\begin{aligned} A_{\text{total}} &= (A_1 + A_2 + A_3 + A_4) \\ &\quad - (A_{12} + A_{13} + A_{14} + A_{23} + A_{24} + A_{34}) \\ &\quad + (A_{123} + A_{124} + A_{134} + A_{234}) \\ &\quad - A_{1234}. \end{aligned}$$

This contains six canceling redundant terms: $A_{13} = A_{123}$, $A_{14} = A_{124}$, and $A_{134} = A_{1234}$. Computing these terms would be wasteful. Such redundancy does not occur when we use the alpha complex: the part of the Voronoi regions contained in the respective atom balls for the redundant terms do not intersect. Therefore, the corresponding edges and triangles do not enter the alpha complex. In two dimensions, we have terms of at most three disk intersections, corresponding to triangles in the alpha complex. Similarly, in three dimensions the most complicated terms are intersections of four spherical balls, and they correspond to tetrahedra in the alpha complex.

Voids and pockets: Voids and pockets represent the concave regions of a protein surface. Because shape-complementarity is the basis of many molecular recognition processes, binding and other activities frequently occur in pocket or void regions of protein structures. For example, the majority of enzyme reactions take place in surface pockets or interior voids.

The topological structure of the alpha shape also offers an effective method for computing voids and pockets in proteins. Consider the Delaunay tetrahedra that are not included in the alpha shape. If we repeatedly merge any two such tetrahedra on the condition that they share a 2-simplex triangle, we will end up with discrete sets of tetrahedra. Some of them will be completely isolated from the outside, and some of them are connected to the outside by triangle(s) on the boundary of the alpha shape. The former corresponds to voids (or cavities) in proteins, the latter corresponds to *pockets* and *depressions* in proteins.

A pocket differs from a depression in that it must have an opening that is at least narrower than one interior cross section. Formally, the *discrete flow* (Edelsbrunner et al., 1998) explains the distinction between a depression and a pocket. In a two-dimensional Delaunay triangulation, the empty triangles that are not part of the alpha shape can be classified into obtuse triangles and acute triangles. The largest angle of an obtuse triangle is more than 90 degrees, and the largest angle of an acute triangle is less than 90 degrees. An empty obtuse triangle can be regarded as a "source" of empty space that "flows" to its neighbor, and an empty acute triangle a "sink" that collects flow from its obtuse empty neighboring triangle(s). In Fig. 6.5a, obtuse triangles 1, 3, 4, and 5 flow to the acute triangle 2, which is a sink. Each of the discrete

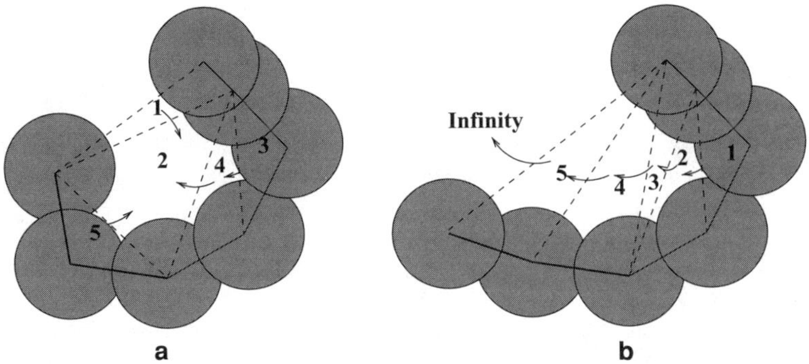

Fig. 6.5 Discrete flow of empty space illustrated for two-dimensional disks. (a) Discrete flow of a pocket. Triangles 1, 3, 4, and 5 are obtuse. The free volume flows to the "sink" triangle 2, which is acute. (b) In a depression, the flow is from obtuse triangles to the outside.

empty spaces on the surface of protein can be organized by the flow systems of the corresponding empty triangles: Those that flow together belong to the same discrete empty space. For a pocket, there is at least one sink among the empty triangles. For a depression, all triangles are obtuse, and the discrete flow goes from one obtuse triangle to another, from the innermost region to outside the convex hull. The discrete flow of a depression therefore goes to infinity. Figure 6.5b gives an example of a depression formed by a set of obtuse triangles.

Once voids and pockets are identified, we can apply the inclusion–exclusion principle based on the simplices to compute the exact size measurement (e.g., volume and area) of each void and pocket (Liang et al., 1998b; Edelsbrunner et al., 1998).

The distinction between voids and pockets depends on the specific set of atomic radii and the solvent radius. When a larger solvent ball is used, the radii of all atoms will be inflated by a larger amount. This could lead to two different outcomes. A void or pocket may become completely filled and disappear. On the other hand, the inflated atoms may not fill the space of a pocket, but may close off the opening of the pocket. In this case, a pocket becomes a void. A widely used practice in the past was to adjust the solvent ball and repeatedly compute voids, in the hope that some pockets will become voids and hence be identified by methods designed for cavity/void computation. The pocket algorithm (Edelsbrunner et al., 1998) and tools such as CASTp (Liang et al., 1998c; Binkowski et al., 2003b) often make this unnecessary.

6.3 Computation and Software

Computing Delaunay tetrahedrization and Voronoi diagram: It is easier to discuss the computation of tetrahedrization first. The incremental algorithm developed in Edelsbrunner and Shah (1996) can be used to compute the weighted tetrahedrization

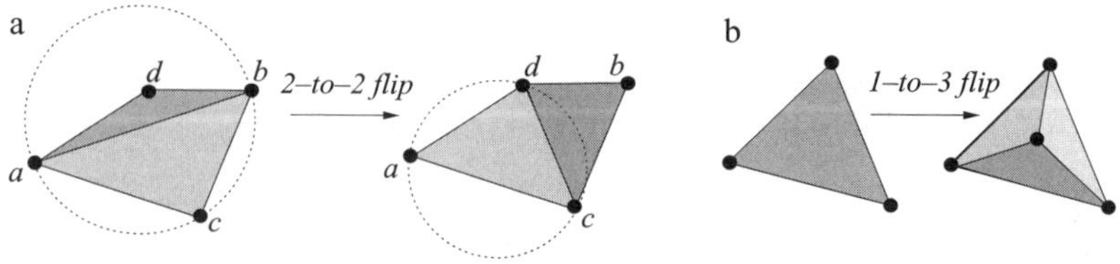

Fig. 6.6 An illustration of *locally Delaunay edge* and *flips*. (a) For the quadrilateral $abcd$, edge ab is not locally Delaunay, as the circumcircle passing through edge ab and a third point c contains a fourth point d. Edge cd is locally Delaunay, as b is outside the circumcircle adc. An *edge-flip* or *2-to-2 flip* replaces edge ab by edge cd, and replace the original two triangles abc and adb with two new triangles acd and bcd. (b) When a new vertex is inserted, we replace the old triangle containing this new vertex with three new triangles. This is called *1-to-3* flips.

for a set of atoms of different radii. For simplicity, we sketch the outline of the algorithm below for two-dimensional unweighted Delaunay triangulation.

The intuitive idea of the algorithm can be traced back to the original observation of Delaunay. For the Delaunay triangulation of a point set, the circumcircle of an edge and a third point forming a Delaunay triangle must not contain a fourth point. Delaunay showed that if all edges in a particular triangulation satisfy this condition, the triangulation is a Delaunay triangulation. It is easy to come up with an arbitrary triangulation for a point set. A simple algorithm to covert this triangulation to the Delaunay triangulation is therefore to go through each of the triangles, and make corrections using "flips" discussed below, if a specific triangle contains an edge violating the above condition. The basic ingredients for computing Delaunay tetrahedrization are generalizations of these observations. We discuss the concept of *locally Delaunay* edge and the *edge-flip* primitive operation below.

Locally Delaunay edge. We say an edge ab is locally Delaunay if either it is on the boundary of the convex hull of the point set, or if it belongs to two triangles abc and abd, and the circumcircle of abc does not contain d (e.g., edge cd in Fig. 6.6a).

Edge-flip. If ab is not locally Delaunay (edge ab in Fig. 6.6a), then the union of the two triangles $abc \cup abd$ is a convex quadrangle $acbd$, and edge cd is locally Delaunay. We can replace edge ab by edge cd. We call this an *edge-flip* or *2-to-2 flip*, as two old triangles are replaced by two new triangles.

We recursively check each boundary edge of the quadrangle $abcd$ to see if it is also locally Delaunay after replacing ab by cd. If not, we recursively edge-flip it.

Incremental algorithm for Delaunay triangulation. Assume we have a finite set of points (namely, atom centers) $\mathcal{S} = \{z_1, z_2, \ldots, z_i, \ldots, z_n\}$. We start with a large auxiliary triangle that contains all these points. We insert the points one by one. At all times, we maintain a Delaunay triangulation $\mathcal{D}_i$ up to insertion of point z_i.

After inserting point z_i, we search for the triangle τ_{i-1} that contains this new point. We then add z_i to the triangulation and split the original triangle τ_{i-1} into three smaller triangles. This split is called *1-to-3 flip*, as it replaces one old triangle with three new triangles. We then check if each of the three edges in τ_{i-1} still satisfies

Algorithm 1 Delaunay triangulation

```
Obtain random ordering of points {z_1, ..., z_n};
for i = 1 to n do
    find τ_{i-1} such z_i ∈ τ_{i-1};
    add z_i, and split τ_{i-1} into three triangles (1-to-3 flip);
    while any edge ab not locally Delaunay do
        flip ab to other diagonal cd (2-to-2 edge flip);
    end while
end for
```

the locally Delaunay requirement. If not, we perform a recursive edge-flip. This algorithm is summarized in Algorithm 1.

In $\mathbb{R}^3$, the algorithm of tetrahedrization becomes more complex, but the same basic ideas apply. In this case, we need to locate a tetrahedron instead of a triangle that contains the newly inserted point. The concept of locally Delaunay is replaced by the concept of *locally convex*, and there are flips different than the 2-to-2 flip in $\mathbb{R}^3$ (Edelsbrunner and Shah, 1996). Although an incremental approach, i.e., sequentially adding points, is not necessary for Delaunay triangulation in $\mathbb{R}^2$, it is necessary in $\mathbb{R}^3$ to avoid nonflippable cases and to guarantee that the algorithm will terminate. This incremental algorithm has excellent expected performance (Edelsbrunner and Shah, 1996).

The computation of Voronoi diagram is conceptually easy once the Delaunay triangulation is available. We can take advantage of the mathematical duality and compute all of the Voronoi vertices, edges, and planar faces from the Delaunay tetrahedra, triangles, and edges. Because one point z_i may be a vertex of many Delaunay tetrahedra, the Voronoi region of z_i therefore may contain many Voronoi vertices, edges, and planar faces. The efficient quad-edge data structure can be used for software implementation (Guibas and Stolfi, 1985).

Volume and area computation: Let V and A denote the volume and area of the molecule, respectively, $\mathcal{K}_0$ for the alpha complex, σ for a simplex in $\mathcal{K}$, i for a vertex, ij for an edge, ijk for a triangle, and $ijkl$ for a tetrahedron. The algorithm for volume and area computation can be written as Algorithm 2. Additional details of volume and area computation can be found in Edelsbrunner et al. (1995), and Liang et al. (1998a).

Software: The software package Delcx for computing weighted Delaunay tetrahedrization, Mkalf for computing the alpha shape, Volbl for computing volume and area of both molecules and interior voids can be found at *www.alphashape.org*. The CASTp webserver for pocket computation can be found at *cast.engr.uic.edu*. There are other studies that compute or use Voronoi diagrams of protein structures

Algorithm 2 Volume and area measurement

$V := A := 0.0;$
for all $\sigma \in \mathcal{K}$ **do**
  if σ is a vertex i **then**
    $V := V + \text{vol}(b_i)\,;\, A := A + \text{area}(b_i);$
  end if
  if σ is an edge ij **then**
    $V := V - \text{vol}(b_i \cap b_j)\,;\, A := A - \text{area}(b_i \cap b_j);$
  end if
  if σ is a triangle ijk **then**
    $V := V + \text{vol}(b_i \cap b_j \cap b_k);\, A := A + \text{area}(b_i \cap b_j \cap b_k);$
  end if
  if σ is a tetrahedron $ijkl$ **then**
    $V := V - \text{vol}(b_i \cap b_j \cap b_k \cap b_l)\,;\, A := A - \text{area}(b_i \cap b_j \cap b_k \cap b_l);$
  end if
end for

(Chakravarty et al., 2002; Goede et al., 1997; Harpaz et al., 1994), although not all computes the weighted version which allows atoms to have different radii.

In this short description of algorithm, we have neglected many details important for geometric computation, for example, the problem of how to handle geometric degeneracy, namely, when three points are colinear, or when four points are coplanar. Interested readers should consult the excellent monograph by Edelsbrunner for a detailed treatise of these and other important topics in computational geometry (Edelsbrunner, 2001).

6.4 Applications: Packing Analysis

An important application of the Voronoi diagram and volume calculation is the measurement of protein packing. Tight packing is an important feature of protein structure (Richards, 1974a, 1977), and is thought to play important roles in protein stability and folding dynamics (Levitt et al., 1997). The packing density of a protein is measured by the ratio of its van der Waals volume and the volume of the space it occupies. One approach is to calculate the packing density of buried residues and atoms using Voronoi diagram (Richards, 1974a, 1977). This approach was also used to derive parameters of radii of atoms (Tsai et al., 1999).

Based on the computation of voids and pockets in proteins, a detailed study surveying major representatives of all known protein structural folds showed that there is a substantial amount of voids and pockets in proteins (Liang and Dill, 2001). On average, every 15 residues introduces a void or a pocket (Fig. 6.7a). For a perfectly solid three-dimensional sphere of radius r, the relationship between volume $V = 4\pi r^3/3$ and surface area $A = 4\pi r^2$ is: $V \propto A^{3/2}$. In contrast, Fig. 6.7b shows that the van der Waals volume scales linearly with the van der Waals surface areas

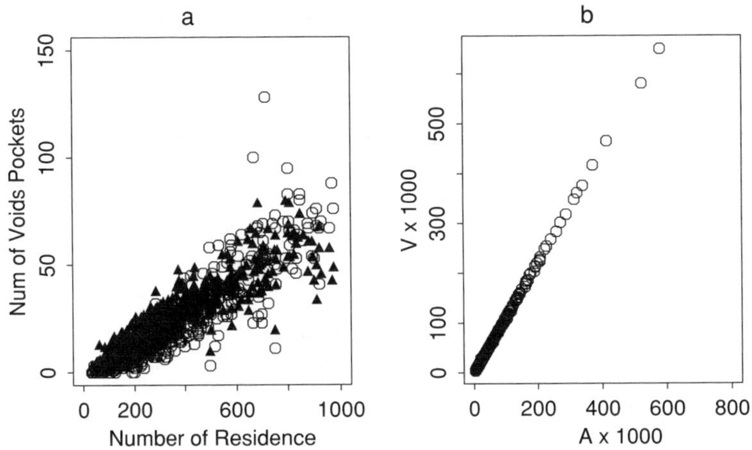

Fig. 6.7 Voids and pockets for a set of 636 proteins representing most of the known protein folds, and the scaling behavior of the geometric properties of proteins. (a) The number of voids and pockets detected with a 1.4 Å probe is linearly correlated with the number of residues in a protein. Only proteins with less than 1,000 residues are shown. Solid triangles and empty circles represent the pockets and the voids, respectively. (b) The van der Waals volume and van der Waals area of proteins scale linearly with each other. Similarly, molecular surface volume also scales linearly with molecular surface area using a probe radius of 1.4 Å. (Data not shown. Figure adapted after Liang and Dill, 2001).

of proteins. The same linear relationship holds irrespective of whether we relate molecular surface volume and molecular surface area, or solvent-accessible volume and solvent-accessible surface area. This and other scaling behavior point out that the protein interior is not packed as tight as solid (Liang and Dill, 2001). Rather, packing defects in the form of voids and pockets are common in proteins.

If voids and pockets are prevalent in proteins, an interesting question is what is the origin of the existence of these voids and pockets. This question was studied by examining the scaling behavior of packing density and coordination number of residues through the computation of voids, pockets, and edge simplices in the alpha shapes of random compact chain polymers (Zhang et al., 2003). For this purpose, a 32-state discrete state model was used to generate a large ensemble of compact self-avoiding walks. This is a difficult task, as it is very challenging to generate a large number of independent conformations of very compact chains that are self-avoiding. The results in Zhang et al. (2003) showed that it is easy for compact random chain polymers to have similar scaling behavior of packing density and coordination number with chain length. This suggests that proteins are not optimized by evolution to eliminate voids and pockets, and the existence of many pockets and voids is random in nature, due to the generic requirement of compact chain polymers. The frequent occurrence and the origin of voids and pockets in protein structures raise a challenging question: How can we distinguish voids and pockets that perform biological functions such as binding from those formed by random chance? This question is related to the general problem of protein function prediction.

6.5 Applications: Protein Function Prediction from Structures

Conservation of protein structures often reveals very distant evolutionary relationships, which are otherwise difficult to detect by sequence analysis (Todd et al., 2001). Comparing protein structures can provide insightful ideas about the biochemical functions of proteins (e.g., active sites, catalytic residues, and substrate interactions) (Holm and Sander, 1997; Martin et al., 1998; Orengo et al., 1999).

A fundamental challenge in inferring protein function from structure is that the functional surface of a protein often involves only a small number of key residues. These interacting residues are dispersed in diverse regions of the primary sequences and are difficult to detect if the only information available is the primary sequence. Discovery of local spatial motifs from structures that are functionally relevant has been the focus of many studies.

Graph-based methods for spatial patterns in proteins: To analyze local spatial patterns in proteins. Artymiuk et al. developed an algorithm based on subgraph isomorphism detection (Artymiuk et al., 1994). By representing residue side chains as simplified pseudo-atoms, a molecular graph is constructed to represent the patterns of side-chain pseudo-atoms and their interatomic distances. A user-defined query pattern can then be searched rapidly against the Protein Data Bank for similarity relationship. Another widely used approach is the method of geometric hashing. By examining spatial patterns of atoms, Fischer et al. developed an algorithm that can detect surface similarity of proteins (Fischer et al., 1982; Norel et al., 1994). This method has also been applied by Wallace et al. for the derivation and matching of spatial templates (Wallace et al., 1997). Russell developed a different algorithm that detects side-chain geometric patterns common to two protein structures (Russell, 1998). With the evaluation of statistical significance of measured root-mean-square distance, several new examples of convergent evolution were discovered, where common patterns of side chains were found to reside on different tertiary folds.

These methods have a number of limitations. Most require a user-defined template motif, restricting their utility for automated database-wide search. In addition, the size of the spatial pattern related to protein function is also often restricted.

Predicting protein functions by matching pocket surfaces: Protein functional surfaces are frequently associated with surface regions of prominent concavity (Laskowski et al., 1996; Liang et al., 1998c). These include pockets and voids, which can be accurately computed as we have discussed. Computationally, one wishes to automatically identify voids and pockets on protein structures where interactions exist with other molecules such as substrates, ions, ligands, or other proteins.

Binkowski et al. developed a method for predicting protein function by matching a surface pocket or void on a protein of unknown or undetermined function to the pocket or void of a protein of known function (Binkowski et al., 2003a, 2005). Initially, the Delaunay tetrahedrization and alpha shapes for almost all of the structures in the PDB databank are computed (Binkowski et al., 2003b). All surface pockets and interior voids for each of the protein structures are then exhaustively computed (Edelsbrunner et al., 1998; Liang et al., 1998b). For each pocket and void, the residues forming the wall are then concatenated to form a short sequence fragment of amino acid residues, while ignoring all intervening residues that do not participate in the formation of the wall of the pocket or void. Two sequence fragments, one from the query protein and another from one of the proteins in the database, both derived from pocket or void surface residues, are then compared using dynamic programming. The similarity score for any observed match is assessed for statistical significance using an empirical randomization model constructed for short sequence patterns.

For promising matches of pocket/void surfaces showing significant sequence similarity, we can further evaluate their similarity in shape and in relative orientation. The former can be obtained by measuring the coordinate root-mean-square distance (RMSD) between the two surfaces. The latter is measured by first placing a unit sphere at the geometric center $z_0 \in \mathbb{R}^3$ of a pocket/void. The location of each residue $z = (x, y, z)^T$ is then projected onto the unit sphere along the direction of the vector from the geometric center: $\boldsymbol{u} = (z - z_0)/||z - z_0||$. The projected pocket is represented by a collection of unit vectors located on the unit sphere, and the original orientation of residues in the pocket is preserved. The RMSD distance of the two sets of unit vectors derived from the two pockets are then measured, which is called the oRMSD for *orientation* RMSD (Binkowski et al., 2003a). This allows similar pockets with only minor conformational changes to be detected (Binkowski et al., 2003a).

The advantage of the method of Binkowski et al. is that it does not assume prior knowledge of functional site residues, and does not require *a priori* any similarity in either the full primary sequence or the backbone fold structures. It has no limitation in the size of the spatially derived motif and can successfully detect patterns small and large. This method has been successfully applied to detect similar functional surfaces among proteins of the same fold but low sequence identities, and among proteins of different fold (Binkowski et al., 2003a, 2004).

Function prediction through models of protein surface evolution: To match local surfaces such as pockets and voids and to assess their sequence similarity, an effective scoring matrix is critically important. In the original study of Binkowski et al., BLOSUM matrix was used. However, this is problematic, as BLOSUM matrices were derived from analysis of precomputed large quantities of sequences, while the information of the particular protein of interest has limited or no influence. In addition, these precomputed sequences include buried residues in the protein core, whose conservation reflects the need to maintain protein stability rather than to maintain protein function. In Tseng and Liang (2005, 2006), a continuous time Markov process was developed to explicitly model the substitution rates of residues in binding pockets. Using a Bayesian Markov chain Monte Carlo method, the residue substitution rates at functional pockets are estimated. The substitution rates are found to be very different for residues in the binding site and residues on the remaining surface of proteins. In addition, substitution rates are also very different for residues in the buried core and residues on the solvent-exposed surfaces.

These rates are then used to generate a set of scoring matrices of different time intervals for residues located in the functional pocket. Application of protein-specific and region-specific scoring matrices in matching protein surfaces results in significantly improved sensitivity and specificity in protein function prediction (Tseng and Liang, 2005, 2006).

In a large-scale study of predicting protein functions from structures, a subset of 100 enzyme families are collected from the total of 286 enzyme families containing between 10 and 50 member protein structures with known Enzyme Classification (E.C.) labels. By estimating the substitution rate matrix for residues on the active site pocket of a query protein, a series of scoring matrices of different evolutionary time is derived. By searching for similar pocket surfaces from a database of 770,466 pockets derived from the CASTp database (with the criterion that each must contain at least 8 residues), this method can recover active site surfaces on enzymes similar to that on the query structure at an accuracy of >92%. Figure 6.8 shows the Receiver Operating Characteristics curve of this study. An example of identifying human amylase using template surfaces from *B. subtilis* and from barley is shown in Fig. 6.9.

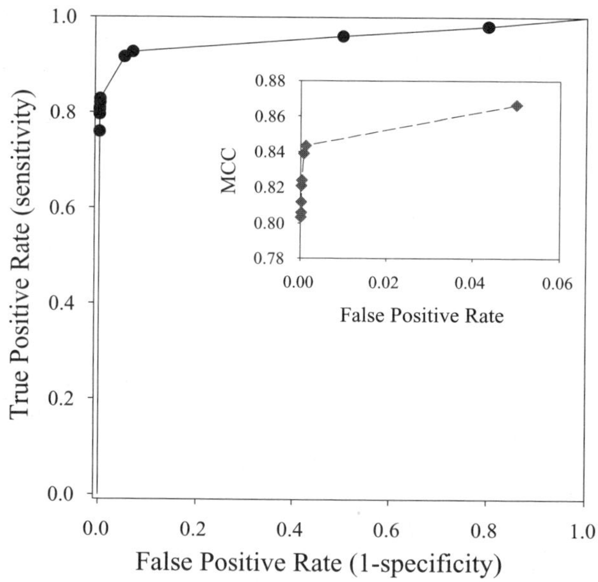

Fig. 6.8 Large scale protein function inference. Results by matching similar functional surfaces for 100 protein families. A correct prediction is made if the matched surface comes from a protein structure with the same Enzyme Classification (E.C.) number (upto the 4-th digit) as that of the query protein. The x-axis of the Receiver Operating Characteristics (ROC) curve reflects the false positive rate (1-specificity) at different statistical significance p-value by the cRMSD measurement, and the y-axis reflects the true positive rate (sensitivity). The inset shows the curve for Mathews's coefficients.

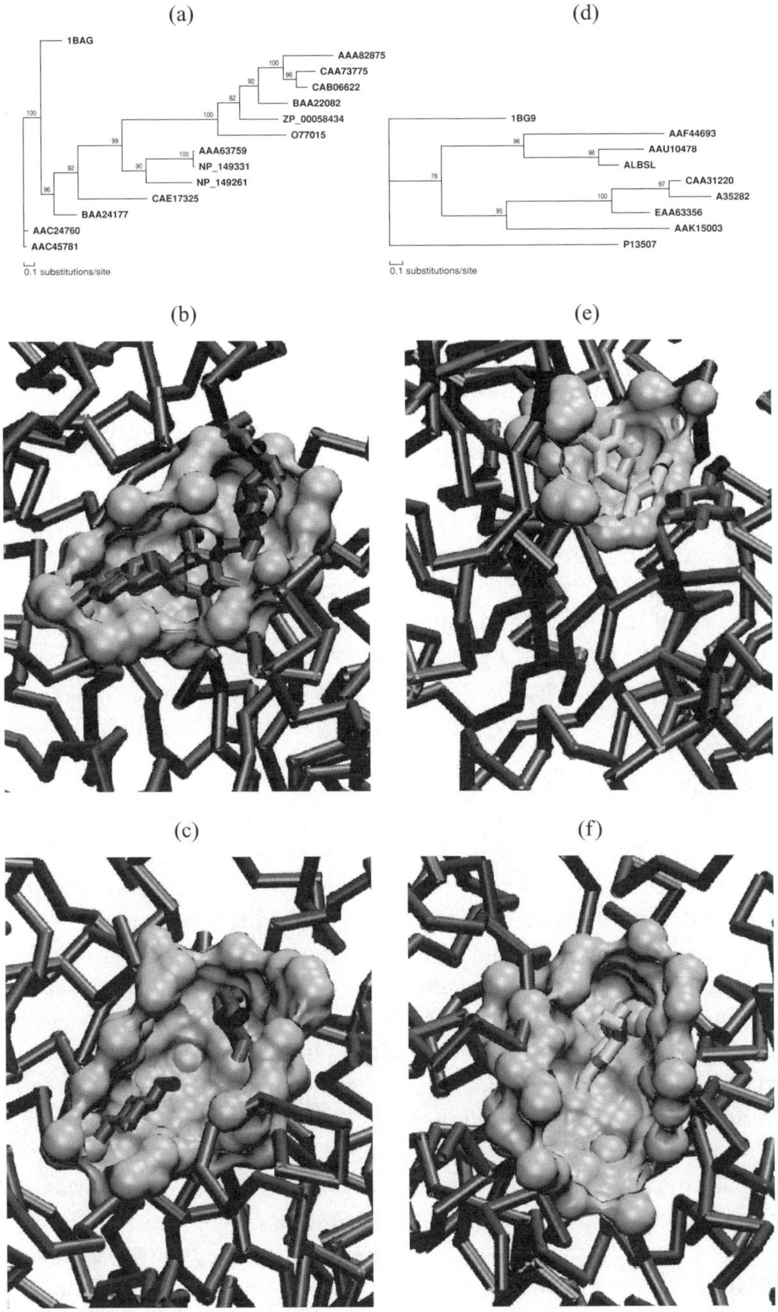

Fig. 6.9 Protein function prediction as illustrated by the example of alpha amylases. Two template binding surfaces are used to search database of protein surfaces to identify protein structures that are of similar functions. (a) The phylogenetic tree for the template PDB structure 1bag from *B. subtilis*. (b) The template binding pocket of alpha amylase on 1bag. (c) A matched binding surface on a different protein structure (1b2y from human, full sequence identity 22%) obtained by querying with 1bag. (d) The phylogenetic tree for the template structure 1bg9 from *H. vulgare*. (e) The template binding pocket on 1bg9. (f) A matched binding surface on a different protein structure (1u2y from human, full sequence identity 23%) obtained by querying with 1bg9. (Adapted from Tseng and Liang, 2006.)

The method of surface matching based on evolutionary model is also especially effective in solving the challenging problems of protein function prediction of orphan structures of unknown function (such as those obtained in structural genomics projects), which have only sequence homologues that are themselves hypothetical proteins with unknown functions.

6.6 Discussion

A major challenge in studying protein geometry is to understand our intuitive notions of various geometric aspects of molecular shapes, and to quantify these notions with mathematical models that are amenable to fast computation. The advent of the union of the ball model of protein structures enabled rigorous definition of important geometric concepts such as solvent-accessible surface and molecular surface. It also led to the development of algorithms for area and volume calculations of proteins. Deep understanding of the topological structure of molecular shapes is also based on the idealized union of ball model (Edelsbrunner, 1995). A success in approaching these problems is exemplified in the development of the pocket algorithm (Edelsbrunner et al., 1998). Another example is the recent development of a rigorous definition of protein–protein binding or interaction interface and algorithm for its computation (Ban et al., 2004).

Perhaps a more fundamental problem we face is to identify important structural and chemical features that are the determinants of biological problems of interest. For example, we would like to know what are the shape features that have significant influences on protein solvation, protein stability, ligand-specific binding, and protein conformational changes. It is not clear whether our current geometric intuitions are sufficient, or are the correct or the most relevant ones. There may still be important unknown shape properties of molecules that elude us at the moment.

An important application of geometric computation of protein structures is to detect patterns important for protein function. The shape of local surface regions on a protein structure and their chemical texture are the basis of its binding interactions with other molecules. Proteins fold into specific native structure to form these local regions for carrying out various biochemical functions. The geometric shape and chemical pattern of the local surface regions, and how they change dynamically are therefore of fundamental importance in computational studies of proteins.

Another important application is the development of geometric potential functions. Potential functions are important for generating conformations, for distinguishing native and near-native conformations from other decoy conformations in protein structure predictions (Singh et al., 1996; Zheng et al., 1997; Li et al., 2003; Li and Liang, 2005b) and in protein–protein docking (Li and Liang, 2005a). They are also important for peptide and protein design (Li and Liang, 2005a; Hu et al., 2004). Chapter 3 describes in detail the development of geometric potential and applications in decoy discrimination and in protein–protein docking prediction.

We have not described in detail the approach of studying protein geometry using graph theory. In addition to side-chain pattern analysis briefly discussed earlier, graph-based protein geometric model also has led to a number of important insights, including the optimal design of model proteins formed by hydrophobic and polar residues (Kleinberg, 2004), and methods for optimal design of side-chain packing (Xu, 2005; Leaver-Fay et al., 2005). Another important topic we did not touch upon is the analysis of the topology of protein backbones. Based on concepts from knot theory, Røgen and Bohr developed a family of global geometric measures for protein structure classification (Røgen and Bohr, 2003). These measures originate from integral formulas of Vassiliev knot invariants. With these measures, Røgen and Fain further constructed a system that can automatically classify protein chains into folds (Røgen and Fain, 2003). This system can reproduce the CATH classification system that requires explicit structural alignment as well as human curation.

Further development of descriptions of geometric shape and topological structure, as well as algorithms for their computation will provide a solid foundation for studying many important biological problems. The other important tasks are then to show how these descriptors may be effectively used to deepen our biological insights and to develop accurate predictive models of biological phenomena. For example, in computing protein–protein interfaces, a challenging task is to discriminate surfaces that are involved in protein binding from other nonbinding surface regions, and to understand in what fashion this depends on the properties of the binding partner protein.

Undoubtedly, evolution plays central roles in shaping the function and stability of protein molecules. The method of analyzing residue substitution rates using a continuous time Markov model (Tseng and Liang, 2005, 2006), and the method of surface mapping of conservation entropy and phylogeny (Lichtarge et al., 1996; Glaser et al., 2003) only scratch the surface of this important issue. Much remains to be done in incorporating evolutionary information in protein shape analysis for understanding biological functions.

6.7 Summary

The accumulation of experimentally solved molecular structures of proteins provides a wealth of information for studying many important biological problems. With the development of a rigorous model of the structure of protein molecules, various shape properties, including surfaces, voids, and pockets, and measurements of their metric properties can be computed. Geometric algorithms have found important applications in protein packing analysis, in developing potential functions, in docking, and in protein function prediction. It is likely further development of geometric models and algorithms will find important applications in answering additional biological questions.

6.8 Further Reading

The original work of Lee and Richards surface can be found in Lee and Richards (1971), where they also formulated the molecular surface model (Richards, 1985). Michael Connolly developed the first method for the computation of the molecular surface (Connolly, 1983). Tsai et al. described a method for obtaining atomic radii parameter (Tsai et al., 1999). The mathematical theory of the union of balls and alpha shape was developed by Herbert Edelsbrunner and colleague (Edelsbrunner, 1995; Edelsbrunner and Mücke, 1994). Algorithm for computing weighted Delaunay tetrahedrization can be found in Edelsbrunner and Shah (1996), or in a concise monograph with in-depth discussion of geometric computing (Edelsbrunner, 2001). Details of area and volume calculations can be found in Edelsbrunner et al. (1995), and Liang et al. (1998a,b). The theory of pocket computation and applications can be found in Edelsbrunner et al. (1998), and Liang et al. (1998c). Richards and Lim offered a comprehensive review on protein packing and protein folding (Richards and Lim, 1994). A detailed packing analysis of proteins can be found in Liang and Dill (2001). The study on inferring protein function by matching surfaces is described in Binkowski et al. (2003a). The study of the evolutionary model of protein binding pocket and its application in protein function prediction can be found in Tseng and Liang (2006).

Acknowledgments

This work is supported by grants from the National Science Foundation (CAREER DBI0133856), the National Institutes of Health (GM68958), the Office of Naval Research (N000140310329), and the Whitaker Foundation (TF-04-0023). The author thanks Jeffrey Tseng for help in preparing this chapter.

References

Artymiuk, P.J., Poirrette, A.R., Grindley, H.M., Rice, D., and Willett, P. 1994. A graph-theoretic approach to the identification of three-dimensional patterns of amino acid side-chains in protein structure. *J. Mol. Biol.* 243:327–344.

Bader, R. 1994. *Atoms in Molecules: A Quantum Theory*. London, Oxford University Press.

Ban, Y., Edelsbrunner, H., and Rudolph, J. 2004. Interface surfaces for protein–protein complexes. In: RECOMB, pp. 205–212.

Binkowski, T.A., Adamian, L., and Liang, J. 2003a. Inferring functional relationship of proteins from local sequence and spatial surface patterns. *J. Mol. Biol.* 332:505–526.

Binkowski, T., Freeman, P., and Liang, J. 2004. pvSOAR: Detecting similar surface patterns of pocket and void surfaces of amino acid residues on proteins. *Nucleic Acids Res.* 32:W555–W558.

Binkowski, T., Joachimiak, A., and Liang, J. 2005. Protein surface analysis for function annotation in high-throughput structural genomics pipeline. *Protein Sci.* 14:2972–2981.

Binkowski, T.A., Naghibzadeh, S., and Liang, J. 2003b. CASTp: Computed atlas of surface topography of proteins. *Nucleic Acids Res.* 31:3352–3355.

Bondi, A. 1964. VDW volumes and radii. *J. Phys. Chem.* 68:441–451.

Chakravarty, S., Bhinge, A., and Varadarajan, R. 2002. A procedure for detection and quantitation of cavity volumes in proteins. Application to measure the strength of the hydrophobic driving force in protein folding. *J. Biol. Chem.* 277:31345–31353.

Connolly, M.L. 1983. Analytical molecular surface calculation. *J. Appl. Crystallogr.* 16:548–558.

Crippen, G.M., and Havel, T.F. 1988. *Distance Geometry and Molecular Conformation*. New York, John Wiley & Sons.

Edelsbrunner, H. 1995. The union of balls and its dual shape. *Discrete Comput. Geom.* 13:415–440.

Edelsbrunner, H. 2001. *Geometry and Topology for Mesh Generation*. London, Cambridge University Press.

Edelsbrunner, H., Facello, M., Fu, P., and Liang, J. 1995. Measuring proteins and voids in proteins. In *Proc. 28th Ann. Hawaii Int. Conf. System Sciences*. Los Alamitos, CA, IEEE Computer Scociety Press, Vol. 5, pp. 256–264.

Edelsbrunner, H., Facello, M., and Liang, J. 1998. On the definition and the construction of pockets in macromolecules. *Discrete Appl. Math.* 88:18–29.

Edelsbrunner, H., and Mücke, E. 1994. Three-dimensional alpha shapes. *ACM Trans. Graphics* 13:43–72.

Edelsbrunner, H., and Shah, N. 1996. Incremental topological flipping works for regular triangulations. *Algorithmica* 15:223–241.

Finney, J.L. 1975. Volume occupation, environment and accessibility in proteins. The problem of the protein surface. *J. Mol. Biol.* 96:721–732.

Fischer, D., Norel, R., Wolfson, H., and Nussinov, R. 1993. Surface motifs by a computer vision technique: Searches, detection, and implications for protein–ligand recognition. *Proteins Struct. Funct. Genet.* 16:278–292.

Gellatly, B.J., and Finney, J. 1982. Calculation of protein volumes: An alternative to the Voronoi procedure. *J. Mol. Biol.* 161:305–322.

Gerstein, M., and Richards, F.M. 1999. *Protein Geometry: Distances, Areas, and Volumes*. International Union of Crystallography, Vol. F, Chapter 22.

Glaser, F., Pupko, T., Paz, I., Bell, R., Shental, D., Martz, E., and Ben-Tal, N. 2003. Consurf: Identification of functional regions in proteins by surface-mapping of phylogenetic information. *Bioinformatics* 19:163–164.

Goede, A., Preissner, R., and Frommel, C. 1997. Voronoi cell: New method for allocation of space among atoms: Elimination of avoidable errors in calculation of atomic volume and density. *J. Comput. Chem.* 18:1113–1123.

Guibas, L., and Stolfi, J. 1985. Primitives for the manipulation of general subdivisions and the computation of Voronoi diagrams. *ACM Trans. Graphics* 4:74–123.

Harpaz, Y., Gerstein, M., and Chothia, C. 1994. Volume changes on protein folding. *Structure* 2:641–649.

Holm, L., and Sander, C. 1997. New structure: Novel fold? *Structure* 5:165–171.

Hu, C., Li, X., and Liang, J. 2004. Developing optimal nonlinear scoring function for protein design. *Bioinformatics* 20:3080–3098.

Kleinberg, J. 2004. Efficient algorithms for protein sequence design and the analysis of certain evolutionary fitness landscapes. In RECOMB, pp. 205–212.

Kratky, K.W. 1981. Intersecting disks (and spheres) and statistical mechanics. I. Mathematical basis. *J. Stat. Phys.* 25:619–634.

Laskowski, R.A., Luscombe, N.M., Swindells, M.B., and Thornton, J.M. 1996. Protein clefts in molecular recognition and function. *Protein Sci.* 5:2438–2452.

Leaver-Fay, A., Kuhlman, B., and Snoeyink, J. 2005. An adaptive dynamic programming algorithm for the side chain placement problem. *Pac. Symp. Biocomput.* pp. 17–28.

Lee, B., and Richards, F.M. 1971. The interpretation of protein structures: Estimation of static accessibility. *J. Mol. Biol.* 55:379–400.

Levitt, M., Gerstein, M., Huang, E., Subbiah, S., and Tsai, J. 1997. Protein folding: The endgame. *Annu. Rev. Biochem.* 66:549–579.

Li, X., Hu, C., and Liang, J. 2003. Simplicial edge representation of protein structures and alpha contact potential with confidence measure. *Proteins* 53:792–805.

Li, X., and Liang, J. 2005a. Computational design of combinatorial peptide library for modulating protein–protein interactions. *Pac. Symp. Biocomput.* pp. 28–39.

Li, X., and Liang, J. 2005b. Geometric cooperativity and anticooperativity of three-body interactions in native proteins. *Proteins* 60:46–65.

Liang, J., and Dill, K.A. 2001. Are proteins well-packed? *Biophys. J.* 81:751–766.

Liang, J., Edelsbrunner, H., Fu, P., Sudhakar, P.V., and Subramaniam, S. 1998a. Analytical shape computing of macromolecules I: Molecular area and volume through alpha-shape. *Proteins* 33:1–17.

Liang, J., Edelsbrunner, H., Fu, P., Sudhakar, P.V., and Subramaniam, S. 1998b. Analytical shape computing of macromolecules II: Identification and computation of inaccessible cavities inside proteins. *Proteins* 33:18–29.

Liang, J., Edelsbrunner, H., and Woodward, C. 1998c. Anatomy of protein pockets and cavities: Measurement of binding site geometry and implications for ligand design. *Protein Sci.* 7:1884–1897.

Lichtarge, O., Bourne, H., and Cohen, F. 1996. An evolutionary trace method defines binding surfaces common to protein families. *J. Mol. Biol.* 257:342–358.

Martin, A.C.R., Orengo, C.A., Hutchinson, E.G., Michie, A.D., Wallace, A.C., Jones, M.L., and Thornton, J.M. 1998. Protein folds and functions. *Structure* 6:875–884.

Norel, R., Fischer, D., Wolfson, H.J., and Nussinov, R. 1994. Molecular surface recognition by a computer vision-based technique. *Protein Eng.* 7:39–46.

Orengo, C.A., Todd, A.E., and Thornton, J.M. 1999. From protein structure to function. *Curr. Opin. Struct. Biol.* 9:374–382.

Petitjean, M. 1994. On the analytical calculation of van der Waals surfaces and volumes: Some numerical aspects. *J. Comput. Chem.* 15:507–523.

Rhodes, G. 1999. *Crystallography Made Crystal Clear: A Guide for Users of Macromolecular Models*. San Diego, Academic Press.

Richards, F.M. 1974a. The interpretation of protein structures: Total volume, group volume distributions and packing density. *J. Mol. Biol.* 82:1–14.

Richards, F.M. 1974b. The interpretation of protein structures: Total volume, group volume distributions and packing density. *J. Mol. Biol.* 82:1–14.

Richards, F.M. 1977. Areas, volumes, packing, and protein structures. *Annu. Rev. Biophys. Bioeng.* 6:151–176.

Richards, F.M. 1985. Calculation of molecular volumes and areas for structures of known geometries. *Methods Enzymol.* 115:440–464.

Richards, F.M., and Lim, W.A. 1994. An analysis of packing in the protein folding problem. *Q. Rev. Biophys.* 26:423–498.

Richmond, T.J. 1984. Solvent accessible surface area and excluded volume in proteins: Analytical equations for overlapping spheres and implications for the hydrophobic effect. *J. Mol. Biol.* 178:63–89.

Rieping, W., Habeck, M., and Nilges, M. 2005. Inferential structure determination. *Science* 309:303–306.

Røgen, P., and Bohr, H. 2003. A new family of global protein shape descriptors. *Math. Biosci.* 182:167–181.

Røgen, P., and Fain, B. 2003. Automatic classification of protein structure by using Gauss integrals. *Proc. Natl. Acad. Sci. USA* 100:119–124.

Russell, R. 1998. Detection of protein three-dimensional side-chain patterns: New examples of convergent evolution. *J. Mol. Biol.* 279:1211–1227.

Singh, R.K., Tropsha, A., and Vaisman, I.I. 1996. Delaunay tessellation of proteins: Four body nearest-neighbor propensities of amino-acid residues. *J. Comput. Biol.* 3:213–221.

Todd, A.E., Orengo, C.A., and Thornton, J.M. 2001. Evolution of function in protein superfamilies, from a structural perspective. *J. Mol. Biol.* 307:1113–1143.

Tsai, J., Taylor, R., Chothia, C., and Gerstein, M. 1999. The packing density in proteins: Standard radii and volumes. *J. Mol. Biol.* 290:253–266.

Tseng, Y., and Liang, J. 2005. Estimating evolutionary rate of local protein binding surfaces: A Bayesian Monte Carlo approach. *Proc. 2005 IEEE EMBC Conf.*

Tseng, Y., and Liang, J. 2006. Estimation of amino acid residue substitution rates at local spatial regions and application in protein function inference: A Bayesian Monte Carlo approach. *Mol. Biol. Evo.* 23:421–436.

Wallace, A.C., Borkakoti, N., and Thornton, J.M. 1997. TESS: A geometric hashing algorithm for deriving 3d coordinate templates for searching structural databases. Application to enzyme active sites. *Protein Sci.* 6:2308–2323.

Word, J., Lovell, S., Richardson, J., and Richardson, D. 1999. Asparagine and glutamine: Using hydrogen atom contacts in the choice of side-chain amide orientation. *J. Mol. Biol.* 285:1735–1747.

Xu, J. 2005. Rapid protein side-chain packing via tree decomposition. In RECOMB, pp. 423–439.

Zhang, J., Chen, R., Tang, C., and Liang, J. 2003. Origin of scaling behavior of protein packing density: A sequential Monte Carlo study of compact long chain polymers. *J. Chem. Phys.* 118:6102–6109.

Zheng, W., Cho, S.J., Vaisman, I.I., and Tropsha, A. 1997. A new approach to protein fold recognition based on Delaunay tessellation of protein structure. In *Pacific Symposium on Biocomputing'97*, (R. Altman, A. Dunker, L. Hunter, and T. Klein, Eds.). Singapore, World Scientific, pp. 486–497.

7 Local Structure Prediction of Proteins

Victor A. Simossis and Jaap Heringa

7.1 Introduction

Protein architecture represents a complex and multilayered hierarchy (Fig. 7.1; Crippen, 1978; Rose, 1979). It starts from a linear chain of amino acid residues (primary structure) that arrange themselves in space to form local structures (secondary structure and supersecondary structure) and extends up to the globular three-dimensional structure of a fully functional folded protein (tertiary and quaternary structure).

This chapter focuses on how the physicochemical properties of the primary structure enable the prediction of the local structural features of a protein, in particular how the secondary (Section 7.2) and supersecondary structure (Section 7.3) of a protein can be predicted from sequence and how disordered regions (Section 7.4) and sequence repeats (Section 7.5) can be detected. In addition, the application of the prediction of these local structures in other fields such as multiple sequence alignment (MSA) (Section 7.6) and tertiary structure prediction (Section 7.7) is discussed. In Section 7.8, a number of currently available software packages that perform these tasks are presented and described in detail. Section 7.9 presents a collection of resources for protein local structure prediction, including online software and databases, with pointers to where they can be used or downloaded.[1] Section 7.10 gives a summary of the chapter's most important points.

7.2 Protein Secondary Structure Prediction

A secondary structure element is a section of consecutive residues in a protein sequence that corresponds to a local region in the associated protein tertiary structure and shows distinct geometrical features. The two basic secondary structure types, *the α-helix* and *β-strand*, are regular and easily distinguishable in protein tertiary structures (Appendix 1: Biological and Chemical Basics Related to Protein Structures), while other types are sometimes harder to classify. For this reason, the majority of secondary structure prediction methods use a three-class alphabet for their

[1] The quoted Internet addresses (URLs) were valid at the time this chapter was written, but may be subject to change in the future.

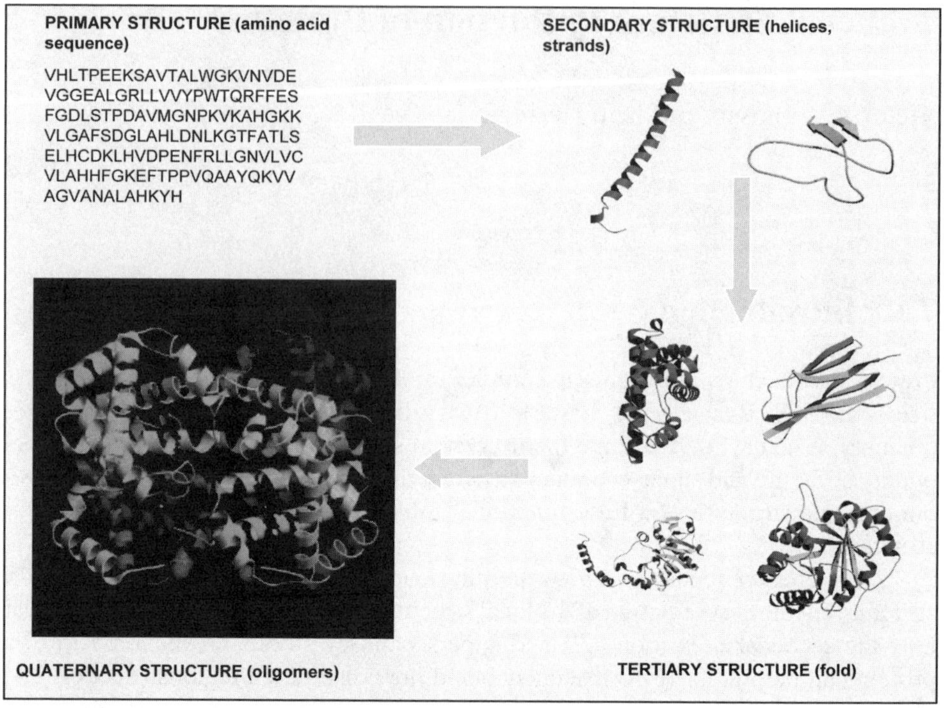

Fig. 7.1 Protein hierarchical classification

predictions: α-helix (H), β-strand (E), and other; the latter are often referred to as *coil* (C).

Approximately 50% of the amino acids in all known proteins are associated with either α-helices or β-strands, while on average the remaining half of protein secondary structure is irregular. The primary reason for the regularity observed in helices and strands is the innate polar nature of the protein backbone, which comprises a polar nitrogen and oxygen atom in each peptide bond between two successive amino acid residues. For a protein to become foldable with an acceptable internal energy, the parts of the backbone buried in the internal protein core need to form hydrogen bonds between these polar atoms. The α-helix and β-strand conformations are optimal for this, since each nitrogen atom can associate with an oxygen partner (and vice versa) within and between both secondary structure types. However, in order to satisfy the hydrogen-bonding constraints, β-strands need to interact with other β-strands, which they can do in a parallel or antiparallel fashion to form a β-*pleated sheet.* As a result, β-strands depend on crucial interactions between residues that are remotely situated in the sequence and therefore are believed to have more pronounced context dependencies than α-helices. Consequently, most prediction methods have greatest difficulty in predicting β-strands correctly.

7.2.1 Biochemical Features of Secondary Structures Used in Prediction

Analyses of secondary structure and related features of the many protein structures deposited in the Protein Data Bank (PDB) (Berman et al., 2000) have resulted in a set of rules about α-helices, β-strands, and coil structures that are important for secondary structure prediction. Most prediction methods, either implicitly or explicitly, make use of these observations when performing their predictions.

7.2.1.1 α-Helices

Considering that ideally one turn of the helical structure is made up of 3.6 residues, the minimum predicted length for an α-helix should be three or four residues. Also, α-helices are often positioned against a buried protein core and have one phase contacting core hydrophobic amino acids, while the opposite phase interacts with the solvent. This results in so-called amphipathic helices (Schiffer and Edmundson, 1967), which show an alternating pattern of three to four hydrophobic residues followed by three to four hydrophilic residues. As an additional rule, proline residues are rare in middle segments as they disrupt the α-helical turn, while they are more frequent in the first two positions of the structure.

7.2.1.2 β-Strands

Normally, two or more β-strands constitute a β-pleated sheet with two strands forming either edge. The hydrophobic nature of such edge strands is different from that of strands that are positioned inside the sheet because they are shielded on both sides. As side chains of constituent residues along a β-strand alternate the direction in which they protrude, edge strands of a β-sheet can show an alternating pattern of hydrophobic–hydrophilic residues, while buried strands typically comprise hydrophobic residues only. The β-strand is the most extended conformation (i.e., consecutive Cα atoms are farthest apart), so that it takes relatively few residues to cross the protein core with a strand. Therefore, the number of residues in a β-strand is usually limited and can be anything from two or three amino acids. Further, β-strands can be disrupted by single residues that induce a kink in the extended structure of the backbone. Such so-called β-bulges consist of relatively hydrophilic residues.

7.2.1.3 Coil Structures

Multiple alignments of protein sequences often display gapped and/or highly variable regions, which would be expected to correspond to loop (coil) regions rather than the other two basic secondary structures. Loop regions contain a high proportion of small polar residues like alanine, glycine, serine, and threonine. Glycine and proline residues are also seen in loop regions, the former due to their inherent flexibility, and the latter for entropic reasons relating to the observed rigidity in their kinking the backbone.

7.2.2 Secondary Structure Prediction: The Beginning

The use of computers to predict protein secondary structure started over 30 years ago (Nagano, 1973). All computational methods devised early on based their predictions on single protein sequences and the average prediction accuracy lingered for a long time in the range between 50 and 60% correctness, i.e., 50–60% of the residues used for predictions were correctly assigned a secondary structure class H, E, or C (Schulz, 1988). A random prediction would yield about 40% correctness given the observed distribution of the three states in globular proteins, i.e., 30% α-helix, 20% β-strand, and 50% coil. Although significantly beyond the random level, the accuracy of the early prediction methods was not sufficient to allow the successful prediction of protein topology, i.e., the folded structural arrangement of protein secondary structures.

The pioneering algorithms of Nagano (Nagano, 1973) and Chou and Fasman (Chou and Fasman, 1974) were aimed at predicting the secondary structure for single sequences and relied on a statistical treatment of compositional information. Lim's method (Lim, 1974) represented the first attempt to incorporate stereochemical rules in prediction. The method relied mainly on conserved hydrophobic patterns in secondary structures such as amphipathicity in helices (Schiffer and Edmundson, 1967). The early and popular GOR method (Garnier et al., 1978; Gibrat et al., 1987) considered the influence and statistics of flanking residues on the conformational state of a selected amino acid to be predicted. The early methods by Nagano (Nagano, 1973), Lim (Lim, 1974), Chou-Fasman (Chou and Fasman, 1974) and the GOR method (Garnier et al., 1978; Gibrat et al., 1987) were reported to perform single sequence secondary structure prediction with accuracies of 50, 54, 56, and 64.4% (GOR IV; Garnier et al., 1996), respectively.

7.2.3 From Early to Recent Prediction: The Key Advances

The first important breakthrough for secondary structure prediction was the use of multiple sequence alignment (MSA) information (Dickerson et al., 1976), which was incorporated into an automatic prediction method for the first time by Zvelebil et al. (1987). The use of the evolutionary information stored in an MSA of a family of homologous proteins, as opposed to using a single sequence, is essential for more accurate predictions and as a result, all current state-of-the-art secondary structure prediction methods use MSAs.

Second, the use of increasingly sensitive machine-learning techniques made the translation process of the evolutionary information in MSAs more accurate. Since the 1990s, methods have employed various complex decision-making techniques including neural networks (NNs), *k*-Nearest-Neighbor analysis (kNN), Example-Based Learning (EBL), Hidden Markov Models (HMMs), and Support Vector Machines (SVMs). An overview of these techniques is provided in Section 7.2.5.

The third element that allowed secondary structure prediction methods to rapidly advance was the dramatic increase in protein sequence and structure data,

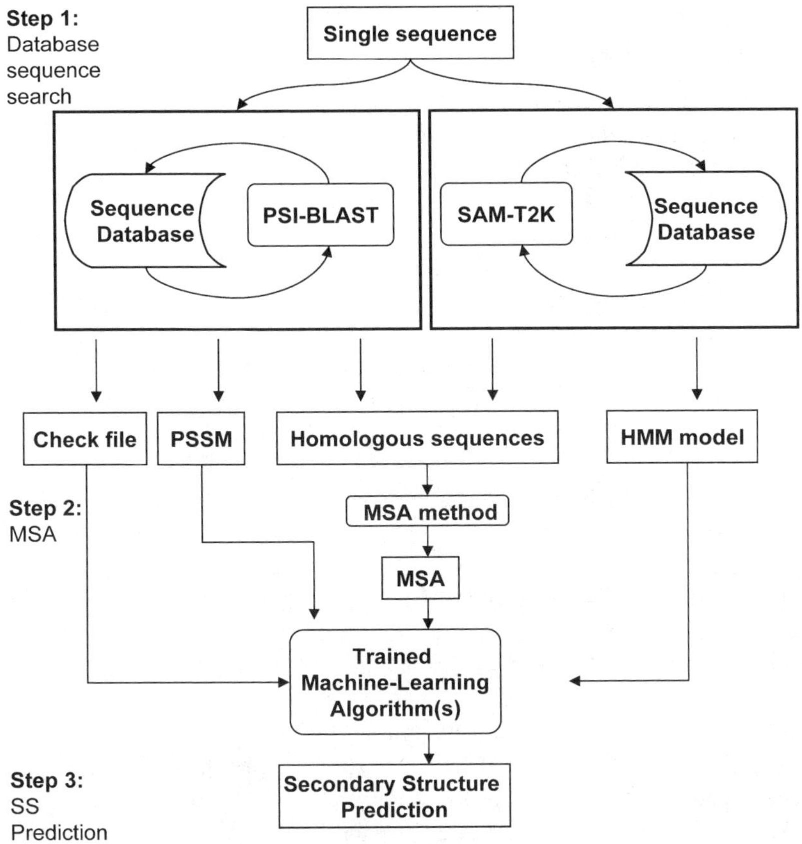

Fig. 7.2 The currently employed three-step process that leads to secondary structure prediction. Step 1: sequence database searching (here we show the currently top methods PSI-BLAST and SAM-T2K). Step 2: multiple sequence alignment (MSA) of the selected sequences either in the possible output formats of the database search methods or by separately employed MSA methods. Step 3: secondary structure prediction based on one of the MSA types of Step 2.

combined with the enhanced sensitivity of automatic database searching tools (Altschul et al., 1997; Friedberg et al., 2000). This allowed the correct identification of more divergent homologues and subsequently the creation of larger structural family profiles that encapsulate more divergent information. In addition, this increase in information also allowed the training of machine-learning algorithms on larger data sets, resulting in higher method accuracy and sensitivity.

As a result, the three standard steps used by almost all current secondary structure prediction methods are: (1) detecting homologues from a database for the sequence to be used as input, (2) aligning these sequences, and (3) using the position-specific information in the MSA to predict the secondary structure of the input sequence (Fig. 7.2).

7.2.4 Multiple Sequence Alignment Becomes a Secondary Structure Prediction Standard

Obtaining the sequences to compile an MSA is done in two main ways: either the MSA is made from already selected homologous sequences or a database homology search engine is used with the query sequence as input to identify homologous sequences in sequence databases. In the latter case, an MSA method is then used to align the query and homologous sequences.

Since the successful first use of the PSI-BLAST database search tool (Altschul et al., 1997; Altschul and Koonin, 1998) in the prediction method PSIPRED (Jones, 1999), most newly developed, but also older prediction methods [such as PHD (Rost and Sander, 1993) that was updated to PHDpsi (Przybylski and Rost, 2002)], have followed in the same footsteps and use PSI-BLAST to produce their input MSAs. More details about how individual prediction methods do this are discussed later in Section 7.8.

PSI-BLAST is an iterative database search technique that searches a preset sequence database [e.g., the protein sequence database SWISS-PROT (Bairoch and Boeckmann, 1991); for recent update see Boeckmann et al. 2003)], in multiple separate iterative steps. The resulting PSI-BLAST MSA, also expressed as a position-specific scoring matrix (PSSM), does not represent a global alignment, but typically contains information from the most similar fragment of each selected homologous sequence.

Concurrent to the PSIPRED innovation, HMMs (Section 7.2.5.3) were introduced as a means to search a preset sequence database for the identification of more divergent sequence homologues in the iterative SAM-T99 method (Karplus et al., 1998, 1999). The difference between this method and PSI-BLAST is that the query sequence is aligned against sequences in the database using HMMs. In addition, an HMM model of the generated MSA can be used instead of a profile (PSSM) to iterate the search or search other databases. Initially, the alignments produced by the SAM-T99 search engine were better in that they more accurately detected remote homologues compared to other search engines including PSI-BLAST (Park et al., 1998). However, at present both search engines perform equally well due to greatly reduced unrelated-hit contaminations in PSI-BLAST (Schaffer et al., 2001). At the moment, the search engine of SAM-T99 has been updated to a newer version SAM-T2K (Karplus et al., 2001), which also incorporates secondary structure information in its scoring. SAM-T2K is an integral part of the SAM-T02 (Karplus et al., 2002, 2003) structure prediction method (Section 7.8).

In more recent years, the detection of homologies between distant sequences has been significantly improved through profile–profile local alignment (Rychlewski et al., 2000; Ginalski et al., 2003, 2004; Mittelman et al., 2003; von Ohsen et al., 2003, 2004; Capriotti et al., 2004; Edgar and Sjolander, 2004; Tomii and Akiyama, 2004; Wang and Dunbrack, 2004; Soding, 2005). In these latter approaches, single sequence input is enriched with homologous position-specific information, mainly by using PSI-BLAST. This enriched information can be represented as either a profile or an HMM and two profiles or HMMs or a combination of the two can be aligned using different profile–profile scoring schemes. Recent comparison studies of such scoring schemes (Rychlewski et al., 2000; Edgar and Sjolander, 2004; Ohlson et al., 2004) suggest that the scoring scheme based on information theory used in *prof_sim* (Yona and Levitt, 2002) is the most sensitive. However, comparing HMM profile pairs appears to produce even better results (Soding, 2005). Many of these profile comparison methods have now moved on to the incorporation of predicted secondary structure information into their profile-scoring schemes to further sensitize the detection of homologies (Ginalski et al., 2003, 2004; Chung and Yona, 2004; von Ohsen et al., 2004; Soding, 2005). A more comprehensive account of this intertwining of sequence alignment and secondary structure prediction is discussed in Section 7.6.

To date, these new-generation homology detection methods have not been explicitly applied to the secondary structure prediction problem, but attempts have been made for the *ab initio* prediction of tertiary structure (von Ohsen et al., 2004).

7.2.5 State-of-the-Art Secondary Structure Prediction Techniques

The most popular machine learning approaches used in current secondary structure prediction methods include *k*-nearest-neighbor analysis, artificial neural networks, hidden Markov models, and support vector machines. Each technique offers unique advantages and also has associated drawbacks in tackling complex problems such as pattern recognition, which for our purpose is the identification of structural classes from consecutive residue patterns. In the descriptions to follow, we give a basic overview of each technique and discuss their strengths and weaknesses.

7.2.5.1 *k*-Nearest-Neighbor

The *k*-nearest-neighbor (kNN) technique is an instance-based machine learning technique. In order to predict the secondary structure of a protein, for which the structure is unknown, the technique extrapolates from already existing information of related proteins. As a result, the performance of this approach is directly dependent on whether closely related examples of known secondary structure are available. Assuming enough "related" information is available, kNN has distinct advantages over other methods: it is easy to program and can deal with complex problems using low-complexity approximations; it can deal with noisy data; it involves no training or retraining with new data and never loses information content because all learning material is explicitly used every time the method is run.

The prediction main steps involve the creation of a library of protein "fragments" of known secondary structure, the creation of a distance representation scheme for relating the library fragments to the query and a decision scheme for discerning between multiple matching possibilities. The various *k*-nearest-neighbor prediction methods approach these steps differently. As a simple example, let *X* be a protein sequence for which we want to do a prediction (Fig. 7.3). Let our example

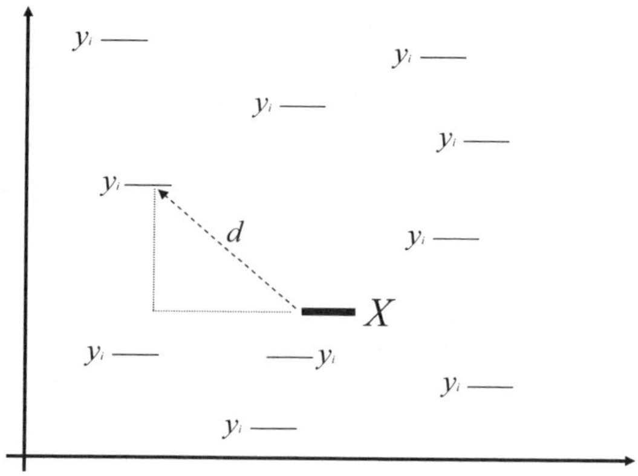

Fig. 7.3 The kNN approach to classifying the secondary structure of a sequence fragment based on a database of other fragments with known secondary structures. The fragment X under consideration in this case is represented by a thick black line and the surrounding lines represent the database sequences being assessed for relatedness (long arrows correspond to small relatedness).

method have access to a large, nonredundant database of variable-length protein sequence fragments (y_i) with their corresponding secondary structures. Our method would take sequence X and use it to scan the database for related fragments (neighbors). Now, let our measure of "relatedness" be the local alignment score between our sequence X and each fragment y_i. The higher the score, the more related the fragment is and therefore the lower the Euclidean distance d from X. After identifying potential neighboring fragments, we would sort them and use only the k nearest ones for the prediction to minimize errors and processing time, as long as we ensured that k allowed good coverage across the whole length of X. Finally, for all sequence positions where more than one possible secondary structure was present, our decision scheme could be a "majority vote" consensus, where the most prominent secondary structure is assigned. Here, each possibility could be further weighed in relation to the fragment's distance from X, making the closest fragments count more than less related ones. The string of these decisions would be our prediction. Again, at this point we could apply a filter to "tidy up" the prediction, for example correcting impossible structures such as a single-residue helix. In any case, the possibilities are many and this was merely intended as a guide example for how a kNN prediction works.

kNN methods outperform classical neural network (NN) methods when closely related examples to a query are available, but their success is highly restricted as they perform poorly in all other cases. The huge increase in data availability has provided the kNN approaches with larger, more diverse sets of examples to train on, thus increasing the space in which they accurately perform. Nonetheless, the data sets that are currently available are still far from covering the entire protein universe. As a result, the NN approach is still the best way to predict secondary structure in a wide range of cases. As a solution to this, methods have been developed over

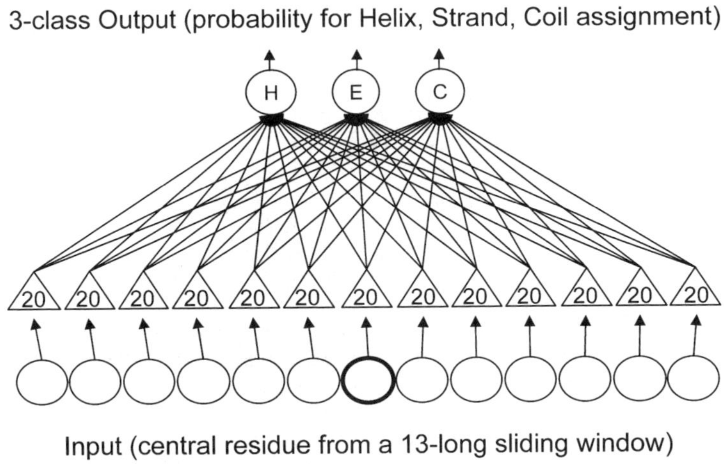

Fig. 7.4 A generic schematic representation for secondary structure prediction NN using a window of 13 residues. The number 20 in the hidden layer positions represents the 20 amino acid possibilities for each position of the 13-long window with respect to the central residue. The path through the network outputs a value for the central residue being either helix (H), strand (E), or coil (C).

the years that attempt to combine the best of both worlds, an example of which we describe in Section 7.8 [APSSP2 (Raghava, 2002a)].

7.2.5.2 Neural Networks

NNs are complex machine-learning systems that are based on nonlinear statistics. They consist of multiple interconnected layers of input and output units, and can also contain intermediate (or "hidden") unit layers (for a review, see Minsky and Papert, 1988). Each unit in a layer receives information from one or more other connected units and determines its output signal based on the weights of the input signals (Fig. 7.4). The weights of an NN are chosen depending on the training procedure and the training set. The training procedure is done by adjusting the weights of the internal connections to optimize the grouping of a set of input patterns into a set of output patterns. In other words, an NN tries to encapsulate the basic trends of the training set (usually a large number of nonredundant examples) and apply them to unknown cases. NNs are powerful learning tools, but there is a risk of overtraining the network, which leads to proper recognition of those patterns the NN has been confronted with during training, but much less successful recognition of patterns that have not been seen. For this reason, training sets must be large in number and nonredundant so they can capture a representative sample and thus decrease their bias toward specific cases. More importantly, when testing a trained NN, the test set must be absolutely separate from the training set and as divergent as possible so that the testing is objective and as unbiased as possible. NNs are very common in secondary structure prediction and are used by all top-performing

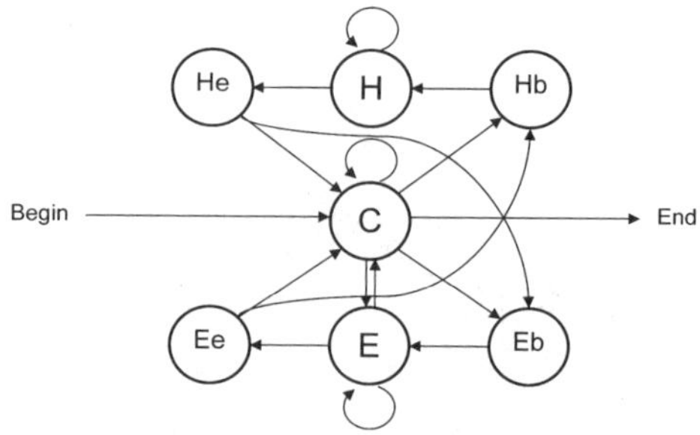

Fig. 7.5 An HMM for secondary structure prediction. He and Ee are helix and strand end positions, respectively; Hb and Eb are helix and strand beginning positions; and H and E are all other helix and strand positions. The original figure from Lin et al. (2005) is reproduced with permission from Bioinformatics, Oxford University Press.

methods, whether in combination with other systems, like YASPIN (Lin et al., 2005) or APSSP2 (Raghava, 2002a), or on their own (more on specific NN methods in Section 7.8).

7.2.5.3 Hidden Markov Models

HMMs are a class of probabilistic models usually applied to time series or linear sequences (for reviews see Eddy, 1996; Durbin, 1998; Durbin et al., 2000). They were first introduced to Bioinformatics in the 1980s (Churchill, 1989) and have been applied as protein profile models in the last decade (Krogh et al., 1994). The basic structure of an HMM is a series of *states* that are linked together through *state transitions*. Each of these states also has a symbol emission probability distribution for generating a symbol in the alphabet. For example, let us consider a sequence modeling HMM that describes three possible amino acid states, according to what secondary structure element (SSE) they are in (Fig. 7.5). In any point after the HMM is initialized, we are in a state X (helix, strand or coil) and have the possibility of either switching to a different state Y or remaining in the same state X. The decision for this is governed by the relation between state transition probability from state X to state Y, and from state X to X. In addition, when the transition is made the HMM will emit (generate) a character from the alphabet (in this case the SSE symbols H, E, or C) with the probability linked to that state. This process is repeated until an end state is reached. In the end there are two layers in our HMM, a hidden *state sequence* that we do not see and a *symbol sequence* that we do see.

An HMM can be parameterized by either training or building procedures. In the training procedure of a sequence–structure prediction HMM, a set of unaligned sequences would be used, while in the building procedure, a set of prealigned sequences would be used. It is generally advisable to build HMMs whenever there is reference information possible.

The HMMs that are used in sequence database searching and structure prediction are profile HMMs. A profile HMM is a strictly linear, left-to-right model that comprises a series of nodes, each corresponding to a column in a multiple alignment

(Krogh et al., 1994). Each node has a match state, insert state, and delete state. Each sequence uses a series of these states to traverse the model from start to end. Using a match state indicates that the sequence has a character in that column, while using a delete state indicates that the sequence does not. Insert states allow sequences to have additional characters between columns. In many ways, these models correspond to profiles. The primary advantage of these models over standard methods of sequence search is their ability to characterize an entire family of sequences.

7.2.5.4 Support Vector Machines

The SVM, first introduced by Vladimir Vapnik in 1992, is a linear learning machine based on recent advances in statistical theory (Vapnik, 1995, 1998) In other words, the main function of SVMs is to classify input patterns by first being trained on labeled data sets (supervised learning). SVMs have been shown to be a significant enhancement in function compared to other commonly used machine learning algorithms such as the perceptron algorithm (see Section 7.2.5.2, Neural Networks) and have been applied to many areas such as handwriting, face, voice, and object recognition and text characterization (for a comprehensive description of SVMs see Cristianini and Shawe-Taylor, 2000). With the turn of the millennium, SVMs were extensively applied to classification and pattern recognition problems in bioinformatics (for reviews see Byvatov and Schneider, 2003; Noble, 2004).

The power of SVMs lies in their use of nonlinear kernel (similarity) functions. When a linear algorithm such as the SVM uses a dot product, replacing it with a nonlinear kernel function allows it to operate in different space. Hence, the kernel functions used in SVMs implicitly map the input (training or test data) into high-dimensional feature spaces. In the high-dimensional feature spaces, linear classifications of the data are possible (each classifier is a separate dimension); they become nonlinear in following steps where they are transformed back to the original input space. As a result, although SVMs are linear learning machines regarding the high-dimensional feature spaces, in fact they act as nonlinear classifiers.

The key is to carefully design the kernel (similarity) criteria during training so as to best discriminate each class (for more information on kernels used in computational biology see Schoelkopf et al., 2004).Ultimately, the kernel function generates a maximum-margin hyperplane between two classes and resides somewhere in space (Fig. 7.6). For example, if we were training an SVM for helix prediction, given training examples labeled either "helix" or "nonhelix," our kernel function would generate a maximum-margin hyperplane that would split the "helix" and "nonhelix" training examples so that the distance from the closest examples (the margin) to the hyperplane would be maximized (Fig. 7.6). If the hyperplane is not able to fully separate the "helix" and "nonhelix" examples, the SVM will choose a hyperplane that splits the examples as cleanly as possible, while still maximizing the distance to the nearest cleanly split examples. The parameters of the maximum-margin hyperplane are derived by solving a quadratic programming (QP) optimization problem. The examples closest to the hyperplane (decision boundary) are "support vectors,"

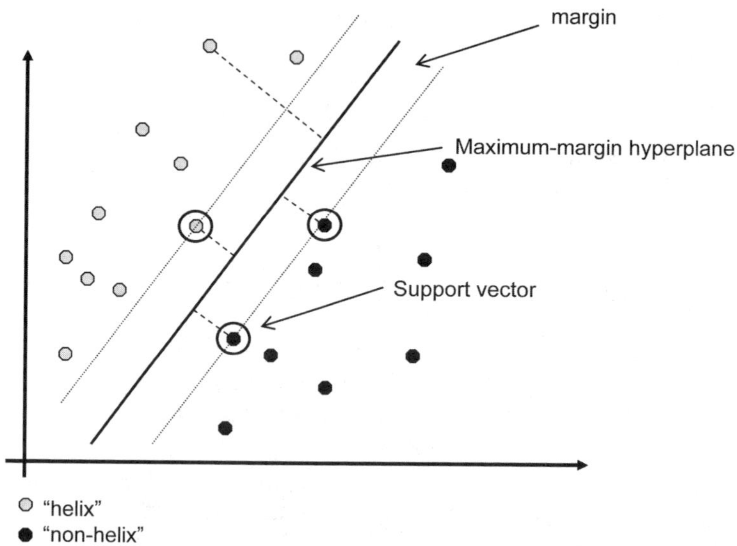

Fig. 7.6 SVM diagram illustrating an optimal separation of "helix" (red dots) and "nonhelix" (black dots) elements showing the position of the hyperplane.

while the ones far from it have no effect (Fig. 7.6). After training, any "unknown" input for which we want to decide whether it is helix or not is mapped into the high-dimensional space and the SVM decides whether it is "helix" or "nonhelix." However, since secondary structure elements are usually classified in three states [helix (H), strand (E), and coil (C)], the actual recognition challenge is not binary (helix or nonhelix), but multiclass and therefore the prediction is still incomplete. The multiclass recognition problem is tackled differently across SVM prediction methods (Hua and Sun, 2001; Kim and Park, 2003; Ward et al., 2003; Guo et al., 2004; Hu et al., 2004), an example of which is described in Section 7.8.

7.2.6 Consensus Secondary Structure Prediction

The majority of secondary structure prediction methods are trained using information from proteins of known 3D structure. In modern studies, training is performed on large data sets, thus avoiding overfitting, and the training data sets do not include any of the proteins used to assess the final version of the method (jack-knife testing). However, each method is trained on different sets of proteins and as a consequence this introduces a bias to the prediction performance, depending on the type of proteins used in the training set.

An early attempt to minimize these biasing effects was to combine predictions from various methods to produce a single consensus (Cuff et al., 1998; Cuff and Barton, 1999). The consensus was derived by majority voting, where the per-residue predicted states from each method were each given an equal "vote" and the consensus kept the prediction that got the majority of the "votes." The philosophy of deriving a consensus prediction is similar to that of having three clocks on a boat: if one clock shows the wrong time there are always the other two to check for consistency and since the probability that two out of three clocks will go wrong at the same time and in a similar way is very low, it is a safe assumption to go with the majority. During the same time, other strategies for consensus prediction were developed such as the combination of different NN outputs (Chandonia and Karplus, 1999; Cuff and Barton, 2000; King et al., 2000; Petersen et al., 2000); optimal method choice for the consensus scheme by linear regression statistics (Guermeur et al., 1999) and decision trees (Selbig et al., 1999); deriving a consensus from cascaded multiple secondary structure classifiers (Ouali and King, 2000); and expressing the consensus as a composite predicted secondary structure, where the variation in prediction is not resolved but used as extended information for the successive database searching steps for fold recognition (An and Friesner, 2002).

From these consensus-deriving strategies, the "majority voting" consensus-deriving scheme has been employed in recent investigations using more state-of-the-art predictions methods and the results have consistently shown that a consensus prediction is better than any of the single predictions produced by the methods used for deriving the consensus (Albrecht et al., 2003; McGuffin and Jones, 2003; Ward et al., 2003). Recently, an extension to the "majority voting" dimensionality and segmentation capabilities was introduced by using dynamic programming (DP) to produce an optimally segmented consensus in an investigation of the effect of alignment gaps in secondary structure prediction (Simossis and Heringa, 2004a). The DP approach for generating a consensus has also very recently been applied to β-barrel protein prediction (Bagos et al., 2005) and the generation of a consensus from multiple secondary structure prediction methods (Simossis and Heringa, 2005). In Fig. 7.7 we illustrate the relation between the "majority voting" and DP approaches. The original "majority voting" strategy is illustrated as a small part of the DP strategy, such that the window length of 1 (first row of the search matrix) represents the original "majority voting" and the remaining window lengths represent the added information used for the derivation of the consensus.

7.2.7 Tertiary Structure Feedback for Secondary Structure Prediction

In the prediction techniques we have described up to now, predicting the secondary structure of a protein from its amino acid sequence has mainly involved using adjacent information. However, when a protein folds, the secondary structure elements that were initially formed can be influenced by the dynamics of formerly distant regions, which now have been brought closer due to the structural rearrangement in three-dimensional space (Blanco et al., 1994; Ramirez-Alvarado et al., 1997; Reymond et al., 1997). Although many initial conformations remain unchanged in the folded protein, there are regions that undergo transitions from one SSE type to

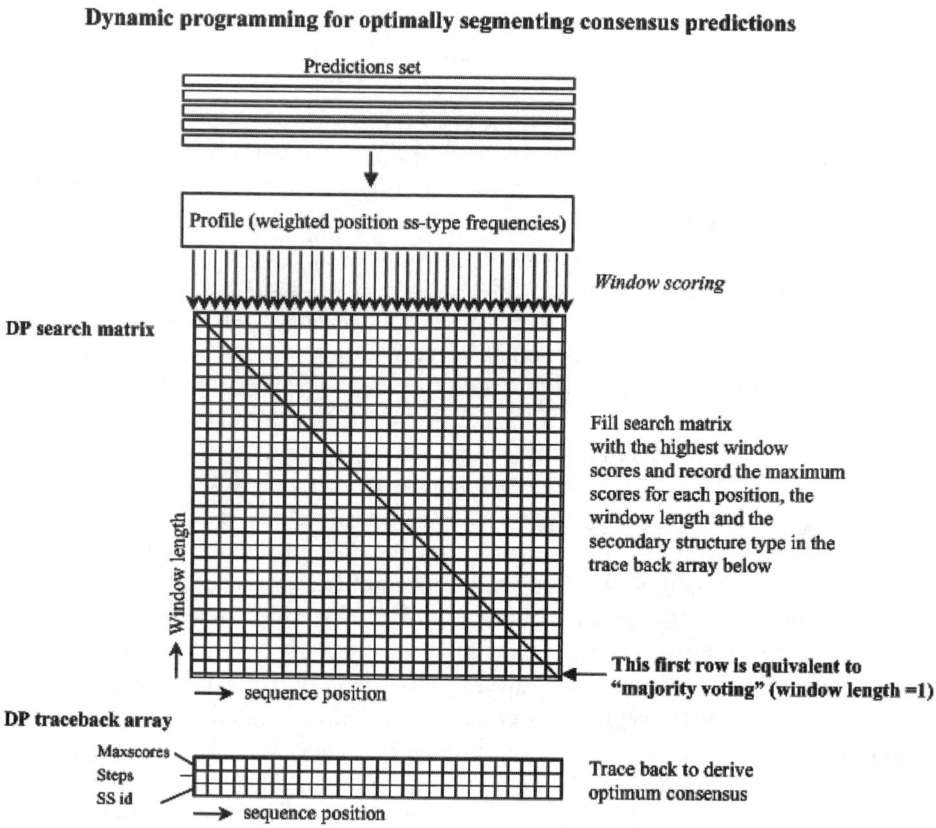

Fig. 7.7 The dynamic programming optimal segmentation strategy for deriving a consensus secondary structure prediction from multiple methods. The majority voting approach is limited to the first row of the search matrix, while the use of all possible segmentations of the information allows further optimization by dynamic programming.

another as a result of different types of interactions (Minor and Kim, 1996; Cregut et al., 1999; Luisi et al., 1999; Derreumaux, 2001; Macdonald and Johnson, 2001). As a result, even the best prediction methods make wrong predictions for these cases because the transition changes only happen as a result of tertiary structure interactions and have not yet occurred in the unfolded state.

Meiler and Baker (2003) used low-resolution tertiary structure models to feed back three-dimensional information to the predictions and successfully raised the quality of the predictions, particularly in β-strands (Meiler and Baker, 2003). However, the applicability of the method is limited since it is only applicable to single-domain proteins and is not able to account for interdomain interactions.

In another approach, surface turns that change the overall direction of the chain ("U" turns) were predicted using multiple alignments and predicted secondary structure propensities to improve the quality of the predictions (Hu et al., 1997; Kolinski et al., 1997).

7.3 Protein Supersecondary Structure Prediction

As a globular protein folds, different regions of the peptide backbone often come together (Wetlaufer, 1973; Unger and Moult, 1993). The compact combinations of two or more adjacent β-strand and α-helical structures, irrespective of the sequence similarity, form frequently recurring structural motifs that are known as supersecondary structures (Rao and Rossmann, 1973). Almost two-thirds of all residues in secondary structures are part of some type of supersecondary structure motif (Salem et al., 1999) and one-third of all known proteins can be classified into ten super folds, which are made up of different combinations between three basic supersecondary structures: the α-hairpin, the β-hairpin, and the β-α-β motif (Salem et al., 1999).

The three basic types of supersecondary structure that are the most frequently observed include the α-α motifs (α-hairpins, α-corners, and helix-turn-helix), the β-β motif (β-hairpin), and the β-α-β motif (two parallel β-strands, separated by an α-helix antiparallel to them, with two hairpins separating the three secondary structures). Other simple combinations of secondary structure types include the α-β-β and β-β-α motifs (Chothia, 1984). Some repetitions or combinations of the above simple supersecondary structures are also predominant in protein structures, such as the β-β-β (β-meander), which is formed by two β-hairpins sharing the middle strand. More elaborate supersecondary structure combination motifs include the *Greek key* (jellyroll) motif (Hutchinson and Thornton, 1993), the *four-helix bundle* (two α-α units connected by a loop), and the *Rossman fold* (effectively two β-α-β units that each form one-half of an open twisted parallel β-sheet). Although the majority of protein folds consist of several supersecondary structures, they can also be constituted by secondary structures in other contexts. An example of the latter is the *globin* fold, six of whose helices cannot be assigned to any of the aforementioned α-α supersecondary structures.

An interesting α-α motif is formed when a pair of α-helices adopt a superhelical twist, resulting in a coiled-coil conformation. The usual left-handed coiled-coil interaction involves a repeated motif of seven helical residues (*abcdefg*), where the *a* and *d* positions are normally occupied by hydrophobic residues constituting the hydrophobic core of the helix–helix interface, while the other positions display a high likelihood to comprise polar residues. Another feature is that the heptad *e* and *g* positions are often charged and can form salt bridges.

An example of a more complex and higher-order supersecondary structure is the WD repeat (tryptophan-aspartate repeat), which is associated with a sequence motif approximately 31 amino acids long that encodes a structural repeat and usually ends with tryptophan-aspartic acid (WD). WD-repeat-containing proteins are thought to contain at least four copies of the WD repeat because all WD-repeat proteins are speculated to form a circular β-propeller structure. This is demonstrated by the crystal structure of the G protein β-subunit, which is the only WD-repeat-containing structure available. It contains seven WD repeats, each of which folds into a small antiparallel β-sheet. WD-repeat proteins have critical roles in many essential biological functions ranging from signal transduction, transcription regulation, to apoptosis, but are probably best known due to their association with several human diseases.

7.3.1 Fundamentals of Supersecondary Structure Prediction

The identification of supersecondary structure motifs is largely, but not entirely, based on secondary structure prediction, as already discussed in Section 7.2. However, secondary structure information is a flat version of the protein structure, in contrast to supersecondary structure motifs that are recurring three-dimensional units that are formed when a protein folds. As a result, observing a pattern in secondary structure, e.g., strand–coil–strand, does not necessarily mean that it signifies a β-hairpin supersecondary structure. On the contrary, such a basic secondary structure pattern could belong to a variety of different supersecondary structure motifs (Rost et al., 1997; de la Cruz and Thornton, 1999). Therefore, in order to extend from a collapsed secondary structure prediction to three-dimensional supersecondary structure prediction, a more detailed description of its properties is needed.

As described earlier, the secondary structure elements that make up supersecondary structure units are joined together by flexible regions that in the three-state classification of secondary structure are referred to as coil (C). Unlike the helix (H) and strand (E) classes, coil adopts a wide range of conformations and is able to change the protein backbone into the different supersecondary structure motifs. As a result, identifying the properties of joining coil regions between helices and strands is the key to identifying a supersecondary structure motif type.

7.3.2 Predicting Protein Supersecondary Structure

The methods that have been developed for the prediction of supersecondary structures have mainly employed machine learning techniques similar to those used in secondary structure prediction, namely, HMMs (Bystroff et al., 2000), NNs (Sun et al., 1997; Kuhn et al., 2004) and SVMs (Cai et al., 2003). In addition to these, some methods have also employed statistical regression techniques such as Monte Carlo-based simulations (Forcellino and Derreumaux, 2001) and a combination of secondary structure prediction and threading against known tertiary motifs (de la Cruz et al., 2002).

7.3.2.1 Neural Networks

Refer to Section 7.2.5.2 for an overview of NNs.

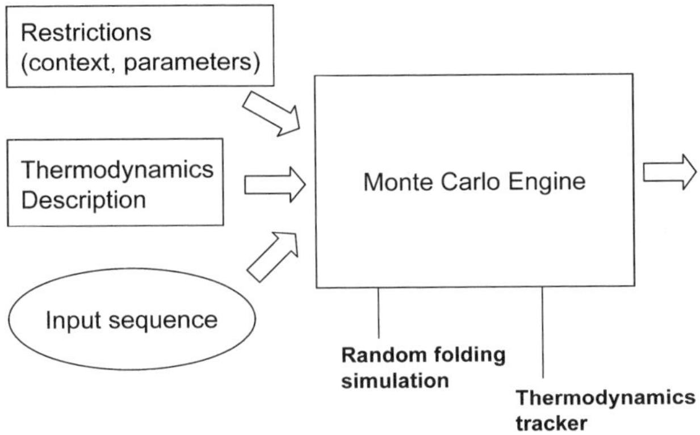

Fig. 7.8 A simplified representation of the Monte Carlo method for predicting supersecondary structure from sequence. This process is repeated enough times so that both the runtime and the accuracy are at acceptable levels.

7.3.2.2 Hidden Markov Models

Refer to Section 7.2.5.3 for an overview of HMMs.

7.3.2.3 Support Vector Machines

Refer to Section 7.2.5.4 for an overview of SVMs.

7.3.2.4 Monte Carlo Simulations

A Monte Carlo (MC) simulation is a stochastic technique, i.e. it uses random numbers and probability statistics to investigate problems (Fig. 7.8; for a comprehensive account see Frenkel and Smit, 2002). The invention of the MC method is often accredited to Stanislaw Ulam, a Polish mathematician who is primarily known for the design of the hydrogen bomb with Edward Teller in 1951. However, Ulam did not invent the concept of statistical sampling, but was the first to use computers to automate it. Together with John von Neumann and Nicholas Metropolis, he developed algorithms for computer implementations of the method, as well as means of transforming nonrandom problems into random forms that would facilitate their solution via statistical sampling. The method was first published in 1949 .()(Metropolis and Ulam, 1949). Nicholas Metropolis named the method after the casinos of Monte Carlo.

The strength and usefulness of MC methods is that they allow us to perform computations that would otherwise be impossible. For example, solving equations that describe the interactions between two atoms is fairly simple, but when attempting to solve the same equation for a fold or a whole protein (hundreds or thousands of atoms), the task is impossible. Basically, an MC simulation samples a large system in a number of random configurations. When selecting the number of random configurations, one way to minimize the standard error is to maximize the sample size. However, due to the fact that this will be computationally expensive, a better solution is to restrict the variance of the random sample. As a result, depending on the restrictions applied to the MC simulation, configurations are accepted or rejected. Standard techniques of variance reduction include antithetic variates, control variates, importance sampling, and stratified sampling (see Frenkel and Smit, 2002).

In the case of protein structures, atoms are randomly moved in a predefined space so that the thermodynamics of the protein in a folded state are respected. The energies of these randomly folded proteins are calculated and according to a predefined selection criterion, they are kept as candidate structures or thrown away. The resulting set of candidate structures generated from this random sampling can be used to approximate the proteins folded state. The same principles apply to supersecondary structure prediction, where only the knowledge of the primary structure of short peptides is needed to sample a number of feasible conformations (Derreumaux, 2001).

7.4 Protein Disordered Region Detection

Disordered regions are regions of proteins or entire proteins, which lack a fixed tertiary structure, essentially because they are partially or fully unfolded. Disordered regions have been shown to be involved in a variety of functions, including DNA recognition, modulation of specificity/affinity of protein binding, molecular threading, activation by cleavage, and control of protein lifetimes. In a recent survey, Dunker et al. (2002) classified the functions of approximately 100 disordered regions into four broad categories: molecular recognition, molecular assembly/disassembly, protein modification, and entropic chains, the latter including flexible linkers, bristles, and springs (Dunker et al., 2002).

Although disordered regions lack a defined 3D structure in their native states, they frequently undergo disorder-to-order transitions upon binding to their partners. As it is known that the amino acid sequence determines a protein's 3D structure, it is appropriate to assume that the amino acid sequence determines the lack of fixed 3D structure as well. Disordered proteins are found throughout the three kingdoms, but are predicted to be more common in eukaryotes than in archaea or eubacteria (Dunker et al., 2000). This would imply that intrinsic disorder is widespread but might be increasingly required for more complex protein functions.

Disordered proteins are gaining increased attention in the biological community (Wright and Dyson, 1999). Following the earlier work on protein folding by Ptitsyn (1994) on molten globule structures, native proteins have been divided in three folding states: ordered (fully folded), collapsed (molten globule-like), or extended (random coil-like). These three forms can occur in localized regions of proteins or comprise entire sequences. Protein function may arise from any of the three forms or from structural transitions between these forms resulting from changes in environmental conditions. The collapsed and extended forms correspond to intrinsic disorder while the fully folded, ordered form is generally comprised of three secondary structure types: α-helix, β-sheet, and coil. However, since the collapsed (molten globule-like) state is known to have secondary structure, the presence or absence of secondary structure cannot be used to distinguish between ordered and disordered proteins. Another feature of disordered proteins or protein regions is that these are intrinsically dynamic and thus have relative coordinates and Ramachandran angles that vary significantly over time, while those in ordered proteins generally are comparatively invariant over time.

For proteins whose X-ray structures are known, the existence of disordered stretches can be identified directly by looking for amino acids that are missing from the electron density maps. A number of disordered regions in proteins have been directly characterized by NMR-based structure elucidation (Bracken, 2001; Dyson and Wright, 2002). Other sources of experimental evidence for disorder include a random coil-type circular dichroism spectrum and an extended hydrodynamic radius, while also limited, time-resolved proteolysis can provide useful information (Dunker et al., 2001).

As the accumulated experimental evidence of disordered regions is still limited and likely to cover only a small fraction of these regions existing in nature, alternative information-based prediction approaches have been developed. In accordance with the hypothesis that a protein's structure and function are determined by its amino acid sequence, it is possible in principle to predict long stretches of 30 or more consecutive disordered residues from the primary structure. Distinguishing features for disordered regions include a higher average flexibility index value (Vihinen et al., 1994), a lower sequence complexity (Romero et al., 2001) as estimated by the popular NSEG method (Wootton and Federhen, 1996), a lower aromatic content (Xie et al., 1998), and different patterns regarding charge and hydrophobicity (Xie et al., 1998; Uversky et al., 2000). The state-of-the-art disorder prediction methods (Section 7.9.3) are thus generally based on the assumption that different types of disordered sequences are more similar to each other than to ordered sequences and vice versa, although protein regions in both the ordered and disordered class can display local variations in flexibility.

Young et al. (1999) successfully predicted regions likely to undergo structural change by using secondary structure prediction techniques. The authors examined protein regions for which secondary structure prediction methods gave equally strong preferences for two different states (Young et al., 1999). Such regions were then further processed combining simple statistics and expert rules. The final method was tested on 16 proteins known to undergo structural rearrangements, and on a number of other proteins. The authors reported no false positives, and identified most known disordered regions. The Young et al. method was further applied to the myosin family (Kirshenbaum et al., 1999), which led to the prediction of likely disordered regions that were previously unidentified, even though the tertiary structure of myosin was known.

In Section 7.9.3, a number of state-of-the-art methods are identified that have been developed especially for the prediction of protein disorder. These methods include the PONDR suite (Obradovic et al., 2003), and the methods FoldIndex (Uversky et al., 2000), DISEMBL (Linding et al., 2003a), GLOBPLOT (Linding et al., 2003b), DISOPRED2 (Ward et al., 2004), PDISORDER (unpublished), and DISpro (Cheng et al., 2005). The highest prediction accuracies reported are currently beyond 90% (Cheng et al., 2005).

7.5 Internal Repeats Detection

7.5.1 Genomic Repeats

An important characteristic of genomes, and particularly for those of eukaryotes, is the high frequency of internal sequence repeats. For example, the human genome is estimated to contain more than 50% of reiterated sequences (e.g., Heringa, 1998). One of the main evolutionary mechanisms for repeat duplication is recombination (Marcotte et al., 1999), which favors additional duplication after initial repeat copies have been made. In the case of tandem repeats, there is believed to be a pronounced correlation between copy number of repeats and further gene duplication (Heringa, 1994) due to gene slippage. Gene duplication can ease the selection pressure on an individual gene and thus lead to an accelerated divergence of the duplicated genes, thereby increasing the scope for evolution toward novel functions.

7.5.2 Protein Repeats

Given widespread duplication and rearrangement of genomic DNA and subsequent gene fusion events, also at the protein level internal sequence repeats are abundant and found in numerous proteins. Gene duplication may enhance the expression of an associated protein or result in a pseudogene where less stringent selection of mutations can quickly lead to divergence resulting in an improved protein. An advantage of duplication followed by gene fusion at the protein level is that the protein resulting from the new single gene complex shows a more complex and often symmetrical architecture, conferring the advantages of multiple, regulated and spatially localized functionality. Many protein repeats comprise regular secondary structures and form multirepeat assemblies in three dimensions of diverse sizes and functions. In general, internal repetition affords a protein enhanced evolutionary prospects due to an enlargement of its available binding surface area. Constraints on sequence conservation appear to be relatively lax, for example due to binding functions ensuing from multiple, rather than, single repeats. Repeat proteins often fulfill important cellular roles, such as zinc-finger proteins that bind DNA, the β-propeller domain of integrin α-subunits implicated in cell–cell and cell–extracellular matrix interactions, or titin in muscle contraction, which consists of many repeated Ig and Fn3 domains. It is interesting in this regard that Marcotte et al. (1999) estimated that eukaryotic proteins are three times more likely to have internal repeats than prokaryotic proteins. The similarities found within sets of internal repeats can be 100% in the case of identical repeats, down to the level where any discernible sequence similarity has been lost as a result of mutation and insertion/deletion events. A classical example of this is chymotrypsin, where fusion of two duplicated genes, each coding for a separate β-barrel domain, has resulted in a two-domain enzyme. The active site consists of amino acids of both domains and shows a greatly enhanced activity as compared to a suspected ancestral active center within an individual ancestral barrel (Heringa, 1994). The amino acid sequences of the two barrels have diverged so much that the duplication event had to be inferred from the structural similarity (McLachlan, 1979).

7.5.3 Protein Repeats Detection

Considerable sequence divergence as well as the short lengths of many sequence repeats imply that repeats detection can be a particularly arduous task. The problem of recognizing internal sequence repeats in proteins has been tackled by many researchers. One of the pioneers in the automatic detection of repeats was McLachlan, who devised the first methods over three decades ago (McLachlan, 1972). These methods relied on Fourier analysis (McLachlan and Stewart, 1976; McLachlan, 1977) and this technique remained popular (Kolaskar and Kulkarni-Kale, 1992; Taylor et al., 2002). Although Fourier transforms are designed to detect periodic behavior, the application to protein sequence signals is compromised by the fact that many repeats are distant as a result of mutations and insertions/deletions, and can be intervened by different irregular sequence stretches. Moreover, proteins can contain multiple repeat types, all with different base periodicities, which decrease the periodic signal for any one type. Finally, Fourier techniques require a relatively large number of repetitions, whereas many proteins contain only few repeats.

Another approach to delineate repeats in protein sequences was made by exploring DP. First attempts were made by McLachlan (1983) who used the DP technique over fixed window lengths on myosin rod repeats (McLachlan, 1983). Boswell and McLachlan (1984) elaborated the method by incorporating dampening factors and allowing the occurrence of gaps (Boswell and McLachlan, 1984). Argos (1987) also adopted the window technique but exploited physicochemical properties of amino acids in addition to the PAM250 residue exchange matrix (Dayhoff et al., 1983), and used the technique to detect repeats in, for example, frog transcription factor IIIA (TFIIIA), human hemopexin, and chick tropoelastin (Argos, 1987). Huang et al. (1990) used local alignments (Smith and Waterman, 1981) to find the repeats in rabbit globin genes (Huang et al., 1990). Their method SIM is a memory optimized implementation of the approach introduced by Waterman and Eggert (1987), which calculates a list of top-scoring nonintersecting local alignments, meaning that no alignment has a given matched amino acid pair in common.

Following these initial developments, a number of methods of delineating internal repeats in protein sequences were reported. These include the early and popular REPRO method (Heringa and Argos, 1993), the fast but inexact method of Pellegrini et al. (1999), the RADAR method (Heger and Holm, 2000), and the TRUST method (Szklarczyk and Heringa, 2004). Taylor et al. (2002) devised a method based on

Fourier analysis to automatically annotate repeating local 3D fragments in protein tertiary structures. These methods are discussed in more detail in Section 7.8.4. Links to various web interfaces to these methods are provided in Section 7.9.3.

7.6 Applications to Multiple Sequence Alignment

The prediction of protein local structure elements has increasingly infiltrated the field of sequence alignment in recent years. In this section we will discuss how structural features such as secondary structure, supersecondary structure, and repeats can enrich the information used in sequence-based alignment methods toward a more accurate detection of similarity.

7.6.1 Structure Is More Conserved Than Sequence

Most alignment methods, irrespective of whether they align two or more sequences, rely entirely on the residue information provided by the sequences they align. As would be expected, the detection of similarities between sequences becomes harder as the level of mutational change that has occurred through evolution increases (Rost et al., 1994). It has been known for many years that alignment quality suffers when the sequence identity of two sequences drops below 30%, the so-called "twilight zone".

Unlike primary structure, the higher structural levels are more conserved through evolution (Chothia and Lesk, 1986). The reason for this is that function is mostly connected to the structure of a protein rather than its residue composition. Therefore, although mutations may alter individual residues in a protein, the structure remains relatively unchanged so that functionality is not lost or inhibited. As a result, structure is a better candidate for detection of homology in distant relatives. Consequently, the use of structural information has been integrated into many alignment methods (Heringa, 1999, 2000, 2002; Ginalski et al., 2003; Chung and Yona, 2004; Ginalski et al., 2004; Simossis and Heringa, 2004b; von Ohsen et al., 2004; Soding, 2005). However, the number of known structures compared to the number of protein sequences remains limiting because the rate at which sequences are added to databases is much faster than that at which protein structures are solved. As a result, in the absence of a crystal structure, predictions of protein local structures can be used to fill the gap.

7.6.2 Integrating Predicted Local Structure Information into an Alignment

The integration of known or predicted secondary structure information into an alignment algorithm can be done in several ways. Early on, the approach simply involved increasing the gap penalties in helical or strand regions in order to bias the algorithm to insert gaps in between SSE regions (Sander and Schneider, 1991). In more recent approaches, in addition to the generalized exchange matrices, secondary-structure specific exchange matrices [e.g., the Lüthy series (Luthy et al., 1994)] have been used for scoring those sequence or profile positions that belong to the same secondary structure class (Heringa, 1999, 2000, 2002). Other researchers have combined the two types of matrices in different schemes (Yu et al., 1998; Hedman et al., 2002; Ginalski et al., 2003; Chung and Yona, 2004; Teodorescu et al., 2004).

In terms of other predicted local structure elements such as repeats and supersecondary structure, no systematic analysis has been done on how its incorporation might aid alignment quality. However, it is conceivable that the identification of repeats prior to alignment would greatly aid the correct positioning of repeated regions and avoid incorrect shifts of the alignment. Similarly and probably more importantly, the assignment of basic supersecondary structures to the known or predicted secondary structure would also greatly improve the alignment of sequences. At least, it would help discern between secondary structure patterns that although similar in two dimensions, do not actually fold the same way in the 3D structure and therefore may not align as tightly as would be assumed by current secondary structure-guided methods.

7.6.3 Local Structure Prediction and Alignment Interdependence

The majority of the current secondary structure-integrating strategies are limited to pairwise local alignment strategies implemented for homology detection (Ginalski et al., 2003, 2004; Chung and Yona, 2004; von Ohsen et al., 2004; Soding, 2005). Conversely, the use of predicted secondary structure to guide MSA has not been exhaustively investigated. Early on, Heringa used PREDATOR (Frishman and Argos, 1996, 1997) predictions to guide the alignments of the DP method PRALINE (Heringa, 1999) and found improvements in alignment quality when aligning 13 flavodoxins with cheY, a distant signal transduction protein that has very low sequence similarity but shares the same fold as the flavodoxins (Heringa, 2000). In this case, the secondary structure prediction program used did not depend on the MSA quality.

Later, Heringa (2002) also extended the MSA–secondary structure prediction interrelationship to an iterative scheme using SSPRED (Mehta et al., 1995), a more advanced MSA-dependent method of the time. In this scenario, an initial MSA is used for the prediction of the secondary structures of the sequences to be aligned and then these predictions are reintroduced to produce a new secondary structure-guided alignment. The new, more correct alignment is then used in the next iteration step to derive new, more accurate secondary structure predictions and so on. Simossis and Heringa have recently re-designed PRALINE (Heringa, 1999) to use the MSA-dependent secondary structure prediction methods PHD (Rost and Sander, 1993), PROFsec (Rost, personal communication), JNET (Cuff and Barton, 2000), and SSPro2 (Pollastri et al., 2002) in this iterative approach. Preliminary results show that for the alignment of the 13 flavodoxin sequences and cheY, the initial

PHD prediction for the most difficult sequence (cheY) is vastly improved by this iterative scheme.

7.7 Applications to Local Protein Tertiary Structure Prediction

Protein tertiary structure prediction is a vast and intense area of research. The ability to predict a protein's 3D structure from the amino acid sequence is one of the outstanding grand challenges in molecular biology, despite almost 40 years of computational research on the subject. A multitude of approaches have been attempted over the years to predict tertiary structure, ranging from simplified lattice models to full-scale energy-based atomic modeling using complex force fields. These can be grouped into two fundamentally different classes of methods to predict 3D structure from amino acid sequence. The first is *ab initio* prediction, which attempts to predict the folding of an amino acid sequence without any direct reference to other known protein structures. Computer-based calculations are employed that attempt to minimize the free energy of a structure with a given amino acid sequence or to simulate the folding process. The utility of these methods is limited by the vast number of possible conformations, the marginal stability of proteins, and the subtle energetics of weak interactions in aqueous solution. For a detailed account of *ab initio* prediction, see Chapter 13. The second group of methods takes advantage of our growing knowledge of 3D structures of proteins. In these *knowledge-based methods,* an amino acid sequence of unknown structure is examined for compatibility with any known protein structures. These techniques are also referred to as threading (see Chapter 12). If a significant match is detected, the known structure can be used as an initial model. Knowledge-based methods have led to many insights into the 3D conformation of proteins of known sequence but unknown structure. To date, the most reliable way to predict a protein structure is homology modeling, where the sequence of an unknown protein is aligned to another homologous protein sequence for which the tertiary structure is known. Typically, for those parts of the query sequence that are aligned with core secondary structures of the template structure, the backbone topologies of these structures are taken. It is clear that for this transfer of information the quality of the alignment between query and template sequence is crucial. A recent survey suggested that the recent improvements in scope and quality of comparative models largely come from the increased number of available protein sequences, resulting in better multiple sequence alignments (Cozzetto and Tramontano, 2005). Techniques have also been created to optimize the alignment of query and template sequence(s) by incorporating information from the template structure (e.g., Kleinjung et al., 2004). Two tasks then remain: one is to model the sidechains of the core elements, while the other is to model the loops connecting the core SSEs. Loop modeling has been defined as finding the ensemble of possible backbone structures, associated with the sequence segment corresponding to the loop, that are geometrically consistent with preceding and following parts of the loop whose 3D structures are given. The latter, also referred to as loop closure, is a complicated chore to achieve.

A vast number of folding experiments suggest that two conformational states are present to any significant extent, folded and unfolded. Such observations demonstrate that protein folding and unfolding result from a *cooperative transition* (Bystroff et al., 2000). The ultimate consequence of cooperativity is that if a protein is placed in conditions under which some part of the protein structure is rendered thermodynamically unstable, the interactions between it and the remainder of the protein will be lost. The loss of these interactions, in turn, will then destabilize the remainder of the structure. However, the conclusion that conditions leading to the disruption of any part of a protein structure will unravel the protein completely, cannot be generally maintained given the recent observations of natively disordered protein regions or even complete proteins (see Section 7.4).

Applications such as homology modeling or protein docking are based on the assumption that a protein's inner core is less prone to movement than surface residues. This notion is supported by the fact that within homologous families, variations of the basic 3D topology associated with a given family are normally located at loop regions, ranging from the extension of a loop by one or a few extra residues, via additional SSEs, to complete domain insertions.

The most important applications of local 3D structure modeling are recognizing and modeling protein ligand-binding sites (An et al., 2005) and protein–protein interaction (PPI) sites, where the ability to model the conformation of surface residues is a crucial issue. Particularly, PPI sites are notoriously difficult to predict (Bordner and Abagyan, 2005). Furthermore, the computational methods designed for these tasks are computationally intensive, such that web interfaces to available programs are largely absent.

Another use of local 3D protein structures is by using local segments as found in the PDB database of tertiary structures (Dutta and Berman, 2005). An important example of this is the Robetta server (Kim et al., 2004) for homology or *ab initio* modeling, which makes use of fragment libraries. Fragment libraries are the pieces of experimentally determined structures that Robetta uses to guide the search of conformational space when predicting structures using its *ab initio* protocol, as well as longer loop conformations in homology models.

Apart from improvements in force fields leading to enhanced and flexible docking approaches, further developments might come from new mesoscopic modeling approaches, in which protein structures are not described at the atomic level, but by means of mesoscopic quantities like the number of effective particles ("beads") in a polymer and an effective potential between these particles. Such approaches aim to be more computationally efficient, allowing genomic pipeline screening modes, while preserving or even enhancing accuracy.

Another avenue to future improvements will be to utilize the ever-growing genomic sequence database and exploit evolutionary comparison methods, bridging for example multiple alignment information and structural descriptions of known binding sites and/or ligands. In such a knowledge-based approach, protein–ligand and protein–protein interactions might be delineated in the absence of three-dimensional modeling scenarios.

7.8 Software Packages

In the following section we describe local structure prediction software packages that at the time this chapter was written were either available for use as a web service or downloadable for local use. The examples cover secondary and supersecondary structure prediction tools using all machine learning approaches discussed (Sections 7.9.1 and 7.9.2), disordered region (Section 7.9.3), and repeats detection (Section 7.9.4).

7.8.1 Secondary Structure Prediction

7.8.1.1 *k*-Nearest-neighbor

Software Package 1. PSSP/ APSSP/ APSSP2
In the original PSSP secondary structure prediction method (Raghava, 2000) the authors introduced the combination of an NN and a customised kNN technique on single sequence prediction. The principle behind this combination was that on the one hand, protein queries that had closely related examples of known secondary structure would get a better prediction using the kNN technique than solely using an NN, but on the other hand, in the event of example absence, the NN would provide a better prediction than the kNN approach. The combination of the predictions of the two techniques was based on per-residue state-specific probabilities of correct prediction calculated by the NN rather than a binary (one or the other) use of the techniques according to the query. In addition, the final prediction was further filtered using an extra NN, much like that operating in PHD, where single-residue strands and helices were corrected.

The customization of the kNN technique used in PSSP was first the extending of the existing database of known examples that had earlier been derived from 126 proteins (Raghava, 2000) to a much larger training dataset by using all of the proteins in the 1998 version of the PDB database (Berman et al., 2000). This way, the number of possible examples was greatly increased; leading to an increase in the number of proteins the technique could correctly handle. Second, due to the increase in examples, the authors developed a way to minimize the computational time of comparing the query to the example database, reporting an 800-fold increase in computational speed (Raghava, 2000).

The most recent versions of PSSP are APSSP (Raghava, 2002b) and APSSP2 (Raghava, 2002a). APSSP and APSSP2 are both three-step methods that use an MSA as input to a combined NN and Example-Based Learning (EBL) system. The main difference between the two is the way the initial step is carried out. In APSSP, the first step is performed automatically by the external secondary structure prediction method Jnet (Cuff and Barton, 2000). The Jnet method is described in detail later on in the consensus approach section. Conversely, in its first step APSSP2 generates an MSA using the PSIBLAST search tool (Altschul et al., 1997) and an initial prediction is produced using the same standard NN as PSSP (Raghava, 2000). In the second step, a customized EBL technique has replaced the kNN approach and is used to generate a second separate prediction. As in PSSP, in the third step, the secondary structures predicted from the first two steps are combined based on prediction reliability scores.

The PSSP and APSSP methods have now been replaced by APSSP2, which is available online as an automatic prediction server and as part of the EVA assessment server (Koh et al., 2003) (see Section 7.9.1).

7.8.1.2 Neural Networks

Software Package 1. PHD/PHDpsi/PROFsec

The secondary structure prediction method PHD (Rost and Sander, 1993) was the first algorithm to employ NNs and database searching. At a time when prediction was stuck in the high ends of 60%, it gave a groundbreaking boost to about 73% (Rost and Sander, 1993). The modus operandi of PHD was the search of the SWISS-PROT (Bairoch and Boeckmann, 1991; Boeckmann et al., 2003) database using the MAXHOM MSA method (Sander and Schneider, 1991). The resulting MSA was passed into the PHD three-layer network and generated its prediction [more details can be found in Rost and Sander (1993) and a review in Heringa (2000)]. In later years, PHD was updated to PHDpsi (Przybylski and Rost, 2002) that used the iterative homology search engine PSI-BLAST on the much larger BIG database, which is a nonredundant merge of the PDB (Berman et al., 2000), TrEMBL, and SWISS-PROT. More recently, although still unpublished, PHD has evolved even more and now takes advantage of bidirectional recurrent NNs (BRNNs) in PROFsec (Rost, personal communication).

All three PHD flavors can be used online as part of the Predict Protein Server (Rost, personal communication), which is one of the first members of the EVA assessment server (Koh et al., 2003) (see Section 7.9.1).

Software Package 2. PSIPRED

The PSIPRED method incorporates MSA information and NNs (Jones, 1999). The alignment information used is represented by a position-specific scoring matrix (PSSM) generated by the PSI-BLAST algorithm (Altschul et al., 1997; Altschul and Koonin, 1998) and is inputted to a two-layered NN.

The accuracy of the PSIPRED method is 76.5%, as evaluated by the author (Jones, 1999) and continues to rank among the best methods according to the EVA assessment server (Koh et al., 2003) (see Section 7.9.1).

Software Package 3. SSPro

SSPro is an NN prediction method that employs 11 BRNNs (bidirectional recurrent neural networks) to generate its predictions, instead of the commonly used feedforward networks. The first version of SSPro (Baldi et al., 1999) used BLAST (Altschul et al., 1997) to generate multiple alignments as input, while in the second and current SSPro version (Pollastri et al., 2002), multiple alignments of homologue sequences are obtained using PSI-BLAST (Altschul et al., 1997; Altschul and Koonin, 1998).

The authors have quoted SSPro2 to have an average prediction accuracy (Q3) of 78% (Pollastri et al., 2002). In addition, the SSPro algorithm has also been experimentally implemented to predict eight-state secondary structure (H: α-helix, G: 3/10-helix, I: π-helix, E: extended strand, B: β-bridge, T: turn, S: bend, C: coil) from primary sequence. In the same paper as SSPro2, the authors present SSpro8 trained using BLAST and PSI-BLAST profiles. The quoted overall performance (Q8) of SSpro8 was 62–63%. An automatic server is available online and is part of the EVA assessment server (Koh et al., 2003) (see Section 7.9.1).

Software Package 4. YASPIN

YASPIN (Lin et al., 2005) uses a feedforward perceptron NN with one hidden layer to predict the SSEs (Bishop, 1995). These predictions are then filtered by an HMM.

The YASPIN NN uses the soft-max transition function (Bishop, 1995) with a window of 15 residues. For each residue in that window, 20 units are used for the scores in the PSSM and 1 unit is used to mark where the window spans termini of protein chains. In total, the input layer has 315 units (21×15). For the hidden layer we have used 15 units. The output layer has 7 units, corresponding to 7 local structure states: helix beginning (Hb), helix (H), helix end (He), beta beginning (Eb), beta (E), beta end (Ee), and coil (C).

The seven-state output of the NN is then filtered through an HMM, which uses the Viterbi algorithm (Durbin, 1998) to optimally segment the seven-state predictions. The HMM defines the transition probabilities between the seven local structure states. The final output is a three-state secondary structure prediction (H: helix, E: beta strand, C: coil). The YASPIN server has recently been added to the EVA assessment server (Koh et al., 2003) (see Section 7.9.1).

7.8.1.3 Hidden Markov Models

Software Package 1. SAM-T99/SAM-T02

The SAM (Sequence Alignment and Modelling system) software package (Hughey and Krogh, 1996) is a collection of tools that use linear HMMs for sequence analysis. Integrated into the package are the SAM-T99 (Karplus et al., 1998; 1999) and SAM-T02 (Karplus et al., 2002; 2003) structure prediction methods. Both methods predict the fold and secondary structure of a target protein sequence using multitrack HMMs and NNs.

The SAM-T02 method is an updated version of SAM-T99 and also incorporates secondary structure information into the scoring schemes it uses. Its HMMs and NNs have been trained on MSAs generated by the SAM-T2K iterated search procedure (Karplus et al., 2001).

The procedure SAM-T02 follows is to build an MSA of homologues to the target sequence using SAM-T2K and then employ NNs to make local structure predictions. The refinement and analysis of the HMM alignments returned can be performed by additional software found in the SAM package. Online servers for both SAM-T99 and SAM-T02 are available, but the authors recommend the use of the most updated SAM-T02 method (see Section 7.9.1). Although SAM-T02 has not yet been added to the EVA assessment server (Koh et al., 2003), SAM-T99 is still one of the highest performing methods (Section 7.9.1).

7.8.1.4 Support Vector Machines

Software Package 1. Hua and Sun

Hua and Sun (2001) were the first to apply SVMs to predict the secondary structure at each location along a protein strand. In their method SSEs fall into three categories: helix (H), sheet (E), or coil. Accordingly, this multiclass recognition problem was addressed by training three separate SVMs, one per SSE. The protein sequence is encoded in redundant binary fashion, using an 11-residue sliding window. The final classification of a given amino acid is the label associated with the SVM that assigns the discriminant score that is farthest from zero. The per-residue accuracy (Q3) and segment overlap (SOV) (Zemla et al., 1999) scores quoted by the authors are 73.5 and 76.2%, respectively, which are comparable to existing NN-based methods. Unfortunately, no SVM methods are currently part of an automated assessment server.

7.8.1.5 Consensus Prediction

Software Package 1. Jpred/Jnet

The initial implementation of Jpred was a purely majority voting consensus-deriving method (Cuff et al., 1998; Cuff and Barton, 1999) (for review see Heringa, 2000). The current Jpred (Jnet) server (Cuff and Barton, 2000) uses a refined and processed PSI-BLAST-generated alignment, the PSI-BLAST and HMM profile of that alignment and performs its predictions through two fully connected three-layer NNs.

The alignments Jnet uses are generated using PSI-BLAST to scan a Seg- (Wootton and Federhen, 1996) and helixfilt- (D.Jones, unpublished) filtered version of the combined SWISS-PROT (Bairoch and Boeckmann, 1991) and TrEMBL protein sequence database. After three iterations of PSI-BLAST, all sequence pairs in the generated alignment are compared and the sequence percentage identities are used to cluster them. All sequences in the alignment with more than 75% identity are removed. The alignment is then further processed by removing all gaps from the target sequence including the corresponding column beneath that gap. This type of alignment processing is also observed in PHD, PHDpsi, and PROFsec and it is essential for the NN to work.

In addition to the alignment, Jnet also uses the PSI-BLAST-generated PSSM (PSI-BLAST profile) and the HMM profile of the alignment (using HMMER). The three input files are each used as input to the two neural networks. The first neural network uses a 17-residue sliding window and predicts the per-residue propensity of helix, strand, and coil. The second network acts as a filter for each per-residue prediction from the first network, using a 19-residue sliding window. Finally, the three predictions are used to generate a per-residue consensus, which is the final prediction.

The authors of Jnet have quoted an average prediction accuracy of 76.4% when using all three input types, 71.6% when only using the refined and processed PSI-BLAST alignment, 74.4% when using the alignment, and HMM profile of the alignment, and 75.2% when using the alignment and the PSI-BLAST profile of the alignment. The EVA server has stopped the assessment of the Jpred method due to a move in URLs. The quoted Q3 of 72.8% over 167 proteins refers to the original Jpred method (see Section 7.9.1).

Software Package 2. SymSSP/SymPRED

The original "majority voting" technique was improved upon by the use of dynamic programming (DP) (Needleman and Wunsch, 1970) in the SymSSP (Simossis and Heringa, 2004a) and soon after in the SymPRED (Simossis and Heringa, submitted) methods. In both applications, an alignment of secondary structures is reduced to a weighted profile that describes helix, strand, or coil content of each position. The profile is scanned in windows of increasing length, from single position (window of 1) up to the length of the whole sequence, for each secondary structure type. Each secondary structure type segment is scored as the sum of all of its positions. These equivalent secondary structure-specific window scores are compared and the highest one is used to fill a search matrix. Finally, the DP routine finds the optimal path through the search matrix and thus provides an optimally segmented consensus prediction.

In the SymSSP method, the strategy was applied to the alignment-based predictions of a single method that preprocesses the input alignment prior to prediction by removing whole alignment positions that show a gap in the top sequence. When most of the discarded information was recovered for the methods PHD (Rost and Sander, 1993), PROFsec (B. Rost, personal communication), and SSPro (Pollastri et al., 2002), the results showed consistent modest improvement of the prediction quality of these methods based on Q3 and SOV (Zemla et al., 1999) score results . In the case of SymPRED, the DP strategy was applied to the predictions of various prediction methods and the results were compared to recent simple "majority voting" investigations (Albrecht et al., 2003; McGuffin and Jones, 2003; Ward et al., 2003). In both investigations the predictions produced were of higher quality than those produced by the simple "majority voting" strategy.

7.8.2 Supersecondary Structure Prediction

7.8.2.1 Software Package 1. COILS2

Closely related to secondary structure prediction is the prediction of coiled-coil structures. If a soluble protein is predicted to contain α-helices, higher-order information as well as increased confidence in predictions made could be gained from testing the possibility of it containing a coiled-coil supersecondary structure.

The program COILS2 (Lupas et al., 1991; Lupas, 1996) compares a query sequence with a database of known parallel two-stranded coiled coils. A similarity score is derived and compared to two score distributions, one for globular proteins (without coiled coils) and one for known coiled-coil structures. The two scores are then converted to a probability for the query sequence to adopt a coiled-coil conformation. Since the program assumes the presence of heptad repeats, probabilities are derived using default window lengths of 14, 21, and 28 amino acids. The program can also use user-defined window lengths for the prediction of extreme coiled-coil lengths. A recently updated scoring matrix, based on data from recent coiled-coil structures and containing amino acid type propensities for various positions in the heptad repeat, shows improved recognition of coiled-coil elements. The COILS2 method accurately recognizes left-handed two-stranded coiled coils but loses sensitivity for coiled-coil structures consisting of more than two strands. Also, it is not able to recognize right-handed or buried coiled–coil helices and therefore is not applicable to transmembrane coiled-coil structures known to show basically similar coiled-coil conformations as soluble proteins, albeit with dramatically different and more hydrophobic constituent amino acids (Langosch and Heringa, 1998).

7.8.2.2 Software Package 2. WD-repeats Prediction

The server "WD repeat Family of Proteins" (see http://bmerc-www.bu.edu/wdrepeat/) is able to recognize putative WD-repeat sequences associated with 4- to 9-bladed 3D WD-repeat structures. These models combine a particular so-called Type-1 structural model with sequence-specific pattern information. Multidomain proteins can be handed to the server intact; the region containing the WD-repeat domain will be identified by the server automatically.

The analysis algorithm is based on probabilistic Discrete State-space Models (DSMs), and optimal filtering and smoothing algorithms (Stultz et al., 1993). The mathematical basis for the models and algorithms is described in White et al. (1994).

A protein sequence submitted to the server is first classified as "generic" or "wd repeat." The class "generic" is designed for proteins not containing WD repeats. Superclass "wd repeat" is designed for the WD-repeat family of proteins. Under this superclass, there are six macroclasses for WD-repeat proteins, each of which contains a different number of WD repeats. Sequences containing fewer than four WD repeats will not be reported as a WD-repeat protein. This is due to the assumption that all WD-repeat proteins adopt a β-propeller fold, which must have at least four blades to form a circular structure. The 4- to 9-bladed (WD4 to WD9) models that can be produced by the server correspond to sequence length ranges of 187–279, 233–332, 278–385, 323–437, 368–489, and 413–541 residues, respectively. To handle longer sequences, the algorithm is able to add leader and trailer to the models on the fly. Therefore, all models can recognize WD repeats within sequences longer than its maximum domain length up to an upper limit on sequence length of 1000 residues.

Each WD repeat has two conserved profiles denoted "profile 1" and "profile 2" (which may be approximated as "GHXXXVXXVXFX" and "XLASGSXDXTIKVWD," respectively, as shown at http://bmerc-www.bu.edu/wdrepeat/) that are used in the DSM prediction. The probabilities of occurrence of each of these profiles will be reported if WD repeats are identified in the sequence. In addition, the β strands within each of the aligned putative WD repeats will be designated, although individual β-strand probabilities will not be reported. To provide the user insight in the 3D orientation of the WD repeats, a skeleton coordinate file in PDB format is included.

7.8.3 Disordered Region Prediction

7.8.3.1 Software Package 1. PONDR

The PONDR suite contains several disorder prediction methods (Obradovic et al., 2003). The predictions from the methods VL2 and VLXT in the PONDR suite (Obradovic et al., 2003) come from ensembles of feedforward NNs trained on combinations of amino acid composition, flexibility, and sequence complexity. Sequence information is parsed using windows of generally 21 amino acids. The amino acid attributes are calculated over this window, and these values are used as inputs for the NNs, which calculate a value for the central amino acid in each window. These prediction values are then smoothed over a sliding window of 9 amino acids. If a residue value exceeds a threshold, the residue is declared disordered. Another predictor VL3 was trained using ordinary least squares regression with partitioning of the training set to cluster various "flavors" of disorder (Vucetic et al., 2003). Recently, a new disorder predictor VSL1 was added to the PONDR suite. The VSL1 predictor obtained the best results in a comparison including 20 different disorder prediction methods presented at the CASP6 structure prediction meeting in December 2004. The methods in the PONDR suite are not freely available.

7.8.3.2 Software Package 2. FoldIndex

The FoldIndex program is based on the calculations developed by Uversky et al. (2000) and predicts whether a sequence will fold by computing its mean net charge and hydrophobicity (Uversky et al., 2000). The window parameter for the FoldIndex classifier was set to 31 residues as this value achieved the highest accuracy on a validation set. The resulting data show that the combination of low mean hydrophobicity and relatively high net charge represents an important prerequisite for the absence of regular structure in proteins under physiologic conditions, thus leading to "natively unfolded" proteins.

7.8.3.3 Software Package 3. DisEMBL

Linding et al. (2003a) developed the NN-based method DisEMBL. The authors carefully selected a number of protein sets—including a coil and a "hot loop" set—to train the neural nets using 5-fold cross validation, while the best parameter settings were selected based on ROC curves. The optimal network architecture was a window size of 19 residues and 30 hidden units. The coil and hot loop NN ensembles, the score distributions of positive and negative test examples were estimated using Gaussian kernel density estimation. Based on these distributions, a calibration curve for converting NN output scores to probabilities was constructed. To predict disorder for an unknown query sequence, the network output is smoothed and the resulting amino acid disorder probabilities are plotted.

7.8.3.4 Software Package 4. GLOBPLOT

The GLOBPLOT method (Linding et al., 2003b) is based on the hypothesis that the tendency for disorder can be expressed as $P = RC - SS$ where RC and SS are the propensity for a given amino acid to be in "random coil" and regular "secondary structure," respectively. The RC and SS propensity values were derived by the authors employing a data set using a single representative of each superfamily in the SCOP database (version 1.59). The two types of propensities were then combined in a single "Russel/Linding" amino acid propensity set, which is able to discriminate between disorder and globular packing.

7.8.3.5 Software Package 5. DISOPRED

The DISOPRED2 method (Ward et al., 2004) exploits an SVM classifier based on a linear kernel function and compares favorably to the above methods across the range of decision thresholds. Ward et al. (2004) also noted that using homologous sequences improves disorder prediction slightly as compared to single sequence prediction, but the beneficial effect is clearly lower than that for secondary structure prediction.

7.8.3.6 Software Package 6. PDISORDER

The PDISORDER method (Softberry, Inc.) exploits a combination of machine learning techniques comprising NNs, linear discriminant functions, and an acute smoothing procedure. At the recent CASP6 prediction assessment workshop, the method scored high in terms of the correlations it yields with crystallographic B-factors, which are included as evidence for disorder.

7.8.3.7 Software Package 7. DISpro

Cheng et al. (2005) reported a state-of-the-art disorder prediction accuracy of 92.8% with a false positive rate of 5% on large cross-validated tests. Their method DISpro uses evolutionary information in the form of profiles, predicted secondary structure and relative solvent accessibility, and ensembles of 1D-recursive NNs. The method shows an improved performance over previous methods, for example using the CASP5 data set (Cheng et al., 2005).

7.8.4 Internal Repeats Recognition

7.8.4.1 Software Package 1. REPRO

Heringa and Argos (1993) adapted the basic Waterman and Eggert algorithm to repeat situations within a single protein by demanding, in addition to top-scoring alignments being nonintersecting, that locally aligned fragments do not overlap. They introduced a graph-based iterative clustering mechanism, which takes the thus produced list of top-scoring nonoverlapping local alignments for a single query sequence, declares the N-terminal matched amino acid pair in each top alignment as start sites of a repeats pair, and then attempts to delineate associated start-sites within the top alignments (i.e., find more repeats internal to the top alignment) that match the repeat type based on alignment consistency with already clustered members of the repeat type. If such new repeats are found, the cluster procedure is iterated. The cluster consistency criterion assesses the number of established repeats that align with a putative repeat, and selects it only if three or more of such top-scoring alignments can be found and if at least one of these associated alignments has already contributed one or more repeat members to the current repeat type and therefore can be trusted to be "in phase" with that repeat type. After the clustering phase, the repeats can be multiply aligned and turned into a profile, which can then be slid over the query sequence to verify the repeats already found and possibly detect new incarnations missed by the preceding algorithmic steps (Heringa and Argos, 1993): If new repeats are found, the profile can be updated and the procedure iterated. The REPRO algorithm is able to detect multiple repeat types independently, and is a sensitive but slow technique. A web server for the REPRO algorithm is available at http://ibivu.cs.vu.nl (George and Heringa, 2000).

7.8.4.2 Software Package 2. Pellegrini et al.

A quick algorithm for calculating the length and copy number of internal repeat sets has been devised by Pellegrini et al. (1999). The method uses the Waterman and Eggert algorithm and converts the scores of the selected top alignments to probabilities. An $N \times N$ path matrix, where N is the length of the protein sequence, is then filled with ones for matrix cells corresponding to local nonintersecting alignments that score above a preset threshold value for the probabilities, and zero values elsewhere. Two simple summing protocols are then applied to this matrix to obtain an approximate notion of the repeat length and copy number, albeit the repeat boundaries are not determined. Marcotte et al. (1999) used the algorithm to derive a general census of repeats in proteins using the SWISS-PROT protein sequence database.

7.8.4.3 Software Package 3. RADAR

The method RADAR (Heger and Holm, 2000) basically follows the algorithmic steps of the REPRO method (Heringa and Argos, 1993). It calculates nonintersecting

Table 7.1 A list of all prediction methods independently assessed by the EVA server and their corresponding overall scores and test set sizes, until the end of 2004. Methods whose names are in boldface have been covered in Section 7.8

Method	Test set	Score	Server URL (assume "http://" at the start of each address)
APSSP2	393	75.1	*www.imtech.res.in/raghava/apssp2/*
Jpred	167	72.8	*www.compbio.dundee.ac.uk/~www-jpred/submit.html*
JUFO	133	68.9	*www.jens-meiler.de/jufo.html*
PHD	446	72.2	*cubic.bioc.columbia.edu/predictprotein/*
PHDpsi	440	73.3	*cubic.bioc.columbia.edu/predictprotein/*
PROF_king	443	72.7	*www.aber.ac.uk/~phiwww/prof/*
PROFsec	443	75.3	*cubic.bioc.columbia.edu/predictprotein/*
Prospect	315	71.7	*compbio.ornl.gov/cgi-bin/PROSPECT/*
PSIPRED	443	76.2	*bioinf.cs.ucl.ac.uk/psipred/psiform.html*
SABLE	156	76.0	*sable.cchmc.org/*
SABLE2	99	76.9	*sable.cchmc.org/*
SAM-T99sec	396	76.0	*www.cse.ucsc.edu/research/compbio/HMM-apps/T99-query.html*
SCRATCH	217	75.7	*www.igb.uci.edu/tools/scratch/*
SSPRO2	257	74.3	*www.igb.uci.edu/tools/scratch/*
SSPRO4	68	78.7	*www.igb.uci.edu/tools/scratch/*
YASPIN	80	71.0	*ibivu.cs.vu.nl/programs/yaspinwww/*

local alignments, and then uses these in an iterative procedure to determine the shortest nonreducible repeat unit and determine the associated boundaries. A profile is constructed from a multiple alignment of a repeat set, and slid over the query sequence to capture more repeats. The whole procedure is then iterated in an attempt to find multiple repeat types. The RADAR step to find the shortest possible repeat unit, includes an iterative wraparound DP algorithm to detect the smallest repeat unit within a potentially reducible set of repeats. The RADAR method is sensitive and sufficiently fast for genomic application.

7.8.4.4 Software Package 4. REP

Andrade et al. (2000) produced a supervised repeats detection method REP, which searches the query sequence using a number of profiles, each profile containing

Table 7.2 A list of methods for predicting coiled-coil and WD-repeat protein regions from sequence

Method	Server URL (assume "http://" at the start of each address)	Reference
COILS2	*iubio.bio.indiana.edu:7780/archive/00000527/*	Lupas et al., 1991
WD-repeat Prediction	*bmerc-www.bu.edu/psa/request.htm*	The same URL

Table 7.3 A list of methods for predicting disordered protein regions from sequence

Method	Server URL (assume "http://" at the start of each address)	Methodology	Reference
FoldIndex	*bioportal.weizmann.ac.il/fldbin/findex*	Charge-hydrophobicity patterns	Priluski et al., 2005
DISpro	*www.ics.uci.edu/~baldig/diso.html*	Neural net	Cheng et al., 2005
DISEMBL	*dis.embl.de/*	Neural net	Linding et al., 2003a
GLOBPLOT	*globplot.embl.de/*	Amino acid propensities	Linding et al., 2003b
DISOPRED2	*bioinf.cs.ucl.ac.uk/disopred/disopred.html*	SVM	Ward et al., 2004
PONDR	*www.pondr.com*	Neural net	Obradovic et al., 2003
DRIPPRED	*sbcweb.pdc.kth.se/cgi-bin/maccallr/disorder/submit.pl*	Kohonen self-organizing maps	*http://www.forcasp.org/paper2127.html*
PDISORDER	*www.softberry.com/berry.phtml?topic=pdisorder&group=programs&subgroup=propt*	Neural network Linear discriminant function Acute smoothing procedure	*http://www.forcasp.org/upload/2197.28.pdf*

the information of a multiple alignment of a known repeats family. The user can scan the query sequence for the following repeat types: Ankyrin, Armadillo, HAT, HEAT, HEAT_AAA, HEAT_ADB, HEAT_IMB, Kelch, Leucin-e-Rich Repeats, PFTA, PFTB, RCC1, TPR, and WD40.

Table 7.4 List of methods for internal repeats recognition

Method	Server URL (assume "http://" at the start of each address)	Methodology	Reference
Pellegrini et al.	*www.doe-mbi.ucla.edu/Services/Repeats/*	Waterman and Eggert algorithm	Pellegrini et al., 1999
RADAR	*www.ebi.ac.uk/Radar/*	Local alignment	Heger and Holm, 2000
REP	*www.embl-heidelberg.de/~andrade/papers/rep/search.html*	Checking known repeat types	Andrade et al., 2000
REPRO	*ibivu.cs.vu.nl/programs/reprowww/*	Local alignment, graph clustering	George and Heringa, 2000
TRUST	*ibivu.cs.vu.nl/programs/trustwww/*	Transitivity	Szklarczyk and Heringa, 2004

7.8.4.5 Software Package 5. TRUST

Szklarczyk and Heringa (2004) developed a method TRUST for protein internal repeats detection based on transitivity of repeats. The authors reported an increased sensitivity and accuracy of the method. This is achieved by exploiting the concept of transitivity of alignments, which relies on mutual reinforcement (or attenuation) of repeat signals, and thus can be used as a noise filter. Starting from local suboptimal alignments, the application of transitivity allows (1) identification of distant repeat homologues for which no alignments were found; (2) gaining confidence about consistently well-aligned regions; and (3) reducing the contribution of nonhomologous repeats. The thus obtained increased consistency generally leads to a virtually noise-free profile representing a generalized repeat with high fidelity. The TRUST method also employs a rigid statistical test for self-sequence and profile-sequence alignments.

7.9 Resources

This section contains useful resources available at the time this chapter was written for online software applications and other useful material.

7.9.1 Secondary Structure Prediction

7.9.2 Supersecondary Structure Prediction

7.10 Summary

This chapter presents an overview of issues in predicting local structural features of proteins. The inherent hierarchical order of protein structure is discussed in a bottom-up fashion, from secondary structure via supersecondary structure to prediction aspects of local three-dimensional structure, the latter including protein disordered region detection and internal repeats recognition. Some approaches to use these structural features in multiple sequence alignment are also discussed. State-of-the-art prediction methods are described and the addresses of their web interfaces, if available, are provided.

References

Albrecht, M., Tosatto, S.C., Lengauer, T., and Valle, G. 2003. Simple consensus procedures are effective and sufficient in secondary structure prediction. *Protein Eng.* 16:459–462.

Altschul, S.F., and Koonin, E.V. 1998. Iterated profile searches with PSI-BLAST—A tool for discovery in protein databases. *Trends Biochem. Sci.* 23:444–447.

Altschul, S.F., Madden, T.L., Schaffer, A.A., Zhang, J., Zhang, Z., Miller, W., and Lipman, D.J. 1997. Gapped BLAST and PSI-BLAST: A new generation of protein database search programs. *Nucleic Acids Res.* 25:3389–3402.

An, J., Totrov, M., and Abagyan, R. 2005. Pocketome via comprehensive identification and classification of ligand binding envelopes. *Mol. Cell. Proteomics* 4:752–761.

An, Y., and Friesner, R.A. 2002. A novel fold recognition method using composite predicted secondary structures. *Proteins* 48:352–366.

Andrade, M.A., Ponting, C.P., Gibson, T.J., and Bork, P. 2000. Homology-based method for identification of protein repeats using statistical significance estimates. *J. Mol. Biol.* 298:521–537.

Argos, P. 1987. Analysis of sequence-similar pentapeptides in unrelated protein tertiary structures. Strategies for protein folding and a guide for site-directed mutagenesis. *J. Mol. Biol.* 197:331–348.

Bagos, P.G., Liakopoulos, T.D., and Hamodrakas, S.J. 2005. Evaluation of methods for predicting the topology of beta-barrel outer membrane proteins and a consensus prediction method. *BMC Bioinformatics* 6:7.

Bairoch, A., and Boeckmann, B. 1991. The SWISS-PROT protein sequence data bank. *Nucleic Acids Res.* 19 (Suppl.):2247–2249.

Baldi, P., Brunak, S., Frasconi, P., Soda, G., and Pollastri, G. 1999. Exploiting the past and the future in protein secondary structure prediction. *Bioinformatics* 15:937–946.

Berman, H.M., Westbrook, J., Feng, Z., Gilliland, G., Bhat, T.N., Weissig, H., Shindyalov, I.N., and Bourne, P.E. 2000. The Protein Data Bank. *Nucleic Acids Res.* 28:235–242.

Bishop, C.M. 1995. *Neural Networks for Pattern Recognition.* Oxford, Clarendon Press.

Blanco, F.J., Rivas, G., and Serrano, L. 1994. A short linear peptide that folds into a native stable beta-hairpin in aqueous solution. *Nat. Struct. Biol.* 1:584–590.

Boeckmann, B., Bairoch, A., Apweiler, R., Blatter, M.C., Estreicher, A., Gasteiger, E., Martin, M.J., Michoud, K., O'Donovan, C., Phan, I., Pilbout, S., and Schneider, M. 2003. The SWISS-PROT protein knowledgebase and its supplement TrEMBL in 2003. *Nucleic Acids Res.* 31:365–370.

Bordner, A.J., and Abagyan, R. 2005. Statistical analysis and prediction of protein–protein interfaces. *Proteins Struct. Funct. Bioinf.* 60:353–366.

Boswell, D.R., and McLachlan, A.D. 1984. Sequence comparison by exponentially-damped alignment. *Nucleic Acids Res.* 12:457–464.

Bracken, C. 2001. NMR spin relaxation methods for characterization of disorder and folding in proteins. *J. Mol. Graph. Model* 19:3–12.

Bystroff, C., Thorsson, V., and Baker, D. 2000. HMMSTR: A hidden Markov model for local sequence–structure correlations in proteins. *J. Mol. Biol.* 301:173–190.

Byvatov, E., and Schneider, G. 2003. Support vector machine applications in bioinformatics. *Appl. Bioinf.* 2:67–77.

Cai, Y. D., Feng, K.Y., Li, Y.X., and Chou, K.C. 2003. Support vector machine for predicting alpha-turn types. *Peptides* 24:629–630.

Capriotti, E., Fariselli, P., Rossi, I., and Casadio, R. 2004. A Shannon entropy-based filter detects high-quality profile–profile alignments in searches for remote homologues. *Proteins* 54:351–360.

Chandonia, J.M., and Karplus, M. 1999. New methods for accurate prediction of protein secondary structure. *Proteins* 35:293–306.

Cheng, J., Sweredoski, M.J., and Baldi, P. 2005. Accurate prediction of protein disordered regions by mining protein structure data. *Data Mining Knowledge Discovery* 11:213–222.

Chothia, C. 1984. Principles that determine the structure of proteins. *Annu. Rev. Biochem.* 53:537–572.

Chothia, C., and Lesk, A.M. 1986. The relation between the divergence of sequence and structure in proteins. *EMBO J* 5:823–826.

Chou, P.Y., and Fasman, G.D. 1974. Prediction of protein conformation. *Biochemistry* 13:222–245.

Chung, R., and Yona, G. 2004. Protein family comparison using statistical models and predicted structural information. *BMC Bioinformatics* 5:183.

Churchill, G.A. 1989. Stochastic models for heterogeneous DNA sequences. *Bull. Math. Biol.* 51:79–94.

Cozzetto, D., and Tramontano, A. 2005. Relationship between multiple sequence alignments and quality of protein comparative models. *Proteins* 58:151–157.

Cregut, D., Civera, C., Macias, M.J., Wallon, G., and Serrano, L. 1999. A tale of two secondary structure elements: When a beta-hairpin becomes an alpha-helix. *J. Mol. Biol.* 292:389–401.

Crippen, G.M. 1978. The tree structural organization of proteins. *J. Mol. Biol.* 126:315–332.

Cristianini, N., and Shawe-Taylor, J. 2000. *An Introduction to Support Vector Machines and Other Kernel-Based Learning Methods.* New York, Cambridge University Press.

Cuff, J.A., and Barton, G.J. 1999. Evaluation and improvement of multiple sequence methods for protein secondary structure prediction. *Proteins* 34:508–519.

Cuff, J.A., and Barton, G.J. 2000. Application of multiple sequence alignment profiles to improve protein secondary structure prediction. *Proteins* 40:502–511.

Cuff, J.A., Clamp, M.E., Siddiqui, A.S., Finlay, M., and Barton, G.J. 1998. JPred: A consensus secondary structure prediction server. *Bioinformatics* 14:892–893.

Dayhoff, M.O., Barker, W.C., and Hunt, L.T. 1983. Establishing homologies in protein sequences. *Methods Enzymol.* 91:524–545.

de la Cruz, X., Hutchinson, E.G., Shepherd, A., and Thornton, J.M. 2002. Toward predicting protein topology: An approach to identifying beta hairpins. *Proc. Natl. Acad. Sci. USA.* 99:11157–11162.

de la Cruz, X., and Thornton, J.M. 1999. Factors limiting the performance of prediction-based fold recognition methods. *Protein Sci.* 8:750–759.

Derreumaux, P. 2001. Evidence that the 127–164 region of prion proteins has two equi-energetic conformations with beta or alpha features. *Biophys. J.* 81:1657–1665.

Dickerson, R.E., Timkovich, R., and Almassy, R.J. 1976. The cytochrome fold and the evolution of bacterial energy metabolism. *J. Mol. Biol.* 100:473–491.

Dunker, A.K., Brown, C.J., Lawson, J.D., Iakoucheva, L.M., and Obradovic, Z. 2002. Intrinsic disorder and protein function. *Biochemistry* 41:6573–6582.

Dunker, A.K., Lawson, J.D., Brown, C.J., Williams, R.M., Romero, P., Oh, J.S., Oldfield, C.J., Campen, A.M., Ratliff, C.M., Hipps, K.W., Ausio, J., Nissen, M.S., Reeves, R., Kang, C., Kissinger, C.R., Bailey, R.W., Griswold, M.D., Chiu, W., Garner, E.C., and Obradovic, Z. 2001. Intrinsically disordered protein. *J. Mol. Graph. Model* 19:26–59.

Dunker, A.K., Obradovic, Z., Romero, P., Garner, E.C., and Brown, C.J. 2000. Intrinsic protein disorder in complete genomes. *Genome Inform. Ser. Workshop Genome Inform.* 11:161–171.

Durbin, R. 1998. *Biological Sequence Analysis: Probabilistic Models of Proteins and Nucleic Acids.* New York,Cambridge University Press.

Durbin, R., Eddy, S., Krogh, A., and Mitchison, G. 2000. Markov chains and hidden Markov models. In *Biological Sequence Analysis: Probalistic Models of Proteins and Nucleic Acids*. New York, Cambridge University Press, pp. 46–79.

Dutta, S., and Berman, H.M. 2005. Large macromolecular complexes in the Protein Data Bank: A status report. *Structure* 13:381–388.

Dyson, H.J., and Wright, P.E. 2002. Insights into the structure and dynamics of unfolded proteins from nuclear magnetic resonance. *Adv. Protein Chem.* 62:311–340.

Eddy, S.R. 1996. Hidden Markov models. *Curr. Opin. Struct. Biol.* 6:361–365.

Edgar, R.C., and Sjolander, K. 2004. COACH: Profile–profile alignment of protein families using hidden Markov models. *Bioinformatics* 20:1309–1318.

Forcellino, F., and Derreumaux, P. 2001. Computer simulations aimed at structure prediction of supersecondary motifs in proteins. *Proteins* 45:159–166.

Frenkel, D., and Smit, B. 2002. Monte Carlo simulations. In: *Understanding Molecular Simulation: From Algorithms to Applications* (D. Frenkel, M. Klein, M. Parrinello, and B. Smit, Eds.). San Diego, Academic Press, pp. 23–58.

Friedberg, I., Kaplan, T., and Margalit, H. 2000. Evaluation of PSI-BLAST alignment accuracy in comparison to structural alignments. *Protein Sci.* 9:2278–2284.

Frishman, D., and Argos, P. 1996. Incorporation of non-local interactions in protein secondary structure prediction from the amino acid sequence. *Protein Eng.* 9:133–142.

Frishman, D., and Argos, P. 1997. Seventy-five percent accuracy in protein secondary structure prediction. *Proteins* 27:329–335.

Garnier, J., Gibrat, J.F., and Robson, B. 1996. GOR method for predicting protein secondary structure from amino acid sequence. *Methods Enzymol* 266:540–553.

Garnier, J., Osguthorpe, D.J., and Robson, B. 1978. Analysis of the accuracy and implications of simple methods for predicting the secondary structure of globular proteins. *J. Mol. Biol.* 120:97–120.

George, R.A., and Heringa, J. 2000. The REPRO server: Finding protein internal sequence repeats through the Web. *Trends Biochem. Sci.* 25:515–517.

Gibrat, J.F., Garnier, J., and Robson, B. 1987. Further developments of protein secondary structure prediction using information theory. New parameters and consideration of residue pairs. *J. Mol. Biol.* 198:425–443.

Ginalski, K., Pas, J., Wyrwicz, L.S., von Grotthuss, M., Bujnicki, J.M., and Rychlewski, L. 2003. ORFeus: Detection of distant homology using sequence profiles and predicted secondary structure. *Nucleic Acids Res.* 31:3804–3807.

Ginalski, K., von Grotthuss, M., Grishin, N.V., and Rychlewski, L. 2004. Detecting distant homology with Meta-BASIC. *Nucleic Acids Res.* 32:W576–581.

Guermeur, Y., Geourjon, C., Gallinari, P., and Deleage, G. 1999. Improved performance in protein secondary structure prediction by inhomogeneous score combination. *Bioinformatics* 15:413–421.

Guo, J., Chen, H., Sun, Z., and Lin, Y. 2004. A novel method for protein secondary structure prediction using dual-layer SVM and profiles. *Proteins* 54:738–743.

Hedman, M., Deloof, H., Von Heijne, G., and Elofsson, A. 2002. Improved detection of homologous membrane proteins by inclusion of information from topology predictions. *Protein Sci.* 11:652–658.

Heger, A., and Holm, L. 2000. Rapid automatic detection and alignment of repeats in protein sequences. *Proteins* 41:224–237.

Heringa, J. 1994. The evolution and recognition of protein sequence repeats. *Comput. Chem.* 18:233–243.

Heringa, J. 1998. Detection of internal repeats: How common are they? *Curr. Opin. Struct. Biol.* 8:338–345.

Heringa, J. 1999. Two strategies for sequence comparison: Profile-preprocessed and secondary structure-induced multiple alignment. *Comput. Chem.* 23:341–364.

Heringa, J. 2000. Computational methods for protein secondary structure prediction using multiple sequence alignments. *Curr. Protein Pept. Sci.* 1:273–301.

Heringa, J. 2002. Local weighting schemes for protein multiple sequence alignment. *Comput. Chem.* 26:459–477.

Heringa, J., and Argos, P. 1993. A method to recognize distant repeats in protein sequences. *Proteins* 17:391–341.

Hu, H.J., Pan, Y., Harrison, R., and Tai, P.C. 2004. Improved protein secondary structure prediction using support vector machine with a new encoding scheme and an advanced tertiary classifier. *IEEE Trans. Nanobiosci.* 3:265–271.

Hu, W.P., Kolinski, A., and Skolnick, J. 1997. Improved method for prediction of protein backbone U-turn positions and major secondary structural elements between U-turns. *Proteins* 29:443–460.

Hua, S., and Sun, Z. 2001. A novel method of protein secondary structure prediction with high segment overlap measure: Support vector machine approach. *J. Mol. Biol.* 308:397–407.

Huang, C.H., Lin, Y.S., Yang, Y.L., Huang, S.W., and Chen, C.W. 1998. The telomeres of Streptomyces chromosomes contain conserved palindromic sequences with potential to form complex secondary structures. *Mol. Microbiol.* 28:905–916.

Huang, X.Q., Hardison, R.C., and Miller, W. 1990. A space-efficient algorithm for local similarities. *Comput. Appl. Biosci.* 6:373–381.

Hughey, R., and Krogh, A. 1996. Hidden Markov models for sequence analysis: Extension and analysis of the basic method. *Comput. Appl. Biosci.* 12:95–107.

Hutchinson, E.G., and Thornton, J.M. 1993. The Greek key motif: Extraction, classification and analysis. *Protein Eng.* 6:233–245.

Jones, D.T. 1999. Protein secondary structure prediction based on position-specific scoring matrices. *J. Mol. Biol.* 292:195–202.

Karplus, K., Barrett, C., Cline, M., Diekhans, M., Grate, L., and Hughey, R. 1999. Predicting protein structure using only sequence information. *Proteins* Suppl. 3:121–125.

Karplus, K., Barrett, C., and Hughey, R. 1998. Hidden Markov models for detecting remote protein homologies. *Bioinformatics* 14:846–856.

Karplus, K., Karchin, R., Barrett, C., Tu, S., Cline, M., Diekhans, M., Grate, L., Casper, J., and Hughey, R. 2001. What is the value added by human intervention in protein structure prediction? *Proteins* Suppl. 5:86–91.

Karplus, K., Karchin, R., Draper, J., Casper, J., Mandel-Gutfreund, Y., Diekhans, M., and Hughey, R. 2003. Combining local-structure, fold-recognition, and new fold methods for protein structure prediction. *Proteins* 53 (Suppl.6):491–496.

Karplus, K., Karchin, R., Hughey, R., Draper, J., Mandel-Gutfreund, Y., Casper, J., and Diekhans, M. 2002. SAM-T02: Protein structure prediction with neural nets, hidden Markov models, and fragment packing. CASP 5.

Kim, D.E., Chivian, D., and Baker, D. 2004. Protein structure prediction and analysis using the Robetta server. *Nucleic Acids Res.* 32:W526–531.

Kim, H., and Park, H. 2003. Protein secondary structure prediction based on an improved support vector machines approach. *Protein Eng.* 16:553–560.

King, R.D., Ouali, M., Strong, A.T., Aly, A., Elmaghraby, A., Kantardzic, M., and Page, D. 2000. Is it better to combine predictions? *Protein Eng.* 13:15–19.

Kirshenbaum, K., Young, M., and Highsmith, S. 1999. Predicting allosteric switches in myosins. *Protein Sci.* 8:1806–1815.

Kleinjung, J., Romein, J., Lin, K., and Heringa, J. 2004. Contact-based sequence alignment. *Nucleic Acids Res.* 32:2464–2473.

Koh, I.Y., Eyrich, V.A., Marti-Renom, M.A., Przybylski, D., Madhusudhan, M.S., Eswar, N., Grana, O., Pazos, F., Valencia, A., Sali, A., and Rost, B. 2003. EVA: Evaluation of protein structure prediction servers. *Nucleic Acids Res.* 31:3311–3315.

Kolaskar, A.S., and Kulkarni-Kale, U. 1992. Sequence alignment approach to pick up conformationally similar protein fragments. *J. Mol. Biol.* 223:1053–1061.

Kolinski, A., Skolnick, J., Godzik, A., and Hu, W.P. 1997. A method for the prediction of surface "U"-turns and transglobular connections in small proteins. *Proteins* 27:290–308.

Krogh, A., Brown, M., Mian, I.S., Sjolander, K., Haussler, D. 1994. Hidden Markov models in computational biology. Applications to protein modeling. *J. Mol. Biol.* 235:1501–1531.

Kuhn, M., Meiler, J., and Baker, D. 2004. Strand–loop–strand motifs: Prediction of hairpins and diverging turns in proteins. *Proteins* 54:282–288.

Kurtz, S., and Schleiermacher, C. 1999. REPuter: Fast computation of maximal repeats in complete genomes. *Bioinformatics* 15:426–427.

Langosch, D., and Heringa, J. 1998. Interaction of transmembrane helices by a knobs-into-holes packing characteristic of soluble coiled coils, *Proteins: Struct. Func. and Gen.* 31:150–159.

Lim, V.I. 1974. Structural principles of the globular organization of protein chains. A stereochemical theory of globular protein secondary structure. *J. Mol. Biol.* 88:857–872.

Lin, K., Simossis, V.A., Taylor, W.R., and Heringa, J. 2005. A simple and fast secondary structure prediction method using hidden neural networks. *Bioinformatics* 21:152–159.

Linding, R., Jensen, L.J., Diella, F., Bork, P., Gibson, T.J., and Russell, R.B. 2003a. Protein disorder prediction: Implications for structural proteomics. *Structure* 11:1453–1459.

Linding, R., Russell, R.B., Neduva, V., and Gibson, T.J. 2003b. GlobPlot: Exploring protein sequences for globularity and disorder. *Nucleic Acids Res.* 31:3701–3708.

Luisi, D.L., Wu, W.J., and Raleigh, D.P. 1999. Conformational analysis of a set of peptides corresponding to the entire primary sequence of the N-terminal domain of the ribosomal protein L9: Evidence for stable native-like secondary structure in the unfolded state. *J. Mol. Biol.* 287:395–407.

Lupas, A. 1996. Prediction and analysis of coiled-coil structures. *Methods Enzymol* 266:513–525.

Lupas, A., Van Dyke, M., and Stock, J. 1991. Predicting coiled coils from protein sequences, *Science* 252:1162–1164.

Luthy, R., Xenarios, I., and Bucher, P. 1994. Improving the sensitivity of the sequence profile method. *Protein Sci.* 3:139–146.

Macdonald, J.R., and Johnson, W.C., Jr. 2001. Environmental features are important in determining protein secondary structure. *Protein Sci.* 10:1172–1177.

Marcotte, E.M., Pellegrini, M., Yeates, T.O., and Eisenberg, D. 1999. A census of protein repeats. *J. Mol. Biol.* 293:151–160.

McGuffin, L.J., and Jones, D.T. 2003. Benchmarking secondary structure prediction for fold recognition. *Proteins* 52:166–175.

McLachlan, A.D. 1972. Repeating sequences and gene duplication in proteins. *J. Mol. Biol.* 64:417–437.

McLachlan, A.D. 1977. Analysis of periodic patterns in amino acid sequences: Collagen. *Biopolymers* 16:1271–1297.

McLachlan, A.D. 1979. Gene duplications in the structural evolution of chymotrypsin. *J. Mol. Biol.* 128:49–79.

McLachlan, A.D. 1983. Analysis of gene duplication repeats in the myosin rod. *J. Mol. Biol.* 169:15–30.

McLachlan, A.D., and Stewart, M. 1976. The 14-fold periodicity in alpha-tropomyosin and the interaction with actin. *J. Mol. Biol.* 103:271–298.

Mehta, P.K., Heringa, J., and Argos, P. 1995. A simple and fast approach to prediction of protein secondary structure from multiply aligned sequences with accuracy above 70%. *Protein Sci.* 4:2517–2525.

Meiler, J., and Baker, D. 2003. Coupled prediction of protein secondary and tertiary structure. *Proc. Natl. Acad. Sci. USA* 100:12105–12110.

Metropolis, N., and Ulam, S. 1949. The Monte Carlo method. *J. Am. Stat. Assoc.* 44:335–341.

Minor, D.L., Jr., and Kim, P.S. 1996. Context-dependent secondary structure formation of a designed protein sequence. *Nature* 380:730–734.

Minsky, M.L., and Papert, S. 1988. *Perceptrons: An Introduction to Computational Geometry.* Cambridge, Mass., MIT Press.

Mittelman, D., Sadreyev, R., and Grishin, N. 2003. Probabilistic scoring measures for profile–profile comparison yield more accurate short seed alignments. *Bioinformatics* 19:1531–1539.

Nagano, K. 1973. Logical analysis of the mechanism of protein folding. I. Predictions of helices, loops and beta-structures from primary structure. *J. Mol. Biol.* 75:401–420.

Needleman, S.B., and Wunsch, C.D. 1970. A general method applicable to the search for similarities in the amino acid sequence of two proteins. *J. Mol. Biol.* 48:443–453.

Noble, W.S. 2004. Support vector machine applications in computational biology. In *Kernel Methods in Computational Biology* (J.-p. Vert, B. Schoelkopf, and K. Tsuda, Eds.). Cambridge, Mass., MIT Press, pp. 71–92.

Obradovic, Z., Peng, K., Vucetic, S., Radivojac, P., Brown, C.J., and Dunker, A.K. 2003. Predicting intrinsic disorder from amino acid sequence. *Proteins* 53 (Suppl. 6):566–572.

Ohlson, T., Wallner, B., and Elofsson, A. 2004. Profile–profile methods provide improved fold-recognition: A study of different profile–profile alignment methods. *Proteins* 57:188–197.

Ouali, M., and King, R.D. 2000. Cascaded multiple classifiers for secondary structure prediction. *Protein Sci.* 9:1162–1176.

Park, J., Karplus, K., Barrett, C., Hughey, R., Haussler, D., Hubbard, T., and Chothia, C. 1998. Sequence comparisons using multiple sequences detect three times as many remote homologues as pairwise methods. *J. Mol. Biol.* 284:1201–1210.

Pellegrini, M., Marcotte, E.M., and Yeates, T.O. 1999. A fast algorithm for genome-wide analysis of proteins with repeated sequences. *Proteins* 35:440–446.

Petersen, T.N., Lundegaard, C., Nielsen, M., Bohr, H., Bohr, J., Brunak, S., Gippert, G.P., and Lund, O. 2000. Prediction of protein secondary structure at 80% accuracy. *Proteins* 41:17–20.

Pollastri, G., Przybylski, D., Rost, B., and Baldi, P. 2002. Improving the prediction of protein secondary structure in three and eight classes using recurrent neural networks and profiles. *Proteins* 47:228–235.

Prilusky, J., Felder, C.E., Zeev-Ben-Mordehai, T., Rydberg, E., Man, O., Beckmann, J.S., Silman, I., and Sussman, J.L. 2005. FoldIndex: A simple tool to predict whether a given protein sequence is intrinsically unfolded. *Bioinformatics* 21:3435–3438.

Przybylski, D., and Rost, B. 2002. Alignments grow, secondary structure prediction improves. *Proteins* 46:197–205.

Ptitsyn, O.B. 1994. Kinetic and equilibrium intermediates in protein folding. *Protein Eng* 7:593–596.

Raghava, G.P.S. 2000. Protein secondary structure prediction using nearest neighbor and neural network approach. CASP 4, 75-76.

Raghava, G.P.S. 2002a. APSSP2: A combination method for protein secondary structure prediction based on neural network and example based learning. CASP 5. URL: http://www.imtech.res.in/raghava/apssp2/

Raghava, G.P.S. 2002b. APSSP: Automatic method for protein secondary structure prediction. CASP 5. URL: http://www.imtech.res.in/raghava/apssp2/

Ramirez-Alvarado, M., Serrano, L., and Blanco, F.J. 1997. Conformational analysis of peptides corresponding to all the secondary structure elements of protein L B1 domain: Secondary structure propensities are not conserved in proteins with the same fold. *Protein Sci.* 6:162–174.

Rao, S.T., and Rossmann, M.G. 1973. Comparison of super secondary structures in proteins. *J. Mol. Biol.* 76:241–256.

Reymond, M.T., Merutka, G., Dyson, H.J., and Wright, P.E. 1997. Folding propensities of peptide fragments of myoglobin. *Protein Sci.* 6:706–716.

Romero, P., Obradovic, Z., Li, X., Garner, E.C., Brown, C.J., and Dunker, A.K. 2001. Sequence complexity of disordered protein. *Proteins* 42:38–48.

Rose, G.D. 1979. Hierarchic organization of domains in globular proteins. *J. Mol. Biol.* 134:447–470.

Rost, B., and Sander, C. 1993. Prediction of protein secondary structure at better than 70% accuracy. *J. Mol. Biol.* 232:584–599.

Rost, B., Sander, C., and Schneider, R. 1994. Redefining the goals of protein secondary structure prediction. *J. Mol. Biol.* 235:13–26.

Rost, B., Schneider, R., and Sander, C. 1997. Protein fold recognition by prediction-based threading. *J. Mol. Biol.* 270:471–480.

Rychlewski, L., Jaroszewski, L., Li, W., and Godzik, A. 2000. Comparison of sequence profiles. Strategies for structural predictions using sequence information. *Protein Sci.* 9:232–241.

Salem, G.M., Hutchinson, E.G., Orengo, C.A., and Thornton, J.M. 1999. Correlation of observed fold frequency with the occurrence of local structural motifs. *J. Mol. Biol.* 287:969–981.

Sander, C., and Schneider, R. 1991. Database of homology-derived protein structures and the structural meaning of sequence alignment. *Proteins* 9: 56–68.

Schaffer, A.A., Aravind, L., Madden, T.L., Shavirin, S., Spouge, J.L., Wolf, Y.I., Koonin, E.V., and Altschul, S.F. 2001. Improving the accuracy of PSI-BLAST protein database searches with composition-based statistics and other refinements. *Nucleic Acids Res.* 29:2994–3005.

Schiffer, M., and Edmundson, A.B. 1967. Use of helical wheels to represent the structures of proteins and to identify segments with helical potential. *Biophys. J.* 7:121–135.

Schoelkopf, B., Tsuda, K., and Vert, J.-P.(Eds.). 2004. *Kernel Methods in Computational Biology.* Cambridge, Mass., MIT Press.

Schulz, G.E. 1988. A critical evaluation of methods for prediction of protein secondary structures. *Annu. Rev. Biophys. Biophys. Chem.* 17:1–21.

Selbig, J., Mevissen, T., and Lengauer, T. 1999. Decision tree-based formation of consensus protein secondary structure prediction. *Bioinformatics* 15:1039–1046.

Simossis, V.A., and Heringa, J. 2004a. The influence of gapped positions in multiple sequence alignments on secondary structure prediction methods. *Comput. Biol. Chem.* 28(5-6:351–366.

Simossis, V.A., and Heringa, J. 2004b. Integrating protein secondary structure prediction and multiple sequence alignment. *Curr. Protein Pept. Sci.* 5:249–266.

Simossis, V.A., and Heringa, J. 2005. SYMPRED consensus secondary structure prediction. http://ibi.vu.nl/programs/sympredwww/

Smit, A., Hubley, R., and Green, P. 2004. RepeatMasker open-3.0. 1996–2004. http://www.repeatmasker.org.

Smith, T.F., and Waterman, M.S. 1981. Identification of common molecular subsequences. *J. Mol.. Biol.* 147:195–197.

Soding, J. 2005. Protein homology detection by HMM–HMM comparison. *Bioinformatics* 21:951–960.

Stultz, C.M., White, J.V., and Smith, T.F. 1993. Structural analysis based on state-space modeling.*Protein Sci.* 2:305–314.

Sun, Z., Rao, X., Peng, L., and Xu, D. 1997. Prediction of protein supersecondary structures based on the artificial neural network method. *Protein Eng.* 10:763–769.

Szklarczyk, R., and Heringa, J. 2004. Tracking repeats using significance and transitivity. *Bioinformatics* 20 (Suppl. 1):I311–I317.

Taylor, W.R., Heringa, J., Baud, F., and Flores, T.P. 2002. A Fourier analysis of symmetry in protein structure. *Protein Eng.* 15:79–89.

Teodorescu, O., Galor, T., Pillardy, J., and Elber, R. 2004. Enriching the sequence substitution matrix by structural information. *Proteins* 54:41–48.

Tomii, K., and Akiyama, Y. 2004. FORTE: A profile–profile comparison tool for protein fold recognition. *Bioinformatics* 20:594–595.

Unger, R., and Moult, J. 1993. Finding the lowest free energy conformation of a protein is an NP-hard problem: Proof and implications. *Bull. Math. Biol.* 55:1183–1198.

Uversky, V.N., Gillespie, J.R., and Fink, A.L. 2000. Why are “natively unfolded” proteins unstructured under physiologic conditions? *Proteins* 41:415–427.

van Belkum, A., Scherer, S., van Alphen, L., and Verbrugh, H. 1998. Short-sequence DNA repeats in prokaryotic genomes. *Microbiol. Mol. Biol. Rev.* 62:275–293.

Vapnik, V.N. 1995. *The Nature of Statistical Learning Theory.* New York, Springer.

Vapnik, V.N. 1998. *Statistical Learning Theory.* New York, Wiley.

Vihinen, M., Torkkila, E., and Riikonen, P. 1994. Accuracy of protein flexibility predictions. *Proteins* 19:141–149.

von Ohsen, N., Sommer, I., and Zimmer, R. 2003. Profile–profile alignment: A powerful tool for protein structure prediction. *Pac. Symp. Biocomput.* 252–263.

von Ohsen, N., Sommer, I., Zimmer, R., and Lengauer, T. 2004. Arby: Automatic protein structure prediction using profile–profile alignment and confidence measures. *Bioinformatics* 20:2228–2235.

Vucetic, S., Brown, C.J., Dunker, A.K., and Obradovic, Z. 2003. Flavors of protein disorder. *Proteins* 52:573–584.

Wang, G., and Dunbrack, R.L., Jr. 2004. Scoring profile-to-profile sequence alignments. *Protein Sci.* 13:1612–1626.

Ward, J.J., McGuffin, L.J., Bryson, K., Buxton, B.F., and Jones, D.T. 2004. The DISOPRED server for the prediction of protein disorder. *Bioinformatics* 20:2138–2139.

Ward, J.J., McGuffin, L.J., Buxton, B.F., and Jones, D.T. 2003. Secondary structure prediction with support vector machines. *Bioinformatics* 19:1650–1655.

Waterman, M.S., and Eggert, M. 1987. A new algorithm for best subsequence alignments with application to tRNA–rRNA comparisons. *J. Mol. Biol.* 197: 723–728.

Wetlaufer, D.B. 1973. Nucleation, rapid folding, and globular intrachain regions in proteins. *Proc. Natl. Acad. Sci. USA* 70:697–701.

White, J.V., Stultz, C.M., and Smith, T.F. 1994. Protein classification by stochastic modeling and optimal filtering of amino-acid sequences. *Math. Biosci.* 119:35–75.

Wootton, J.C., and Federhen, S. 1996. Analysis of compositionally biased regions in sequence databases. *Methods Enzymol.* 266:554–571.

Wright, P.E., and Dyson, H.J. 1999. Intrinsically unstructured proteins: Re-assessing the protein structure–function paradigm. *J. Mol. Biol.* 293:321–331.

Xie, Q., Arnold, G.E., Romero, P., Obradovic, Z., Garner, E., and Dunker, A.K. 1998. The sequence attribute method for determining relationships between sequence and protein disorder. *Genome Inform. Ser. Workshop Genome Inform.* 9:193–200.

Yona, G., and Levitt, M. 2002. Within the twilight zone: A sensitive profile–profile comparison tool based on information theory. *J. Mol. Biol.* 315:1257–1275.

Young, M., Kirshenbaum, K., Dill, K.A., and Highsmith, S. 1999. Predicting conformational switches in proteins. *Protein Sci.* 8:1752–1764.

Yu, L., White, J.V., and Smith, T.F. 1998. A homology identification method that combines protein sequence and structure information. *Protein Sci.* 7:2499–2510.

Zemla, A., Venclovas, C., Fidelis, K., and Rost, B. 1999. A modified definition of Sov, a segment-based measure for protein secondary structure prediction assessment. *Proteins* 34:220–223.

Zvelebil, M.J., Barton, G.J., Taylor, W.R., and Sternberg, M.J. 1987. Prediction of protein secondary structure and active sites using the alignment of homologous sequences. *J. Mol. Biol.* 195:957–961.

Further Reading

Jones, N.C., and Pevzner P.A. 2004. *An Introduction to Bioinformatics Algorithms.* Cambridge, MA, MIT Press.

Konopka, A.K., and Crabbe, M.J.C. (Eds.). 2004. *Compact Handbook of Computational Biology*. New York, Dekker.

8 Protein Contact Map Prediction

Xin Yuan and Christopher Bystroff

8.1 Introduction

Proteins are linear chains that fold into characteristic shapes and features. To understand proteins and protein folding, we try to represent the protein molecule in such a way that its features are easy to see and manipulate. A simple representation facilitates algorithm design for structure prediction. The simplicity of the three-state character string representation of secondary structure is part of the reason for secondary structure prediction receiving so much attention early in the era of computational biology. One-dimensional strings are easily understood, parsed, mined, and manipulated. But secondary structure alone does not tell us enough about the overall shapes and features of a protein. We need a simple way to represent the overall tertiary structure of a protein.

Here we explore a two-dimensional Boolean matrix representation of protein structure, where each dimension is the residue number and each value is true if the residues are spatial neighbors and false otherwise—called a "contact map." A contact map is the simplest representation of a protein that can be faithfully projected back into three dimensions. As such it has received increased attention in recent years from bioinformaticists, who see this as a data structure that is readily amenable to data mining and machine learning.

The first goal of this chapter is to introduce the contact map data structure, how it is calculated from the three-dimensional structure and how it is transformed from two dimensions into three. Then we will explore a series of computational methods that have attempted to predict contact maps directly from the primary sequence, with or without the help of template structures from the protein database. Next, we will discuss the various ways that contact map predictions may be evaluated for accuracy. Finally, we will present some of the other ways contact maps have proven useful.

8.2 Definition of Interresidue Contacts and Contact Maps

Interresidue contacts have been defined in various ways. Routinely, a contact is said to exist when a certain distance is below a threshold. The distance may be that between Cα atoms (Vendruscolo et al., 1997), between the Cβ atoms (Lund et al., 1997; Thomas et al., 1996; Olmea and Valencia, 1997; Fariselli et al., 2001a,b), or it may be the minimum distance between any pair of atoms belonging to the side chain or to the backbone of the two residues (Fariselli and Casadio, 1999; Mirny and Domany, 1996). The following definitions of the interresidue distance D_{ij} have appeared at some time in the literature, each associated with a threshold distance:

1. The distance between alpha carbons (CA)
2. The distance between beta carbons (CB), using alpha carbon for glycine
3. The minimum distance between the van der Waals (vdW) spheres of heavy atoms (HS)
4. The minimum distance between the vdW spheres of backbone heavy atoms (BB)
5. The minimum distance between the vdW spheres of side-chain heavy atoms or alpha carbons (SC)

By one informed account, the best definition for interresidue distance is SC, the minimum distance between side-chain or alpha-carbon atoms (Berrera et al., 2003). Using a cutoff distance of around 1.0 Å between vdW spheres, SC-based contact maps efficiently recognized homologue sequences in a "threading" experiment, where a query sequence is assigned an energy score for every possible alignment of the sequence to a set of template structures. In a threading experiment, the definition of a contact determines which residues in the sequence are used to sum the energy. Using the minimum distance between residues and using a short distance cutoff makes the definition of a contact more energetically realistic, and this makes the sum of amino acid contact potentials a better approximation of the true energy. Contact potentials are energy functions that measure the pairwise side-chain-dependent free energy of residue–residue contacts, irrespective of the side-chain conformations. Contact potentials may be derived from contact maps statistics, as described in a later section.

Simpler, backbone atom-based definitions (CA or CB) with longer distance cutoffs are more readily projected into three dimensions, since the atomic positions used to calculate the distance depend only on the backbone angles. CB is slightly more meaningful than CA in an energetic sense, since side chains that point toward each other, and therefore have a shorter CB distance than CA distance, are more likely to make an actual physical contact. Using the minimum distance measures (HS, BB, or SC) can make projection into three dimensions more difficult because we have not saved the precise identity of the contacting atoms in the Boolean matrix. CB distances with a cutoff of 8 Å were chosen for use in the Critical Assessment of Structure Prediction (CASP) experiments (Moult et al., 2003), and this is the definition that we will discuss here, unless otherwise specified.

Having defined what we mean by the distance D_{ij}, the definition of a contact map is a straightforward distance threshold. For a protein of N amino acids the contact map is an $N \times N$ matrix C whose elements are given, for all $i, j = 1, \ldots, N$, by

$$C_{ij} = \begin{cases} 1 & \text{if } D_{ij} < Dcutoff \\ 0 & \text{otherwise} \end{cases} \tag{8.1}$$

The set of all D_{ij} is commonly referred to as the "distance matrix." Therefore, we can think of C_{ij} as a *thresholded distance matrix*. The mean difference between two distance matrices is sometimes called the "distance matrix error" (DME), as follows:

$$DME\,(a,b) = \frac{\sum_{i=i}^{N-loc} \sum_{j=i+loc}^{N} \left| D_{ij}^{a} - D_{ij}^{b} \right|}{0.5\,(N - loc - 1)\,(N - loc)} \tag{8.2}$$

DME is variously defined as the average of absolute differences or the root-mean-square distance difference, often with a cutoff (*loc*) to exclude local distances. The DME can be shown to correlate with the root-mean-square deviation (RMSD) in atomic positions if both numbers are derived from the same structures:

$$RMSD\,(a,b) = \sqrt{\frac{\sum_{i=1,N} \left(x_i^a - x_i^b\right)^2}{N}} \tag{8.3}$$

By association, since the contact map error (CME) is a crude approximation of the DME, we can say that the sum of differences between two contact maps is a crude approximation of the RMSD between the two proteins they represent:

$$CME\,(a,b) = \frac{\sum_{i=1}^{N-loc} \sum_{j=i+loc}^{N} \left| C_{ij}^{a} - C_{ij}^{b} \right|}{0.5\,(N - loc - 1)\,(N - loc)} \tag{8.4}$$

But this is at best a rough correlation, and then only under the special constraint that each contact map C_{ij} is derived from a 3D structure. As we will discuss later, a simple measure such as CME by itself is usually not a good indicator of structural prediction accuracy. This topic is discussed again in Section 8.5.

8.3 Features of a Contact Map

Contact maps and distance matrices are "internal coordinates," and as such are independent of the reference frame of the Cartesian atomic coordinates. This frame invariance, plus the Boolean property, makes contact maps attractive to practitioners of machine learning and data mining techniques. Patterns within contact maps are meaningful even when taken out-of-context.

It is well-known that the number of contacts scales linearly with the chain length (Thomas et al., 1996; Vendruscolo et al., 1997; Fariselli and Casadio, 1999). The slope of the linear dependence depends only on how a contact is defined. Using CB distances and a cutoff distance of 8 Å, and ignoring local contacts with $|i - j| < 3$, the number of contacts in a compact globular protein is approximately 3.0 times the length of the protein, with a relatively small standard deviation of ±0.4. Since every contact involves two residues, this number implies an average of about 6 (±0.8)

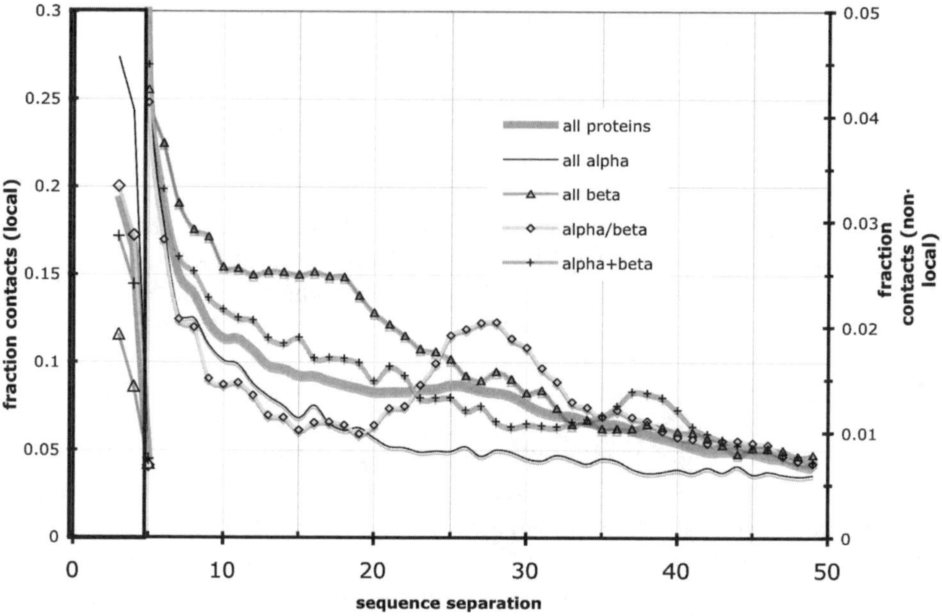

Fig. 8.1 Fraction of all CB contacts with cutoff distance 8.0 Å as a function of sequence separation distance for the four main SCOP classes of proteins. About half of all contacts are local ($3 \leq |i - j| \leq 5$, left axis). Different fold classes have significant differences in the contact profile. The peaks at around 28 in alpha/beta proteins correspond to the sequence distance where parallel strands are separated by one alpha helix, called βαβ-units.

contacts per residue. These numbers are consistent across all fold classes, probably reflecting the invariant packing density and size of amino acids. Parallel α/β proteins deviate most from this average, with an average of 3.3 contacts per residue, but this difference is less than one standard deviation. There are many protein chains with far fewer than three contacts per residue but these are generally not globular domains. Instead they are often parts of larger complexes which, when taken together, also average six contacts per residue.

Most interresidue contacts in proteins are local, and the likelihood of finding a contact drops quickly as the sequence distance between residues increases. There are interesting and obvious class-dependent differences in the sequence separation profile of contacts (Fig. 8.1). This distribution is important to consider when assessing the accuracy of contact map predictions, since local contacts are easier to predict than nonlocal ones. The "contact order" of a protein is defined as the average sequence distance between contacting residues, and this number has been shown to correlate with the folding rate for many small proteins (Plaxco et al., 1998). Some studies use the contact order as a measure of the topological compexity of the fold (Kuznetsov and Rackovsky, 2004; Punta and Rost, 2005). Recently, the notion of contact order has been refined to take nested loop closures into account, giving an "effective contact order" which is probably a much better measure of fold complexity (Chavez

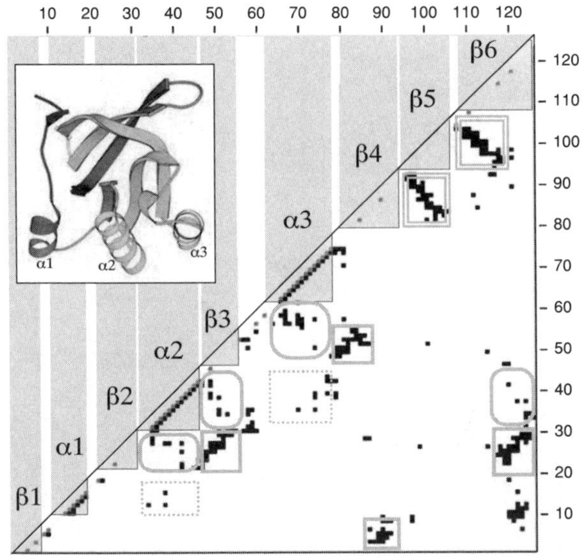

Fig. 8.2 Contact map of glutathione reductase, domain 2 (PDB code 3GRS, residues 166–290). Black boxes are contacts, gray boxes: $i, i+3$ contacts, shaded triangles: contacts within secondary structure elements, gray rectangles: parallel beta-strands, double rectangles: antiparallel beta-strands, dotted rectangles: helix–helix contacts, rounded rectangles: helix–strand contacts. Inset: Molscript (Kraulis, 1991) drawing of 3GRS structure.

et al., 2004). In this study it was understood that the contact order should reflect the configuration entropy lost on the formation of a contact. The effective contact order is the entropy of the closure of a loop that may already contain contacts within it.

A trained eye can identify secondary structure elements in a contact map by looking at the local contacts, i.e., those near the diagonal of the matrix. A helix has an unbroken row of contacts between $i, i \pm 4$ pairs. Extended strands have no local contacts with $3 < |i - j| < 5$, although occasional $i, i+3$ contacts occur in β strands where β-bulges or β-bends occur. Loops have some local contacts but never an unbroken row. Figure 8.2 shows images of common contact patterns that are found between secondary structure elements. Antiparallel and parallel β strands give rise to unbroken rows of contacts in the off-diagonal region. A row of contacts that is perpendicular to the diagonal of the matrix represents a pair of antiparallel strands. These are contacts between residues $i + k$ and $j - k$, where k goes from zero through the length of the strand pairing. Similarly, a row of contacts that is parallel to the diagonal represents a pair of parallel strands, with contacts between $i + k$ and $j + k$. Consequently, β sheets appear as a set of perpendicular or parallel rows of contacts. The strand order can be determined by tracing the pairing interactions (gray rectangles in Fig. 8.2). Contacts between α-helices and other secondary structure elements appear as broken rows or “tire tracks.” If the two contacting elements are both helices, then the contacts appear every three or four residues in both directions, following the periodicity of the helix. If one of the elements is a strand, then we see

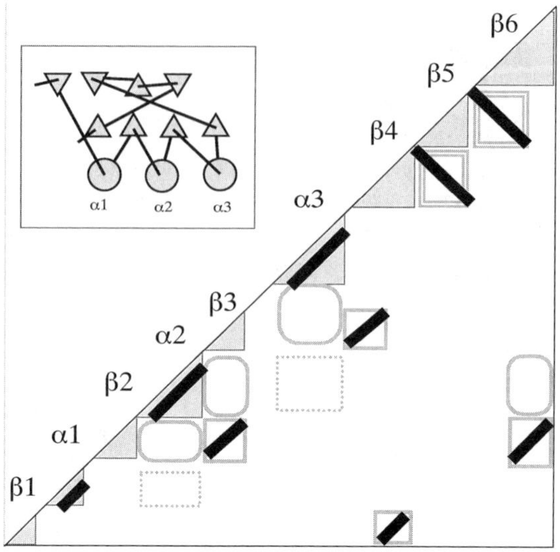

Fig. 8.3 Idealized features in contact maps (thick bars) may be converted to a topological cartoon (Michalopoulos et al., 2004) using simple drawing conventions.

a periodicity of two in the contacts in that direction, since the side chains in a strand alternate sides of the sheet. Domains can be seen as regions of the chain that have dense contacts, since intradomain contacts outnumber interdomain contacts.

If there is additional knowledge to resolve the ambiguity in overall handedness, then the entire molecule can be reconstructed by hand from a contact map. For example, for α/β proteins we can assume that any parallel β-α-β unit has a right-handed crossover (more than 99% of all parallel β-α-β supersecondary structure units are right-handed). If our assumption is right, then we know on which side of the sheet to place the helix. The presence or absence of helix–helix contacts can be used to resolve the placement of any additional helices with respect to the sheet. However, without some external information about either the overall handedness or the handedness of any substructure, two mirror-image reconstructions are possible. Figure 8.3 shows an idealized contact map, the same one shown in Fig. 8.2, and the corresponding protein topology (TOPS) cartoon (Michalopoulos et al. 2004) that can be drawn using only the simplified contact map. Although TOPS cartoons such as this one cannot be accurately projected to three dimensions without additional information such as key contacts, the TOPS graph structures allow the easy visualization of common topological features in proteins.

8.4 From Contact Map Prediction to 3D Structure

Contact maps that are derived from 3D protein structures can be mapped back to their corresponding structures by taking advantage of the known stereochemistry of amino acids and proteinlike backbone angles. But not all square, symmetrical Boolean matrices map to 3D objects, much less to proteinlike objects.

In mathematical terms, a contact map is an undirected graph, where the vertices are the residues and the edges are the residue–residue contacts. But contact maps that have been derived from true protein structures, or from any other set of points in three dimensions, are a special subset of all undirected graphs called "sphere intersection graphs" or "sphere of influence graphs" (SIGs) (Michael and Quint, 1999). In a SIG the edges represent the intersections of fixed-radius spheres. If a graph is a SIG, then at least one solution exists for the positions of the vertices in 3D. The thresholded distances from the solution configuration must correspond to the contacts in the contact map exactly, or the graph is not a SIG. If there is no solution, then the contact map without modification cannot represent a protein, or for that matter, any set of points in 3D! However, there may exist a subset of the contacts that can potentially represent a protein. The problem of mapping a predicted contact map to 3D is the problem of finding the best SIG within a contact map.

Determining whether or not a contact map is a SIG remains an open problem for the general case (Michael and Quint, 1999). But heuristic methods can be applied that use additional information about proteins, including the key facts that (1) adjacent vertices are linked with their distance fixed at 3.8 Å and (2) that all nodes are self-avoiding (i.e., no two nodes can be closer than 3.8 Å). Figure 8.4 illustrates the key constraints on a proteinlike SIG. In addition to these constraints, proteins have

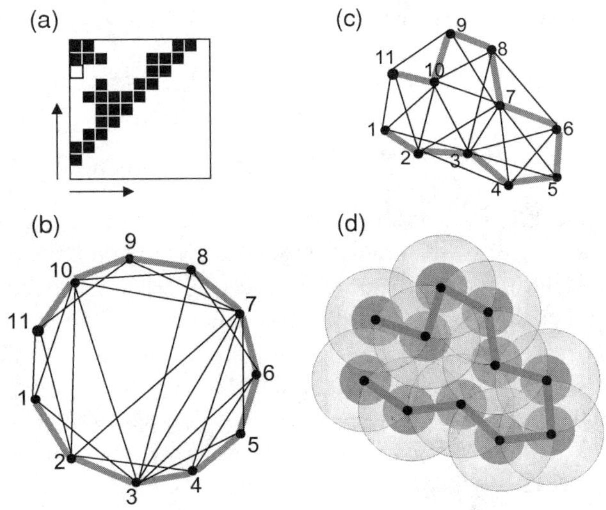

Fig. 8.4 A proteinlike sphere intersection graph (SIG). For a contact map (a) can always be projected to an undirected graph where the vertex positions satisfy nearest neighbor distance constraints and self-avoidance (b). This contact map is a proteinlike SIG because vertex positions are possible (c) such that each edge distance corresponds to a sphere intersection (d, large circles) and all vertices are mutually avoiding (dark circles). The addition of a single contact between 1 and 9 [white box in (a)] breaks the SIG.

characteristic secondary structures and turns that are sequence dependent and restrict the way nodes can be arranged locally along the chain. So while there is still no general solution for finding a SIG within a contact map, the problem of finding a "proteinlike SIG" seems tractable and it is likely it will be solved in the near future.

If the contact map is a proteinlike SIG, then it is possible to reproduce, with considerable accuracy, the 3D structure of the protein's backbone from its contact map (Havel et al., 1979; Saitoh et al., 1993). And at least one heuristic approach has been shown to work in the presence of "noise" contacts, accurately excluding random physically impossible contacts that were added to a true protein contact map (Vendruscolo et al., 1997; Vendruscolo and Domany, 1998). Vendruscolo's method works by minimizing a cost function that contains only geometric constraints, nothing resembling the true energies of the polypeptide chain. The task of predicting the tertiary structure of a protein is split into two steps, making it a crude pathway model. First, a reliable prediction of secondary structure must be realized, then a coarse-grained contact map is used to select contacts between the secondary structure elements. The method succeeds even when up to 10% of the contacts are "noise." Interestingly, it is now possible to reconstruct a contact map from a 1D representation consisting of principal eigenvectors (PE) derived from HS contact maps (Porto et al., 2004). The PE reconstruction of the contact combined with 3D projection using Vendruscolo's method builds models that are typically within RMSD 2.0 Å of the original structure. Unfortunately, there is still a large gap between the prediction accuracy necessary for a good 3D reconstruction and the prediction accuracy possible using today's methods. Worse than that, the distribution of erroneous contact predictions in real cases is probably not random, as this reconstruction algorithm assumes.

8.5 Contact Map Prediction

Contact prediction offers a possible shortcut to predict protein tertiary structure. Over the years, a variety of different approaches have been developed for contact map prediction including neural networks (Fariselli et al., 2001a,b; Pollastri and Baldi, 2002; Lund et al., 1997), support vector machines (Zhao and Karypis, 2003), and association rules (Zaki et al., 2000). Statistical approaches have also been tried, including correlated mutations (Olmea and Valencia, 1997; Thomas et al., 1996; Singer et al., 2002), knowledge-based potentials (Sippl, 1990; Park et al., 2000), and hidden Markov models (Shao and Bystroff, 2003). Statistical pair potentials do not produce sufficiently specific contact predictions. More specific information appears to come from neighboring residues and patterns of mutation, sequence conservation, and predicted secondary structure, all obtainable from multiple sequence alignments. The various features include contacts from patterns of conserved hydrophobic amino acids (Aszodi et al., 1995), sequence profiles derived from multiple sequence alignment (Fariselli et al., 2001a,b; Pollastri and Baldi, 2002; MacCallum, 2004; Hamilton et al., 2004; Shao and Bystroff, 2003), distribution of distances in

Table 8.1 Available servers for contact map predictions

Server	URL	Reference(s)
CORNET	gpcr.biocomp.unibo.it/cgi/predictors/cornet/pred_cmapcgi.cgi	Olmea & Valencia, 1997, Fariselli & Casadio, 1999
PDG	www.pdg.cnb.uam.es:8081/pdg_contact_pred.html	Pazos et al., 1997
HMMSTR	www.bioinfo.rpi.edu/~bystrc/hmmstr/server.php	Shao & Bystroff, 2003
GPCPRED	sbcweb.pdc.kth.se/cgi-bin/maccallr/gpcpred/submit.pl	MacCallum, 2004
PoCM	foo.acmc.uq.edu.au/~nick/Protein/contact.html	Hamilton et al., 2004
CMAPpro	www.ics.uci.edu/~baldig/	Cheng et al., 2005

proteins with known structures (Tanaka and Scheraga, 1976; Wako and Scheraga, 1982; Huang et al., 1995; Mirny and Domany, 1996; Maiorov and Crippen, 1992), correlated mutation and/or combination with other features (Olmea and Valencia, 1997; Fariselli et al., 2001a,b; Pollastri and Baldi, 2002; Hamilton et al., 2004; Göbel et al., 1994; Neher, 1994; Shindyalov et al., 1994), secondary structure information (Shao and Bystroff, 2003; Zaki et al., 2000; Hamilton et al., 2004; Fariselli et al., 2001a,b; Olmea and Valencia, 1997; Zhang and Kim, 2000; Hu et al., 2002). Beyond ones and zeros of a contact map, knowledge-based estimates of residue–residue distance have been used to determine the approximate structure of proteins (Skolnick et al., 1997; Wako and Scheraga, 1982; Monge et al., 1994; Aszodi et al., 1995).

The results in CASP5 (Aloy et al., 2003) and CASP6 (Graña et al., 2005) suggest that there has been at best a very limited improvement for *de novo* contact prediction methods. In the following sections we summarize a few of these approaches to contact map prediction in detail, with an eye toward possible improvements. Table 8.1 lists the currently available web servers for contact map prediction.

8.5.1 Contact Prediction Using Statistical Models

In sequence alignments, some pairs of positions appear to covary in a physicochemically plausible manner, i.e., a "loss of function" point mutation may be rescued by an additional mutation that compensates for the change (Altschuh et al., 1987). Compensating mutations would be most effective if the mutated residues were spatial neighbors; therefore, "correlated mutations" across evolutionary distance should imply spacial proximity. Attempts have been made to quantify this hypothesis and to use it for contact predictions (Neher, 1994; Göbel et al., 1994; Taylor and Hatrick, 1994).

Direct statistical methods require pairwise scoring matrices to compute the contact scores. The scoring matrices are based on *a priori* models of noncovalent residue interactions and/or protein evolution. In various approaches, the matrices have been based on amino acid identity (Shindyalov et al., 1994), amino acid substitution probabilities (Göbel et al., 1994), contact substitution probabilities (Rodionov and Johnson, 1994), biophysical complementarity of electrostatic charge and side chain volume (Neher, 1994), or statistics from evolutionary models (Singer et al., 2002). In the latter case, the energetic value of a contact was estimated as a likelihood matrix, using a large set of proteins of known structure. Mutations are correlated because side-chain interactions have an energetic value, and this energetic value is therefore reflected in the database contact statistics (Fig. 8.5). The predicted target contact energies were calculated by first generating a multiple sequence alignment and then summing the likelihood of all residue pairs in the corresponding columns. The likelihood approach performed better when contacts were local in the sequence, but tended to perform poorly on nonlocal contacts. If combined with other features, the method could give better predictions.

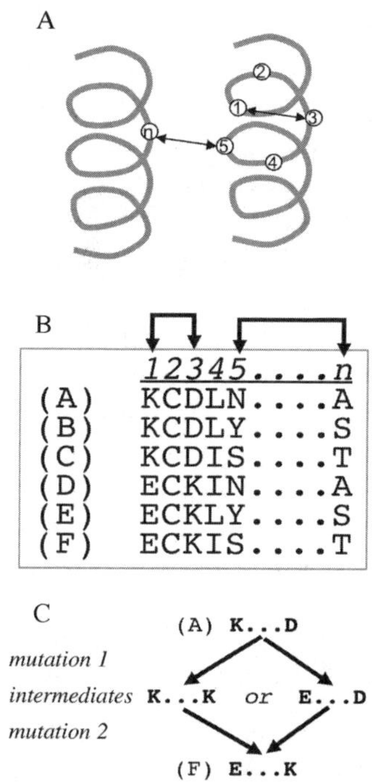

Fig. 8.5 Illustration of correlated mutation theory and application. (A) Several residues are shown in their structure context, in this example, two nearby α-helices. (B) For these, six sequences (A–F) are shown as a multiple alignment. Positions 1 and 3 show correlated substitutions (connected by arrows), as do positions 5 and *n*. (C) The most parsimonious evolutionary pathways are between sequences A and F, for positions 1 and 3. Correlated mutation detects pairs of residue positions that show correlated substitutions without intermediates. The theory is that when a mutation occurs in a structurally important residue (mutation 1), the intermediate has structural instability. Compensatory mutations are then selected (mutation 2) and the structural interaction is restored. Any intermediates are eventually eliminated from the sequence record due to reduced fitness. (Based on a figure from Singer et al., 2002.)

The correlated (or compensatory) mutation information is generally weak. Contact prediction can be improved by combining correlated mutations with other data such as sequence conservation and contact density information (Hamilton et al., 2004). The principle behind contact density is simple. If two nonadjacent residues are in contact, then we expect that the residues adjacent to them will also be in contact with a high probability. Correlated mutations have been combined with other sources of information in some of the methods described in the following sections.

A simpler statistical method is the sequence conservation at single positions. The success of the evolutionary trace method (Lichtarge et al., 1996) in identifying localized side chains based on functional conservation in protein sequence families shows that sequence conservation is both biologically and statistically significant when combined with known structure. In this method, conserved positions are mapped to the surface of a known protein and clustered to find functional sites. In practice, sequence conservation is not used alone but rather as a component of the training data from neural networks, described in the next section.

8.5.2 Contact Maps from Neural Networks

Both the correlated mutation and likelihood approaches performed best on local contacts, but tended to perform poorly on longer sequences where many contacts were nonlocal. Another approach to the problem has been to train neural networks with various encodings of multiple sequence alignments with other inputs such as predicted secondary structure (Fariselli and Casadio, 1999; Fariselli, 2001a,b). These tended to perform better over a wide range of sequence lengths. Fariselli's CORNET predictor claims to have the best contact prediction results to date. It was specifically designed to include evolutionary information in the form of a sequence profile, sequence conservation, correlated mutations, and predicted secondary structures. Sequence conservation was taken from the HSSP database (Dodge et al., 1998). Correlated mutations were calculated as previously described (Olmea and Valencia, 1997; Göbel et al., 1994). This neural network approach involved encoding frequencies of residues in columns of a multiple sequence alignment, as well as having inputs based on predicted secondary structures, length of input sequence, and residue separation. Briefly, each position in the alignment has a distance array that contains the interresidue distances between all of the possible pairs of sequences at that position. The distance between residues is defined using an early amino acid scoring function (McLachlan, 1971). The correlation value between each pair of positions in the alignment is computed as the correlation of the two arrays for each possible residue pair. The network was trained by using the back-propagation algorithm, with a single output neuron coding for contact (1) and noncontact (0). Contacts were defined using Cβ atoms (CB) with an 8-Å cutoff, and only those separated by at least six residues were used. The hidden layer consisted of eight neurons. Each residue pair in the sequence was coded as an input vector of 210 elements ($20 \times (20 + 1)/2$), representing all possible pairs of amino acids. CORNET has an average off-diagonal (nonlocal) contact accuracy of 21%. While this result is more than six times better than a chance prediction, it is still far from providing sufficient accuracy for a reliable 3D reconstruction.

In GIOHMM (Pollastri and Baldi, 2002), a new neural net architecture was introduced. The contact matrix was represented as a 2D graph. It is implemented in two steps. The first step is the construction of a statistical graphical model (Bayesian network) for contact maps, where the states are arranged in one input plane, one output plane, and four hidden planes. The parameters of the Bayesian network are the local conditional probability distributions. The second step is the reparameterization of the graphical model using artificial recurrent neural networks. In the training of the neural net, the input includes the information for the contact, secondary structure, and solvent accessibility. The authors cite a prediction accuracy of 60.5% for CB contacts with an 8-Å cutoff and 45% for CB contacts with a 10-Å cutoff, but only local contacts were considered ($|i - j| < 7$). While intriguing, these numbers cannot be compared directly with those mentioned above. Prediction of local contacts is intermediate between secondary structure prediction, for which the highest three-state prediction accuracies average 75–80% (Jones, 1999), and nonlocal contact map prediction, for which a highest accuracy of 21% has been reported. The same group (P. Baldi) has recently released a new contact map predictor, CMAPpro, as part of a battery of tools for protein feature prediction (Cheng et al., 2005). The innovation in this neural net architecture is a heirarchical scheme where the output of local contact predictions is used as the input for predicting nonlocal contacts.

Lund et al. combined two independent data driven methods (Lund et al., 1997). The first used statistically derived probability distributions of the pairwise distance between two residues, similar to the knowledge-based pair potentials of Sippl (1990). The second consisted of a neural network with a single hidden layer connected to two three-residue windows a defined distance apart on the sequence. For both of these functions, the underlying physical determinants of the statistics are the various chemical affinities between short sequence patterns of amino acid side chains. Nonpolar side chains attract through the hydrophobic effect, polar side chains through hydrogen bonds and salt bridges. This affinity alone does not determine the likelihood of a contact but is combined with sequence separation distance, since the polypeptide chain has a certain degree of stiffness that limits the ways the side chains can come together when the loop is short. Their results showed that prediction by neural networks is more accurate than predictions by probability density functions. The accuracy of the prediction can be increased by using sequence profiles instead of single sequences.

As mentioned earlier, patterns of contacts form when an α-helix is in contact with a strand, a helix with a helix, or when two strands are paired in a β sheet. A recent study used a neural network approach to find patterns of correlated mutations (Hamilton et al., 2004). The main input to the neural network was a matrix of 25 mutational correlation values for a pair of five-residue windows centered on the residues of interest. Each entry in the matrix is the correlation between two residues (Göbel et al., 1994). This information was combined with other inputs such as predicted secondary structure using Psi-Pred (Jones, 1999; McGuffin et al.,

2000), the type of amino acids, and the input sequence length. Using this method an average prediction accuracy of 21.7% was obtained. The accuracy was found to be relatively consistent across different sequence lengths, but to vary widely with the secondary structure. As with previous studies, contact predictions were found to be particularly difficult for α-helical proteins (Fariselli and Casadio, 1999, Fariselli et al., 2001b). Fariselli suggested that the poor predictions from their methods on this subset of proteins might be a result of the underrepresentation of α-type proteins in the training set. But in Hamilton et al., even if they trained the model on proteins of α-type to predict an α-type protein, no improvement in prediction accuracy was obtained. It might indicate that the patterns of contact are less locally defined in α-helical proteins and may require the window size to be larger. Alternatively, the predictions could be improved by finding a better measure of correlated mutations, and perhaps by applying the contact occupancy filtering as described in Olmea and Valencia (1997).

8.5.3 A Genetic Algorithm for β-Strand Contacts

MacCallum has noted that protein architectures impose regularities in local sequence environments (MacCallum, 2004). Based on the fact that many proteins have pairs of neighboring strands with similar sequence patterns, the GPCPRED algorithm used only sequence profile and residue separation information as input to a genetic programming approach to contact prediction. Sequence profiles are classified using a self-organizing map algorithm (SOM), and the new classes reveal a distinctive "striping" pattern across facing strand pairs. The predictions were equal to or better than existing automated contact predictors that use more fitting parameters. Predictions of sets of "$L/10$" contacts (i.e., number of contacts predicted equals length of protein over 10), each between positions separated by at least eight residues, were 27% correct for proteins up to length $L = 400$. As they suggest, the predictions could be improved if they included additional information such as sequence conservation and correlated mutations. As good as they are, the predictions cannot be uniquely mapped to three dimensions, but with an additional postprocessing step based on the packing rules, this could be remedied.

8.5.4 Contact Prediction Using Support Vector Machine

A support vector machine (SVM) is a method for binary classification in an arbitrary feature space and as such is well-suited for the contact map problem. In one study (Zhao and Karypis, 2003), contact and noncontact residue pairs were treated as positive and negative instances in a feature space comprised of position-dependent information for amino acid content, physicochemical environment, secondary structure, and evolutionary correlation. SVM was used to define an optimal multidimensional hyperplane for dividing contacts and noncontacts in the space of the features. The model was trained on all classes of protein structure in the CATH database (Orengo et al., 1997). The results indicated that the secondary structure feature is most helpful for contact prediction in proteins containing β strands. On the other hand, correlated mutations and sequence profile methods performed the best for proteins containing α-helices. Models learned separately for different protein classes might result in better performance in contact prediction.

8.5.5 Prediction Using Association Rules

Data mining was used to extract valuable information from true contact maps (Hu et al., 2002; Zaki et al., 2000) in the form of recurrent nonlocal contact patterns and sequence–contact association rules. Zaki et al. developed a string encoding and hashing technique to extract all of the nonlocal contact patterns for a sliding window across all contact maps of existing structures. The contact patterns were clustered based on their similarities, and sequence-to-contact relationships were expressed as logical statements, or association rules. By applying association rules to the output of the hidden Markov model HMMSTR (Bystroff et al., 2000), their contact map predictions had about 20% accuracy for $L/2$ contacts with $|i - j| \geq 4$, corresponding to about 20% coverage. Even with low coverage the predictions contained physically impossible combinations of contacts (see Fig. 8.6). To make the predictions more meaningful, there is a need to filter out the physically impossible contacts.

8.5.6 Prediction Using Pathway Models

A contact prediction method that makes use of sequence profiles, fragment templates, and pathway models was used for the first time in the CASP5 experiment (Shao and Bystroff, 2003), with accuracies comparable to or higher than previous approaches, depending on how accuracy is measured. In this prediction method, the first step is to assign a probability to each potential contact. The probability in this case is the database-derived likelihood of contact between any two local structure motifs.

Local structure motifs were predicted probabilistically as Markov states from the HMMSTR model (Bystroff et al., 2000). A matrix γ expresses the probability of each motif at each sequence position, solved using the Forward/Backward algorithm (Rabiner, 1989):

$$\gamma\,(i, q) = P(q|i) \tag{8.5}$$

Then the contact potential $G(p, q, s)$ between any two HMMSTR states p and q, given a sequence separation s, was calculated as the negative log of the sum over all joint probabilities as follows:

$$G(p, q, s) = -\log \frac{\sum\limits_{CATH} \sum\limits_{i \ni D_{i,i+s} < 8\mathring{A}}}{\gamma(i, p)\gamma(i+s, q)} \sum_{CATH} \sum_{i} \gamma(i, p)\gamma(i+s, q) \tag{8.6}$$

In the numerator, the sum is over all residue pairs $(i, i + s)$ that are in contact in all CATH proteins. In the denominator, the sum is over all residue pairs $(i, i + s)$. To predict contacts, we first calculate the contact potential E_{ij}, by summing $G(p, q, s)$ over all states p at i and all states q at j. E_{ij} may be thresholded to give a contact map prediction.

This algorithm implies that the local structure motif folds first, followed by motif–motif condensation to form larger units. Rule-based filtering techniques were applied to remove contacts that were impossible given a previously defined set of contacts. "Common sense" rules were applied. For example, any one β-strand residue may pair with at most two other β strands, not three of course. Other rules enforced the physically possible density of contacts and mutual contacts, and the triangle inequality. In addition, contacts were assigned only if they had an effective sequence separation of 8 or less after "loop closure," similar to the effective contact order (Chavez et al., 2004). This gave local contacts opportunity to form before assigning nonlocal contacts (Fig. 8.6). This simple folding pathway model was sufficient to extract the correct set of contacts for some but not all of the CASP5 targets (Shao and Bystroff, 2003; Bystroff and Shao, 2003). The most common error was the wrong

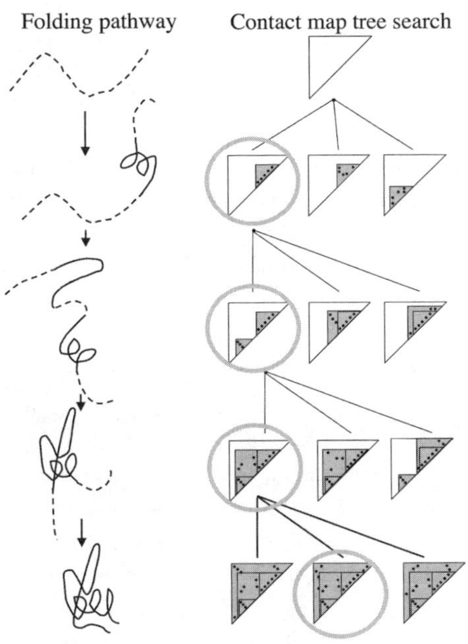

Fig. 8.6 In the HMMSTRCM ("hamster CM") method, a folding pathway is expressed as a tree search in contact map space where each branch represents the addition of new contacts to the previous set of contacts (shaded triangles, thick lines). An energy function may be applied to select among alternative sets of contacts. Local contacts (shaded triangles) form first. These come together (larger shaded triangles), subject to a set of simple rules. HMMSTRCM succeeded in cases where the initial contacts were correctly assigned, but could not overcome bad initial assignments.

choice of the nucleation site, since early errors propagated to further errors. A better means to choose the folding nucleation site would remedy this problem.

8.6 Evaluation of Contact Map Predictions

The current evaluation criteria used for contact map predictions include (Graña et al., 2005; Aloy et al., 2003; Koh et al., 2003): (1) accuracy: the number of correctly predicted contacts divided by the total number of predicted contacts; (2) coverage: the number of correctly predicted contacts divided by the total number of contacts; (3) improvement over random: the calculated accuracy divided by the random accuracy; and (4) the delta evaluation: the percentage of correctly predicted contacts that are within a certain number (delta) of residues of the experimental contact, measured along the sequence.

Another useful measure is the distance distribution of predicted contacts, X_d:

$$X_d = \sum_{i=1}^{15} \frac{P_{ip} - P_{ia}}{15d_i} \tag{8.7}$$

where the sum runs over 15 distance bins covering the range from 0 to 60 Å. d_i is the distance representing each bin. P_{ip} is the percentage of predicted contacts whose true distance is in bin i. P_{ia} is the same percentage but for all of the residue pairs, not just contacts. Defined in this way, $X_d > 0$ indicates that more of the predicted contacts are either true contacts or close to being true contacts. $X_d \leq 0$ indicates that the contacts are random (Pazos et al. 1997).

Each of these criteria is well-behaved when the prediction is close to perfect, but they diverge to different extents from good behavior when the prediction strays from the path of perfection. A good predicted contact map should map uniquely to the correct 3D structure. Contact maps may be divided into blocks representing contacts between secondary structure elements (Figs. 8.2 and 8.3). If a prediction identifies most of the contact blocks but the overall accuracy is low, we may still recognize it as a good prediction that maps to a single correct structure. On the other hand, if the accuracy is high but correct contacts concentrated in the local region or in one block of the molecule, then it is less meaningful as a 3D structure. In our opinion, we might have included the “block count” as another evaluation criterion. It would be defined as the total number of counts of true contact blocks, with one count for each block. The higher the block count, the better the prediction is. Since the nonlocal contacts are harder to predict than the local ones, we might give more weight to the nonlocal contacts.

If we look at the parallels between contact map prediction and the solution of protein structures by NMR, we can see a more logical way of defining accuracy in contact map prediction. NMR structures are solved by applying distance geometry methods while minimizing a cost function that seeks to satisfy as many of the experimental distance constraints as possible. The result of distance geometry is an ensemble of possible structures where each structure is a local minimum of the distance geometry cost function and each structure satisfies the distance constraints to about the same extent. If the distance constraints from the NMR experiment are more self-consistent and mutually re-enforcing, then the ensemble is more tightly clustered, and the average pairwise RMSD between members of the ensemble is small. Some of the best NMR structures have ensembles with RMSDs around 1.0 Å, the worst have very high RMSDs approaching random. Most often this reflects the disorder in the polypeptide rather than the quality of the NMR data.

A contact map prediction represents an ensemble of states in the same way and for the same reason as a set of NMR distance constraints represents an ensemble of states. Therefore, it makes sense to measure the quality of a contact map prediction in the same way as we measure the quality of an NMR structure, by sampling an ensemble of 3D solutions and then measuring the diversity of the ensemble. If the ensemble is mostly disordered, then the DME metric might make more sense, or the size of the largest fragment with an average RMSD below a cutoff. In any case, a measure of contact map accuracy in 3D would alleviate the problems associated with 2D accuracy assessments, and the accuracy would better correlate with the usefulness of the prediction.

8.7 Other Applications of Contact Maps

Up to this point, we have confined the discussion to contact prediction in globular proteins, but the prediction of membrane protein structures is potentially much more valuable. Membrane protein structures are harder to characterize experimentally, since membranes interfere with both crystallization and NMR experiments. Electron microscopy, sometimes using monoclonal antibodies, and fluorescent resonance energy transfer (Eisenhawer et al., 2001) have been used with some success to obtain the gross layout of the transmembrane parts of membrane proteins, but in general these experiments are not sufficient to build a detailed model. Molecular simulations have been used successfully to refine the structure of the transmembrane regions (Enosh et al., 2004). One way contact map predictions can potentially be used is to assign a contact or noncontact value to residues in the soluble part of a membrane protein and use that information to predict contacts within the membrane. This would work because transmembrane regions are either helices or strands, and these generally pass directly through the membrane without any turns. Thus, contacts on the membrane surface imply contacts within the membrane and on the other side.

Contact maps have been used to align sequences to structures and to align structures to structure, even nonsequentially. The correlated mutation metric described in Fig. 8.5 ignores the identity of the amino acids involved. A more specific model for correlated mutations has been constructed in which a score for each of the possible amino acid substitution pairs was stored in a 400×400 contact substitution matrix called CAO (Contact Accepted mutatiOn) (Lin et al., 2003; Kleinjung et al., 2004). Each matrix element expresses the degree, positive or negative, to which the two mutations were observed in tandem in known structures. For example, the score for F_Y:I_V would be positive if a contact between an F and a Y was frequently observed to mutate F $\rightarrow$ I and Y $\rightarrow$ V as compared to random chance. CAO is not used, like correlated mutations, to predict contacts directly, but instead it is used to score sequence alignments to proteins of known structure. The greater sensitivity of this method allowed the authors to assign functional annotations to previously uncharacterized sequences with improved confidence.

We have used contact maps to align structures nonsequentially, and have applied the contact map alignment to search for conserved packing arrangements in protein cores (Yuan and Bystroff, 2005). The program SCALI (Structural Core ALIgnment) assembles a nonsequential alignment from a pairs-list of short gapless local alignments. Each of these short alignments relates some of the contacts in the target to some contacts in the template. The contact map score determines which segments to keep in a search through alignment space. The resulting alignment is nonsequential if the aligned segments are ordered differently in the two proteins. Nonsequential alignments do not imply homology, but may be used to find structural motifs. We used nonsequential alignments to find recurrent multibody interactions in protein cores.

8.8 Conclusions

Contact maps represent a useful and easily manipulated data structure for protein structure prediction by statistical, machine learning, and simulation methods. Progress is being made toward building predictive models that use this data structure, and new insights are being discovered about the nature of protein folding. Contact maps are bridging the gap between accurate 1D structure predictions and 3D structure predictions, but much work remains to be done. Here is a short list of open problems in contact map prediction as discussed in this chapter.

- *Scoring and error correction*. The impact of this representation on the field of protein structure prediction depends on advances in methods for correcting imperfect contact map predictions.
- *Nonlocal contacts*. Most methods are far more accurate on local contacts, but the global topology is defined by the nonlocal, or long range, contacts.
- *Proteinlike SIG recognition*. A general solution for the problem of recognizing a sphere intersection graph given proteinlike constraints remains an open problem.
- *Evaluation*. As in all areas of structure prediction, methods for evaluating success in contact map prediction need to correlate with usefulness, otherwise interative training of any sort will not converge on the truth.
- *Reconstruction of HS and SC contact maps*. Contact maps based on side chains work the best for fold recognition, but projecting maps into 3D is problematic.

If a fool-proof method can be established for converting side-chain contacts to a 3D ensemble, it will eventually unleash the power of machine learning methods to attack the protein folding problem.

Recommended Reading

Baker, D. 2000. A surprising simplicity to protein folding. *Nature* 405:39–42.

Vendruscolo, M., Najmanovich, R., and Domany, E. 1999. Protein folding in contact map space. *Phys. Rev. Lett.* 82:656–659.

References

Aloy, P., Stark, A., Hadley, C., and Russell, R.B. 2003. Predictions without templates: New folds, secondary structure, and contacts in CASP5. *Proteins* 53 (Suppl. 6):436–456.

Altschuh, D., Lesk, A.M., Bloomer, A.C., and Klug, A. 1987. Correlation of coordinated amino acid substitutions with function in viruses related to tobacco mosaic virus. *J. Mol. Biol.* 193:693–707.

Aszodi, A., Gradwell, M.J., and Taylor, W.R. 1995. Global fold determination from a small number of distance restraints. *J. Mol. Biol.* 251:308–326.

Berrera, M., Molinari, H., and Fogolari, F. 2003. Amino acid empirical contact energy definitions for fold recognition in the space of contact maps. *BMC Bioinformatics* 4:8.

Bystroff, C., and Shao, Y. 2003. Modeling protein folding pathways. In *Practical Bioinformatics* (J.M. Bujnicki, Ed.). Berlin, Springer-Verlag.

Bystroff, C., Thorsson, V., and Baker, D. 2000. HMMSTR: A hidden Markov model for local sequence–structure correlations in proteins. *J. Mol. Biol.* 301:173–190.

Chavez, L.L., Onuchic, J.N., and Clementi, C. 2004. Quantifying the roughness on the free energy landscape: Entropic bottlenecks and protein folding rates. *J. Am. Chem. Soc.* 126:8426–8432.

Cheng, J., Randall, A., Sweredoski, M., and Baldi, P. 2005. SCRATCH: A protein structure and structural feature prediction server. *Nucleic Acids Res.* 33: 72–76.

Dodge, C., Schneider, R., and Sander, C. 1998. The HSSP database of protein structure–sequence alignments and family profiles. *Nucleic Acids Res.* 26:313–315.

Dosztanyi, Z., Fiser, A., and Simon, I. 1997. Stabilization centers in proteins: Identification, characterization and predictions. *J. Mol. Biol.* 272:597–612.

Eisenhawer, M., Cattarinussi, S., Kuhn, A., and Vogel, H. 2001. Fluorescence resonance energy transfer shows a close helix–helix distance in the transmembrane M13 procoat protein. *Biochemistry* 40:12321–12328.

Enosh, A., Fleishman, S.J., Ben-Tal, N., and Halperin, D. 2004. Assigning transmembrane segments to helices in intermediate-resolution structures. *Bioinformatics* 20 (Suppl. 1):I122–I129.

Fariselli, P., and Casadio, R. 1999. A neural network based predictor of residue contacts in proteins. *Protein Eng.* 12:15–21.

Fariselli, P., Olmea, O., Valencia, A., and Casadio, R. 2001a. Prediction of contact maps with neural networks and correlated mutations. *Protein Eng.* 14:835–843.

Fariselli, P., Olmea, O., Valencia, A., and Casadio, R. 2001b. Progress in predicting inter-residue contacts of proteins with neural networks and correlated mutations. *Proteins* Suppl. 5:157–62.

Göbel, U., Sander, C., Schneider, R., and Valencia, A. 1994. Correlated mutations and residue contacts in proteins. *Proteins* 18:309–317.

Graña, O., Baker, D., Maccallum, R.M., Meiler, J., Punta, M., Rost, B., Tress, M.L., and Valencia, A. 2005. CASP6 assessment of contact prediction. *Proteins* [Epub 26 Sep 2005].

Hamilton, N., Burrage, K., Ragan, M.A., and Huber, T. 2004. Protein contact prediction using patterns of correlation. *Proteins* 56:679–684.

Havel, T.F., Crippen, G.M., and Kuntz, I.D. 1979. Effects of distance constraints on macromolecular conformation. II. Simulation of experimental results and theoretical predictions. *Biopolymers* 18:73–81.

Hu, J., Shen, X., Shao, Y., Bystroff, C., and Zaki, M.J. 2002. Mining protein contact maps. *BIOKDD 2002*, Edmonton, Canada.

Huang, E.S., Subbiah, S., and Levitt, M. 1995. Recognizing native folds by the arrangement of hydrophobic and polar residues. *J. Mol. Biol.* 252:709–720.

Jones, D.T. 1999. Protein secondary structure prediction based on position-specific scoring matrices, *J. Mol. Biol.* 292:195–202.

Kleinjung, J., Romein, J., Lin, K., and Heringa, J. 2004. Contact-based sequence alignment. *Nucleic Acids Res.* 32:2464–2473.

Koh, I.Y., Eyrich, V.A., Marti-Renom, M.A., Przybylski, D., Madhusudhan, M.S., Eswar, N., Grana, O., Pazos, F., Valencia, A., Sali, A., and Rost, B. 2003. EVA: Evaluation of protein structure prediction servers. *Nucleic Acids Res.* 31:3311–3315.

Kraulis, P.J. 1991. MOLSCRIPT: A program to produce both detailed and schematic plots of protein structures. *J. App. Crystallogr.* 24:946–950.

Kuznetsov, I.B., and Rackovsky, S. 2004. Class-specific correlations between protein folding rate, structure-derived, and sequence-derived descriptors. *Proteins* 54:333–334.

Lichtarge, O., Bourne, H.R., and Cohen, F.E. 1996. An evolutionary trace method defines binding surfaces common to protein families. *J. Mol. Biol.* 257:342–358.

Lin, K., Kleinjung, J., Taylor, W., and Heringa, J. 2003. Testing homology with CAO: A contact-based Markov model of protein evolution. *Comp. Biol. Chem.* 27:93–102.

Lund, O., Frimand, K., Gorodkin, J., Bohr, H., Bohr, J., Hansen, J., and Brunak, S. 1997. Protein distance constraints predicted by neural networks and probability density functions. *Protein Eng.* 10:1241–1248.

MacCallum, R.M. 2004. Striped sheets and protein contact prediction. *Bioinformatics* 20(Suppl. 1):I224–I231.

Maiorov, V.N., and Crippen, G.M. 1992. Contact potential that recognizes the correct folding of globular proteins. *J. Mol. Biol.* 227:876–888.

McGuffin, L.J., Bryson, K., and Jones, D.T. 2000. The PSIPRED protein structure prediction server. *Bioinformatics* 16:404–405.

McLachlan, A.D. 1971. Tests for comparing related amino-acid sequences. Cytochrome c and cytochrome c 551. *J. Mol. Biol.* 61:409–424.

Michael, T.S., and Quint, T. 1999. Sphere of influence graphs in general metric spaces. *Math. Comput. Model.* 29:45–53.

Michalopoulos, I., Torrance, G.M., Gilbert, D.R., and Westhead, D.R. 2004. TOPS: An enhanced database of protein structural topology. *Nucleic Acids Res.* 32:D251–D254.

Mirny, L., and Domany, E. 1996. Protein fold recognition and dynamics in the space of contact maps. *Proteins* 26:391–410.

Monge, A., Friesner, R.A., and Honig, B. 1994. An algorithm to generate low-resolution protein tertiary structures from knowledge of secondary structure. *Proc. Natl. Acad. Sci. USA* 91:5027–5029.

Moult, J., Fidelis, K., Zemla, A., and Hubbard, T. 2003. Critical assessment of methods of protein structure prediction (CASP)–round V. *Proteins* 53 (Suppl. 6):334–339.

Neher, E. 1994. How frequent are correlated changes in families of protein sequences? *Proc. Natl. Acad. Sci. USA* 91:98–102.

Olmea, O., and Valencia, A. 1997. Improving contact predictions by the combination of correlated mutations and other sources of sequence information. *Fold Des.* 2:S25–S32.

Orengo, C.A., Michie, A.D., Jones, S., Jones, D.T., Swindells, M.B., and Thornton, J.M. 1997. CATH—A hierarchic classification of protein domain structures. *Structure* 5:1093–1108.

Park, K., Vendruscolo, M., and Domany, E. 2000. Toward an energy function for the contact map representation of proteins. *Proteins* 40:237–248.

Pazos, F., Helmer-Citterich, M., Ausiello, G., and Valencia, A. 1997. Correlated mutations contain information about protein–protein interaction. *J. Mol. Biol.* 271:511–523.

Plaxco, K.W., Simons, K.T., and Baker, D. 1998. Contact order, transition state placement and the refolding rates of single domain proteins. *J. Mol. Biol.* 277:985–994.

Pollastri, G., and Baldi, P. 2002. Prediction of contact maps by GIOHMMs and recurrent neural networks using lateral propagation from all four cardinal corners. *Bioinformatics* 18(Suppl. 1):S62–S70.

Porto, M., Bastolla, U., Roman, H.E., and Vendruscolo, M. 2004. Reconstruction of protein structures from a vectorial representation. *Phys. Rev. Lett.* 92:218101–218104.

Punta, M., and Rost, B. 2005. Protein folding rates estimated from contact predictions. *J. Mol. Biol.* 348:507–512.

Rabiner, L.R. 1989. A tutorial on hidden Markov models and selected applications in speech recognition. *Proc. IEEE* 77:257–286.

Rodionov, M.A., and Johnson, M.S. 1994. Residue–residue contact substitution probabilities derived from aligned three-dimensional structures and the identification of common folds. *Protein Sci.* 3:2366–2377.

Saitoh, S., Nakai, T., and Nishikawa, K. 1993. A geometrical constraint approach for reproducing the native backbone conformation of a protein. *Proteins* 15:191–204.

Shao, Y., and Bystroff, C. 2003. Predicting interresidue contacts using templates and pathways. *Proteins* 53(Suppl. 6):497–502.

Shindyalov, I.N., Kolchanov, N.A., and Sander, C. 1994. Can three-dimensional contacts in protein structures be predicted by analysis of correlated mutations? *Protein Eng.* 7:349–358.

Singer, M.S., Vriend, G., and Bywater, R.P. 2002. Prediction of protein residue contacts with a PDB-derived likelihood matrix. *Protein Eng.* 15:721–725.

Sippl, M.J. 1990. Calculation of conformational ensembles from potentials of mean force. An approach to the knowledge-based prediction of local structures in globular proteins. *J. Mol. Biol.* 213:859–883.

Skolnick, J., Kolinski, A., and Ortiz, A.R. 1997. MONSSTER: A method for folding globular proteins with a small number of distance restraints. *J. Mol. Biol.* 265:217–241.

Tanaka, S., and Scheraga, H.A. 1976. Medium- and long-range interaction parameters between amino acids for predicting three-dimensional structures of proteins. *Macromolecules* 9:945–950.

Taylor, W.R., and Hatrick, K. 1994. Compensating changes in protein multiple sequence alignments. *Protein Eng.* 7:341–348.

Thomas, D.J., Casari, G., and Sander, C. 1996. The prediction of protein contacts from multiple sequence alignments. *Protein Eng.* 9:941–948.

Vendruscolo, M., and Domany, E. 1998. Efficient dynamics in the space of contact maps. *Fold Des.* 3:329–336.

Vendruscolo, M., Kussell, E., and Domany, E. 1997. Recovery of protein structure from contact maps. *Fold Des.* 2:295–306.

Wako, H., and Scheraga, H.A. 1982. Visualization of the nature of protein folding by a study of a distance constraint approach in two-dimensional models. *Biopolymers* 21:611–632.

Yuan, X., and Bystroff, C. 2005. Non-sequential structure-based alignments reveal topology-independent core packing arrangements in proteins. *Bioinformatics* 27:1010–1019.

Zaki, M.J., Shan, J., and Bystroff, C. 2000. Mining residue contacts in proteins using local structure predictions. *Proceedings IEEE International Symposium on Bio-Informatics and Biomedical Engineering*, Arlington, VA.

Zhang, C., and Kim, S.H. 2000. Environment-dependent residue contact energies for proteins. *Proc. Natl. Acad. Sci. USA* 97:2550–2555.

Zhao, Y., and Karypis, G. 2003. Prediction of contact maps using support vector machines. *BIBE 2003*, Bethesda, MD. IEEE Computer Society, pp. 26–36.

9 Modeling Protein Aggregate Assembly and Structure

Jun-tao Guo, Carol K. Hall, Ying Xu, and Ronald Wetzel

9.1 Introduction

One might say that "protein science" got its start in the domestic arts, built around the abilities of proteins to aggregate in response to environmental stresses such as heating (boiled eggs), heating and cooling (gelatin), and pH (cheese). Characterization of proteins in the late nineteenth century likewise focused on the ability of proteins to precipitate in response to certain salts and to aggregate in response to heating. Investigations by Chick and Martin (Chick and Martin, 1910) showed that the inactivating response of proteins to heat or solvent treatment is a two-step process involving separate denaturation and precipitation steps. Monitoring the coagulation and flocculation responses of proteins to heat and other stresses remained a major approach to understanding protein structure for decades, with solubility, or susceptibility to aggregation, serving as a kind of benchmark against which results of other methods, such as viscosity, chemical susceptibility, immune activity, crystallizability, and susceptibility to proteolysis, were compared (Mirsky and Pauling, 1936; Wu, 1931). Toward the middle of the last century, protein aggregation studies were largely left behind, as improved methods allowed elucidation of the primary sequence of proteins, reversible unfolding studies, and ultimately high-resolution structures. Curiously, the field of protein science, and in particular protein folding, is now gravitating back to a closer look at protein aggregation and protein aggregates. Unfortunately, the means developed during the second half of the twentieth century for studying native, globular proteins have not proved immediately amenable to the study of aggregate structures. Great progress is being made, however, to modify classical methods, including NMR and X-ray diffraction, as well as to develop newer techniques, that together should continue to expand our picture of aggregate structure (Kheterpal and Wetzel, 2006; Wetzel, 1999).

The current situation presents opportunities and challenges for computational methods, but with an ironic contrast to the history of the development of these methods for solvated globular proteins. Computational modeling methods for globular proteins were developed against a background in which many protein structures and properties were well known and reasonably well understood, and could therefore provide target structures and fundamental design concepts. In contrast, there are no complete high-resolution structures of protein aggregates like amyloid. Furthermore, at least some aggregate structures appear to be hierarchical, and it is not well understood how much the higher degrees of order, such as bundling of protofilaments into amyloid fibrils, might contribute to the overall stability of the fibril. This introduces an awkward uncertainty when attempting to model a fibril substructure; it is not clear how much stability should be expected within the substructure, and how much might be provided only through its superassembly into a higher order of structure. It does not seem unreasonable to assume that the folding and stabilization of aggregates is guided by the same general relationships that guide the folding and stabilization of globular proteins (Williams et al., 2004). At the same time, it is possible that protein aggregates might be stabilized by some factors that are not so well understood, owing to our ignorance of both aggregate structure and assembly characteristics.

This chapter provides an overview of this challenging field, at a time when computational approaches are just beginning to be utilized to model aggregate structures and assembly. The first part of the chapter gives an overview of folding and aggregation and a survey of the process and products of misfolding and aggregation. The second segment provides a brief description of some of the major physical techniques currently being used to characterize various aspects of aggregate structure. The final part describes computational methods for approaching aggregate structure and aggregate assembly.

9.2 Folding and Misfolding

The recognition that simple proteins in ideal circumstances achieve their native folded states through thermodynamic control, influenced simply by their amino acid sequences and necessary cofactors (Anfinsen, 1973), introduced the conundrum of how it might be possible for all possible folded states to be explored during protein folding, to identify and secure the free energy minimum, in a realistic time frame (Levinthal, 1969). The solution to this paradox is now thought to be a combination of two factors: the lack of complete randomness in the starting, unfolded state (Fleming and Rose, 2005), and a restricted folding surface, often characterized as a rough-textured funnel, that features many (but far from infinite) parallel and interconnected pathways leading from the large number of unfolded conformations to the native state (Dill and Chan, 1997). There are exceptions to the general rule of thermodynamic control, however. Fifteen years ago two proteins were described that fold to a kinetically controlled local free energy minimum, producing a relatively stable, isolatable folded state that can be induced to reengage the folding pathway to produce a more stably folded monomeric structure (Creighton, 1992). These are alpha-lytic protease (Baker et al., 1992) and the serpin PAI-1 (Levin and Santell, 1987).

Besides these rarely reported alternative folded states, some globular proteins exhibit additional alternative states: misfolded aggregates. Although the ability of proteins to convert, under stress, to irreversibly aggregated, insoluble structures has long been appreciated, protein aggregates have never seriously been included in debates over whether protein folding is controlled kinetically or thermodynamically. This may be because aggregates were considered an unnatural aberration, or because it has never been clear whether aggregate formation itself is under kinetic or thermodynamic control.

Over the past two to three decades, the biological importance of protein aggregates as nontrivial, alternative folded states has been established. Misfolding/aggregation has been revealed to be a major side reaction during normal protein folding in the cell (Turner and Varshavsky, 2000), which is consistent with the huge investment paid by molecular and cellular evolution in developing pathways to manage aggregation such as molecular chaperones (Hartl and Hayer-Hartl, 2002) and the ubiquitin proteasome system (Petrucelli and Dawson, 2004; Vigouroux et al., 2004). Aggregates are associated with a wide variety of human diseases, including Alzheimer's disease, Creutzfeldt-Jakob disease, Huntington disease, Type II diabetes, and Parkinson disease (Martin, 1999; Merlini and Bellotti, 2003). These manifestations of aggregation can be added to the more biotechnological challenges of protein stability (often synonymous with avoidance of aggregation), inclusion body formation during recombinant expression, and aggregate formation during refolding (Wetzel, 1994; Wetzel and Goeddel, 1983). Given the ubiquitous nature of protein aggregation, the importance of learning more about aggregate structure is clear.

The question of kinetic versus thermodynamic control of the formation of amyloid and other aggregates is only in the early stages of being addressed. For simple peptides that have no highly stable solution conformation, it is sometimes possible to establish an equilibrium between bulk phase monomer and amyloid, and to actually determine an equilibrium constant and free energy (O'Nuallain et al., 2005; Williams et al., 2004, 2006). For stably folded, globular proteins, which only form amyloid under conditions that disfavor the native state (Colon and Kelly, 1992; Hurle et al., 1994; McCutchen et al., 1993), it may be difficult to identify conditions where both native and fibril species can be populated simultaneously so that relative stabilities might be directly assessed. Kinetic partitioning appears to play a big role in aggregation during folding reactions of some proteins, where rapid transformation to an aggregation-committed intermediate can effectively take molecules out of the productive folding pathway (Finke et al., 2000; Goldberg et al., 1991; Haase-Pettingell and King, 1988). The above considerations are important for two major reasons. First, without having some idea of the relative thermodynamic stability of amyloid fibrils, it is difficult to validate computational models. Second, an awareness of both the thermodynamics and kinetics of aggregation will be important when evaluating computational simulations of the aggregation process.

9.2.1 Types of Aggregates

Inclusion body formation in bacteria: Bacterial inclusion bodies (IBs) rich in protein were first observed when bacteria were fed amino acid analogues that were incorporated into protein (Prouty et al., 1975) and were later observed in recombinant expression of proteins, where they require elaborate methodologies for either suppressing their formation or facilitating recovery of native protein (De Bernardez Clark et al., 1999; Marston and Hartley, 1990). Except for a general enrichment in β-sheet, we have very little knowledge of the intimate details of protein structure within IBs.

In spite of their different appearances under the EM, IBs and amyloid fibrils share quite a few similarities in their formation and properties (Wetzel, 1992), including a shared role of native state destabilization in their formation (Chan et al., 1996) and a shared susceptibility toward mutations (Wetzel, 1994). Monomeric nucleation, only recently described in the aggregation kinetics of polyglutamine amyloid formation (Chen et al., 2002), was very recently also observed in protein aggregation in bacteria (Ignatova and Gierasch, 2005). IBs and amyloid, along with other protein aggregates, including thermally induced aggregates and aggregates formed during folding *in vitro*, appear to consist largely of β-sheet structure (Oberg et al., 1994; Sunde and Blake, 1997). The ultrastructure of neuronal IBs formed in Huntington's disease tissue shows that these large aggregates are collections of fibrillar structures (DiFiglia et al., 1997), presumably polyglutamine amyloid fibrils (Scherzinger et al., 1997).

Amyloid and other disease-related aggregation: Over the past 10–20 years, amyloid fibrils and related pathological protein aggregates have become increasingly important subjects of research. This is primarily because of the evidence that these aggregates are associated with human disease (Martin, 1999; Merlini and Bellotti, 2003). Relative to other aggregates, we know a considerable amount about amyloid structure. Some of these details will be discussed below in the review of biophysical and biochemical techniques for analyzing fibril structure. Nonamyloid aggregates, such as spherical oligomers and protofibrils (Caughey and Lansbury, 2003), may be even more important to human disease than are mature fibrils.

Amyloid fibrils are normally found outside the cell, but there are clearly aberrant aggregation reactions that occur inside the cell as well. This is not surprising, given the complex biochemical systems that have evolved to deal, at least in part, with protein misfolding and aggregation, including molecular chaperones (Hartl and Hayer-Hartl, 2002), the ubiquitin proteasome system (Petrucelli and Dawson, 2004; Vigouroux et al., 2004), aggresomes (Kopito, 2000), and autophagy (Glickman, 2000). We know much less about intracellular aggregates than we do about extracellular amyloid. Aggregation in the presence of chaperones, even when they can not completely block aggregation, can lead to aggregates with substantially altered properties (Muchowski et al., 2000). The important cross-talk between fundamental biophysics and evolved biochemical pathways in the intracellular aggregation of proteins is little understood.

Aggregation during refolding: The denaturing conditions required for recovery of material from IBs yield proteins that must be refolded. Some proteins can be easily refolded from the denatured state *in vitro*; these include classical models for protein folding studies like ribonuclease and staph nuclease. More typically, proteins refold inefficiently, leading to formation of aggregates that limit the yields of the folding reaction. An important area of biotechnology is the development of methods for improving refolding yields (De Bernardez Clark et al., 1999).

Aggregation of stressed native proteins: Aggregation *in vitro* induced by exposure of the native protein to stresses like heat or denaturing solvents is also important in biotechnological applications such as enzyme reactors. This is also the classic protein aggregation first studied as one of the few established means of characterizing protein structure (Chick and Martin, 1910). Aggregation often takes place in the thermal unfolding transition zone. If aggregates form at an elevated temperature, triggered by unfolding of the native state, they tend to remain stable and insoluble when the solution is returned to lower temperatures. This essentially irreversible removal of native protein from solution, either at an elevated temperature or during the return to the normal temperature range (which many times can occur without detectible chemical changes), is the basis for the thermal inactivation of many proteins. The ability of a particular protein to aggregate under defined conditions depends broadly on two features: (1) the stability of the native fold and (2) the aggregation tendency of the unfolded or misfolded states, including folding intermediates (Wetzel, 1994).

The above type of aggregate is presumably related to another major aggregation problem of the pharmaceutical industry, the aggregates that form in protein solutions during processing or during storage in the vial. Aggregates in injected proteins are a serious problem for several reasons, including (1) diminution of active molecules, (2) inflammation at the injection site, and (3) stimulation of an antibody response (Cleland et al., 1993; Hermeling et al., 2004; Shire et al., 2004). These aggregates can form in more subtle ways, not only during long-term storage, but also under the influence of shear forces (for example, during injection) or due to an inadequate lyophilization process. The study of how these aggregates form, their structures, their biological effects, and most importantly how to avoid them, is of critical importance to the pharmaceutical industry.

9.2.2 Structural Hierarchy of Amyloid and Conformational Isomerism

An added complication to thinking about amyloid structure is the number of levels of structure involved. Like the coiled coils of collagenlike molecules, amyloid fibrils exhibit a quaternary structure that is both a distinguishing feature of amyloid and at the same time a source of great variation in morphological forms (Goldsbury et al., 2000; Sunde and Blake, 1997). The fundamental assembly unit of amyloid is the protofilament, a structure normally not seen in isolation but rather as a component of higher structure in fibrils. The classical amyloid fibril appears in the EM as a twisted rope; the protofilaments are the major strands making up the rope. There can be two to six protofilaments bundled together to make an amyloid fibril, with the number tending to be characteristic of a particular protein fibril. Protofilaments appear to be capable of assembling into other types of superstructures as well, such as ribbons containing multiple protofilaments in a flat array, or twisted bundles composed of multiple-protofilament ribbons (Goldsbury et al., 2000). These alternatively assembled states, often viewed in the same preparation of fibrils (Goldsbury et al., 2000), may arise from, or at least be associated with, different conformations of the component protein as it is packed into the fibril structure (Petkova et al., 2005). If, as it appears, all of these forms are based on the protofilament, it would appear that the simplest target of structure determination and prediction should be the protofilament.

It remains to be seen whether differences in quaternary structure (that is, protofilament packing) are the basis for self-propagating conformational variants of amyloid fibrils, or result from underlying differences in secondary and tertiary structure (such as formation of the amyloid filament core). Whatever their source in the assembly pathways and/or structural nuances of amyloid, different superassembled states of protofilaments can be very important biologically. Conformational variation among amyloid fibrils is now considered to be the underlying factor controlling strain phenomena and species barriers in mammalian (Horiuchi et al., 2000) and yeast (Tanaka et al., 2004) prion biology.

The recent recognition that a single amyloidogenic polypeptide can make multiple amyloid conformational states that can exhibit tangible structural differences (Petkova et al., 2005; Tanaka et al., 2004) introduces a further complication into the goal of approaching amyloid structure through a combined and iterative program of experiment and computational modeling. A necessary prerequisite to successful improvement of modeling approaches is the ability to validate methods through comparison of modeled structures with experimental data. Yet different approaches to characterizing some amyloid fibril structures have led to conflicting results. For example, solid-state NMR suggests that the C-terminal amino acids of Aβ(1–40) are involved in H-bonded β-sheet (Petkova et al., 2005) in the fibril, while scanning proline mutagenesis (Williams et al., 2004) and hydrogen exchange (Whittemore et al., 2005; Kheterpal et al., 2006) suggest that the C-terminal residues are disordered. It now seems possible that both results are correct: the solid-state NMR result was obtained on fibrils grown under conditions of agitation, while the proline and hydrogen exchange results were obtained on fibrils grown without stirring; these two methods of preparation are linked to fibrils with substantially differing structures (Petkova et al., 2005).

While the existence of amyloid isomerism is, in the short term, an unwelcome complication, in the long term it is an advantage for computational studies since it provides an opportunity to predict and account for relatively subtle variations among multiple fibril conformations.

9.3 Experimental Approaches to Aggregate Structure

X-ray diffraction: The difficult challenges of measuring aggregate structure owe mainly to the difficulty of crystallizing intact fibrils, so that detailed crystal diffraction patterns cannot be obtained. Particularly well-ordered aggregates can exhibit a few key reflections in powder diffraction experiments consistent with dominant repeating structure. In particular, amyloid fibrils aligned in the X-ray beam typically generate what is called the cross-β pattern, in which the reflections associated with strand–strand repeats in the hydrogen-bonding direction and in the β-sheet packing direction are at 90° to each other. The pattern indicates that the β-extended chains are oriented in the fibril (and protofilament) such that they are perpendicular to the fibril axis, and held together by hydrogen bonds that are oriented parallel to the fibril axis (Sunde and Blake, 1997). In rare cases, high-resolution data on crystalline forms of short, amyloidogenic peptides have provided a number of potential insights into amyloid structure, such as extremely low water content (Balbirnie et al., 2001; Nelson et al., 2005) and intimate side chain packing interactions (Makin et al., 2005; Nelson et al., 2005). While these and other (Elam et al., 2003; Schiffer et al., 1985) examples of infinite β-sheet formation in crystal structures may offer important insights into amyloid fibril structure, it must be kept in mind that protein crystals and amyloid fibrils, while related, are different, for example in the hierarchical structure of fibrils discussed above. The question of whether crystals of amyloidogenic peptides accurately reflect fine details of amyloid structure has not been addressed experimentally.

Electron microscopy (EM): Our impressions of the hierarchical, filamentous structure of amyloid, and more globular structures of some other aggregates, come primarily from EM studies. Scanning transmission EM can be used to determine the mass per length of fibrils and other aggregates (Goldsbury et al., 2000). This analysis confirms objectively and quantitatively what is already apparent by visual inspection, that most fibril preparations are heterogeneous, containing a number of distinct types of fibrils that vary both by the amount of twist and by the number of protofilaments per fibril (Goldsbury et al., 2000). These morphological variations may, at least in some cases, correspond to different self-replicating structural variants that can exhibit different stabilities (Tanaka et al., 2004) and molecular substructures (Petkova et al., 2005). In particularly favorable cases, especially in cryo-EM, image averaging has provided significantly enhanced detail, such as the number of protofilaments and the overall cross-sectional density within the fibril (Jimenez et al., 1999, 2002; Wille et al., 2002).

Atomic force microscopy (AFM): Scanning probe techniques can provide striking images that offer similar resolution to EM but with different strengths and sometimes superior results (Stine et al., 1996). Staining is not necessary in AFM, and precise height information can be obtained. The protofilament-based structural hierarchy of amyloid fibrils can be studied using AFM (Kanno et al., 2005). Nonfibrillar, potentially biologically relevant oligomeric states can also be visualized (Harper et al., 1997). In principle, scanning probe imaging also offers a wealth of probe types that might provide information on the chemical nature of the imaged surfaces. Force measurements can elucidate segmental stability (Kellermayer et al., 2005).

While technically challenging, another advantage of AFM over EM is the ability to monitor temporal changes in structure in situ (Goldsbury et al., 1999).

Circular dichroism (CD): Like most other optical spectroscopies, CD is generally considered to be ineffective in analysis of large, light-scattering aggregates such as most fibril preparations. In rare cases the technique works remarkably well, not only demonstrating the high β-sheet content expected of an amyloid preparation, but giving fibril formation kinetics that track with other measures of amyloid assembly (Chen et al., 2002). Conformational variants can also be distinguished by CD in favorable cases (Yamaguchi et al., 2005).

Vibrational spectroscopy: Fourier transform infrared spectroscopy (FTIR) is one of the few techniques available for analysis of the protein solid state, and has been used to advantage in amyloid studies to demonstrate the existence of β-sheet, other secondary structural elements, and the parallel/antiparallel nature of the sheet. The method is useful as a qualitative tool to follow changes in secondary structure as an assembly reaction proceeds, and, with appropriate cautions, can be used to extract secondary structure content. There are a number of technical problems that must be sorted out for the effective use of this method in analyzing protein aggregates (Nilsson, 2004).

Solid-state NMR: Great strides have been made in recent years in the collection and analysis of NMR data on protein aggregates in the solid state. Chemical shift information can be interpreted in terms of the ϕ,ψ angles of the alpha carbons, and interatomic distances up to about 6 Å can be deduced from dipole–dipole couplings (Tycko, 2000). In favorable cases, the collection of a large number of distance restraints can allow piecing together of a fairly high resolution structure within a single extended chain element (Jaroniec et al., 2004). Other key structural information obtained using solid-state NMR includes the parallel/antiparallel nature of adjacent strands in protofilaments (Benzinger et al., 1998), the identification of long-range, nonbonded interactions (Petkova et al., 2002), and the demonstration of altered intrastrand folding in self-propagating fibrils exhibiting different morphologies in EMs (Petkova et al., 2005). Since solid-state NMR data can be collected on bona fide amyloid fibrils, when high-resolution data can be obtained (Jaroniec et al., 2004) this method appears to offer the most complete insights available into amyloid fibril structure.

Electron spin resonance: Judiciously chosen probes coupled to the thiol side chain of mutationally introduced Cys residues have been used to sense local environment in globular proteins (Hubbell et al., 2000), and this technique is now being applied to amyloid fibrils. In principle, one can observe distances between spin labels, and the mobility and solvent accessibility of the label at different positions in a protein structure. The method has been used to map out regions of parallelism and side-chain mobility in amyloid fibrils of Aβ(1–40) (Torok et al., 2002) and a number of other proteins.

Hydrogen–deuterium exchange: Hydrogen exchange (HX) has been used with great success to map structural features of globular proteins, in particular backbone secondary structure and sterically inaccessible sites. For globular proteins where data acquisition by real-time NMR and mass spectrometry (MS) is possible, methods can be adapted to examine the structure of the native state as well as dynamic aspects of protein structure and folding (Ferraro et al., 2004; Li and Woodward, 1999). Analysis of exchange into aggregates is significantly complicated by the requirement to break aggregate structure and restore the soluble, monomeric state to the precursor protein before it can be analyzed. The challenge in studies of aggregate structure has been to minimize and adjust for the loss of exchange information occurring during the sample processing. Loss of information due to HX during sample processing can be minimized in an MS approach using an in-line T-tube that makes sample dissolution and presentation to the MS probe one seamless operation (Kheterpal et al., 2000). Effective schemes for correcting data to account for exchange during sample preparation have been described (Kheterpal et al., 2003b). The HX-MS technique is rapid and sensitive enough to collect exchange protection data on the oligomeric/protofibrillar intermediates of Aβ(1–40) amyloid assembly; interestingly, this work showed that there is a class of highly protected H-bonds even in metastable protofibrils (Kheterpal et al., 2003a). Proteolytic fragmentation coincident with fibril dissolution can be used to obtain protection information on segments of the peptide in the fibril (Wang et al., 2003, Kheterpal et al., 2006). By using MS to analyze a large collection of overlapping proteolytic fragments from fibrils exposed to D_2O, it is possible to assign protection factors at the single residue level (Del Mar et al., 2005).

When signal dispersion allows, and individual hydrogens have been assigned, NMR can be used to follow the fate of each hydrogen as signal loss follows deuterium exchange. The trick to obtaining exchange information by NMR is to identify an aprotic NMR solvent that dissolves the protein, minimizes artifactual exchange during data collection, and allows good dispersion of the amide proton resonances. An effective solvent has been dichloroacetic acid/dimethylsulfoxide mixtures, and this has allowed the NMR method to map exchange protection in amyloid fibrils composed of peptides (Kuwata et al., 2003) and even the small protein β_2-microglobulin (Hoshino et al., 2002). Identical Aβ(1–40) fibrils have been exposed to HD exchange monitored by both MS (Kheterpal et al., 2000, 2003b, 2006) and NMR (Whittemore et al., 2005), and the results from the two methods are in very good agreement.

Limited proteolysis: Limited proteolysis has been used to understand globular protein structure, motility, and folding (Fontana et al., 1997). The major advantage of the technique is the ability to provide structural information on systems that are not readily amenable to other techniques. Major disadvantages are low resolution and a dependence for success on factors, such as proteolysis kinetics, often out of the control of the experimenter. The basic principle is that solvent-accessible regions of the polypeptide backbone are revealed by their ability to be cleaved by an endoproteinase such as trypsin. Potential limitations of the use of the method are: (1) proteolysis

```
         10         20         30         40
DAEFRHDSGY EVHHQKLVFF AEDVGSNKGA IIGLMVGGVV
```

Fig. 9.1 The amino acid sequence of Aβ(1–40).

requires at least transient exposure of a six- to eight-residue segment containing the protease site; (2) the protein may unfold and undergo multiple cleavages after the first cleavage event so rapidly that it becomes difficult to interpret any of the data in terms of native structure; and (3) accessible segments lacking a cleavage site will not be detected. The latter can be detected as a problem by comparing the cleavage kinetics of the folded, target state and the unfolded state; it can potentially be overcome by exploring multiple proteases.

Limited proteolysis has been used to effectively demonstrate the lack of stable structure in the N-terminal 10–14 residues of Aβ(1–40) (Fig. 9.1) when it assembles into an amyloid fibril (Kheterpal et al., 2001). Interestingly, a percentage of the Aβ(1–40) molecules in the aggregate was not cleaved, suggesting either a second class of aggregate or a second class of folded peptides in the Aβ fibril. [Similar results have been obtained in HX experiments on α-synuclein fibrils (Del Mar et al., 2005).] Somewhat surprisingly, the rates of cleavage at sites in the exposed N-terminus were comparable for monomer and fibril, suggesting efficient proteolysis of the fibril despite its expected poor diffusion rate. Limited proteolysis has also been used to characterize amyloidogenic intermediates (Polverino de Laureto et al., 2003). This is even more challenging than analysis of fibrils, and is only possible if the molecular species under investigation is highly populated and relatively stable with respect to the time course of proteolysis and analysis.

Side chain accessibility analysis: Group-specific chemical modification reactions have historically been used to probe for the location of particular residues in globular protein structure (Means and Feeney, 1971). To some extent, this is possible for amyloid fibril analysis as well. For example, an amine-specific reagent provided evidence that Lys28 in some molecules of Aβ(1–40) in the amyloid fibril has an exposed side chain (Iwata et al., 2001). Overall, however, exploitation of naturally occurring amino acid reactivities suffers from the relative chemical inertness of many of the 20 standard amino acids. One can overcome this limitation by preparing single Cys mutants of the amyloidogenic peptide. When built into fibrils, the chemical accessibility of the reactive sulfhydryl side chain of Cys provides structural details otherwise difficult to obtain (Shivaprasad and Wetzel, 2006).

Cross-linking analysis: Covalent cross-linking analysis is another chemical approach to protein structure determination that has been used historically for characterizing the interaction sites for protein ligands and protein–protein interactions. The trick in a rigorous analysis is to adjust the timing of the appearance and disappearance of the reactive species so that it is faster than the time scale for appreciable structural rearrangement or dissociation in the target. Otherwise, cross-linking may be guided more by chemical feasibility than by spatial proximity.

Photoaffinity labeling, in principle, can overcome this problem, but the lifetime of some photoactive species can be long, and their ultimate mode of reaction can be by relatively slow nucleophilic displacement rather than by rapid bond insertion (Chowdhry and Westheimer, 1979). One interesting approach to photo-cross-linking is to use the natural photochemical activity of the normal amino acid side chains to mediate cross-linking (Bitan and Teplow, 2004). This has the potential advantage of avoiding the use of chemical analogues, but, as discussed above, the precision of the method may sometimes be limited by strong biases in the reactivity of different amino acids. Interestingly, features of Aβ(1–42) oligomer formation determined using this method have been replicated in a simulation of Aβ aggregate (Urbanc et al., 2004b).

With appropriate controls, a chemical cross-linking approach can also provide useful information. Exposure to oxidizing conditions of amyloid fibrils grown from double Cys mutants appears to introduce intrapeptide cross-links only when the side chains are in contact within the fibril, providing important distance restraints for fibril structure model building (Shivaprasad and Wetzel, 2004).

Kinetic analysis of fibril assembly: In principle, assembly kinetics can inform us about the structure of a reaction product but only to the extent that the assembly mechanism is well defined and the transition state associated with the rate-limiting step resembles the final product. For protein aggregates these assumptions are often difficult to make. Even so, assembly kinetics has been informative in some of the cases where it has been used, especially in mutational analysis of amyloid structure. For example, the difficult question of polyglutamine aggregate structure was approached by an analysis of the kinetics of aggregation of polyglutamine sequences containing proline–glycine (PG) pairs at regular intervals. Optimal aggregation kinetics were observed when the number of glutamine residues between PG pairs was nine, corresponding to extended chains in the range of seven or eight Gln residues (Thakur and Wetzel, 2002). This is consistent with suggestions that the most stable β-sheets are about seven residues wide (Stanger et al., 2001). The potential value of using assembly kinetics as a link to structure is illustrated by the fact that subsequent X-ray fiber diffraction studies on aggregates of normal, unbroken polyglutamine sequences confirmed the width of the β-sheets in the aggregate structure suggested by the PG mutational analysis (Sharma et al., 2005).

The above polyglutamine kinetics analysis was fruitful because the aggregation reaction appears to be a relatively simple case of nucleated growth polymerization (Chen et al., 2002). In contrast, spontaneous amyloid fibril formation under native conditions by most other peptides, including yeast prions (Collins et al., 2004; Serio et al., 2000), appears much more complex. Oligomeric intermediates are often observed, and it is not always clear whether they are on-pathway or off-pathway. In spite of this uncertainty, the use of fibril formation kinetics to score the effects of proline substitutions on fibril formation by Aβ(1–42) gave information that in many ways is in agreement with a similar analysis of Aβ(1–40) fibrils scored by fibril stability (see below). A number of differences were observed (Morimoto et al., 2004), however, and it is not clear whether these are to be attributed to the use of kinetics information or the difference in peptide structure.

As difficult as fibrils and other aggregates are to investigate structurally, an even more challenging problem is the structure of the aggregation nucleus. Yet because it holds the key to the kinetics of aggregate formation and is relatively small, the nucleus is a particularly attractive assembly to try to model computationally. Due to the complexity of many amyloid assembly reactions mentioned above, nucleation kinetics analysis has proved very difficult. An exception to this is polyglutamine, which does not seem to form significant off-pathway oligomeric aggregates under normal aggregation conditions. Treating the nucleus as a thermodynamic entity and the least stable species on the reaction pathway allows one to model the nucleation kinetics by placing the nucleus in a preequilibrium with bulk phase monomer (Ferrone, 1999). Applying the resulting kinetics expression to data from a sedimentation assay (Wetzel, 2005) yielded the surprising result that the nucleus for polyglutamine aggregation is a high-energy form of the monomer (Chen et al., 2002). Subsequent studies allowed calculation of K_{N*}, the equilibrium constant describing the structural interconversion of bulk phase monomer and nucleus, yielding a value of about 10^{-9} for a disease-associated Q_{47} repeat polyglutamine (Bhattacharyya et al., 2005). This very low K_{eq} suggests that it will be difficult indeed to obtain any direct structural information on the aggregation nucleus, emphasizing the importance of simulations. One model for the polyglutamine nucleus was recently proposed based on computational studies (Khare et al., 2005).

Analysis of elongation kinetics can also be useful in evaluating amyloid structures, in particular in probing compatibility between mutational or conformational variants of fibrils. Cross-seeding experiments where seeding is sufficiently heavy that the lag phase is eliminated can be particularly helpful in gauging the compatibility between two or more monomers for adding to a preexisting fibril (O'Nuallain et al., 2004). The possible linkage between fibril assembly and disassembly kinetics, on the one hand, and fibril stability, on the other, is illustrated by the ability to recapitulate the fibril elongation equilibrium constant by propagating the microscopic forward and reverse rate constants from detailed analysis of fibril elongation (O'Nuallain et al., 2005).

Thermodynamic analysis of fibril structure and dynamics: Growth of some amyloid fibrils stops short of complete aggregation, and the endpoint can be shown to reflect a dynamic equilibrium (Jarrett et al., 1994; O'Nuallain et al., 2005). This allows calculation of the free energy of elongation, which further allows many of the kinds of analysis typically done on globular proteins, such as mutational analysis of stability (Williams et al., 2004, 2006). A significant amount of information about amyloid fibril structure has been gleaned using scanning mutational analysis of the Aβ(1–40) peptide as the changes affect the peptide's ability to engage the amyloid structure. Insertion of prolines (Williams et al., 2004), alanines (Williams et al., 2006), cysteines (Shivaprasad and Wetzel, 2006), modified cysteines (Shivaprasad and Wetzel, 2006), and disulfide bonds (Shivaprasad and Wetzel, 2004) has generated information that has deepened our understanding of fibril structure and how it is stabilized. The energetic consequences of certain mutations of residues thought to be in β-sheet in the amyloid fibril were shown to be remarkably similar (Williams et al., 2006) to the effect of the same mutation in a parallel β-sheet in a globular protein (Merkel et al., 1999), providing important validation of this approach and suggesting a fundamental similarity in the way that amyloid fibrils and globular proteins are stabilized.

At the same time, amyloid appears to be somewhat more plastic than globular proteins, sometimes disseminating the destabilizing effects of a lesion through structural distortions at a considerable distance from the site of the mutation (Williams et al., 2004, 2006). Thus, fibrils appear to be capable of adjusting their β-sheet networks in response to some disruptive mutations, for example resulting in fibrils that exhibit decreased stability even while featuring a greater number of backbone H-bonds (Williams et al., 2004). Lack of additivity in some double Ala mutants also may indicate a more plastic structure in amyloid (Williams et al., 2006).

The aforementioned experimental approaches, such as fiber diffraction, electron microscopy, HD exchange, solid-state NMR, limited proteolysis, electron paramagnetic resonance spectroscopy (EPR), and various chemical approaches, have yielded valuable information about aggregate structure. But they are not sufficient to derive high-resolution structure of protein aggregates. Computational techniques can offer a complementary alternative to experimental methods in building structural models of protein aggregates including amyloid fibrils, testing the stabilities of the model structures and studying the aggregate assembly process.

9.4 Computational Approaches to Aggregate Structure

Computational methods, especially protein structure prediction and molecular dynamics (MD) simulations, have been widely used for modeling protein structures and studying their dynamic behaviors. For example, the first three-dimensional working model of human plasma vitronectin was predicted through a combination of computational methods, specifically protein threading and domain docking, and experimental observations (Xu et al., 2001). The predicted model is consistent with all known experimental observations, including positioning of the ligand binding sites, accessibility of protease cleavage sites (Xu et al., 2001), and data from small-angle scattering experiments (Lynn et al., 2005). MD simulations have become an important tool in studying the physical basis of the structure and function of biomolecules since the first simulation work was published about three decades ago (Karplus and McCammon, 2002). One of the examples that illustrate the power of MD simulations to obtain functionally relevant information, which has been impossible using experimental techniques, is the study of conformational changes of GroEL (Ma et al., 2000). GroEL consists of two rings, each of which has seven identical subunits stacked back to back (Xu et al., 1997). MD simulations have been used successfully to demonstrate the conformational changes between the open and closed states. The simulation results have also shown that the subunits adopt an intermediate conformation with ATP bound, which is supported by cryo-EM results (Karplus and McCammon, 2002).

Whereas computational structure predictions have been used extensively for normally folded proteins, their application to misfolded structures and protein aggregates has been limited. This is not surprising, given that amyloid peptides or proteins adopt different conformations in soluble and fibril states. The soluble monomer structures for some of the amyloid precursor proteins have been solved and deposited into the Protein Data Bank (PDB) (Berman et al., 2000), such as Aβ(1–40) (Coles et al., 1998; Sticht et al., 1995), insulin (Hua et al., 1995), prion (Riek et al., 1996), and transthyretin (TTR) (Blake et al., 1978). In most cases, however, the soluble monomer structures provide very little insight into the possible conformations of the molecules in the amyloid fibrils. For example, Aβ(1–40), prion, and insulin have predominantly α-helical structures in physiological conditions or in organic solvents while amyloid fibrils formed by these peptides or proteins have predominantly β-sheet structures. A conformational transition from α-helix to β-sheet has been suggested as the key step in the formation of an ordered structure upon aggregation in these cases. Other protein aggregates, such as β_2-microglobulin and TTR, appear to result from the assembly of the states that have both amyloid- and native-like structures, suggesting a role for native structure in amyloid assembly. At the same time, amyloid fibril formation is not restricted to the relatively small number of proteins associated with well-recognized clinical disorders. Experiments have shown that many proteins, including such a well-known molecule as myoglobin, under suitable conditions, can form amyloid fibrils, which suggests that the ability to form such fibrils may be a generic property of polypeptide chains (Dobson, 1999, 2003). These observations, along with the aforementioned lack of information on the contribution of protofilament packing to the stability of the fibril, make it very challenging to model misfolded protein structures.

There are two possible atomic resolution computational approaches to modeling amyloid fibril structure. The first is to simulate the fibril formation process including conformational changes from the native globular protein, seed formation, and protofilament packing. However, the atomic-resolution simulation methods that are favored by the protein folding community cannot be applied to the study of amyloid fibril formation due to the long time scales involved in seed formation and in the fibrillation process. The large system sizes also present problems to computational simulation. The other atomic resolution approach bypasses the fibril formation process and studies the chemical interactions that stabilize the fibril structure. The rationale is that amyloid fibrils are formed from some regularly repeating building blocks revealed by X-ray diffraction patterns (Sunde and Blake, 1998). Therefore, the problem of modeling amyloid fibril structures can be partitioned into two steps: modeling the monomer structural features observed in amyloid fibrils and the packing of monomer structures in oligomers. The stabilities of the proposed oligomer models can then be tested using MD simulations.

We will discuss atomic-level structure modeling of the amyloid fibril cores and the low- to intermediate-resolution models of aggregate assembly.

9.4.1 Atomic Resolution Computational Approaches

MD simulations of small amyloid forming peptides: Amyloid fibril models can be constructed from scratch based solely on experimental observations and MD simulations can then be applied to test the validity of the models. Nussinov and colleagues have done extensive studies on short amyloid peptides in an attempt to obtain the underlying chemical principles of the atomic interactions involved in amyloid formation. These peptides include a fragment (residues 113–120) derived from the Syrian hamster prion protein (Ma and Nussinov, 2002a), two peptides (residues 22–27 and 22–29) from the human islet amyloid polypeptide (Zanuy et al., 2003; Zanuy and Nussinov, 2003), several fragments from Aβ amyloid protein (Ma and Nussinov, 2002b), and a peptide (residues 15–19) from human calcitonin (Haspel et al., 2005). The structure of each strand was constructed using standard modeling software, such as Insight II's biopolymer module (http://www.accelrys.com/). Then the peptide chains were placed with predefined parallel or antiparallel orientations. The hydrogen-bonded chains were placed at a distance of ∼5.0 Å from each other. The distance between the sheets was set to ∼10 Å, which corresponds to the average distance in a cross-β structure (Ma and Nussinov, 2002a; Sunde and Blake, 1998). The stabilities of these supramolecular structures and the contribution of the key residues to the stability were tested using MD simulations. The assumption in using a simulation approach to test the stabilities of oligomers is that if the peptides within the model oligomers can survive high-temperature MD simulations, then the oligomers are considered stable. For example, simulations of three strands of $A\beta_{16-22}$ revealed that antiparallel β-sheet structure is preferred while the parallel β-sheet structure is less stable, which is consistent with solid-state NMR data. Using a similar approach, Zanuy and Nussinov studied every possible amyloid organization of a segment (residues 22–27) of human islet amyloid polypeptide (hIAPP), such as peptide conformations within sheets and the lateral arrangements between sheets. They found that this short segment prefers an antiparallel arrangement of strands within sheets and a parallel lateral association. In the lateral association, the aromatic side chains play an important role in intersheet interactions (Zanuy and Nussinov, 2003).

Protein threading approaches for modeling amyloid fibril structures: Protein threading seems to be a feasible approach and a natural fit for modeling monomer structures within amyloid fibrils. First, protein structures in the PDB (Berman et al., 2000) with cross-β features might hold the key to understanding the folding pattern in amyloid fibrils (Jenkins and Pickersgill, 2001; Wetzel, 2002). For example, the parallel β-helical fold fulfills the basic requirements for an underlying primordial structure of amyloid fibrils, such as intrinsic cross-β structure and main chain hydrogen bonding. If the amyloid fibril folding pattern is present in solved globular structures, one obvious question is why these proteins such as parallel β-helical proteins do not oligomerize. As addressed by Richardson and Richardson (2002), proteins with β-sheet element are prevented from oligomerizing by N- and C-terminal caps, while in amyloid fibrils an indefinite number of β-strands may be hydrogen-bonded together into a very stable assembly due to an inherent propensity of β-structure to form sheetlike structures. Second, protein threading, one of the three popular structure prediction methods, identifies a structural homologue or analogue through aligning the query sequence onto template structures and finds the best possible template through evaluating sequence–structure fitness using empirical energy functions. Threading is a valuable method for finding structural analogs as it in principle does not rely on sequence similarity. The assumption is that some intrinsic interaction patterns between the residues of stable protein structures contribute to the specific folding pattern. Given recent suggestions that fibrils are stabilized by forces common to all proteins, hydrophobic interactions and hydrogen bonding, and not by forces particular to a specific sequence (Bucciantini et al., 2002; Kayed et al., 2003; Williams et al., 2005), threading should be a useful approach to predicting amyloid structure as amyloid proteins do not share any detectable sequence similarity though they share a number of structural features.

Currently, two approaches have been applied in amyloid fibril structure modeling, implicit threading and explicit threading. In implicit threading, the peptide sequences can be mapped to a known structure that fits the proposed model. For example, Li et al. (1999) used an implicit threading method to construct their twisted model of Aβ amyloid protofilaments based on limited experimental observation that Aβ may form an antiparallel β-sheet with a turn located around residues 25–28. The basic building block, a dimer of an antiparallel β-sheet with a turn located at residues 25–28 for Aβ(12–42), was constructed using the high-resolution structure of TTR (PDB ID: 2pab) (Blake et al., 1978) as a template. In their model, 48 monomers of Aβ(12–42) stack with four monomers per layer to form a twisted helical turn of β-sheet. MD simulations were applied to the model in explicit aqueous solution to test the stability of the protofilament model. Their simulation result suggests that the twist observed in synchrotron X-ray studies might be the result of protofilament packing, rather than from the structure of individual protofilaments. Using the threading algorithm "TOPITS," Chaney et al. (1998) identified three possible templates for Aβ(1–42) structure. All three proteins share an antiparallel β-sheet structure. The resulting model of Aβ(1–42) from threading studies displays a Greek key motif with four antiparallel β-strands (1–6, 9–15, 18–24, and 29–36). They also proposed that two Aβ molecules should form a dimer in order to shield unfavorable hydrophobic domains from the aqueous environment. In their Aβ protofilament model, the C-terminal domain (residues 30–42) of each Aβ molecule of the dimer extends toward the center to form an antiparallel β-sheet with the other Aβ dimer. In their protofilament model, the twisted β-sheet is highly hydrophobic yet is exposed to an aqueous environment. To resolve this thermodynamically unfavorable situation, a fibril model with three protofilaments was constructed, which has a compact and thermodynamically favorable structure with hydrophobic β-sheets buried inside and the hydrophilic β-barrels made of residues 1–28 exposed to aqueous environments.

The aforementioned Aβ amyloid protofilament or fibril models have one common feature, a core structure containing antiparallel β-sheets, either intermolecular or intramolecular. Compelling evidence from solid-state NMR and liquid suspension EPR studies on full-length Aβ fibrils suggests that the peptides in the fibril core are in-register, parallel arrangements (Benzinger et al., 1998; Petkova et al., 2002; Torok et al., 2002). Though a number of models involving parallel β-sheet have been proposed, there is no consensus on a unique structure model due to the uncertainty of the number and the location(s) of turns in the Aβ peptide (Lakdawala et al., 2002; Petkova et al., 2002). Recently, proline scanning mutagenesis experiments on Aβ(1–40) and Aβ(1–42), a technique used to search for regions involved in turns and disordered structure, have provided valuable information regarding the possible turn regions. Experimental data from Williams et al. (2004) suggest that the 15–36 sequence of Aβ(1–40) is involved in the amyloid core formation with three β strands separated by two turns at residues 22–23 and 29–30, which resembles an existing folding pattern of the parallel β-helical proteins. In fact, this β-helical-like model for amyloid fibrils has previously been suggested as possible folding motif in Aβ, insulin fibrils, and polyglutamine fibrils (Jimenez et al., 2002; Perutz et al., 2002; Wetzel, 2002). Based on recent experimental observations, Guo et al. constructed a structural model for the Aβ amyloid fibril core structure using a threading technique and MD simulations (Fig. 9.2) (Guo et al., 2004). In their approach, Aβ(15–36) was threaded against the representative parallel β-helical proteins and several non-β-helical all-β proteins as controls. The sequence–structure alignments with top threading scores were consistent with proline scanning mutagenesis data with respect to the locations of turns and β-strands. The non-β-helical templates did not score as well as β-helical proteins. Using the highest scoring alignments from the threading analysis, and the strong evidence from solid-state NMR and EPR studies that Aβ monomers are in an in-register, parallel β-sheet organization in Aβ

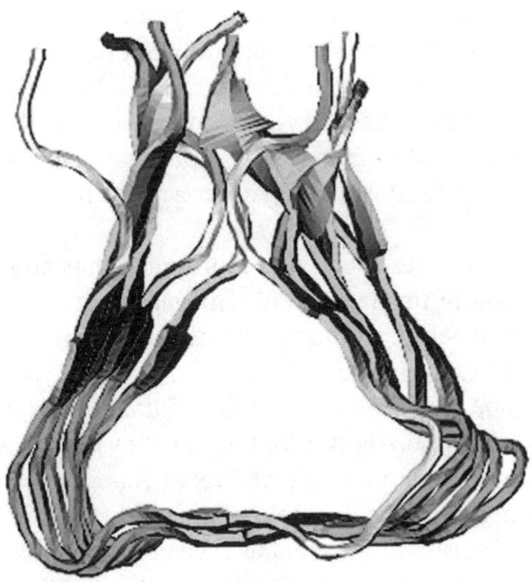

Fig. 9.2 Structural model of the core of Aβ amyloid fibril.

fibrils, both left-handed and right-handed 6-mer models were generated as the core of protofilaments and were subjected to MD simulations. The simulation results revealed that the left-handed model is more stable than the right-handed model. The total number of hydrogen bonds in the left-handed model during simulation is in agreement with the HD exchange experiments (Guo et al., 2004).

Govaerts and colleagues also applied threading approaches to the modeling of prion fibril structures (Govaerts et al., 2004). Their studies suggest that the sequence of PrP27–30 is compatible with a parallel left-handed β-helical fold. In their study, residues 89–174 are threaded onto the structure of the β-helical domain of uridyltransferase (1G97), which results in four rungs of β-helices. The α-helical region of residues 177–227 is packed onto the β-helix in an arrangement appearing in known β-helical protein structures. The exact position of the α-helices is optimized to fit the densities observed in the projection maps of the 2D crystals (Govaerts et al., 2004). The trimeric model from the packing of three parallel left-handed β-helical monomers matches the structural constraints of the PrP27–30 crystals.

It will be interesting to see if other amyloid forming sequences can be threaded reasonably well onto β-helical structures although it should be noted that there is in fact no unequivocal evidence that the strand arrangements in amyloid fibrils formed by a particular sequence are independent of the length of the fragment studied. For example, shorter Aβ peptides may form antiparallel β-sheets while the full-length Aβ peptides adopt a parallel organization (Ma and Nussinov, 2002b).

Computational docking approach: Computational docking approach has been used to predict the fibril structure of β_2-microglobulin (Benyamini et al., 2003). Traditionally, computational docking methods are used to predict protein–protein or protein–ligand interactions. Since fibril formation is a polymerization process, docking methods should be useful in examining the building blocks of fibrils (Zanuy et al., 2004). However, docking methods are only applicable in cases satisfying the following requirements: (1) the monomer structure is known; (2) the segments involved in fibril formation have been revealed by experiments; and (3) there is little change in the monomer structure between the globular and fibrillar states. Because of these constraints, as of now, the docking approach has limited applications in fibril structure modeling.

Benyamini and colleagues found that this approach is applicable to the modeling of the β_2-microglobulin fibril structure (Benyamini et al., 2003). The basic idea is that if the monomer structure is known and experimental data have suggested the possible segments and structural changes involved in the amyloid formation process, the fibril structure can be constructed by "guided" docking experiments. In their sequence and structure analysis on β_2-microglobulin, they proposed that less conserved regions are more likely to undergo conformational change that may lead to amyloid fiber formation. β_2-microglobulin is a seven-stranded β-sandwich structure (Saper et al., 1991). Sequence conservation analysis revealed that unlike the conserved interior strands, strands A (residues 6–12), D (residues 53–60), and G (residues 91–94) are less conserved, suggesting they are prone to conformational changes, which is consistent with many experimental observations. HD exchange experiments showed that strands A and G are not involved in fibril formation (Hoshino et al., 2002). Limited proteolysis studies revealed that strands A, D, and G are protected in the globular form but are not protected in the fibrillar form of β_2-microglobulin (Monti et al., 2002). They proposed that only the interior strands of β_2-microglobulin structure (without strands A and G) are involved in fibril structure. The docked fibril structures using the "core" β_2-microglobulin continuous β-sheet structure with the cross-β pattern are in agreement with the structural features of some amyloid fibrils. However, as discussed earlier, the specific requirements of this approach, which are that the fibril state monomer structure be known, have limited applications of the docking approach in modeling other amyloid fibril structures.

9.4.2 Low-Resolution Models

While the atomic resolution approaches reviewed thus far offer insights on the stability of postulated amyloid fibril structures, they do not tell us much about the assembly process. The problem is, as mentioned earlier, that the atomic detail that makes high-resolution models so realistic also makes them extremely computationally intensive, precluding their application to problems involving large conformational changes or long time scales. A more promising approach for the study of aggregate assembly is the class of models known as low-resolution models.

Low-resolution models, also called simplified folding models, rely on a coarse-grained representation of protein geometry and energetics. They typically account for the motion of groups of atoms along the protein and ignore the motion of the solvent atoms in order to enhance computational efficiency. The absence of solvent atoms in low-resolution models means that effective potentials, or potentials of mean force, must be used to describe the interactions between residues. There are two types of low resolution models: lattice models which represent a protein as a linear chain of residues confined to a lattice, and off-lattice models which represent a protein as a chain of residues or groups of residues moving through continuous space. Low-resolution models have provided valuable insights into the basic principles of protein folding due to their ability to monitor large conformational changes and long time scales (Chan and Dill, 1990; Dill, 1990; Go and Taketomi, 1978, 1979; Kolinski et al., 1986; Kuntz et al., 1976; Levitt, 1976; Levitt and Warshel, 1975; Skolnick and Kolinski, 1990; Taketomi et al., 1975; Tanaka and Scheraga, 1976). Their weakness is their inability to make definitive statements about the folding of specific proteins.

While the nonspecificity of low-resolution protein models is a serious disadvantage when trying to locate a protein's native state structure using simulations, it is much less of a disadvantage in simulating protein aggregation. The reason for this is that fibrillization seems to be less sensitive to the details of the inter- and intramolecular potentials than protein folding is. Support for this idea is the observation that the basic crossed-β protofilament structure is the same for many proteins with different sequences. In fact, some investigators believe that fibrillization is an intrinsic property of proteins under slightly denatured concentrated conditions stemming from the interplay between protein geometry, backbone hydrogen bonding, and hydrophobicity (Dobson, 1999, 2003). The weak dependence of protofilament structure on sequence and the great speed of low-resolution model simulations make this approach well suited to examinations of aggregation kinetics on a molecular level.

Low-resolution model studies of protein aggregation can be differentiated based on several characteristics: on-lattice versus off-lattice, two-dimensional versus three-dimensional, few chains versus many chains, short chains versus long chains, simple interaction potential versus complex interaction potential, amorphous aggregate versus ordered (fibrillar) aggregate, and whether or not the structure of the monomer in the fibrillar state differs or is the same as the structure of the monomer in the isolated native state. As will be seen from the discussion below, the models have gotten more complex and more realistic as time has progressed.

Low-resolution lattice models: The earliest low-resolution model of protein aggregation was a lattice model introduced in 1994 by Patro and Przybycien (Patro and Przybycien, 1994; Patro et al., 1996). They modeled a system of proteins as a collection of hexagons with polar and nonpolar surface sites moving on a two-dimensional lattice. Their aim was to learn how surface characteristics influence the formation kinetics and structure of the observed aggregates. Since their monomers were essentially in the folded state, they could not explore how folding and unfolding impacts aggregation. This early work spurred other groups to apply lattice models in the study of protein aggregation, but with the caveat that the model proteins were allowed to fold and unfold.

The lattice models can be distinguished according to the size of the "alphabet" used to represent the various residues. The simplest of the lattice models is the two-letter-alphabet "HP model" introduced by Lau and Dill (1989), in which a protein is modeled as a chain of hydrophobic (H) and polar (P) residues arranged in a specific sequence. Nonbonded H beads attract each other with strength ε to account for the hydrophobic effect while nonbonded P–P and H–P interactions are set equal to zero. This mimics the tendency of hydrophobic residues to bury themselves in the protein interior in order to avoid contact with water. Even this simple "two-letter-alphabet" model allows investigators to extract (via Monte Carlo simulation or exact enumeration) the general theoretical principles that underlie the connection between a protein's sequence and its native structure, folding pathways, kinetics, and the like.

A common theme running through many of the lattice simulations is the nature of the monomer structure within the ordered aggregate. The question is whether the protein remains soluble in its native state conformation under all conditions, adopts its native state structure within the ordered aggregate, or adopts an alternate structure (the so-called prionlike structure). This theme was explored by Giugliarelli et al. (2000) using a two-dimensional HP model; they found that the answer to the previous question depends sensitively on the amino acid composition. In fact, the most stable proteins were those whose fraction of hydrophobic residues is similar to that found in naturally occurring proteins. Harrison et al. studied the formation of dimers using two- and three-dimensional HP lattice models (Harrison et al., 1999) and a simple two-dimensional four-letter-alphabet lattice model (Harrison et al., 2001)with residues of types H (hydrophobic), P (polar), A, and B, where A and B have a particular affinity for each other. They observed that the protein sequences that were marginally stable as monomers were more likely to be stabilized in an alternate conformation by the multimeric interactions in a dimer aggregate, which was at an energy minimum. The dimers that were rich in β-sheet structure were more likely to propagate their conformations onto other chains, hence the term "prionlike."

The HP model has been used by a number of groups to learn how the structure of the ordered aggregate depends on the protein sequence, concentration, and temperature. Istrail et al. (1999) studied the dependence of the aggregation of two model proteins on the hydrophobic/hydrophilic sequence and composition along the chain as well as chain packing fraction (essentially the concentration). Not surprisingly, the higher the number of hydrophilic residues, the lower the aggregation propensity is. Dima and Thirumalai (2002) used a three-dimensional HP lattice model containing two proteins to probe how the conformational change from a compact monomeric state to an oligomeric β-sheet state depends on temperature and concentration. They observed three distinct ordered states, only one of which contained the native state.

Other investigators have adopted much larger alphabets to describe their lattice proteins. The 20-letter alphabet proposed by Miyazawa and Jernigan (1985) is quite popular. In this model the interresidue interaction potentials are estimated from the numbers of interresidue contacts observed in crystal structures in the PDB. Broglia et al. (1998), Bratko and Blanch (2001), Cellmer et al. (2005), and Leonhard et al. (2003) have all used this alphabet in their Monte Carlo simulation explorations of how protein sequence and concentration influence the aggregation kinetics and thermodynamics of systems containing a small number of chains.

All of the work mentioned above was limited to only a few chains which is not enough to fully explore the competition between protein folding and aggregation. In contrast, the simulations of Combe and Frenkel (2003), Toma and Toma (2000), and Hall and co-workers (Gupta and Hall, 1997; Gupta et al., 1998; Nguyen and Hall, 2002) were truly multichain systems. Combe and Frenkel performed Monte Carlo simulations on a system containing twenty 8-mer peptides whose interactions were modeled using the "Go model," a model in which the interactions are chosen to favor the known native state. Even though Go models introduce a strong bias toward the isolated chain's native state, they can still be used profitably to explore the kinetics of protein aggregation and the competition between folding and aggregation. Toma and Toma conducted lattice Monte Carlo simulations on systems containing as many as twenty 12-mer HP peptides of three different sequences in an effort to learn how the sequence and concentration affect the formation of an ordered state.

Hall and co-workers (Gupta and Hall, 1997; Gupta et al., 1998; Nguyen and Hall, 2002) conducted simulations on systems containing as many as 40 two-dimensional 16-mer HP chains of a single sequence in an effort to learn how the protein concentration, denaturant concentration, and temperature affected the refolding yield and the kinetics of the aggregation pathway. Denaturant concentration was modeled implicitly; the stronger the interaction between the hydrophobic residues, the weaker the denaturant concentration. Since their aim was to learn how to optimize protein folding yield during recovery from inclusion bodies, they focused primarily on aggregation into nonstructured states. Nguyen and Hall (2002) performed simulations that mimicked four methods of thermal protein renaturation used in the lab: dialysis, dilution (or diafiltration), quenching, and pulse renaturation (fed-batch operation). Based on the simulation results, a strategy for rapidly obtaining high refolding yields was suggested which involved instantaneous removal to intermediate denaturant concentrations followed by dialysis to the final state.

Low-resolution models—off lattice: Off–lattice low-resolution protein models have been used extensively in the past decade to simulate the folding of an isolated protein. They are just beginning to be used in studies of aggregation. In the simplest of these models, a protein is represented as a flexible chain of spheres (think pearl necklace), with each sphere representing a single residue. The interactions between the spheres are represented by energy functions that can be divided roughly into three categories: (1) Go-type potentials in which the parameters are chosen to favor the protein's known native state (Go and Taketomi, 1978; Taketomi et al., 1975), (2) potentials based on the relative hydrophobicity of the side chains as measured by various hydrophobicity scales (and sometimes on the partial charge), and (3) knowledge-based potentials like the Miyazawa–Jernigan potential (Miyazawa and Jernigan, 1985) in which statistical data on residue–residue contacts from the PDB are used to infer side chain/side chain potentials. Local interaction potentials are also used to maintain steric and angular constraints.

Jang et al. (2004a,b) used an off-lattice low-resolution protein model to investigate the thermodynamics and kinetics associated with the folding, aggregation, and fibrillization of β-strand peptides. The goal was to learn why multiprotein systems sometimes form ordered aggregates and sometimes form amorphous aggregates. Their off-lattice protein model consisted of 39 single-sphere residues interacting intramolecularly via a square–well Go potential (Go and Taketomi, 1978; Taketomi et al., 1975) that favored the four-strand antiparallel β-sheet native state and intermolecularly via a second square–well Go potential that favored the formation of a tetrameric β-sheet complex (a model fibril). In fact, the intra- and intermolecular interactions could be interpreted as mimicking hydrogen bonding and hydrophobic interactions, respectively. The ratio of the strengths of the intra- and intermolecular potentials was varied. Discontinuous MD simulations (described in a later section) were performed on systems containing a single protein and four proteins to see how the equilibrium properties, folding pathways, and kinetics varied as a function of the ratio of the intra- and intermolecular interactions and the temperature. A phase diagram was constructed showing which monomer states and which tetrameric complex states were stable at various temperatures and interaction strength ratios. At high temperatures the four-peptide system formed monomers but at low temperatures the system assembled into tetrameric β-sheet complexes that were either partially ordered , ordered, or highly ordered, depending on the relative strength of the inter- and intramolecular interactions. The most ordered structures were found at intermediate values of this ratio. This implies that for ordered aggregates to form, there must be a balance between the hydrogen bonding interactions that hold the β-strands together in a β-sheet and the hydrophobic interactions that hold the sheets together. If either is too large, disordered rather than ordered aggregates will form at high concentrations.

Ding et al. (2002a,b) also used an off-lattice low-resolution model to study the aggregation of a system of eight model Src SH3 domain proteins. Each amino acid residue along the flexible backbone chain was represented by a single backbone sphere and a single side-chain sphere interacting via a Go potential designed to recover the protein's native state. A fibrillar double β-sheet structure was observed with inter-β-strand spacing and inter-β-sheet spacing similar to those observed in experiments.

Intermediate-resolution protein models: In recent years a new class of protein folding models has been introduced, called intermediate-resolution folding models (Derreumaux, 1999; Liwo et al., 1997; Takada et al., 1999; Wallqvist and Ullner, 1994). The idea here was to add more realistic features to the low-resolution protein folding models in the hopes that this would allow a priori prediction of the native state structure of specific proteins based solely on their amino acid sequence. The number of spheres used to represent protein geometry was increased from one to as many as seven. The energy functions were expanded to include not only the three categories described for the low-resolution off-lattice models but also hydrogen-bonding potentials, multibody terms, burial terms (in which the strength of the hydrophobic interaction depends on the extent of burial), and special potentials for disulfide bonds and proline.

Intermediate-resolution protein models are now being used to study aggregation and fibril formation. In 2001, Hall and co-workers (Smith and Hall, 2001a,b,c) introduced an implicit-solvent intermediate-resolution protein model, which they subsequently named PRIME (Nguyen et al., 2004). This model is simple enough to allow the simulation of systems containing many proteins over long time scales, yet contains sufficient molecular detail to mimic real protein dynamics. The level of molecular detail in the protein representation and interaction potential is reduced just to the point at which the key physical features governing protein fibrillization remain and the other features are neglected. In PRIME, each amino acid is composed of four spheres: a three-sphere backbone comprised of united atom NH, C_α, and C=O and a single-sphere side chain (CH_3 for alanine), all with realistic diameters and bond lengths. Ideal backbone bond lengths, bond angles, C_α–C_α distances, and residue L- isomerization are maintained by imposing a series of pseudo bonds whose lengths fluctuate within a tolerance of 2% about the specified values.

All forces in PRIME are modeled by either hard-sphere or square-well potentials with realistic diameters. This was done so as to allow the use of discontinuous MD (DMD) simulation, a very fast alternative to traditional MD simulation that is applicable to systems of molecules interacting via discontinuous potentials, e.g., hard-sphere and square-well potentials (Rapaport, 1978, 1979). Instead of solving Newton's equation of motion at regular spaced time intervals, as in traditional MD, DMD is event-driven. Since discontinuous potentials exert forces only when particles collide (unlike continuous potentials such as the Lennard–Jones potential), the position and velocity of each molecule after a collision can be determined exactly, as opposed to numerically. This imparts great speed to the algorithm, allowing sampling of much wider regions of conformational space, longer time scales, and larger systems than in traditional MD.

The solvent in PRIME is modeled implicitly by factoring its effect into the energy function as a potential of mean force. Interactions between hydrophobic side chains are represented by a square-well potential; interactions between polar side chains or between polar and hydrophobic side chains are represented by a hard sphere interaction. Hydrogen bonding between amide hydrogen atoms and carbonyl oxygen atoms is represented by a directionally dependent square-well attraction of strength between NH and C=O united atoms. (The angle between the "virtual" N–H and C=O vectors can be determined from knowledge of the locations of the adjacent united atoms along the chain.) The strength of the hydrophobic interaction is fixed at a fraction of the strength of the hydrogen bonding interaction; this fraction is the only adjustable parameter in the model.

By combining PRIME with DMD, Nguyen and Hall (2004a, 2005) were able to simulate the spontaneous formation of ordered aggregates, essentially protofilaments, in model systems containing 48 to 96 polyalanine (Ac-$KA_{14}K$-NH_2) peptides (Fig. 9.3). Polyalanine was chosen for study because synthetic polyalanine-based peptides, which form α-helical structures at low temperatures and low peptide concentrations, had been found experimentally to form β-sheet complexes (fibrils) *in vitro* at high temperatures and high peptide concentrations. All simulations started from a random coil configuration equilibrated at a high temperature and then slowly cooled to the temperature of interest so as to minimize kinetic trapping. These simulations took approximately 40 hours on an AMD Athlon MP 2200+ single-processor workstation. The simulation results showed that the populations of α-helices, amorphous aggregates, β-sheets, and fibrils were highly dependent

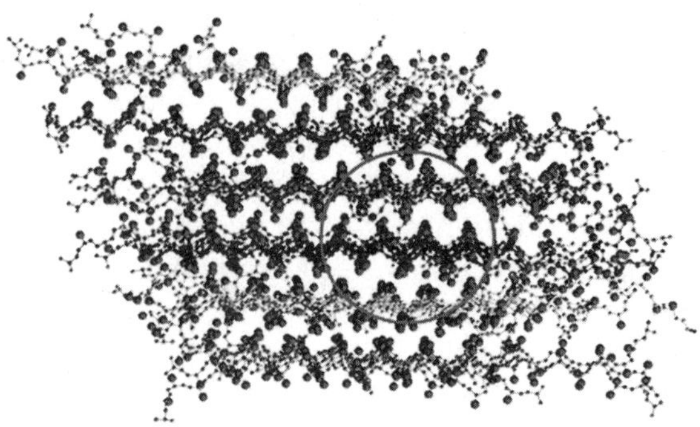

Fig. 9.3 A closeup snapshot of the 96-peptide fibrillar structure of polyalanine.

on temperature and peptide concentration, in qualitative agreement with the experimental results of Blondelle and co-workers (Blondelle et al., 1997; Perez-Paya et al., 1996) on Ac-$KA_{14}K$-NH_2 peptides. The fibrils observed in the simulations mimicked the structural characteristics observed in experiments in that most of the fibrillar peptides were arranged in an in-register parallel orientation, with intra- and intersheet distances similar to those observed in experiments. The simulations revealed that Ac-$KA_{14}K$-NH_2 fibril formation is nucleation dependent, which is similarly observed in experimental studies. The formation of small fibrils was preceded by the appearance of small amorphous aggregates, then β-sheets, and finally rapid growth of a stable fibril. A phase diagram in the temperature–concentration plane was mapped out delineating the regions where random coils, α-helices, β-sheets, fibrils, and amorphous aggregates are stable (Nguyen and Hall, 2004b).

Models similar to PRIME have been developed by Stanley, Ding, Dokholyan, Teplow (Ding et al., 2003; Urbanc et al., 2004a,b) and co-workers for use with DMD (which they call "discrete molecular dynamics" as opposed to "discontinuous molecular dynamics"). Their main focus has been on the aggregation of β amyloid, Aβ, the protein whose oligomerization and fibrillization have been linked to Alzheimer's disease. The Aβ peptide is represented with a four bead per residue model as in PRIME. Their side chain representation and energy function are more complex than used in PRIME for polyalanine in order to account for the differences between all 20 amino acid residues. There are four types of side chains: neutral, charged hydrophilic, hydrophilic, and hydrophobic. The hydropathy of the various side chains is assigned according to the scale of Kyte and Doolittle (1982). In addition to having an attraction between hydrophobic side chains, they include a repulsive interaction between uncharged hydrophilic side chains and either charged or uncharged side chains. The side chain for glycine residues is absent.

This model is being used to tackle the difficult question of why Aβ(1–42) is so much more amyloidogenic than Aβ(1–40). Aβ(1–42) is more likely to be associated with the early onset forms of Alzheimer's, with increased risk for getting Alzheimer's disease, with enhanced neurotoxicity, and with faster formation of fibrils *in vitro*. Urbanc et al. (2004a,b) conducted DMD simulations of systems containing 32 Aβ (1–40) and 32 Aβ(1–42) peptides starting from a system of random coil Aβ monomers. Although they did not observe ordered structures in their simulations, they did observe important early events in the aggregation process including monomer folding and assembly into disordered oligomers of various sizes. Analysis of their results indicates that there are significant differences between the oligomer size distributions of Aβ(1–40) and Aβ(1–42), with Aβ(1–40) more likely to form dimers and Aβ(1–42) more likely to form pentamers, in agreement with *in vitro* size distribution studies. The Aβ(1–42) peptide was likely to form a turn at Gly37–Gly38, whereas the Aβ(1–40) was not. The structural differences between the conformations of the Aβ(1–40) and Aβ(1–42) oligomers suggest that the hydrophobic core of the Aβ(1–42) pentamer is more exposed than that of the Aβ(1–40) pentamer, and is therefore likely to form larger oligomers. This may have implications for the biology of Alzheimer's disease since some believe that it is at the oligomer/protofibril level that Aβ is most toxic with Aβ(1–42) being more toxic than Aβ(1–40) (Caughey and Lansbury, 2003).

9.5 Summary

In summary, high-resolution structural characterization of protein aggregates using classical approaches, such as X-ray crystallography or solution NMR, has been hampered due to the insolubility and noncrystalline nature of the aggregates. Encouraging results have begun to emerge, however. Two high-resolution, detailed structures, both involving crystal packing with "infinite β-sheet" (Schiffer et al., 1985) characteristics, have been obtained from crystals of amyloidogenic peptides (Makin et al., 2005; Nelson et al., 2005). One peptide is a seven-residue fragment of a yeast prion known as Sup35 (Nelson et al., 2005). The other one is a designed 12mer peptide containing two KFFE motifs separated by an AAAK motif (Makin et al., 2005). Experimental approaches as discussed in Section 9.3, including fiber diffraction, electron microscopy, hydrogen–deuterium exchange, solid-state NMR, limited proteolysis, electron paramagnetic resonance spectroscopy, and various chemical approaches, have yielded valuable information about the possible conformations of aggregate structure. But these low-resolution data are not sufficient to establish a high-resolution structure of protein aggregates, without which it will be difficult to address some of the fundamental questions regarding the molecular mechanism of aggregate assembly and the detailed inter- and intramolecular interactions that stabilize protein aggregates. Computational approaches can use the low-resolution experimental data to advance our understanding of the structure of protein aggregates and amyloid fibrils. All of the models constructed so far using computational approaches are motivated by or rely on experimental observations, and are validated using molecular dynamic simulations, a common technique for testing the stabilities of the models.

Considering that the insoluble nature of amyloid fibrils makes it hard to obtain high-resolution, detailed structural information, computational approaches should play a significant role in our efforts to solve the aggregate structure. Although computational methods are making strides in deciphering the mechanism of fibrillogenesis and the fibril core structure, many challenging issues need to be addressed in the future. First, computational studies of the aggregation process currently only apply low- to intermediate-resolution models. Novel ideas are clearly needed to investigate the process at a higher level given current computation capability. Second, it is well known that all amyloid fibrils share the common cross-β structure, but the structural details might vary from sequence to sequence. Both antiparallel and parallel β-sheet organizations have been suggested for the core structure from solid-state NMR studies. Moreover, recent studies have revealed that different growth conditions applied to the same peptide molecule can yield fibrils with distinct morphologies and possibly with different neurotoxicities (Petkova et al., 2005). Most importantly, amyloid fibril morphology correlates with internal architecture of the fibril, such as side-chain packing arrangements, and the sequences involved in the β-structure. Therefore, it is highly likely that there exist variations of amyloid folding motifs. Computational methods should be able to simulate the various structures under different conditions. In addition, comparative studies of amyloid fibril models formed by different amyloid proteins should be done in the future to elucidate the general principles regarding how specific interactions stabilize the fibril structures. Probably the most challenging issue in amyloid fibril structure modeling is the modeling of the packing patterns and detailed interactions among protofilaments, which might vary from fibril to fibril (Jimenez et al., 2002). Last but not least, we should also improve structural modeling techniques in such a way that new experimental data can be incorporated into the model as constraints. With the advance in computation speed and capability and the help from experimental observations, in the near future, we should be able to combine modeling studies with peptide assembling simulations for a better understanding of the process of amyloid fibril formation and detailed structure of the fibrils.

Suggested Further Reading

A special issue of the review journal *Accounts of Chemical Research* on amyloid (Vol 89, Issue 9, Sept., 2006) contains a number of articles on aspects of amyloid structure. Recent methods for analysis of amyloid structure, as well as some computational methods, are described in detail in *Methods in Enzymology, Amyloid, Prions, and Other Protein Aggregates* (R. Wetzel and I. Kheterpal, Eds.), Academic Press, San Diego, 2006, Volumes 412 and 413. The review by Zanuy et al. (2004) provides insights on amyloid structural formation and assembly through computational approaches.

Acknowledgments

The authors acknowledge support from the National Institutes of Health (R01 AG18927 to R.W., Y.X., and J.T.G.; AG18416 to R.W.; and R01 GM056766 to C.K.H.), National Science Foundation (DBI-0354771 to Y.X. and J.T.G.), and the Georgia Cancer Coalition (to Y.X. and J.T.G.).

References

Anfinsen, C.B. 1973. Principles that govern the folding of protein chains. *Science* 181:223–230.

Baker, D., Sohl, J.L., and Agard, D.A. 1992. A protein-folding reaction under kinetic control. *Nature* 356:263–265.

Balbirnie, M., Grothe, R., and Eisenberg, D.S. 2001. An amyloid-forming peptide from the yeast prion Sup35 reveals a dehydrated beta-sheet structure for amyloid. *Proc. Natl. Acad. Sci. USA* 98:2375–2380.

Benyamini, H., Gunasekaran, K., Wolfson, H., and Nussinov, R. 2003. Beta2-microglobulin amyloidosis: Insights from conservation analysis and fibril modelling by protein docking techniques. *J. Mol. Biol.* 330:159–174.

Benzinger, T.L., Gregory, D.M., Burkoth, T.S., Miller-Auer, H., Lynn, D.G., Botto, R.E., and Meredith, S.C. 1998. Propagating structure of Alzheimer's beta-amyloid(10-35) is parallel beta-sheet with residues in exact register. *Proc. Natl. Acad. Sci. USA* 95:13407–13412.

Berman, H.M., Westbrook, J., Feng, Z., Gilliland, G., Bhat, T.N., Weissig, H., Shindyalov, I.N., and Bourne, P.E. 2000. The Protein Data Bank. *Nucleic Acids Res.* 28:235–242.

Bhattacharyya, A.M., Thakur, A., and Wetzel, R. 2005. Polyglutamine aggregation nucleation: Thermodynamics of a highly unfavorable protein folding reaction. *Proc. Natl. Acad. Sci USA* 102:15400–15405.

Bitan, G., and Teplow, D.B. 2004. Rapid photochemical cross-linking—A new tool for studies of metastable, amyloidogenic protein assemblies. *Acc. Chem. Res.* 37:357–364.

Blake, C.C., Geisow, M.J., Oatley, S.J., Rerat, B., and Rerat, C. 1978. Structure of prealbumin: Secondary, tertiary and quaternary interactions determined by Fourier refinement at 1.8 Å. *J. Mol. Biol.* 121:339–356.

Blondelle, S.E., Forood, B., Houghten, R.A., and Perez-Paya, E. 1997. Polyalanine-based peptides as models for self-associated beta-pleated-sheet complexes. *Biochemistry* 36:8393–8400.

Bratko, D., and Blanch, H.W. 2001. Competition between protein folding and aggregation: A three-dimensional lattice-model simulation. *J. Chem. Phys.* 114:561–569.

Broglia, R.A., Tiana, G., Pasquali, S., Roman, H.E., and Vigezzi, E. 1998. Folding and aggregation of designed proteins. *Proc. Nat. Acad. Sci. USA* 95:12930–12933.

Bucciantini, M., Giannoni, E., Chiti, F., Baroni, F., Formigli, L., Zurdo, J., Taddei, N., Ramponi, G., Dobson, C.M., and Stefani, M. 2002. Inherent toxicity of aggregates implies a common mechanism for protein misfolding diseases. *Nature* 416:507–511.

Caughey, B., and Lansbury, P.T. 2003. Protofibrils, pores, fibrils, and neurodegeneration: Separating the responsible protein aggregates from the innocent bystanders. *Annu. Rev. Neurosci.* 26:267–298.

Cellmer, T., Bratko, D., Prausnitz, J.M., and Blanch, H. 2005. Thermodynamics of folding and association of lattice-model proteins. *J. Chem. Phys.* 122:174908.

Chan, H.S., and Dill, K.A. 1990. Origins of structure in globular-proteins. *Proc. Nat. Acad. Sci. USA* 87:6388–6392.

Chan, W., Helms, L.R., Brooks, I., Lee, G., Ngola, S., McNulty, D., Maleeff, B., Hensley, P., and Wetzel, R. 1996. Mutational effects on inclusion body formation in the periplasmic expression of the immunoglobulin V_L domain REI. *Fold. Des.* 1:77–89.

Chaney, M.O., Webster, S.D., Kuo, Y.M., and Roher, A.E. 1998. Molecular modeling of the Abeta1-42 peptide from Alzheimer's disease. *Protein Eng.* 11:761–767.

Chen. S., Ferrone, F., and Wetzel, R. 2002. Huntington's disease age-of-onset linked to polyglutamine aggregation nucleation. *Proc. Natl. Acad. Sci. USA* 99:11884–11889.

Chick, H., and Martin, C.J. 1910. On the "heat coagulation" of proteins. *J. Physiol.* 40:404–430.

Chowdhry, V., and Westheimer, F.H. 1979. Photoaffinity labeling of biological systems. *Annu. Rev. Biochem.* 48:293–325.

Cleland, J.L., Powell, M.F., and Shire, S.J. 1993. The development of stable protein formulations: A close look at protein aggregation, deamidation, and oxidation. *Crit. Rev. Ther. Drug Carrier Syst.* 10:307–377.

Coles, M., Bicknell, W., Watson, A.A., Fairlie, D.P., and Craik, D.J. 1998. Solution structure of amyloid beta-peptide(1-40) in a water-micelle environment. Is the membrane-spanning domain where we think it is? *Biochemistry* 37:11064–11077.

Collins, S.R., Douglass, A., Vale, R.D., and Weissman, J.S. 2004. Mechanism of prion propagation: Efficient amyloid growth in the absence of oligomeric intermediates. PLoS 2:1582–1590.

Colon, W., and Kelly, J.W. 1992. Partial denaturation of transthyretin is sufficient for amyloid fibril formation *in vitro*. *Biochemistry* 31:8654–8660.

Combe, N., and Frenkel, D. 2003. Phase behavior of a lattice protein model. *J. Chem. Phys.* 118:9015–9022.

Creighton, T.E. 1992. Protein folding. Up the kinetic pathway [news; comment]. *Nature* 356:194–195.

De Bernardez Clark, E., Schwarz, E., and Rudolph, R. 1999. Inhibition of aggregation side reactions during *in vitro* protein folding. *Methods Enzymol* 309:217–236.

Del Mar, C., Greenbaum, E., Mayne, L., Englander, S.W., and Woods, V.L., Jr. 2005. Amyloid structure: alpha-synuclein studied by hydrogen exchange and mass spectrometry. *Proc. Natl. Acad. Sci. USA* 102: 15477–15482.

Derreumaux, P. 1999. From polypeptide sequences to structures using Monte Carlo simulations and an optimized potential. *J. Chem. Phys.* 111:2301–2310.

DiFiglia, M., Sapp, E., Chase, K.O., Davies, S.W., Bates, G.P., Vonsattel, J.P., and Aronin, N. 1997. Aggregation of huntingtin in neuronal intranuclear inclusions and dystrophic neurites in brain. *Science* 277:1990–1993.

Dill, K.A. 1990. Dominant forces in protein folding. *Biochemistry* 29:7133–7155.

Dill, K.A., and Chan, H.S. 1997. From Levinthal to pathways to funnels. *Nat. Struct. Biol.* 4:10–19.

Dima, R.I., and Thirumalai, D. 2002. Exploring protein aggregation and self-propagation using lattice models: Phase diagram and kinetics. *Protein Sci.* 11:1036–1049.

Ding, F., Borreguero, J.M., Buldyrey, S.V., Stanley, H.E., and Dokholyan, N.V. 2003. Mechanism for the alpha-helix to beta-hairpin transition. *Proteins Struct. Funct. Genet.* 53:220–228.

Ding, F., Dokholyan, N.V., Buldyrev, S.V., Stanley, H.E., and Shakhnovich, E.I. 2002a. Direct molecular dynamics observation of protein folding transition state ensemble. *Biophys. J.* 83:3525–3532.

Ding, F., Dokholyan, N.V., Buldyrev, S.V., Stanley, H.E., and Shakhnovich, E.I. 2002b. Molecular dynamics simulation of the SH3 domain aggregation suggests a generic amyloidogenesis mechanism. *J. Mol. Biol.* 324:851–857.

Dobson, C.M. 1999. Protein misfolding, evolution and disease. *Trends Biochem. Sci.* 24:329–332.

Dobson, C.M. 2003. Protein folding and misfolding. *Nature* 426:884–890.

Elam, J.S., Taylor, A.B., Strange, R., Antonyuk, S., Doucette, P.A., Rodriguez, J.A., Hasnain, S.S., Hayward, L.J., Valentine, J.S., Yeates, T.O., and Hart, P.J. 2003. Amyloid-like filaments and water-filled nanotubes formed by SOD1 mutant proteins linked to familial ALS. *Nat. Struct. Biol.* 10:461–467.

Ferraro, D.M., Lazo, N.D., and Robertson, A.D. 2004. EX1 hydrogen exchange and protein folding. *Biochemistry* 43:587–594.

Ferrone, F. 1999. Analysis of protein aggregation kinetics. *Methods Enzymol.* 309:256–274.

Finke, J.M., Gross, L.A., Ho, H.M., Sept, D., Zimm, B.H., and Jennings, P.A. 2000. Commitment to folded and aggregated states occurs late in interleukin-1 beta folding. *Biochemistry* 39:15633–15642.

Fleming, P.J., and Rose, G.D. 2005. Conformational properties of unfolded proteins. In *Protein Folding Handbook, Part I* (J. Buchner and T. Kiefhaber, Eds.). Weinheim, Wiley-VCH, pp. 710–736.

Fontana, A., Polverino de Laureto, P., De Filippis, V., Scaramella, E., and Zambonin, M. 1997. Probing the partly folded states of proteins by limited proteolysis. *Fold. Des.* 2:R17–26.

Giugliarelli, G., Micheletti, C., Banavar, J.R., and Maritan, A. 2000. Compactness, aggregation, and prionlike behavior of protein: A lattice model study. *J. Chem. Phys.* 113:5072–5077.

Glickman, M.H. 2000. Getting in and out of the proteasome. *Semin. Cell Dev. Biol.* 11:149–158.

Go, N., and Taketomi, H. 1978. Respective roles of short- and long-range interactions in protein folding. *Proc. Natl. Acad. Sci. USA* 75:559–563.

Go, N., and Taketomi, H. 1979. Studies on protein folding, unfolding and fluctuations by computer simulation. III. Effect of short-range interactions. *Int. J. Pept. Protein Res.* 13:235–252.

Goldberg, M.E., Rudolph, R., and Jaenicke, R. 1991. A kinetic study of the competition between renaturation and aggregation during the refolding of denatured-reduced egg white lysozyme. *Biochemistry* 30:2790–2797.

Goldsbury, C., Kistler, J., Aebi, U., Arvinte, T., and Cooper, G.J. 1999. Watching amyloid fibrils grow by time-lapse atomic force microscopy. *J. Mol. Biol.* 285:33–39.

Goldsbury, C.S., Wirtz, S., Muller, S.A., Sunderji, S., Wicki, P., Aebi, U., and Frey, P. 2000. Studies on the *in vitro* assembly of A beta 1-40: Implications for the search for A beta fibril formation inhibitors. *J. Struct. Biol.* 130:217–231.

Govaerts, C., Wille, H., Prusiner, S.B., and Cohen, F.E. 2004. Evidence for assembly of prions with left-handed beta-helices into trimers. *Proc. Natl. Acad. Sci. USA* 101:8342–8347.

Guo, J.T., Wetzel, R., and Xu, Y. 2004. Molecular modeling of the core of Abeta amyloid fibrils. *Proteins* 57:357–364.

Gupta, P., and Hall, C.K. 1997. Effect of solvent conditions upon refolding pathways and intermediates for a simple lattice protein. *Biopolymers* 42:399–409.

Gupta, P., Hall, C.K., and Voegler, A.C. 1998. Effect of denaturant and protein concentrations upon protein refolding and aggregation: A simple lattice model. *Protein Sci.* 7:2642–2652.

Haase-Pettingell, C.A., and King, J. 1988. Formation of aggregates from a thermolabile *in vivo* folding intermediate in P22 tailspike maturation: A model for inclusion body formation. *J. Biol. Chem.* 263:4977–4983.

Harper, J.D., Lieber, C.M., and Lansbury, P.T., Jr. 1997. Atomic force microscopic imaging of seeded fibril formation and fibril branching by the Alzheimer's disease amyloid-beta protein. *Chem. Biol.* 4:951–959.

Harrison, P.M., Chan, H.S., Prusiner, S.B., and Cohen, F.E. 1999. Thermodynamics of model prions and its implications for the problem of prion protein folding. *J. Mol. Biol.* 286:593–606.

Harrison, P.M., Chan, H.S., Prusiner, S.B., and Cohen, F.E. 2001. Conformational propagation with prion-like characteristics in a simple model of protein folding. *Protein Sci.* 10:819–835.

Hartl, F.U., and Hayer-Hartl, M. 2002. Molecular chaperones in the cytosol: From nascent chain to folded protein. *Science* 295:1852–1858.

Haspel, N., Zanuy, D., Ma, B., Wolfson, H., and Nussinov, R. 2005. A comparative study of amyloid fibril formation by residues 15–19 of the human calcitonin hormone: A single beta-sheet model with a small hydrophobic core. *J. Mol. Biol.* 345:1213–1227.

Hermeling, S., Crommelin, D.J., Schellekens, H., and Jiskoot, W. 2004. Structure-immunogenicity relationships of therapeutic proteins. *Pharm. Res.* 21:897–903.

Horiuchi, M., Priola, S.A., Chabry, J., and Caughey, B. 2000. Interactions between heterologous forms of prion protein: Binding, inhibition of conversion, and species barriers. *Proc. Natl. Acad. Sci. USA* 97:5836–5841.

Hoshino, M., Katou, H., Hagihara, Y., Hasegawa, K., Naiki, H., and Goto, Y. 2002. Mapping the core of the beta(2)-microglobulin amyloid fibril by H/D exchange. *Nat. Struct. Biol.* 9:332–336.

Hua, Q.X., Gozani, S.N., Chance, R.E., Hoffmann, J.A., Frank, B.H., and Weiss, M.A. 1995. Structure of a protein in a kinetic trap. *Nat. Struct. Biol.* 2:129–138.

Hubbell, W.L., Cafiso, D.S., and Altenbach, C. 2000. Identifying conformational changes with site-directed spin labeling. *Nat. Struct. Biol.* 7:735–739.

Hurle, M.R., Helms, L.R., Li, L., Chan, W., and Wetzel, R. 1994. A role for destabilizing amino acid replacements in light chain amyloidosis. *Proc. Natl. Acad. Sci. USA* 91:5446–5450.

Ignatova, Z., and Gierasch, L.M. 2005. Aggregation of a slow-folding mutant of a beta-clam protein proceeds through a monomeric nucleus. *Biochemistry* 44:7266–7274.

Istrail, S., Schwartz, R., and King, J. 1999. Lattice simulations of aggregation funnels for protein folding. *J. Comput. Biol.* 6:143–162.

Iwata, K., Eyles, S.J, and Lee, J.P. 2001. Exposing asymmetry between monomers in Alzheimer's amyloid fibrils via reductive alkylation of lysine residues. *J. Am. Chem. Soc.* 123:6728–6729.

Jang, H.B., Hall, C.K., and Zhou, Y.Q. 2004a. Assembly and kinetic folding pathways of a tetrameric beta-sheet complex: Molecular dynamics simulations on simplified off-lattice protein models. *Biophys. J.* 86:31–49.

Jang, H.B., Hall, C.K., and Zhou, Y.Q. 2004b. Thermodynamics and stability of a beta-sheet complex: Molecular dynamics simulations on simplified off-lattice protein models. *Protein Sci* 13:40–53.

Jaroniec, C.P., MacPhee, C.E., Bajaj, V.S., McMahon, M.T., Dobson, C.M., and Griffin, R.G. 2004. High-resolution molecular structure of a peptide in an amyloid fibril determined by magic angle spinning NMR spectroscopy. *Proc. Natl. Acad. Sci. USA* 101:711–716.

Jarrett, J.T., Costa, P.R., Griffin, R.G., and Lansbury, P.T., Jr. 1994. Models of the b protein C-terminus: Differences in amyloid structure may lead to segregation of "long" and "short" fibrils. *J. Am. Chem. Soc.* 116:9741–9742.

Jenkins, J., and Pickersgill, R. 2001. The architecture of parallel beta-helices and related folds. *Prog. Biophys. Mol. Biol.* 77:111–175.

Jimenez, J.L., Guijarro, J.I., Orlova, E., Zurdo, J., Dobson, C.M., Sunde, M., and Saibil, H.R. 1999. Cryo-electron microscopy structure of an SH3 amyloid fibril and model of the molecular packing. EMBO J. 18:815–821.

Jimenez, J.L., Nettleton, E.J., Bouchard, M., Robinson, C.V., Dobson, C.M., and Saibil, H.R. 2002. The protofilament structure of insulin amyloid fibrils. *Proc. Natl. Acad. Sci. USA* 99:9196–9201.

Kanno, T., Yamaguchi, K., Naiki, H., Goto, Y., and Kawai, T. 2005. Association of thin filaments into thick filaments revealing the structural hierarchy of amyloid fibrils. *J. Struct. Biol.* 149:213–218.

Karplus, M., and McCammon, J.A. 2002. Molecular dynamics simulations of biomolecules. *Nat. Struct. Biol.* 9:646–652.

Kayed, R., Head, E., Thompson, J.L., McIntire, T.M., Milton, S.C., Cotman, C.W., and Glabe, C.G. 2003. Common structure of soluble amyloid oligomers implies common mechanism of pathogenesis. *Science* 300:486–489.

Kellermayer, M.S., Grama, L., Karsai, A., Nagy, A., Kahn, A., Datki, Z.L., and Penke, B. 2005. Reversible mechanical unzipping of amyloid beta-fibrils. *J. Biol. Chem.* 280:8464–8470.

Khare, S.D., Ding, F., Gwanmesia, K.N., and Dokholyan, N.V. 2005. Molecular origin of polyglutamine aggregation in neurodegenerative diseases. *PLoS Comput. Biol.* 1:230–235.

Kheterpal, I., Chen, M., Cook, K.D., and Wetzel, R. 2006. Structural differences in Abeta amyloid protofibrils and fibrils mapped by hydrogen exchange-mass spectrometry with on-line pratcolytic fragmentation. *J. Mol. Biol.* 361:785–795.

Kheterpal, I., Lashuel, H.A., Hartley, D.M., Walz, T., Lansbury, P.T., and Jr., Wetzel, R. 2003a. Abeta protofibrils possess a stable core structure resistant to hydrogen exchange. *Biochemistry* 42:14092–8.

Kheterpal, I., and Wetzel, R. 2006. Amyloid, prions, and other protein aggregates II. In *Methods in Enzymology* (J. N. Abelson and M. I. Simon, Eds.), San Diego, Academic Press.

Kheterpal, I., Wetzel, R., and Cook, K.D. 2003b. Enhanced correction methods for hydrogen exchange–mass spectrometric studies of amyloid fibrils. *Protein Sci.* 12:635–643.

Kheterpal, I., Williams, A., Murphy, C., Bledsoe, B., and Wetzel, R. 2001. Structural features of the Abeta amyloid fibril elucidated by limited proteolysis. *Biochemistry* 40:11757–11767.

Kheterpal, I., Zhou, S., Cook, K.D., and Wetzel, R. 2000. Abeta amyloid fibrils possess a core structure highly resistant to hydrogen exchange. *Proc. Natl. Acad. Sci. USA* 97:13597–13601.

Kolinski, A., Skolnick, J., and Yaris, R. 1986. Monte-Carlo simulations on an equilibrium globular protein folding model. *Proc. Nat. Acad. Sci. USA* 83:7267–7271.

Kopito, R.R. 2000. Aggresomes, inclusion bodies and protein aggregation. *Trends Cell Biol.* 10:524–530.

Kuntz, I.D., Crippen, G.M., Kollman, P.A., and Kimelman, D. 1976. Calculation of protein tertiary structure. *J. Mol. Biol.* 106:983–994.

Kuwata, K., Matumoto, T., Cheng, H., Nagayama, K., James, T.L., and Roder, H. 2003. NMR-detected hydrogen exchange and molecular dynamics simulations provide structural insight into fibril formation of prion protein fragment 106–126. *Proc. Natl. Acad. Sci. USA* 100:14790–14795.

Kyte, J., and Doolittle, R.F. 1982. A simple method for displaying the hydropathic character of a protein. *J. Mol. Biol.* 157:105–132.

Lakdawala, A.S., Morgan, D.M., Liotta, D.C., Lynn, D.G., and Snyder, J.P. 2002. Dynamics and fluidity of amyloid fibrils: a model of fibrous protein aggregates. *J. Am. Chem. Soc.* 124:15150–15151.

Lau, K.F., Dill, K.A. 1989. A lattice statistical-mechanics model of the conformational and sequence-spaces of proteins. *Macromolecules* 22:3986–3997.

Leonhard, K., Prausnitz, J.M., and Radke, C.J. 2003. Solvent–amino acid interaction energies in 3-D-lattice MC simulations of model proteins. Aggregation thermodynamics and kinetics. *Phys. Chem. Chem. Phys.* 5:5291–5299.

Levin, E.G., and Santell, L. 1987. Conversion of the active to latent plasminogen activator inhibitor from human endothelial cells. *Blood* 70:1090–1098.

Levinthal, C. 1969. How to fold gratiously. *Univ. Ill. Bull.* 41:22–24.

Levitt, M. 1976. Simplified representation of protein conformations for rapid simulation of protein folding. *J. Mol. Biol.* 104:59–107.

Levitt, M., and Warshel, A. 1975. Computer-simulation of protein folding. *Nature* 253:694–698.

Li, L., Darden, T.A., Bartolotti, L., Kominos, D., and Pedersen, L.G. 1999. An atomic model for the pleated beta-sheet structure of Abeta amyloid protofilaments. *Biophys. J.* 76:2871–2878.

Li, R., and Woodward, C. 1999. The hydrogen exchange core and protein folding. *Protein Sci.* 8:1571–1590.

Liwo, A., Oldziej, S., Kazmierkiewicz, R., Groth, M., Czaplewski, C. 1997. Design of a knowledge-based force field for off-lattice simulations of protein structure. *Acta. Biochim. Pol.* 44:527–547.

Lynn, G.W., Heller, W.T., Mayasundari, A., Minor, K.H., and Peterson, C.B. 2005. A model for the three-dimensional structure of human plasma vitronectin from small-angle scattering measurements. *Biochemistry* 44:565–574.

Ma, B., and Nussinov, R. 2002a. Molecular dynamics simulations of alanine rich beta-sheet oligomers: Insight into amyloid formation. *Protein Sci.* 11:2335–2350.

Ma, B., and Nussinov, R. 2002b. Stabilities and conformations of Alzheimer's beta -amyloid peptide oligomers (Abeta 16-22, Abeta 16-35, and Abeta 10-35): Sequence effects. *Proc. Natl. Acad. Sci. USA* 99:14126–14131.

Ma, J., Sigler, P.B., Xu, Z., and Karplus, M. 2000. A dynamic model for the allosteric mechanism of GroEL. *J. Mol. Biol.* 302:303–313.

Makin, O.S., Atkins, E., Sikorski, P., Johansson, J., and Serpell, L.C. 2005. Molecular basis for amyloid fibril formation and stability. *Proc. Natl. Acad. Sci. USA* 102:315–320.

Marston, F.A., and Hartley, D.L. 1990. Solubilization of protein aggregates. *Methods Enzymol.* 182:264–276.

Martin, J.B. 1999. Molecular basis of the neurodegenerative disorders [published erratum appears in *N. Engl. J. Med.* 1999 Oct 28;341(18):1407]. *N. Engl. J. Med.* 340:1970–1980.

McCutchen, S.L., Colon, W., and Kelly, J.W. 1993. Transthyretin mutation Leu-55-Pro significantly alters tetramer stability and increases amyloidogenicity. *Biochemistry* 32:12119–12127.

Means, G.E., and Feeney, R.E. 1971. *Chemical Modification of Proteins*. San Francisco, Holden–Day.

Merkel, J.S., Sturtevant, J.M., and Regan, L. 1999. Sidechain interactions in parallel beta sheets: The energetics of cross-strand pairings. *Struct. Fold. Des.* 7:1333–1343.

Merlini, G., and Bellotti, V. 2003. Molecular mechanisms of amyloidosis. *N. Engl. J. Med.* 349:583–596.

Mirsky, A.E., and Pauling, L. 1936. On the structure of native, denatured and coagulated protein. *Proc. Natl. Acad. Sci. USA* 22:439–447.

Miyazawa, S., and Jernigan, R.L. 1985. Estimation of effective interresidue contact energies from protein crystal-structures—Quasi-chemical approximation. *Macromolecules* 18:534–552.

Monti, M., Principe, S., Giorgetti, S., Mangione, P., Merlini, G., Clark, A., Bellotti, V., Amoresano, A., and Pucci, P. 2002. Topological investigation of amyloid fibrils obtained from beta2-microglobulin. Protein Sci 11:2362–2369.

Morimoto, A., Irie, K., Murakami, K., Masuda, Y., Ohigashi, H., Nagao, M., Fukuda, H., Shimizu, T., and Shirasawa, T. 2004. Analysis of the secondary structure of beta-amyloid (Abeta42) fibrils by systematic proline replacement. *J. Biol. Chem.* 279:52781–52788.

Muchowski, P.J., Schaffar, G., Sittler, A., Wanker, E.E., Hayer-Hartl, M.K., and Hartl, F.U. 2000. Hsp70 and hsp40 chaperones can inhibit self-assembly of polyglutamine proteins into amyloid-like fibrils. *Proc. Natl. Acad. Sci. USA* 97:7841–7846.

Nelson, R., Sawaya, M.R., Balbirnie, M., Madsen, A.O., Riekel, C., Grothe, R., and Eisenberg, D. 2005. Structure of the cross-beta spine of amyloid-like fibrils. *Nature* 435:773–778.

Nguyen, H.D., Hall, C.K. 2002. Effect of rate of chemical or thermal renaturation on refolding and aggregation of a simple lattice protein. *Biotechnol. Bioeng.* 80:823–834.

Nguyen, H.D., and Hall, C.K. 2004a. Molecular dynamics simulations of spontaneous fibril formation by random-coil peptides. *Proc. Natl. Acad. Sci. USA* 101:16180–16185.

Nguyen, H.D., and Hall, C.K. 2004b. Phase diagrams describing fibrillization by polyalanine peptides. *Biophys. J.* 87:4122–4134.

Nguyen, H.D., Hall, C.K. 2005. Kinetics of fibril formation by polyalanine peptides. *J. Biol. Chem.* 280:9074–9082.

Nguyen, H.D., Marchut, A.J., and Hall, C.K. 2004. Solvent effects on the conformational transition of a model polyalanine peptide. *Protein Sci.* 13:2909–2924.

Nilsson, M.R. 2004. Techniques to study amyloid fibril formation *in vitro*. *Methods* 34:151–160.

Oberg, K., Chrunyk, B.A., Wetzel, R., and Fink, A. 1994. Native-like secondary structure in interleukin-1β inclusion bodies by attenuated total reflectance FTIR. *Biochemistry* 33:2628–2634.

O'Nuallain, B., Shivaprasad, S., Kheterpal, I, and Wetzel, R. 2005. Thermodynamics of Aβ(1–40) amyloid fibril formation. Biochemistry 44:12709–12718.

O'Nuallain, B., Williams, A.D., Westermark, P., and Wetzel, R. 2004. Seeding specificity in amyloid growth induced by heterologous fibrils. *J. Biol. Chem.* 279:17490–17499.

Patro, S.Y., and Przybycien, T.M. 1994. Simulations of kinetically irreversible protein aggregate structure. *Biophys J.* 66:1274–1289.

Patro, S.Y., Przybycien, T.M., and Isermann, H.P. 1996. Simulations of reversible protein-aggregate and crystal structure. *Abstr. Pap. Am. Chem. Soc.* 211:176-Biot.

Perez-Paya, E., Forood, B., Houghten, R.A., and Blondelle, S.E. 1996. Structural characterization and 5'-mononucleotide binding of polyalanine beta-sheet complexes. *J. Mol. Recognit.* 9:488–493.

Perutz, M.F., Finch, J.T., Berriman, J., and Lesk, A. 2002. Amyloid fibers are water-filled nanotubes. *Proc. Natl. Acad. Sci. USA* 99:5591–5595.

Petkova, A.T., Ishii, Y., Balbach, J.J., Antzutkin, O.N., Leapman, R.D., Delaglio, F., and Tycko, R. 2002. A structural model for Alzheimer's beta-amyloid fibrils based on experimental constraints from solid state NMR. *Proc. Natl. Acad. Sci. USA* 99:16742–16747.

Petkova, A.T, Leapman, R.D., Guo, Z., Yau, W.M., Mattson, M.P., and Tycko, R. 2005. Self-propagating, molecular-level polymorphism in Alzheimer's {beta}-amyloid fibrils. Science 307:262–265.

Petrucelli, L., and Dawson, T.M. 2004. Mechanism of neurodegenerative disease: Role of the ubiquitin proteasome system. *Ann. Med.* 36:315–320.

Polverino de Laureto, P., Taddei, N., Frare, E., Capanni, C., Costantini, S., Zurdo, J., Chiti, F., Dobson, C.M., and Fontana, A. 2003. Protein aggregation and amyloid fibril formation by an SH3 domain probed by limited proteolysis. *J. Mol. Biol* 334:129–141.

Prouty, W.F., Karnovsky, M.J., Goldberg, A.L. 1975. Degradation of abnormal proteins in *Escherichia coli*: Formation of protein inclusions in cells exposed to amino acid analogs. *J. Biol. Chem.* 250:1112–1122.

Rapaport, D.C. 1978. Molecular dynamics simulation of polymer chains with excluded volume. *J. Phys. A-Math. Gen.* 11:L213–L217.

Rapaport, D.C. 1979. Molecular dynamics study of a polymer-chain in solution. *J. Chem. Phys.* 71:3299–3303.

Richardson, J.S., and Richardson, D.C. 2002. Natural beta-sheet proteins use negative design to avoid edge-to-edge aggregation. *Proc. Natl. Acad. Sci. USA* 99:2754–2759.

Riek, R., Hornemann, S., Wider, G., Billeter, M., Glockshuber, R., and Wuthrich, K. 1996. NMR structure of the mouse prion protein domain PrP(121–321). *Nature* 382:180–182.

Saper, M.A., Bjorkman, P.J., and Wiley, D.C. 1991. Refined structure of the human histocompatibility antigen HLA-A2 at 2.6 A resolution. *J. Mol. Biol.* 219:277–319.

Scherzinger, E., Lurz, R., Turmaine, M., Mangiarini, L., Hollenbach, B., Hasenbank, R., Bates, G.P., Davies, S.W., Lehrach, H., and Wanker, E.E. 1997. Huntingtin-encoded polyglutamine expansions form amyloid-like protein aggregates *in vitro* and in vivo. *Cell* 90:549–558.

Schiffer, M., Chang, C.H., and Stevens, F.J. 1985. Formation of an infinite beta-sheet arrangement dominates the crystallization behavior of lambda-type antibody light chains. *J. Mol. Biol.* 186:475–478.

Serio, T.R., Cashikar, A.G., Kowal, A.S., Sawicki, G.J., Moslehi, J.J., Serpell, L., Arnsdorf, M.F., and Lindquist, S.L. 2000. Nucleated conformational conversion and the replication of conformational information by a prion determinant. *Science* 289:1317–1321.

Sharma, D., Shinchuk, L., Inouye, H., Wetzel, R., and Kirschner, D.A. 2005. Polyglutamine homopolymers having 8–45 repeats form slablike β-crystallite assemblies. *Proteins Struct. Funct. Bioinf.* 61:398–411.

Shire, S.J., Shahrokh, Z., and Liu, J. 2004. Challenges in the development of high protein concentration formulations. *J. Pharm. Sci.* 93:1390–1402.

Shivaprasad, S., and Wetzel, R. 2004. An intersheet packing interaction in Aβ fibrils mapped by disulfide crosslinking. *Biochemistry* 43:15310–15317.

Shivaprasad, S., and Wetzel, R. 2006. Scanning cysteine mutagenesis analysis of Aβ(1–40) amyloid fibrils. *J. Biol. Chem.* 281:993–1000.

Skolnick, J., and Kolinski, A. 1990. Simulations of the folding of a globular protein. *Science* 250:1121–1125.

Smith, A.V., Hall, C.K. 2001a. Alpha-helix formation: Discontinuous molecular dynamics on an intermediate-resolution protein model. *Proteins* 44:344–360.

Smith, A.V., and Hall, C.K. 2001b. Assembly of a tetrameric alpha-helical bundle: Computer simulations on an intermediate-resolution protein model. *Proteins* 44:376–391.

Smith, A.V., and Hall, C.K. 2001c. Protein refolding versus aggregation: Computer simulations on an intermediate-resolution protein model. *J. Mol. Biol.* 312:187–202.

Stanger, H.E., Syud, F.A., Espinosa, J.F., Giriat, I., Muir, T., and Gellman, S.H. 2001. Length-dependent stability and strand length limits in antiparallel beta -sheet secondary structure. *Proc. Natl. Acad. Sci. USA* 98:12015–12020.

Sticht, H., Bayer, P., Willbold, D., Dames, S., Hilbich, C., Beyreuther, K., Frank, R.W., and Rosch, P. 1995. Structure of amyloid A4-(1-40)-peptide of Alzheimer's disease. *Eur. J. Biochem.* 233:293–298.

Stine, W.B., Jr., Snyder, S.W., Ladror, U.S., Wade, W.S., Miller, M.F., Perun, T.J., Holzman, T.F., and Krafft, G.A. 1996. The nanometer-scale structure of amyloid-beta visualized by atomic force microscopy. *J. Protein. Chem.* 15:193–203.

Sunde, M., and Blake, C. 1997. The structure of amyloid fibrils by electron microscopy and X-ray diffraction. *Adv. Protein. Chem.* 50:123–159.

Sunde, M., and Blake, C.C. 1998. From the globular to the fibrous state: Protein structure and structural conversion in amyloid formation. *Q. Rev. Biophys.* 31:1–39.

Takada, S., Luthey-Schulten, Z., and Wolynes, P.G. 1999. Folding dynamics with nonadditive forces: A simulation study of a designed helical protein and a random heteropolymer. *J. Chem. Phys.* 110:11616–11629.

Taketomi, H., Ueda, Y., and Go, N. 1975. Studies on protein folding, unfolding and fluctuations by computer simulation. I. The effect of specific amino acid sequence represented by specific inter-unit interactions. *Int. J. Pept. Protein. Res.* 7:445–459.

Tanaka, M., Chien, P., Naber, N., Cooke, R., and Weissman, J.S. 2004. Conformational variations in an infectious protein determine prion strain differences. *Nature* 428:323–328.

Tanaka, S., and Scheraga, H.A. 1976. Medium- and long-range interaction parameters between amino acids for predicting three-dimensional structures of proteins. *Macromolecules* 9:945–950.

Thakur, A., Wetzel, R. 2002. Mutational analysis of the structural organization of polyglutamine aggregates. *Proc. Natl. Acad. Sci. USA* 99:17014–17019.

Toma, L., and Toma, S. 2000. A lattice study of multimolecular ensembles of protein models. Effect of sequence on the final state: Globules, aggregates, dimers, fibrillae. *Biomacromolecules* 1:232–238.

Torok, M., Milton, S., Kayed, R., Wu, P., McIntire, T., Glabe, C.G., and Langen, R. 2002. Structural and dynamic features of Alzheimer's Abeta peptide in amyloid fibrils studied by site-directed spin labeling. *J. Biol. Chem.* 277:40810–40815.

Turner, G.C., and Varshavsky, A. 2000. Detecting and measuring cotranslational protein degradation in vivo. *Science* 289:2117–2120.

Tycko, R. 2000. Solid-state NMR as a probe of amyloid fibril structure. *Curr. Opin. Chem. Biol.* 4:500–506.

Urbanc, B., Cruz, L., Ding, F., Sammond, D., Khare, S., Buldyrev, S.V., Stanley, H.E., and Dokholyan, N.V. 2004a. Molecular dynamics simulation of amyloid beta dimer formation. *Biophys. J.* 87:2310–2321.

Urbanc, B., Cruz, L., Yun, S., Buldyrev, S.V., Bitan, G., Teplow, D.B., and Stanley, H.E. 2004b. *In silico* study of amyloid beta-protein folding and oligomerization. *Proc. Natl. Acad. Sci. USA* 101:17345–17350.

Vigouroux, S., Briand, M., and Briand, Y. 2004. Linkage between the proteasome pathway and neurodegenerative diseases and aging. *Mol. Neurobiol.* 30:201–221.

Wallqvist, A., and Ullner, M. 1994. A simplified amino-acid potential for use in structure predictions of proteins. *Proteins-Struct. Funct. Genet.* 18:267–280.

Wang, S.S., Tobler, S.A., Good, T.A., and Fernandez, E.J. 2003. Hydrogen exchange-mass spectrometry analysis of beta-amyloid peptide structure. *Biochemistry* 42:9507–9514.

Wetzel, R. 1992. Protein aggregation *in vivo*: Bacterial inclusion bodies and mammalian amyloid. In *Stability of Protein Pharmaceuticals: In Vivo Pathways of Degradation and Strategies for Protein Stabilization* (T. J. Ahern and M. C. Manning, Eds.). New York, Plenum Press, pp. 43–88.

Wetzel, R. 1994. Mutations and off-pathway aggregation. *Trends Biotechnol.* 12:193–198.

Wetzel, R. 1999. Amyloid, prions, and other protein aggregates. *Methods Enzymol.* 309:820.

Wetzel, R. 2002. Ideas of order for amyloid fibril structure. *Structure* 10:1031–1036.

Wetzel, R. 2005. Protein folding and aggregation in the expanded polyglutamine repeat diseases. In *The Protein Folding Handbook, Part II*. (J. Buchner and T. Kiefhaber, Eds.). Weinheim, Wiley-VCH, pp. 1170–1214.

Wetzel, R., and Goeddel, D.V. 1983. Synthesis of polypeptides by recombinant DNA methods. In *The Peptides: Analysis, Synthesis, Biology* (J. Meienhofer and E. Gross, Eds.). New York, Academic Press, Vol. 5, pp. 1–64.

Whittemore, N.A., Mishra, R., Kheterpal, I., Williams, A.D., Wetzel, R., and Serpersu, E.H. 2005. Hydrogen-deuterium (H/D) exchange mapping of Aβ1-40 amyloid fibril secondary structure using NMR spectroscopy. *Biochemistry* 44:4434–4441.

Wille, H., Michelitsch, M.D., Guenebaut, V., Supattapone, S., Serban, A., Cohen, F.E., Agard, D.A., and Prusiner, S.B. 2002. Structural studies of the scrapie prion protein by electron crystallography. *Proc. Natl. Acad. Sci. USA* 99:3563–3568.

Williams, A., Portelius, E., Kheterpal, I., Guo, J.-T., Cook, K., Xu, Y., and Wetzel, R. 2004. Mapping abeta amyloid fibril secondary structure using scanning proline mutagenesis. *J. Mol. Biol.* 335:833–842.

Williams, A.D., Sega, M., Chen, M., Kheterpal, I., Geva, M., Berthelier, V., Kaleta, D.T., Cook, K.D., and Wetzel, R. 2005. Structural properties of Aβ protofibrils stabilized by a small molecule. *Proc. Natl. Acad. Sci. USA* 102:7115–7120.

Williams, A.D., Shivaprasad, S., and Wetzel, R. 2006. Alanine scanning mutagenesis of Aβ(1–40) amyloid fibril stability. *J. Mol. Biol.* 357:1283–1294.

Wu, H. 1931. Studies on denaturation of proteins. XII. A theory of denaturation. *Chin. J. Physiol.* 5:321–344.

Yamaguchi, K., Takahashi, S., Kawai, T., Naiki, H., and Goto, Y. 2005. Seeding-dependent propagation and maturation of amyloid fibril conformation. *J. Mol. Biol.* 352:952–960.

Xu, D., Baburaj, K., Peterson, C.B., and Xu, Y. 2001. Model for the three-dimensional structure of vitronectin: Predictions for the multi-domain protein from threading and docking. *Proteins* 44:312–320.

Xu, Z., Horwich, A.L., and Sigler, P.B. 1997. The crystal structure of the asymmetric GroEL-GroES-(ADP)7 chaperonin complex. *Nature* 388:741–750.

Zanuy, D., Gunasekaran, K., Ma, B., Tsai, H.H., Tsai, C.J., and Nussinov, R. 2004. Insights into amyloid structural formation and assembly through computational approaches. *Amyloid* 11:143–161.

Zanuy, D., Ma, B., and Nussinov, R. 2003. Short peptide amyloid organization: Stabilities and conformations of the islet amyloid peptide NFGAIL. *Biophys. J.* 84:1884–1894.

Zanuy, D., and Nussinov, R. 2003. The sequence dependence of fiber organization. A comparative molecular dynamics study of the islet amyloid polypeptide segments 22–27 and 22–29. *J. Mol. Biol.* 329:565–584.

10 Homology-Based Modeling of Protein Structure

Zhexin Xiang

10.1 Introduction

10.1.1 Structural Genomics and Homology Modeling

The human genome project has already discovered millions of proteins (http://www.swissprot.com). The potential of the genome project can only be fully realized once we can assign, understand, manipulate, and predict the function of these new proteins (Sanchez and Sali, 1997; Frishman et al., 2000; Domingues et al., 2000). Predicting protein function generally requires knowledge of protein three-dimensional structure (Blundell et al., 1978; Weber, 1990), which is ultimately determined by protein sequence (Anfinsen, 1973). Protein structure determination using experimental methods such as X-ray crystallography or NMR spectroscopy is very time consuming (Johnson et al. 1994). To date, fewer than 2% of the known proteins have had their structures solved experimentally. In 2004, more than half a million new proteins were sequenced that almost doubled the efforts in the previous year, but only 5300 structures were solved. Although the rate of experimental structure determination will continue to increase, the number of newly discovered sequences grows much faster than the number of structures solved (see Fig. 10.1).

Fortunately, many protein sequences are evolutionarily related, and thus can be classified into different families. Proteins in the same families frequently have noticeable similarities and thus share three-dimensional architecture, which allows a structural description of all proteins in a family even when only the structure of a single member is known. This evolutionary relationship provides the rationale for structural genomics, a systematic and large-scale effort toward structural characterization of all proteins, where a representative protein in each family is chosen to be solved experimentally with the rest reliably predicted by a homology modeling method (Goldsmith-Fischman and Honig, 2003; Al-Lazikani et al., 2001a). Fold recognition has also become an important tool that supplements sequence-based methods to detect remote homologues. However, the line between traditional homology modeling and fold recognition has diminished due to the progress in the alignment sensitivity and the increase in database size. *Ab initio* structure methods have made notable progress in recent years and are extremely important, not only for what they can accomplish but also for what they can teach us about protein folding (Bonneau and Baker, 2001). The progress in *ab initio* prediction makes it possible in a few cases to refine homology models to the accuracy of low-resolution X-ray

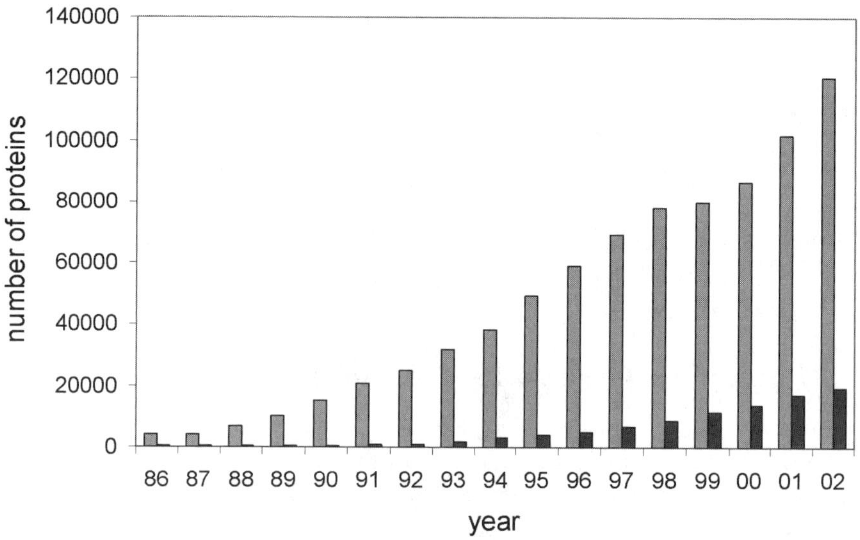

Fig. 10.1 Number of protein sequences and structures available each year. Blue bar denotes the number of protein sequences in SWISS-PROT, red bar is the number of protein structures in PDB.

structures. In fact, if we assume that a native protein structure is at the global free-energy minimum, comparative modeling is a simple scheme to focus the search of conformation space by minimally disturbing those existing solutions, i.e., the experimentally solved structures. The obvious advantage is that the comparative modeling technique relaxes the stringent requirements of force field accuracy and prohibitive conformational searching, because it dispenses with the calculation of a physical chemistry force field and replaces it, in large part, with the counting of identical residues between template and target sequences.

Currently there are about 2 million protein sequences in Swiss-Prot and TrEMBL, but only 7677 protein families have been identified according to the Pfam database (http://pfam.wustl.edu/). This number is strongly dependent on the sequence similarity cutoffs used to cluster the sequence space. If 30% sequence identity cutoff is used, which is generally considered as a threshold for successful homology modeling, statistical estimates place the number somewhere between 10,000 and 30,000 for all proteins in Nature (Liu et al., 2004), but only a fraction of which have distinct spatial arrangements (Brenner et al., 1997). Ninety percent of protein structures deposited today share a similar fold to others already in the PDB. Many of these solved structures are site-directed mutants or inhibitor-bound complexes of previously deposited proteins, and many are related members of families of proteins with similar sequences and closely related three-dimensional structures. Even proteins with no sequence homology can have very similar folds; for example, the sulfate and phosphate binding proteins, the transferrins, and the porphobilinogen deaminase have similar bilobal anion binding structures but not significant sequence identities. Protein topologies such as the four α-helix bundle, the αβ-nucleotide binding motif, the β-jelly roll, the αβ-barrel, and the β-immunoglobulin domain have been found in a wide range of protein structures (Johnson et al., 1994; Al-Lazikani et al., 1997; Efimov, 1993). A recent study has found that roughly 33% of all proteins have complete sequence coverage to a protein with known structure (Ekman et al., 2005). This kind of protein structure redundancy versus protein sequence variability is the cornerstone of homology modeling algorithms. It is quite likely that homology modeling will assume an increasingly important role in both biological and chemical applications with the advent of structural genomics initiatives around the world.

10.1.2 History of Homology Modeling

Homology modeling techniques became important only after 1990 when hundreds of protein structures had already been deposited in the Protein Data Bank. Early modeling studies in the late 1960s and early 1970s frequently relied on the construction of hand-made wire and plastic models and only later depended on computer software (Cox and Bonanou, 1969; Tometsko, 1970). The first homology model was built simply by copying existing coordinates from a homologous protein and those non-identical residues were then substituted by reassembling corresponding side chains (Browne et al., 1969). This approach, called rigid-body assembly, is still widely employed today with considerable success, especially when the proteins have sequence identity above 40% (Greer, 1980, 1981). The homology modeling method was pioneered in two studies by Browne et al. (1969) and Greer (1981). Browne et al. (1969) published the first homology model using an X-ray-derived structure as a template. They modeled bovine α-lactalbumin on the three-dimensional structure of hen egg-white lysozyme, where the pair sequence identity was about 39% and only deletions were considered as the polypeptide chain was shortened in α-lactalbumin. Their prediction was proven generally correct later on when the structure of α-lactalbumin was solved (Acharya et al., 1989).

McLachlan and Shotton (1971) modeled alpha-lytic proteinase of Myxobacter 495 based on the structures of both chymotrypsin and elastase, where the sequence identity between these two proteinases was only 18% and the alignment was fragmented by frequent gaps. When the structure of alpha-lytic proteinase was published (Brayer et al., 1979), it became clear that misalignment of the sequence with those of the known 3D structures led to incorrect regions, but portions of both domains were constructed correctly (Delbaere et al., 1979). This model demonstrated the difficulty in aligning sequences of limited similarity and in modeling variable, mainly loop regions.

Greer was the first to demonstrate the importance of modeling variable regions (1981). By abstracting approximate conformations from a family of homologous proteins of known structures, he could distinguish structurally conserved regions, which contain strong sequence homology, and structurally variable regions, which include all the insertions and deletions. By applying the structural distinction to new sequences, erroneous alignments of the sequences are greatly minimized. For each new aligned sequence, the structurally conserved regions can be constructed from any of the known structures. The construction of variable regions, however, is not straightforward. However, the conformations of loops of just one- or two-residue deletions or insertions can be extrapolated from one of the homologous structures. This approach was further applied to predict the structure of mammalian serine proteases based on a number of proteins from this family, including a variety of blood-serum, intestinal, and pancreatic proteins as well as a closely related bacterial enzyme (Greer, 1981).

Instead of deriving protein backbone structure from only one of its homologues, Taylor (1986) developed a method of generating templates for each part of protein to be modeled based on the conserved patterns observed in the known 3D structures of a family. The conserved templates were derived from a small number of related sequences of the known tertiary structures. The templates were then made more representative by aligning with other sequences of unknown structures. The specificity of the templates was demonstrated by their ability to identify the conserved features in known immunoglobulin and the related sequences but not in other sequences. However, assembling these conserved patterns into a complete structure requires the use of a force field and conformation sampling.

Due to the small number of protein structures available before the 1990s, the comparative modeling technique was not widely successful. The real development of homology modeling began in the mid-1990s with the progress of genome projects and the growth of the number of solved structures in the PDB. With the advent of structural genomics, the importance of homology modeling continues to grow. Although 25 years have passed since Greer's pioneering work on comparative model building of mammalian serine protease in 1981, the basic technique used in today's most advanced modeling programs remains almost the same, i.e., finding the closest homologues as the basis of modeling the query sequence. Recent efforts in comparative modeling have been concentrated on the discovery of distant homologues, the improvement of alignment accuracy, and especially the refinement of models by optimization of empirical energy functions.

10.1.3 Accuracy and Applicability of Homology Modeling

Approximately 57% of all known sequences have at least one domain that is related to at least one protein of known structure (Pieper et al., 2002). The probability of finding a related known structure for a randomly selected sequence from a genome ranges from 30% to 65%, since a few genomes have received more research attention than others (Kelley et al., 2000; Teichmann et al., 1999; Fiser and Sali, 2003). The percentage is steadily increasing because more distinct folds are discovered each year, and because the number of different structural folds that proteins adopt is limited (Irving et al., 2001). Current estimates suggest that there are between 1000 and 5000 folds in the universe of compact globular proteins, with about 200 new folds realized annually from the structure deposition (Brenner et al., 1997). The number of known protein sequences is close to 2 million so far. Over 1.1 million proteins can readily have at least one of their domains reliably predicted with homology modeling methods. Given the rate of experimental structure determination, approximately 6000 proteins each year, it is arguable that homology modeling has already saved up to hundreds of years of human effort, though homology models often have low quality. In the next 10 years, structural genomics will possibly discover all protein distinct folds in Nature, making comparative modeling applicable to almost any protein sequence (Vitkup et al., 2001). The usefulness of comparative modeling is ever increasing as more proteins can be predicted with higher accuracy. The accuracy of homology modeling depends primarily on the sequence similarity between the target sequence and the template structure.

When the sequence identity is above 40%, the alignment is straightforward, there are not many gaps, and 90% of main-chain atoms can be modeled with an RMSD error of about 1 Å (Sanchez and Sali, 1997). In this range of sequence identity, the structural difference between proteins mainly arises from loops and side chains. When the sequence identity is about 30–40%, obtaining correct alignment becomes difficult, where insertions and deletions are frequent. For sequence similarity in this range, 80% of main-chain backbone atoms can be predicted to RMSD 3.5 Å, while the rest of the residues are modeled with larger errors, especially in the insertion and deletion regions (Harrison et al., 1995; Mosimann et al., 1995; Yang and Honig, 2000; Sauder et al., 2000). Even in correctly aligned regions, loop modeling and side-chain placement pose difficulties (Bower et al., 1997; Rapp and Friesner, 1999). When the sequence similarity is below 30%, the main problem becomes the identification of the homologue structures, and alignment becomes much more difficult. For some sequences where the structures in the family are very conserved in evolution (e.g., kinase family), homology modeling can make predictions as accurate as low-resolution X-ray experiments even if the sequence identity is much less than 30% identity to the template (Yang and Honig, 1999; Petrey et al., 2003).

Even if homology modeling is generally much less accurate than experimental methods, it can still be helpful in proposing and testing hypotheses in molecular biology, such as predictions of ligand binding sites (Zhou and Johnson, 1999; Francoijs et al., 2000), substrate specificities (Jung et al., 2000; De Rienzo et al., 2000), function annotation, protein interaction pathways, and drug design (Nugiel et al., 1995; Sanchez and Sali, 1997). It can also provide starting models for solving structures from X-ray crystallography, NMR, and electron microscopy (Talukdar and Wilson, 1999; Ceulemans and Russell, 2004).

10.2 Procedures in Homology Modeling

Given a protein sequence, successful homology modeling usually consists of the following steps as shown in Fig. 10.2: (1) identify the homologue of known structure from the Protein Data Bank; (2) align the query sequence to the template structure; (3) build the model based on the alignment; (4) assess and refine the model. Each step may involve some errors.

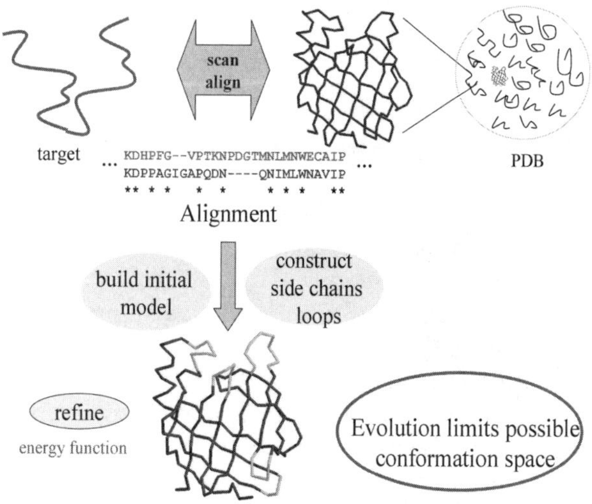

Fig. 10.2 Basic homology modeling protocol. Homology modeling starts by scanning the PDB for sequences similar to the target. The hit of highest sequence similarity is chosen as the template. Model for the target is then built based on the alignment between the target and template sequence, which will be subjected to further refinement.

10.2.1 Homologue Detection and Alignment

Homology modeling starts from selection of homologues with known structures from the PDB. If the query sequence has high sequence identity (>30%) to the structure, the homology detection is quite straightforward which is usually done by comparing the query sequence with all the sequences of the structures in the PDB. This can often be achieved simply with the dynamic programming method (Needleman and Wunsch, 1970) and its derivatives (Smith and Waterman, 1981; Gotoh, 1982). The most popular software is BLAST (Altschul et al., 1997) (http://www.ncbi.nlm.nih.gov/blast/) that searches sequence databases for optimal local alignments to the query. The BLAST program improves the overall speed of searches while retaining good sensitivity by breaking the query and database sequences into fragments, and initially seeking matches between fragments. The matched fragments are then extended in either direction in an attempt to generate an alignment with a score exceeding a particular threshold. The score based on substitution matrices reflects the degree of similarity between the query and the sequence being compared, capable of ranking the quality of each pairwise alignment. The BLAST program functions very well for alignment of sequences with high similarities. But when the sequence identity is well below 30%, homology hits from BLAST are not reliable. A number of alternative strategies have been developed. These include template consensus sequences (Taylor, 1986; Chappey et al., 1991) and profile analysis (Barton and Sternberg, 1990; Suyama et al., 1997; Lolkema and Slotboom, 1998). All these approaches, based on either multiple sequence or structure alignments, are more sensitive because the consensus sequences are more representative of the sequence family, and the profile reflects the conserved structural or functional preferences.

In the past several years, sequence profile methods have emerged as the primary approach in distant homology detection. Position-specific profile search methods such as PSI-BLAST (Altschul et al., 1997) and hidden Markov models (HMMs) (Krogh et al., 1994), as implemented in the SAM (Karplus et al., 1998) and HMMER (http://hmmer.wustl.edu) packages, have vastly improved the accuracy of sequence alignments and have extended the boundaries of detectable sequence similarity. Sequence profiles methods, e.g., PSI-BLAST, start from performing a pairwise search of the database. The significant alignments are then used by the program to construct a position-specific score matrix (PSSM). This matrix replaces the query sequence in the next round of database searching. The procedure may be iterated until no new significant alignments are found. The profile method can be further improved with information from multiple structure alignment, secondary structure prediction, and solvent accessibility. Since the structural information is more conserved than sequence, it may represent the crucial requirement, in the process of evolution, of residues at specific positions with respect to the stability and function of the structure as a whole. Although a major goal of this effort has been remote homologue detection, an important side benefit has been significant improvement in alignment quality, even at levels of sequence identity for which pairwise alignment methods are known not to work. This, in turn, has had a positive impact on the starting alignments used in homology modeling, and thus has the potential to extend the applicability of homology modeling to increasingly lower levels of sequence similarity. Indeed, perhaps the largest part of recent improvements in homology modeling can be traced directly to improvement in sequence alignment algorithms.

If multiple homologues from the PDB have been identified, the next step is to select one or a few templates that are most appropriate for building a model. Sequence similarity between the query and the template is usually considered as the primary criterion used to choose the best template. Higher sequence similarity often suggests a closer relationship in evolution, thus more conservation of structure, and vice versa. On the other hand, gaps (insertion or deletion) in the alignment have a severe impact on the quality of the model to be built, since gaps are regions where no templates are readily available to guide the model building process. Generally, insertions in the alignment are more difficult to handle than deletions, particularly for insertions of more than 10 residues, because modeling inserted residues is a mini *ab initio* problem. Thus, the second criterion is to choose an alignment that has fewer gaps and short insertions. Moreover, since sequences in the same family often share similar structure and function, templates that can be clustered into the same subfamily as the target are often favored. This can usually be achieved with construction of a multiple alignment and phylogenetic tree (Felsenstein, 1981; Retief, 2000). If the function of the protein to be modeled (target protein) is known, templates with similar functions should also be given more consideration. If possible, the environment (e.g., ligands, solvent, temperature, pH) at which the template structure was determined and the native environment of the target protein to be modeled should also be properly taken into account. A protein, such as calcium binding protein, can adopt quite different conformations in solvent under different environments (Mishig-Ochiriin et al., 2005). Thus, structures determined in environments similar to the physiological environment of the target are generally preferred. In addition, structures of higher quality are generally used, such as X-ray structures of high resolution and low R-factor, and NMR structures with sufficient constraints. In the end, structures that are most representative of the family should be used if all other criteria are identical. The trait can easily be calculated as the average RMSD of the structure to all other family members with known structures. The hypothesis is that the target protein is more likely to be similar to the most "typical" structure. Instead of relying on a single template, it can be advantageous to select multiple structures from one or several families. Multiple template structures may be aligned with different domains of the target, thus a composite model can be built with each domain based on the best template. It is also useful for modeling variable regions of a structure family, where the segment, which is not conserved, assumes multiple conformations, and the "best" model is assumed to have the lowest value of some empirical energy function. If the sequence identity is too low and there is no clear hit, a better approach is always to make multiple models with each model based on one template. Thus, the best model is determined by physical chemistry- or statistics-based energy or a combination of both (Sippl, 1995; Petrey et al., 2003; also see Chapters 2 and 3).

In homology modeling, one of the most difficult and important tasks is to improve sequence–template alignment. Although profile methods have significantly improved alignment accuracy, manual inspection is often required to further improve the quality if the alignment is well below 30% identity with frequent gaps. This is because the current alignment software usually seeks an alignment of global optimality with an empirical scoring function that may misalign functionally important residues. Manual inspection of the alignment does not necessarily need to have the model actually built, since residue–residue interaction in the target sequence can easily be identified from their corresponding aligned positions in the template structure. There are several general rules to guide alignment tuning. First, charged residues in the target sequence should not be aligned with a buried residue in the template, unless it will form hydrogen bonds or salt bridges with another residue in the target; second, fragments of predicted secondary structures (alpha helix and beta sheet) in the target sequence should be aligned with the fragments of identical secondary structure characterization from the template; third, residues with known important functions, either for protein activity or structural stability, should be aligned with residues of similar functions in the template; fourth, insertions or deletions in the secondary structure regions should be pushed to the loop regions. Manual editing of the alignment is the most tedious part in homology modeling. A misalignment by only one residue position will result in an error of approximately 4 Å in the model because the current homology-modeling algorithms generally cannot recover from errors in the alignment (Fiser and Sali, 2003).

Table 10.1 Comparative modeling programs

Programs	Availability
NEST	http://trantor.bioc.columbia.edu/programs/jackal/
COMPOSER	http://www-cryst.bioc.cam.ac.uk/
Tripos (COMPOSER)	http://www.tripos.com/
CONGEN	http://www.congenomics.com/congen/congen_toc.html
MODELLER	http://guitar.rockefeller.edu/modeller/modeller.html
InsightII (MODELLER)	http://www.accelrys.com/
SWISS-MODEL	http://www.expasy.ch/swissmod/SWISS-MODEL.html
SCHRODINGER	http://www.schrodinger.com
WHATIF	http://swift.cmbi.kun.nl/whatif/
SEGMOD	http://www.bioinformatics.ucla.edu/genemine/
DRAGON	rmunro@nimr.mrc.ac.uk
ICM	http://www.molsoft.com/
3D-JIGSAW	http://www.bmm.icnet.uk/servers/3djigsaw/
Builder	koehl@csb.stanford.edu
PrISM	http://trantor.bioc.columbia.edu/programs/PrISM/index.html

10.2.2 Model Building

Given the alignment between the query sequence and templates, there are generally four methods in model building depending on how the information in the known structures is transferred to the query sequence. In this section we are going to discuss the first three methods, i.e., rigid body assembly, segment matching, and spatial restraint, and leave our own approach (artificial evolution model building) to Section 10.3 for more detailed description. Table 10.1 shows the most widely used model building programs that are publicly available. Most of the programs were based on the rigid body assembly method, and some have been commercialized, e.g., COMPOSER (Sutcliffe et al., 1987a,b) in Tripos and MODELLER (Sali and Blundell, 1993) in InsightII. In addition to model building protocols, the programs also differ from each other in model refinement.

10.2.2.1 Model Building by Rigid Body Assembly

The simplest and most widely used method is called rigid body assembly (also called cut-and-paste method) as shown in Fig. 10.3. This method was initiated by Greer in 1981 and is still widely used, e.g., in the software packages PriSM (Yang and Honig, 1999), Congen (Bruccoleri, 1993), and COMPOSER (Sutcliffe et al., 1987a,b). It starts from identification of the conserved and variable regions of the templates. The identification can often be achieved from the superimposed template structures. Conserved regions are evident from the multiple structure alignment, that is, the RMSDs (root mean square distance) among the fragments are relatively small, and variable regions are usually in loops with frequent gaps in the structural alignment. A

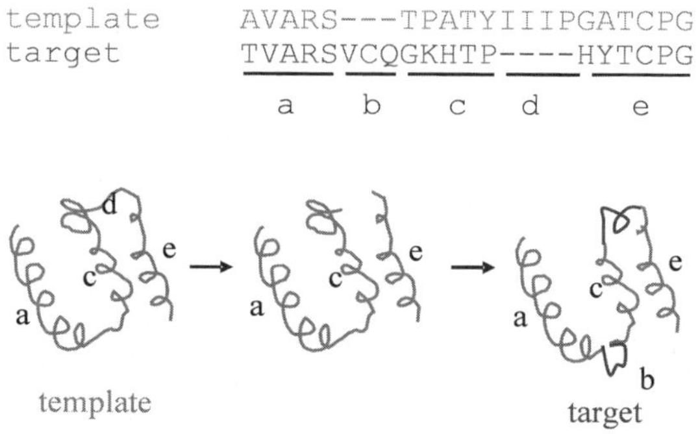

Fig. 10.3 Rigid body assembly with single template. Red and blue denote template and model structure, respectively. Conformations of aligned segments a, c, and e are directly transferred to the model; segment b is a new conformation inserted between a and c, which is obtained from either *ab initio* sampling or database searching; segments c and e are fused following the deletion of segment d.

framework for the superimposed templates can be calculated by averaging the atom coordinates of the structurally conserved regions. The averaging is often weighted based on the sequence similarity of the target sequence to the templates, higher sequence similarity carrying larger weight. The core residues of the target model, i.e., residues aligned with conserved regions of the templates, obtain their main-chain coordinates from the closest conserved segment (in terms of RMSD to the framework) from the template, or from the segment whose template has highest sequence identity to the target. The model is constructed by fitting the core rigid bodies onto the framework. The unconserved, or loop, region is then constructed either with *ab initio* approach, or by searching a database for structures that fit the anchor core regions and have a compatible sequence (Topham et al., 1993). The side chains are modeled based on their intrinsic conformational preferences and on the conformations of the equivalent side chains in the template structures (Sutcliffe et al., 1987a,b). If a single template is chosen, the model construction is straightforward, copying coordinates of aligned residues from the template to the model, and connecting broken segments with database searching or *ab initio* sampling as mentioned earlier. For strong sequence homologues, a single template is sufficient; but for weakly homologous templates, a framework based on weighted averaging over multiple templates is often more reliable.

10.2.2.2 Model Building by Segment Matching

Segment matching (Levitt, 1992), which has been adopted in SegMod software, is based on the finding that most hexapeptide segments of protein structure can be clustered into only 100 structurally different classes (Unger et al., 1989). Homology model construction relies on approximate positions of conserved atoms from the templates as "guiding positions" to calculate the coordinates of other atoms. The guiding positions usually correspond to the atoms of the segments that are conserved in the alignment between the template structures and the target sequence. They can be calculated by averaging the positions of corresponding atoms in all the template structures with weights based on their sequence similarity to the target. The averaged positions are then fitted by all-atom segments that are obtained by scanning databases of short segments of protein structures, or by a conformation search with restraints of potential energies or geometry rules. The segment matching method can construct both side chain and backbone atoms. If the distance between the conserved positions is too large, there may be no proper segments in the database to cover the missing atoms (usually only segments of five residues have their accessible conformations in the databank) (see Fidelis et al., 1994), thus the *ab initio* method may be the only approach. Segment matching could be considered as an extension of the rigid body assembly method since it scans a database of segments not restricted to those in the template structures. Indeed, segment matching has other applications, such as in side chain and loop modeling, where database scanning, instead of *ab initio* conformation sampling, is employed to identify the best conformers for the prediction.

10.2.2.3 Model Building by Satisfaction of Spatial Restraints

The third group of methods, satisfaction of spatial restraints, was proposed by Havel and Snow (1991) and Sali and Blundell (1993). The method was adopted in one of the most widely used homology model building programs, MODELLER (http://salilab.org). The method starts by generating many restraints for the target protein based on its alignment to the template structure. The restraints are generally obtained by assuming that the distance between two residues in the query model is similar to the distance between the two corresponding aligned residues in the template as shown in Fig. 10.4. The restraints are further supplemented with stereochemical constraints on bond angle, bond length, peptide bond dihedral angle, nonbonded van der Waals clashes, and so on. For weak homologues, additional constraints from experiments, if available, should also be used to increase the model accuracy. This additional information can be obtained from experimental data, for instance, distances between atoms of protein residues as measured by mass spectroscopy (MS) (Chapman, 1996), which uses protein cross-linking reagents as molecular rulers, or by nuclear Overhauser effect (NOE) restraints of NMR spectroscopy. In addition to the hard constraints, a lower and upper bound for each restraint is often provided, with a tight bound for stereochemical restraint and a relatively loose bound for longer distances. The bounds can be best estimated from statistical analysis of the relationships between similar protein structures. By scanning a set of related structures, various correlations can be quantified, such as correlations between two corresponding distances, or between corresponding main-chain dihedral angles. These relationships are expressed as conditional probability density functions and can be used directly as spatial restraints. Probabilities for different values of the equivalent distances and main-chain dihedral angles are calculated from the type of residue considered, from main-chain conformation of an equivalent residue, and from sequence similarity between the two proteins. The spatial restraints and the CHARMM22 force field terms

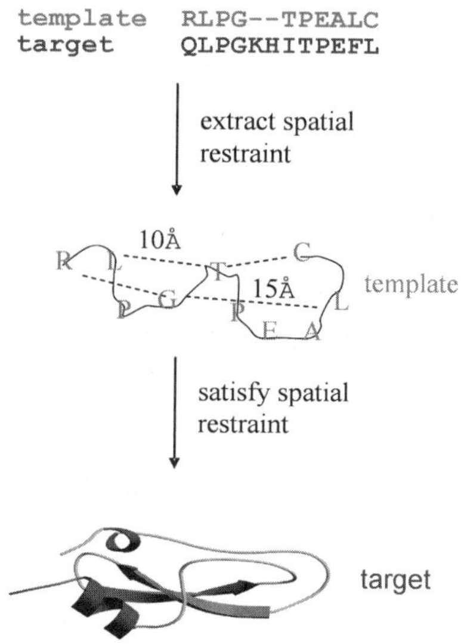

Fig. 10.4 Model building by satisfaction of spatial restraints. Distance restraints to be satisfied by the model are extracted from the template structure based on the alignment between the target sequence and the template. For example, distance between residues L and H (residues 2 and 6) for the model is assumed to be 10 Å, equivalent to the distance between residues L and T (residues 2 and 5) in the template structure.

enforcing proper stereochemistry (Brooks and Karplus, 1983) are combined into an objective function. The model is obtained by optimizing the objective function in Cartesian space. Thus, a proper model should not violate any of the constraints, and have low energy of the objective function. The advantage of the spatial restraint method is that it can use many different types of information about the target sequence including Cα–Cα distance and secondary structure preference. However, for highly homologous sequences, the information is already stored in the template structures, and introducing information derived from other members of the family may degrade the model.

10.2.3 Homology Model Refinement

High-resolution refinement is a difficult task that requires an effective sampling strategy and an accurate energy function. Homology model refinement is primarily focused on tuning alignment and modeling loops and side chains (see Fig. 10.5). Loops are usually the most variable regions of a structure where insertion and deletion often occur. Correct alignment is the most important task for homology modeling, since the errors introduced into the model by misalignment are hard to remove in the later stages of refinement. When the sequence identity is above 40%, errors in the homology structure mainly come from side chains; when the sequence identity is between 30 and 40%, loops and side chains become most problematic. Given a good energy function, loop and side-chain refinement can, in principle, be applied

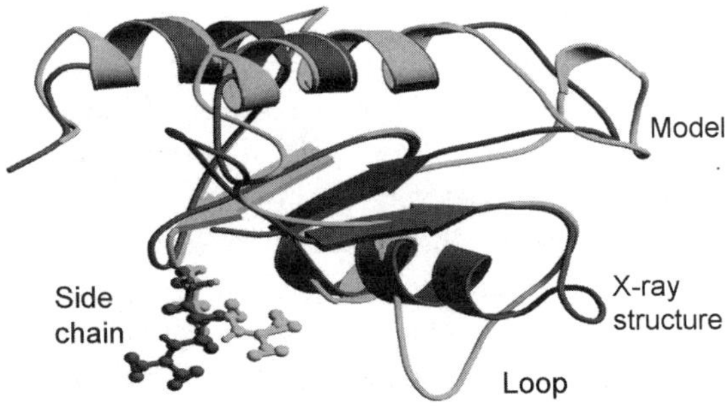

Fig. 10.5 Superimposition of the model for CASP (Critical assessment of techniques for protein structure prediction) Target 113 and its native structure. The root-mean deviation distance is 2.96 Å. The largest RMSDs typically occur in side chains and loops.

repeatedly to relax the backbone closer to native. Refinement on helix and β-sheet can be handled with similar methods as for loops, where proper hydrogen bond constraints should be applied to retain the secondary structure definition (Li et al., 2004). Recent attempts have been made to use physical chemistry energy to refine side chains, loops, and secondary structures, sometimes as a step in choosing the alignment.

10.2.3.1 Loop Prediction

When modeling loops, the basic goal is to predict the conformation of a loop that is fixed at both ends of its protein backbone. A number of methods have been proposed for loop prediction, i.e., *ab initio* methods (Zheng and Kyle, 1996; Rapp and Friesner, 1999; Fiser et al., 2000; Xiang et al., 2002; Jacobson et al., 2004), database-related methods (Li et al., 1999; Wojcik et al., 1999), or a combination of both (Fidelis et al., 1994; van Vlijmen and Karplus, 1997). *Ab initio* methods of loop prediction involve the generation of a large number of randomly chosen candidate conformations and their evaluation with energetic or other criteria. Database methods generate trial conformations based either on sequence relationships to loops of known structure, or on geometric criteria such as the distance between the amino and carboxyl termini of the loop in question. Once loops are generated in this way, energetic criteria are often applied to select the final model.

Clearly, it is important that near-native conformations be present among the trial conformations generated in the first step of loop modeling. Adequate sampling does not appear to be a problem if a large enough number of loops are generated randomly. Indeed, Rapp and Friesner (1999) were able to generate near-native conformations for even a 12-residue loop. However, database methods generate a much smaller number of trial conformations and the lack of a large enough template library to cover the many possible conformations of longer loops (more than five residues not including the stem, or anchoring, residues that are kept fixed) limits their utility for these cases (Fidelis et al., 1994). Using a sequence-dependent database method,

Wojcik et al. (1999) reported an average accuracy of 3.8 Å RMSD for the backbone atoms of an eight-residue loop. Van Vlijmen and Karplus (1997) used CHARMM to optimize initial conformations that were selected from the protein database. They reported improved results for longer loops but their optimization procedure, which involves simulated annealing, effectively extends the range of conformation space searched beyond that provided by the database conformations. In this sense, their approach is closer to *ab initio* loop generation. The accuracy of loop modeling is highly dependent not only on the number of residues in the loop, but also on the distance between the loop stems. Generally, when the distance between the loop stems is shorter, the loop conformation is more like "Ω," and thus has more freedom to move around; therefore, it is more difficult to predict. A database approach is usually more reliable, especially for long loops, if the segment identified from the PDB comes from a protein structure of the same family as the target protein.

Because conformational sampling does not appear to be a problem for loops of less than 12 residues, the quality of the scoring function used to evaluate loop conformations is the major determinant of loop-prediction accuracy. Loop accuracy is usually evaluated in terms of local RMSD (involving the optimal superposition of the predicted and native loop independent of the rest of the structure) or global RMSD (where the RMSD is evaluated with the loop stems kept in place). The latter measure is preferred because the former allows for two loops to be seen as similar, and to have a small RMSD, even if they have very different orientations in the context of the native structure. Rapp and Friesner (1999) used the generalized Born solvation model and the AMBER94 force field to obtain low RMSD values for the two loops they studied. Their approach still needs to be tested on a larger sample size. Fiser et al. (2000) have recently published an extensive *ab initio* study on a data set of 40 loops and also report low RMSD from known structures. Using global RMSD as a criterion, Fiser et al. (2000) reported an accuracy of less than 2 Å for 8-residue loops.

The study of Fiser et al. (2000) utilized a scoring function that included the CHARMM22 force field and statistical preferences taken from protein databases. Scoring functions based entirely on physical chemistry potentials and an accurate solvation model have the potential of identifying the native conformation as lowest in energy, but there are cases where lower energy conformations appear (Smith and Honig, 1994; Steinbach, 2004). One problem may be that most loop prediction approaches seek the lowest energy conformation, thus ignoring conformational entropy effects that will favor broad energy wells. We have recently implemented a procedure called "colony energy" (for detail, see Section 10.3.2) that takes the shape of the energy well into account and yields highly accurate loop prediction (e.g., 1.4 Å global RMSD for eight-residue loops) (Xiang et al., 2002). With crystal environments considered, Jacobson et al. (2004) achieved the best accuracy of 1.0 Å RMSD for eight-residue loops with a computing-intensive approach that combines OPLS all-atom energy function, efficient methods for loop buildup and side-chain optimization, and the hierarchical refinement protocol. Fogolari and Tosatto (2005) demonstrated that molecular mechanics/Poisson–Boltzmann solvent-accessible surface area, if

Table 10.2 Loop modeling program

Programs	Availability
LOOPY	http://trantor.bioc.columbia.edu/programs.html
PLOP	http://francisco.compbio.ucsf.edu/~jacobson/plop_manual/plop_overview.htm
COILS	http://www.ch.embnet.org/software/COILS_form.html
MODELLER (loop module)	http://guitar.rockefeller.edu/modeller/modeller.html
CODA	http://www-cryst.bioc.cam.ac.uk/coda/

combined with the colony energy approach, is very effective in discriminating loop decoys.

Most methodological tests compare predicted loop conformations to known structures, with the backbone conformation of anchoring residues identical to that of the native conformation. This does not properly simulate real modeling conditions under which the backbone of the target protein may not be identical to that of the template. Not surprisingly, loop prediction accuracy degrades as the constraints provided by the loop ends are less accurately defined (Lessel and Schomburg, 1999; Fiser et al., 2000). Table 10.2 shows some loop modeling software that can be easily obtained from the Web. Other loop modeling software only exists as internal components of model building packages listed in Table 10.1. Compared with database scanning methods, most *ab initio* loop prediction programs are very slow.

10.2.3.2 Side-Chain Prediction

The greatest success in the prediction of side-chain conformations has been achieved for core residues where packing constraints significantly simplify the problem. Even for core residues, the accuracy of side-chain prediction degrades when the structure of the backbone is itself not known to a high degree of accuracy. Many side-chain programs are based on rotamer libraries (Ponder and Richard, 1987), which are generally defined in terms of side-chain torsion angles for preferred conformations of a particular side chain. The resolution of rotamer libraries has increased over time and rotamer libraries have been compiled simply by sampling all angles at some given level of resolution (Maeyer et al., 1997). Since backbone conformation changes the frequency of the rotamers, a backbone-dependent rotamer library is often used in side-chain modeling (Dunbrack and Karplus, 1993; Canutescu et al., 2003). The major advantage is to increase computing efficiency, since bad rotamers, e.g., clashing with the backbone, have been automatically removed during construction of the rotamer library. Baker and his co-workers have developed a "solvated rotamer" approach that shows improvement on side-chain packing at protein–protein interfaces (Jiang et al., 2005). This approach extends current side-chain packing methods by using a rotamer library including solvated rotamers with one or more water molecules fixed to polar functional groups in probable hydrogen-bond orientations, together with a simple energetic description of water-mediated hydrogen bonds. As the number of rotamers increases, however, so does the problem of sampling all possible conformations. There have been a variety of approaches developed to deal with the combinatorial problem in side-chain prediction (Lee and Subbiah, 1991; Lee, 1994; Vasquez, 1996; Dahiyat and Mayo, 1997; Gordon and Mayo, 1999; Samudrala et al., 2000; Kingsford et al., 2005).

Accuracies of about 1 Å RMSD have been reported for core residues in known structures where the backbone has been fixed in the native conformation (Koelh and Delarue, 1994; Vasquez, 1996; Bower et al., 1997; Samudrala and Moult, 1998). A number of studies suggest that further improvements may still be possible. Mendes et al. (1999) found, for example, that the use of an intrinsic torsional potential can improve prediction accuracy. Lovell et al. (2000) reported a novel rotamer library in which internal clashes between side chain and backbone are removed. This library could, in principle, be used to improve prediction accuracy. Xiang and Honig (2001) have shown that using a very detailed rotamer library, which is based on rotamers that use Cartesian coordinates taken from known structures rather than idealized bond lengths and angles, yields RMSD values relative to the native of only 0.62 Å for core residues. This appears to constitute a significant improvement over existing procedures and demonstrates that the combinatorial problem, usually assumed to greatly complicate side-chain prediction, may in fact be of little consequence. This was later confirmed in a more detailed study (Desmet et al., 2002), which showed that local minima for all side-chain prediction may be almost as accurate as the global minimum when evaluated against experimentally determined structures. Improvement on side-chain prediction in recent years has mainly come from better energy functions. Eyal et al. (2004) showed that solvent accessibility and contact surface area are important with regard to the accuracy of side-chain prediction, particularly for modeling buried side chains. Liang and Grishin (2002) have developed a new and simple scoring function for side-chain prediction that consists of the following energy terms: contact surface, volume overlap, backbone dependency, electrostatic interactions, and desolvation energy. The weights of these energy terms were optimized to achieve the minimal average root-mean-square deviation between the lowest energy rotamer and the observed side-chain conformation on a training set of high-resolution protein structures. The derived scoring function combined with a Monte Carlo search algorithm was used to place all side chains onto a protein backbone simultaneously. The average prediction accuracy was 87.9 and 73.2% for the first and second torsion angles correctly predicted to within 40 degrees of native. As is the case for loop prediction, side-chain prediction accuracy depends sensitively on the accuracy to which the backbone conformation is known (Huang et al., 1998). This suggests the possibility of developing procedures where side-chain and backbone conformation can be used iteratively to refine homology models.

Table 10.3 lists some publicly available side-chain prediction programs and the methods they used. Earlier side chain predictions, e.g., RAMP (Samudrala and Moult, 1998), SMD (Tuffery et al., 1993), and CONFMAT (Koelh and Delarue, 1994), were usually based on small rotamer libraries; more recent programs use very

Table 10.3 Side-chain modeling program

Programs	Availability
SCAP	http://trantor.bioc.columbia.edu/programs/jackal/
SCWRL	http://dunbrack.fccc.edu/SCWRL3.php
SMOL	Nikolai.Grichine@UTSouthwestern.Edu
SCCOMP	http://atlantis.weizmann.ac.il/~eyale/
RAMP	http://www.ram.org/computing/ramp/ramp.html
SMD	http://condor.urbb.jussieu.fr/Smd.php
CONFMAT	koehl@csb.stanford.edu
MAXSPROUT	http://www.ebi.ac.uk/maxsprout/

detailed rotamer libraries, e.g., SCAP (Xiang and Honig, 2001), SCWRL (Canutescu et al., 2003), SMOL (Liang and Grishin, 2002). In our recent benchmark study of SCAP, SMOL, and SCWRL, SCAP excelled in prediction for core and surface residues (Xiang et al., to be submitted). For partially buried residues, SMOL performed the best, which was due to its more sufficient conformation sampling and optimized scoring function. SCWRL also performed quite well though not as accurately as the other two, but with much less CPU cost. On a 300-MHz SGI machine, SCWRL is very fast, 3 seconds for each protein, while SMOL needs 11,700 seconds and SCAP needs 361 seconds. Since the test was performed on the native protein backbones, their performance may vary with homology models.

10.2.3.3 Other Improvements to Refinement

Recent improvements to refinement have been mainly achieved by increasing alignment accuracy. Almost all alignment software currently in use has to rely on one of the derivatives of dynamic programming. Although dynamic programming can obtain the global optimal alignment for a given scoring matrix, it cannot account for nonlocal residue–residue interactions. For example, double mutant effects can only be properly estimated if their spatial conformations are both available. As such, a cumbersome but effective method of refining alignment is to build multiple models based on the alternative alignments, with the best alignment corresponding to the model of the lowest physical-chemistry energy. The assumption is that a conformation of lower energy is more likely close to the native state. The method becomes possible due to the availability of more discriminatory energy functions and faster model building tools (Petrey et al., 2003). Tens of thousands of models can be built in a short time with Linux clusters, each based on one variation of alignment. An effective scoring energy can be readily applied to the ensemble of models. The energies of these models can be further minimized with an approach similar to genetic algorithm, i.e., shuffling segments among different models by fixing other parts of protein, where the stems of the segment should have identical residues aligned with the template. Similarly, genetic algorithms are also important tools to increase model quality based on multiple templates. The multiple models, each based on one template, will be superimposed. Variable regions identified are exchanged and then optimized among different models. In the optimization process, an RMSD restraint can be applied to restrict sampling to a conformational space close to the averaged framework of the original templates. This method has been utilized in the NEST program and produced satisfactory results in CASP6, which will be discussed in more detail in Section 10.3.

Recent research has attempted to use MD simulation to refine models. Lee et al. (2001) used MD simulations with an explicit solvent model to refine Rosetta models followed by scoring with the Poisson–Boltzman/surface area solvation model. Their results showed that native structures could be distinguished energetically from structurally different low-resolution models. Lu and Skolnick (2003) used a combination of local restraints, knowledge-based potentials, and MD approaches that showed promising improvements over previous studies using standard MD methods. Fan and Mark (2004) used classical MD simulations with explicit water to refine homology models. A significant improvement over the model structures has been observed in a number of cases. The results indicate that homology models could be possibly refined with MD simulations on a time scale of tens to hundreds of nanoseconds. Qian et al. (2004) used the principal components of the variation of backbone structures within a homologous family to define a small number of evolutionarily favored sampling directions and showed that model quality can be improved by energy-based optimization along these directions. Li et al. (2004) developed new hierarchical and multiscale algorithms to sample helices and flanking loops, which were evaluated with an all-atom protein force field (OPLS) and a generalized Born continuum solvent model. This method, integrated with a loop and side-chain modeling technique, can potentially be used to refine homology structures iteratively. The next-generation structure modeling algorithms should be able to refine a protein structure closer to the native conformation. The most critical part is to obtain an energy function that is sensitive enough to discriminate near-native conformations from other nonnative folds. Though conformation sampling is also difficult, computer clusters allow more thorough sampling of states around the original models.

10.2.4 Model Assessment

All models built by homology will have errors as discussed in the previous section. Verification of the model, and estimation of the likelihood and magnitude of errors has become one of the most important steps in advancing the state of the art of homology modeling. Errors of the model are usually estimated either from the energy of the model, or from the resemblance of a given characteristic of the model to real structures. The most critical component is the development of a scoring function that is capable of distinguishing good from bad models.

Scoring functions used for the evaluation of protein models generally fall into two broad categories. "Statistical" effective energy functions (Sippl, 1995) are based on the observed properties of amino acids in known structures, and have been widely used in fold recognition and homology modeling applications. A variety of statistical criteria have been used successfully to discriminate between deliberately misfolded and native structures. Most of them are directly or indirectly based on the analysis of contacts, either interresidue contacts, interatom contacts, or contacts with solvent. For example, preferential distributions of polar and apolar residues inside or outside of a protein can be used to detect completely misfolded models (Baumann et al., 1989); solvation potentials can detect local errors as well as complete misfolds (Holm and Sander, 1992); packing rules have been implemented for structure evaluation (Gregoret and Cohen, 1990). Residue or atom contacts are discriminative because they are energetically favored, and many real structures cannot tolerate too many unfavorable interactions. Thus, for a model to be correct, only a few infrequently observed atomic contacts are allowed. However, bond angles and bond lengths, though powerful in checking the quality of experimental structures, are usually less useful for the evaluation of models because these factors have already been considered appropriately in the model building stage (Fiser and Sali, 2003).

Physical effective energy functions (Lazaridis and Karplus, 1999a) are based on a direct evaluation of the solvation free energy of a protein. It has been demonstrated that such a direct evaluation of the conformational free energy can be at least as successful as statistically based scoring functions in distinguishing the native structure of a protein from an incorrectly folded decoy, although generally at greater computational cost (Janardhan and Vajda, 1998; Vorobjev et al., 1998; Lazaridis and Karplus, 1999b; Petrey and Honig, 2000). A distinct advantage of such physically derived functions is that they are based on well-defined physical interactions, thus making it easier to learn and to gain insight from their performance. Moreover, the success in CASP (Critical Assessment of Protein Structure Prediction) of *ab initio* methods based on purely physical chemistry methods (Lee et al., 1999) suggests that our understanding of the forces that drive protein stability may have reached the point where it can be translated into widely applicable computational tools. One of the major drawbacks of accurate physical chemical description of the folding free energy of a protein is that the treatment of solvation required usually comes at a significant computational expense. Fast solvation models such as the generalized Born (Still et al., 1990) and SCP-ISM (Hassan et al., 2000), together with a variety of simplified scoring schemes (Huang et al., 1995; Petrey and Honig, 2000), may prove to be extremely useful in this regard.

A number of freely available programs can be used to verify homology models as shown in Table 10.4. They generally belong to one of two categories. The first category (e.g., PROCHECK and WHATIF) checks for proper protein stereochemistry, such as symmetry checks, geometry checks (e.g., chirality, bond lengths, bond angles, torsion angles), and structural packing quality; the second category (e.g., VERIFY3D and PROSAII) checks the fitness of sequence to structure, and assigns a score for each residue fitting its current environment. A new graphics software called GRASP2 is also useful in model assessment (Petrey and Honig, 2003). The software can display alignments and template structures simultaneously for assessment of the alignment quality. For example, insertions or deletions can be mapped to the structures to verify that they make sense geometrically. Where residue substitutions

Table 10.4 Model assessment program

Programs	Availability
PROCHECK	http://www.biochem.ucl.ac.uk/~roman/procheck/procheck.html
WHATCHECK	http://www.sander.embl-heidelberg.de/whatcheck/
ProSalI	http://www.came.sbg.ac.at
VERIFY3D	http://www.doe-mbi.ucla.edu/Services/Verify_3D/
ERRAT	http://www.doe-mbi.ucla.edu/Services/Errat.html
ANOLEA	http://www.fundp.ac.be/pub/ANOLEA.html
AQUA	http://www.nmr.chem.uu.nl/users/jurgen/Aqua/server/
Probe	http://kinemage.biochem.duke.edu/software/probe.php
SQUID	http://www.ysbl.york.ac.uk/~oldfield/squid/
PROVE	http://www.ucmb.ulb.ac.be/UCMB/PROVE
ProQ	http://www.sbc.su.se/~bjorn/ProQ
GRASP2	http://trantor.bioc.columbia.edu/programs.html

occur, the user can verify that structural features such as hydrophobic packing are maintained and that active-site residues and other features of the target identified from the literature are conserved. The manual inspection should be combined with existing programs to further identify problems in the model.

10.3 Homology Modeling with JACKAL

A new set of homology modeling tools have been developed that are publicly distributed in the JACKAL package (http://trantor.bioc.columbia.edu/programs.html). JACKAL integrates knowledge-based and physics-based methods for protein structure prediction and refinement. At the heart of our approach to structure prediction and refinement is the use of the colony energy concept (see Section 10.3.2). The purpose of JACKAL is to automate the process of structure prediction, from template identification and alignment tuning to model building, refinement and structure verification. JACKAL contains the following major components: NEST for model building and refinement; SCAP for side-chain modeling (Xiang and Honig, 2001); LOOPY for loop prediction (Xiang et al., 2002); AUTOALIGN for alignment tuning; CONREF for model refinement. The core of JACKAL is the NEST program, which, based on our newly developed artificial evolution algorithm (Fig. 10.6), attempts to build models by simulating the natural process of structural evolution from the template structure to the target model. SCAP and LOOPY are used for residue mutation and insertion/deletion, respectively.

10.3.1 Model Building with Artificial Evolution Algorithm

Given an alignment between the query and template sequence, the alignment can be broken down into a list of operations such as residue mutation, insertion, or deletion

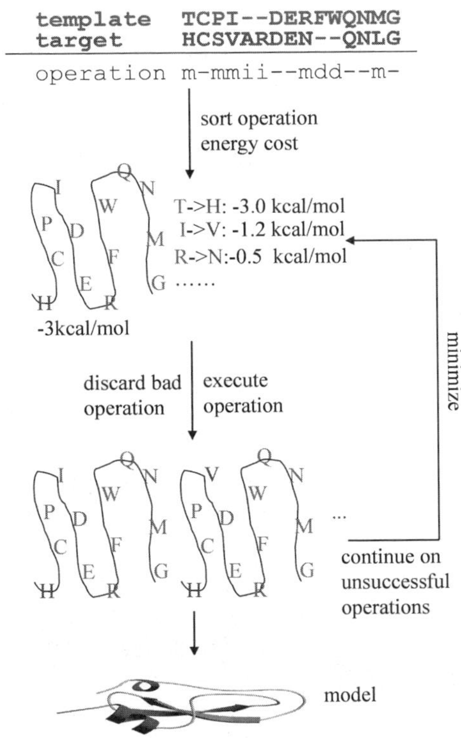

Fig. 10.6 Model building with artificial evolution. m, i, and d denote mutation, insertion, and deletion, respectively.

(Fig. 10.6 shows an alignment of 9 operations, i.e., 5 mutations, 2 insertions, and 2 deletions). Supposing the template to be the "parent structure," it would take Nature billions of years for the template structure to evolve into the target structure. It is unlikely that Nature would finish the daunting task in one step. Instead, a more probable scenario is for Nature to evolve into the target structure via multiple steps with minimal changes to the template structure at each step. Accordingly, building a target model could be considered a process of evolving the template structure based on the alignment so that changes are carried out step by step, each step on one operation. Each operation, i.e., mutation, deletion, or insertion, will disturb the template structure and thus involve an energy cost, either positive or negative. The model building starts from the operation with the least energy cost and so on. Each operation is followed by a slight energy minimization to remove atom clashes. The final structure is then subjected to more thorough energy minimization. The order for the first round of operations does not have to be determined by actually calculating the energy cost for each operation; instead, it can be conveniently estimated empirically. For example, amino-acid mutation is generally easier in evolution than insertion and deletion. As such, mutation operations on residues that are on the protein surface are usually performed first followed by mutation of buried small-sized residues and so on. The operation is considered successful if it does not cause a significant energy penalty (less than 5 kcal/mol) to the structure; otherwise the operation is discarded and will return to the waiting list. Insertion or deletion of multiple residues is considered as a group of operations, each operating on one residue. The operation starts from the middle residue of the segment with deletion preferred over insertion, since the structural effects of deletion are more reliably predicted. Similarly, operations with more than 5 kcal/mol energy cost (an empirical cutoff that can easily be modified) would also be considered unsuccessful and returned to the waiting list for the next round of operations. The next round of operations actually works on the waiting list, starting from the operation of the least energy cost that has been calculated from the previous round, but with a doubled energy cutoff, e.g., operation with energy penalty more than 10 kcal/mol would be considered unsuccessful for the second round. A number of rounds (less than five rounds in total) would finally accomplish the evolution of the template structure to the model, which will be followed by a series of model refinements.

NEST is heavily dependent on our previous progress in side-chain and loop modeling, i.e., the SCAP (Xiang and Honig, 2001) and LOOPY (Xiang et al., 2002) program. Both SCAP and LOOPY have been integrated into the NEST code, though they also exist independently in the JACKAL package. In the case of mutation, the residue in the template structure first has its side chain changed to the corresponding one in the target sequence, followed by several steps of minimization of the new side chain. We have adopted a simple conformational sampling strategy for side-chain modeling. Side-chain modeling is first carried out with all other parts of protein fixed. The complete rotamer conformations for the side chain, which has been compiled from 646 nonredundant high-resolution protein chains, will be assembled onto the backbone. The rotamer with the lowest colony energy (see Section 10.3.2 for a description of the colony energy concept) will be selected as the final conformation. However, if the rotamer of the lowest conformation energy participates in a hydrogen bond, the conformation energy is used instead of the colony energy because entropic effects generally do not favor hydrogen bonding, and an accurate balance between hydrogen-bonding energy and entropy is difficult to achieve in a simplified force field. If the best rotamer has positive energy, neighboring side chains contacting with the rotamer will then be subjected to minimization. For each of the neighboring side chains including the one that has just been mutated, a similar strategy, that is, sampling all possible rotamers with evaluation based on colony energy, will be performed. The minimization procedure starts from the first residue to the last in the neighboring list until all the side-chain conformations retain the same rotamer on further iteration. If the energy of the side chain for the mutant residue is larger than 5 kcal/mol, the mutation will be considered unsuccessful, thus the mutation operation will be returned to the waiting list, and all other affected residues associated with this operation will be restored to their previous configurations.

An algorithm similar to LOOPY is used to minimize regions affected by insertion or deletion. A segment of five to eight residues that covers the residue under consideration is used in the minimization process. In order not to introduce large disturbance to the conserved region, the segment window usually slides to one particular direction depending on the location where the residue has to be handled. For example, a larger part of the segment should be assigned to one side of the residue under consideration if it has lower sequence alignment similarity than the other side; it is advisable for the segment to avoid the helix or β-sheet region in order to keep the regular secondary structure intact. If the insertion or deletion is in the helix or β-sheet, the segment will try to cover as many residues as possible in loops. The segment should overlap with at least one residue on either side of the operation. A shorter segment should be used if the loop has fewer than five residues. The segment is refined by sampling alternative conformations. If the operation is in a loop, random conformations would be generated; otherwise, 50% of the conformations would be generated randomly, and the other 50% would be generated that equivalently extends or shortens the regular secondary structures. In other words, conformational sampling is performed with insertion or deletion pushed to the nearest loop region. The backbone of each conformation is minimized using "direct tweak," a novel energy minimization algorithm that minimizes all torsion angle freedoms of a segment without dislodging the end residues. The "direct tweak" algorithm was achieved by combining conventional energy minimization in torsion-angle space with a set of chain-closure constraints that were based on the random tweak algorithm (Shenkin et al., 1987). Pairs of segments with RMSD greater than 2 Å are then combined (i.e., for an eight-residue segment, the first four residues in one segment joined with the last four in another to form a new segment which is fused in the middle with a segment closure procedure) to generate new segments. This results in a set of the original segments plus all the newly fused segments. Side chains are then assembled onto each of the segments, and the colony energy for each segment is calculated. The lowest 30% survive and the procedure is repeated until a single segment remains. In all of the above steps, no more than 200 segments are retained. The operation is successful only when the energy increase of the segment is less than 5 kcal/mol.

10.3.2 Physical-Chemical Energy and Colony Energy Method

The energy function used in NEST can be expressed as the following terms (for more detailed discussion of energy functions, see Chapters 2 and 3):

$$\Delta E = \Delta E_{\mathrm{vw}} + \Delta E_{\mathrm{torsion}} + \Delta E_{\mathrm{hbond}} + \Delta E_{\mathrm{hydro}}, \tag{10.1}$$

$$\Delta E_{\mathrm{vw}} = 61.66\,\eta\,\exp(-2r^2) * (1/r - 1.12/r^{0.5}), \tag{10.2}$$

$$\Delta E_{\mathrm{hbond}} = \min\left(0, [-16 + 12\Omega]\cos(\theta_{\mathrm{DHA}})\cos(1.5\,\theta_{\mathrm{HAC}})/d_{\mathrm{HA}}^3\right), \tag{10.3}$$

$$\text{if } 2\,\text{Å} < d_{\mathrm{HA}} < 3\,\text{Å},\ \theta_{\mathrm{DHA}} > 90^\circ,\ \text{and } \theta_{\mathrm{HAC}} > 60^\circ;$$

$$\text{else } \Delta E_{\mathrm{hbond}} = 0.$$

ΔE_{vw}, $\Delta E_{\mathrm{torsion}}$, $\Delta E_{\mathrm{hbond}}$, and $\Delta E_{\mathrm{hydro}}$ are van der Waals, torsion, hydrogen bond, and hydrophobic energy, respectively. The van der Waals energy is evaluated with a

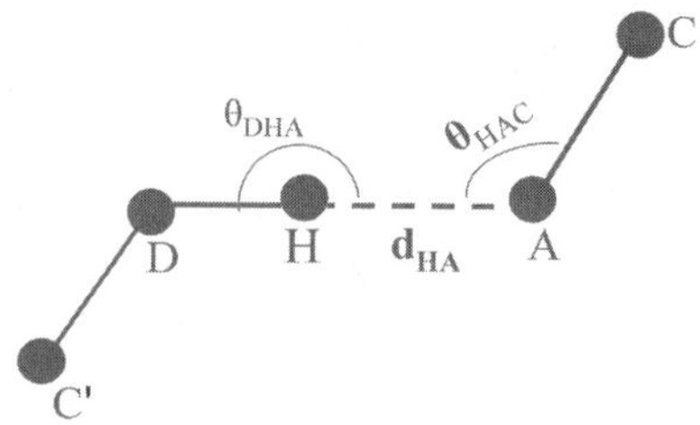

Fig. 10.7 Hydrogen bond. D, A is hydrogen bond donor and acceptor, respectively. H is the proton, and C is carbon atom.

modified expression that fits the CHARMM van der Waals curve but with repulsive term softened to reduce sensitivity to small changes in atomic positions. In Eq. (10.2), η is the energy at the minimum of the potential function and is chosen to correspond to the minimum in the van der Waals potential of the CHARMM22 force field between the two interacting atoms, and r is the ratio of the interatomic distance and the sum of the van der Waals radii of two interacting atoms. The hydrophobic energy is calculated based on the solvent-accessible surface area with the coefficient of 0.025 kcal/mol/Å^2. Here hydrogen-bond energy E_{hbond} was calculated using Eq. (10.3) (see Fig. 10.7), where Ω is the ratio of the solvent-accessible surface area (SASA) of the residue in the protein and the SASA of the same conformation for the residue isolated in solution. D is the hydrogen donor, H is the polar hydrogen, A is the hydrogen acceptor, and C is the carbon atom bonded to A. θ_{DHA} and θ_{HAC} are the angles defined by the coordinates of the respective atoms, and d_{HA} is the distance between atoms H and A. Although the value of θ_{HAC} depends on whether the atomic orbital of the acceptor is sp2 or sp3, the θ_{HAC} angle is nevertheless close to 120°. Since the rotamer library is discretized, we relaxed the standard requirement that θ_{HAC} should be larger than 90° (McDonald and Thornton, 1994). E_{hbond} is defined to assume its minimum value when d_{HA} is 2 Å and θ_{DHA} is 180°. The minimum E_{hbond} values for completely buried and completely exposed side chains are −2 and −0.5 kcal/mol, respectively, representative of experimental data for hydrogen bonds (Efimov and Brazhnikov, 2003).

For each operation (mutation, deletion, and insertion), sufficient conformation sampling is usually performed. The mechanical energy for each sampled conformation is evaluated with Eq. (10.1). NEST does not assume the best prediction to be that of lowest mechanical energy; instead, a new energy term called "colony energy" is used to evaluate all candidates (see Fig. 10.8), and the conformation with the lowest colony energy will be chosen as the prediction (Xiang et al., 2002). For an operation with N sampled conformations, the colony energy of rotamer i, ΔG_i, is calculated as

$$\Delta G_i = -RT * \ln\left[\sum_j \exp(-E_j/(RT) - \beta(\text{RMSD}_{ij}/\text{RMSD}_{\text{avg}})^{\gamma})\right], \qquad (10.4)$$

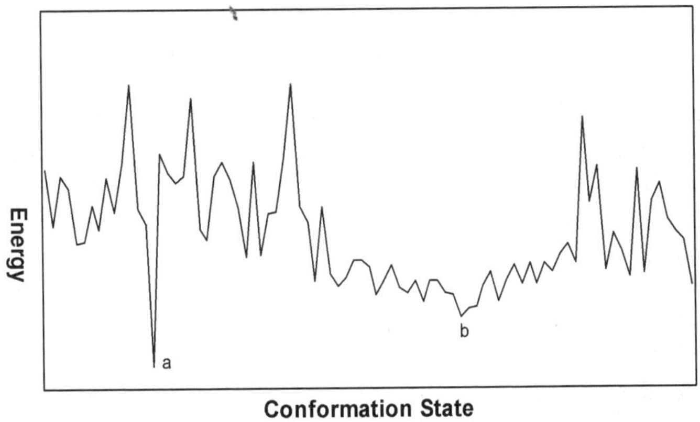

Fig. 10.8 One-dimensional schematic of the sampled conformations. Although conformation a is the global minimum of the mechanical energy, conformation b is structurally similar to many conformations at local minimum, and may possess lower colony energy than conformation a depending on the values of β and γ.

where R is the gas constant, T is absolute temperature, and E_i is the mechanical energy [Eq. (10).1] of the conformation i in the ensemble that has been sampled for the operation. The sum is for all conformations in the ensemble, i.e., j ranges from 1 to N including i. RMSD_{ij} is the root-mean-square distance between conformations i and j. $\mathrm{RMSD}_{\mathrm{avg}}$ is the average of RMSD between any two conformations in the ensemble for a given operation. The parameter β is set to $-\ln(1/2)$. The ranges of conformational energies and three-dimensional structures sampled in a particular application call for the use of γ values that balance the conformational-energy and RMSD-based factors appropriately. Results obtained with the training data set suggested an optimal value for γ would be 1 and 3 for side-chain and loop modeling, respectively, (Xiang et al., 2002). Equation (10.4) approximates entropic effects by favoring those conformations found in regions of configuration space that are visited most frequently.

10.3.3 Model Refinement with JACKAL

Model refinement in JACKAL is performed in two steps. The first step is to increase alignment quality, and the second step is to directly refine the model itself. A meta server is usually used (http://bioinfo.pl/meta/) to identify as many prospective templates as possible. In the absence of a unanimous template identified by all servers, all possible hits will be considered. For example, if multiple templates are identified but all servers point to the same structural family, all structures in the PDB from that family should be used as possible templates. For each template identified, a number of alignments are obtained either from different servers or from alternative alignments based on a particular alignment protocol. Because model building can be done rapidly using NEST, the ensemble of sequence alignments is readily converted to an ensemble of 3D model structures. AUTOALIGN can be used on the ensemble of models to repeatedly improve model quality using genetic algorithms. Specifically, all the models are superimposed based on sequence alignment and the regions of high variability are identified. For each model, AUTOALIGN tries all possible conformations on variable regions with corresponding segments from other models. The resultant candidates are then clustered and ranked using the colony energy. The conformation of the lowest colony energy is chosen as the best choice. The process can be repeated until a stable model is derived.

For each model, unaligned regions corresponding to gaps in the sequence alignment are modeled using the independent LOOPY program with a similar approach discussed above but at a more sufficient conformation sampling and energy minimization. Specifically, 2000 initial conformations are randomly sampled and filtered against the consensus secondary-structure predictions from the meta server. The 2000 conformations are then energy-minimized using our fast "direct tweak" method, and the 300 conformations of lowest energy are kept. An additional 300 are obtained from a fragment database using sequence similarity, secondary structure, and end-point geometry. The 600 conformations are subjected to additional energy minimization, and the conformation of lowest colony energy is selected. Side chains are modeled with the independent SCAP program, where the initial conformation starts from the NEST output. The final model will be further optimized using the CONREF module that refines the model with restraints. The restraints include backbone hydrogen bonds and main-chain framework of the template structure, i.e., an energy penalty would be applied if the sampled structure breaks an existing hydrogen bond or deviates significantly (more than 2 Å) from the original model. This is to guarantee that sampling only visits conformations close to the templates.

10.3.4 Comparison with Other Homology Modeling Software

Homology modeling has been widely used in structure prediction, and many homology modeling tools are available (Table 10.1). Given the same alignment and template, it was generally believed there were no major differences between the best modeling programs. However, a recent study by Wallner and Elofsson (2005) has shown that some programs performed better than others. In their study, a benchmark of six different homology modeling programs—MODELLER, SEGMOD/ENCAD (Levitt, 1992), SWISS-MODEL (Schwede et al., 2003), 3D-JIGSAW (Bates et al., 2001), NEST, and BUILDER (Koehl and Delarue, 1996)—is presented. Their study concluded that no single modeling program consistently outperformed the others in all tests. However, it is quite clear that three modeling programs, MODELLER, NEST, and SEGMOD/ENCAD, perform better than the others. Detailed analysis of these homology modeling programs revealed some interesting differences. For example, using a 1.4-GHz AMD XP processor, NEST needs 17 s on average to build a model, while SEGMOD needs 6 s, and MODELLER needs 43 to 430 s in MODELLER6v2 and MODELLER6v2–10, respectively; MODELLER,

SWISS-MODEL, and BUILDER produce more models that do not converge compared to the other programs; in terms of stereochemistry (bond lengths, bond angles, and side-chain planarity), 3D-JIGSAW, BUILDER, and SWISS-MODEL created more residues with bad chemistry for difficult targets, while the other modeling programs showed a fairly constant number of bad residues at all sequence identities. For sequence identities below 40%, all modeling programs manage to bridge some gaps and build some loops correctly or incorrectly; therefore, accordingly, some models are better or worse than the template. In this region the MODELLER programs, NEST, SEGMOD/ENCAD, and SWISS-MODEL, improved 20% of the models. Only NEST rarely made the models worse, while all other programs deteriorated at least 5% of the models. The authors also found that NEST had more of its models "among best" than the other programs; thus, selecting a model from NEST is almost always a good choice.

10.4 Application of Homology Modeling

Homology modeling is often an efficient way to obtain information about proteins of interest. Compared with *ab initio* protein folding, homology modeling is more accurate and reliable. The quality of a homology model is directly correlated with the sequence similarity between target and template. Though a homology model is not perfect, it is still very useful in a wide spectrum of applications where information about 3D conformation of a protein is required.

Highly homologous models with sequence identity above 50% to the templates often have RMSD from the crystal structure around 2 Å, which is roughly comparable to a medium-resolution X-ray structure except for some gapped regions. Models at this level of accuracy can often be used to study a wide range of biological activities that require the knowledge of conformations of individual residues, such as studying catalytic mechanism (Zhou and Johnson, 1999; Francoijs et al., 2000; Xu et al., 2005; Fischer et al., 2005; Kim et al., 2005), designing and improving ligands (Wang and Hampson, 2005; Niv and Weinstein, 2005), predicting protein partners (Orban et al., 2005), solving X-ray structures with molecular replacement (Cupp-Vickery et al., 2003; Schwarzenbacher et al., 2004), refining NMR structures (Skolnick et al., 1997 and Kim et al., 2004) and defining antibody epitopes (Oakhill et al., 2005). In the middle of the accuracy level are the models based on approximately 35% sequence identity, corresponding to 85% of $C\alpha$ atoms modeled within 3.5 Å of their crystal positions. Though conformations for most side chains have significant errors, fortunately, the active and binding sites are frequently more conserved and are thus modeled more accurately (Sanchez and Sali, 1998; Hassan et al., 2005). Medium-resolution models can be used to improve protein function prediction based on sequence alone (Burley and Bonanno, 2002; Shakhnovich et al., 2003), because ligand binding is more determined by the 3D configurations of active-site residues than by sequence. They can also be used to construct site-directed mutants with altered or destroyed binding capacity, or design proteins with added disulfide bonds for extra stability, which in turn could test hypotheses about the sequence–structure–function relationships (Ivanenkov et al., 2005; Campillo et al., 2005). For models of low accuracy with sequence identity less than 25%, they sometimes have less than 50% of their Cα atoms within 3.5 Å of their correct positions (Fiser and Sali, 2003). Nevertheless, such models still have the correct fold and even knowing only the fold of a protein is frequently sufficient to predict its approximate biochemical function (Al-Lazikani et al., 2001b). Evaluation of models in this low range of accuracy can be used for confirming or rejecting a match between remotely related proteins (Sanchez and Sali, 1998).

Xu et al. (2005) recently used homology modeling to study HSP90 and kinase Erbb1 interaction. The molecular chaperone Hsp90 modulates the function of specific cell signaling proteins. Although targeting Hsp90 with the antibiotic inhibitor geldanamycin (GA) may be a promising approach for cancer treatment, little is known about the determinants of Hsp90 interaction with its client proteins. Previous studies have shown that Erbb1 binds with HSP90 while Erbb2, having 82% sequence identity to Erbb1, does not. The crystal structure of Erbb1 has been solved to a resolution of 2.6 Å, which was used as the template structure to build the homology model for Erbb2. By superimposing the 3D conformations of Erbb1 and Erbb2, a loop within the N lobe of the kinase domain of Erbb2 was identified that determines Hsp90 binding. Further detailed analysis of the Erbb1 crystal structure and Erbb2 model identified a single residue difference (Gly745 on Erbb1 versus Asp778 on Erbb2) that may account for their different interaction with HSP90. The analysis implied that the amino acid sequence of the loop determines the electrostatic and hydrophobic character of the protein's surface, which in turn governs interaction with Hsp90. The hypothesis was later confirmed by a number of carefully designed mutagenesis experiments.

Another study used low-resolution comparative models to annotate protein functions (Al-Lazikani et al., 2001). Janus kinases (JAKs) are a family of nonreceptor protein tyrosine kinases involved in signaling cascades initiated by various cytokines, interferons, and growth factors (Schindler and Darnell, 1995). There are four human JAK proteins: JAK1–3 and TYK2. JAKs share seven main regions of homology, termed JAK-homology domains JH1–7, numbered from the C to the N terminus. JH1 is the C-terminal protein kinase domain, and JH2 is a kinaselike domain whose precise function remains unclear. JH3–7 play a role in receptor interactions. There has been considerable uncertainty as to whether JAKs contain SH2 domains. Application of homology modeling and other sequence profile analysis method strongly indicates that the Janus family of nonreceptor protein tyrosine kinases contains SH2 domains. One of the Janus kinases, human TYK2, has an SH2 domain that contains a histidine instead of the conserved arginine at the key phosphotyrosine-binding position, βB5. Calculations of the pK_a values of the βB5 arginines in a number of SH2 domains and of the βB5 histidine in a homology model of TYK2 suggest that this histidine is likely to be neutral around pH 7, thus indicating that it may have lost the ability to bind phosphotyrosine.

10.5 Summary

Protein-structure prediction has fascinated the scientific community for decades; it is a problem simple to define but difficult to solve. The dream seems more and more attainable with the explosion of sequence and structural information and because of computational advances in many different areas. These include pure sequence analysis, structure-based sequence analysis, conformational analysis of proteins, and the understanding of the energetic determinants of protein stability. Homology modeling has become a widely used tool, and fold recognition has been shown to extend the limits of detection of sequence search methods. The advent of structural genomic initiatives is certain to spur the development of a host of new computational methods aimed at detecting new relationships between sequence, structure, and function. Continued progress in *ab initio* modeling, combined with ever-increasing databases, makes it possible to further refine homology models to higher accuracy. Such models will provide the basis for a more detailed analysis of structure and function relationships than has been available in the past and will provide powerful tools for the analysis of experimental data and for the design of new experiments.

Despite past progress, much remains to be done. A major problem that still plagues structure prediction by homology is that the structure of the target protein may differ significantly from the closest available template. Unlike the rapid advances made in experimental structure determination, progress in homology structure prediction has been incremental as illustrated at the recent CASP (Critical Assessment of Methods for Structure Prediction of Proteins, http://www.forcasp.org) competitions. Reliability of these homology modeling methods depends critically on the level of sequence identity between the modeling target and the template. When sequence identity is 30% or higher, backbone atoms are usually correctly modeled. The majority of the errors come from side-chain and loop placement during refinement with roughly 3–4 Å RMSD compared to high-resolution crystal structures. When the sequence identity drops below 30%, misalignment happens frequently and model quality suffers dramatically. To increase the utilization and value of the computational models in biomedical research, and to reduce the need for still costly experimental structure determination, significant improvement in the reliability and accuracy of modeling techniques is needed by the research community. There are two immediate goals that have to be addressed in the homology modeling community. The first scientific goal is to expand the modeling coverage to more distantly related proteins that exhibit as low as 10% identity to any known structures. The quality of these models should be close to X-ray structures or high-resolution NMR structures with less than 2 Å RMSD for backbone and side-chain atoms. Significant improvement of modeling methods is needed to push the modeling coverage to remote homologues of existing structures without much compromise on quality. This is both an alignment problem and a refinement problem. Future progress on this issue will depend on advances in the energetic evaluation of structures and the evolutionary analysis of sequences, and the integration of these two fields. The second goal is to achieve the standard of high-resolution X-ray crystal structure quality for comparative models that are based on known structures with higher homology (30% sequence identity) to the modeling targets. This is predominantly a high-accuracy refinement problem, although substantial improvement of alignment methods is also required. The aim is to acquire the ability to reliably produce computational models with highly accurate placement of both backbone and side-chain atoms, and to significantly reduce the need for experimental structure determinations for close homologues of known structures.

Further Reading

Bates, P. A., Kelley, L.A., MacCallum, R.M., and Sternberg, M.J.E. 2001. Enhancement of protein modelling by human intervention in applying the automatic programs 3D-JIGSAW and 3D-PSSM. *Proteins Struct. Funct. Genet. Suppl.* 5:39–46.

Fan, H., and Mark, A.E. 2004. Refinement of homology-based protein structures by molecular dynamics simulation techniques. *Protein Sci.* 13:211–220.

Koehl, P., and Delarue, M. 1996. Mean-field minimization methods for biological macromolecules. *Curr. Opin. Struct. Biol.* 6:222–226.

Levitt, M. 1992. Accurate modeling of protein conformation by automatic segment matching. *J. Mol. Biol.* 226:507.

Li, X., Jacobson, M.P., and Friesner, R.A. 2004. High resolution prediction of protein helix positions and orientations. *Proteins* 55:368–382.

Sali, A., and Blundell, T.L. 1993. Comparative protein modelling by satisfaction of spatial restraints. *J. Mol. Biol.* 234:779–815.

Schwede, T., Kopp, J., Guex, N., and Peitsch, M.C. 2003. SWISS-MODEL: An automated protein homology-modeling server. *Nucleic Acids Res.* 31:3381–3385.

Tang, C.L., Xie, L., Koh, I.Y.Y., Posy, S., Alexov, E., and Honig, B. 2003. On the role of structural information in remote homology detection and sequence alignment: New methods using hybrid sequence profiles. *J. Mol. Biol.* 334:1043–1062.

Xiang, Z.X., Csoto, C., and Honig, B. 2002. Evaluating configurational free energies: The colony energy concept and its application to the problem of protein loop prediction. *Proc. Natl. Acad. Sci. USA* 99:7432–7437.

Xiang, Z.X., and Honig, B. 2001. Extending the accuracy limit of side-chain prediction. *J. Mol. Biol.* 311:421–430.

Acknowledgments

I thank Drs. Peter Steinbach, Cinque Csoto, and Jan Norberg for their many useful comments and critical reading of the manuscript. No endorsement by the U.S. Government should be inferred from the mention of trade names, software packages, commercial products, or organizations.

References

Acharya, K.R., Stuart, D.I., Walker, N.P., Lewis, M., and Phillips, D.C. 1989. Refined structure of baboon alpha-lactalbumin at 1.7 A resolution. Comparison with C-type lysozyme. *J. Mol. Biol.* 208:99–127.

Al-Lazikani, A., Jung, J., Xiang, Z.X., and Honig, B. 2001a. Protein structure prediction. *Curr. Opin. Struct. Biol.* 5:51–56.

Al-Lazikani, B., Lesk, A.M., and Chothia, C. 1997. Standard conformations for the canonical structures of immunoglobulins. *J. Mol. Biol.* 273:927–948.

Al-Lazikani, B., Sheinerman, F., and Honig, B. 2001b. Combining multiple structure and sequence alignments to improve sequence detection and alignment: Application to the SH2 domains of Janus Kinases. *Proc. Natl. Acad. Sci. USA* 98:14796–14801.

Altschul, S., Madden, T.L., Schaffer, A.A., Zhang, J., Zhang, Z., Miller, W., and Lipman, D.J. 1997. Gapped BLAST and psi-blast: A new generation of protein database search programs. *Nucleic Acids Res.* 25:3389–3402.

Anfinsen, C.B. 1973. Principles that govern the folding of protein chains. *Science* 181:223–230.

Barton, G.J., and Sternberg, M.J. 1990. Flexible protein sequence patterns. A sensitive method to detect weak structural similarities. *J. Mol. Biol.* 212:389–402.

Bates, P.A., Kelley, L.A., MacCallum, R.M., and Sternberg, M.J.E. 2001. Enhancement of protein modelling by human intervention in applying the automatic programs 3D-JIGSAW and 3D-PSSM. *Proteins Struct. Funct. Genet. Suppl.* 5:39–46.

Baumann, G., Froemmel, C., and Sander, C. 1989. Polarity as a criterion in protein design. *Protein Eng.* 2:329–334.

Blundell, T.L., Bedarkar, S., Rinderknecht, E., and Humble, R.E. 1978. Insulin-like growth factor: A model for tertiary structure accounting for immunoreactivity and receptor binding. *Proc. Natl. Acad. Sci. USA* 75:180–184.

Bonneau, R., and Baker, D. 2001. Ab initio protein structure prediction: Progress and prospects. *Annu. Rev. Biophys. Biomol. Struct.* 30:173–189.

Bower, M., Cohen, F.E., and Dunbrack, R.L., Jr. 1997. Homology modeling with a backbone- dependent rotamer library. *J. Mol. Biol.* 267:170–184.

Brayer, G.D., Delbaere, L.T., and James, M.N. 1979. Molecular structure of the alpha-lytic protease from Myxobacter 495 at 2.8 angstroms resolution. *J. Mol. Biol.* 131:743–775.

Brenner, S.E., Chothia, C., and Hubbard, T.J. 1997. Population statistics of protein structures: Lessons from structural classifications. *Curr. Opin. Struct. Biol.* 7:369–376.

Brooks, B.R., and Karplus, M. 1983. CHARMM: A program for macromolecular energy, minimization, and dynamic calculations. *J. Comput. Chem.* 4:187–217.

Browne, W.J., North, A.C.T., Phillips, D.C., Brew, K., Vanaman, T.C., and Hill, R.C. 1969. A possible three-dimensional structure of bovine alpha-lactalbumin based on that of hen's egg-white lysozyme. *J. Mol. Biol.* 42:65.

Bruccoleri, R.E. 1993. Application of systematic conformational search to protein modeling. *Mol. Simulat.* 10:151–174.

Burley, S.K., and Bonanno, J.B. 2002. Structuring the universe of proteins. *Annu. Rev. Genomics Hum. Genet.* 3:243–262.

Campillo, N.E., Antonio Paez, J., Lagartera, L., and Gonzalez, A. 2005. Homology modelling and active-site-mutagenesis study of the catalytic domain of the pneumococcal phosphorylcholine esterase. *Bioorg. Med. Chem.* 13:6404–6413.

Canutescu, A.A., Shelenkov, A.A., and Dunbrack, R.L., Jr. 2003. A graph-theory algorithm for rapid protein side-chain prediction. *Protein Sci.* 12:2001–2014.

Ceulemans, H., and Russell, R.B. 2004. Fast fitting of atomic structures to low-resolution electron density maps by surface overlap maximization. *J. Mol. Biol.* 338:783–793.

Chapman, J.R. 1996. Mass spectrometry. Ionization methods and instrumentation. *Methods Mol. Biol.* 61:9–28.

Chappey, C., Danckaert, A., Dessen, P., and Hazout, S. 1991. MASH: An interactive program for multiple alignment and consensus sequence construction for biological sequences. *Comput. Appl. Biosci.* 7:195–202.

Cox, R.A., and Bonanou, S.A. 1969. A possible structure of the rabbit reticulocyte ribosome. An exercise in model building. *Biochem. J.* 114:769–774.

Cupp-Vickery, J.R., Urbina, H., and Vickery, L.E. 2003. Crystal structure of IscS, a cysteine desulfurase from Escherichia coli. *J. Mol. Biol.* 330:1049–1059.

Dahiyat, B.I., and Mayo, S.L. 1997. Probing the role of packing specificity in protein design. *Proc. Natl. Acad. Sci. USA* 94:10172–10177.

Delbaere, L.T., Brayer, G.D., and James, M.N. 1979. Comparison of the predicted model of alpha-lytic protease with the x-ray structure. *Nature* 279:165–168.

De Rienzo, F., Fanelli, F., Menziani, M.C., and De Benedetti, P.G. 2000. Theoretical investigation of substrate specificity for cytochromes P450 IA2, P450 IID6 and P450 IIIA4. *J. Comput. Aided. Mol. Des.* 14:93–116.

Desmet, J., Spriet, J., and Lasters, I. 2002. Fast and Accurate Side-chain Topology and Energy Refinement (FASTER) as a new method for protein structure optimization. *Proteins* 48:31–43.

Domingues, F.S., Koppensteiner, W.A., and Sippl, M.J. 2000. The role of protein structure in genomics. *FEBS Lett.* 476:98–102.

Dunbrack, R.L., Jr., and Karplus, M. 1993. Backbone-dependent rotamer library for proteins application to side-chain prediction. *J. Mol. Biol.* 230:543–574.

Efimov, A.V. 1993. Standard structures in proteins. *Prog. Biophys. Mol. Biol.* 60:201–239.

Efimov, A.V., and Brazhnikov, E.V. 2003. Relationship between intramolecular hydrogen bonding and solvent accessibility of side-chain donors and acceptors in proteins. *FEBS Lett.* 554:389–393.

Ekman, D., Bjorklund, A.K., Frey-Skott, J., and Elofsson, A. 2005. Multi-domain proteins in the three kingdoms of life: Orphan domains and other unassigned regions. *J. Mol. Biol.* 348:231–243.

Eyal, E., Najmanovich, R., McConkey, B.J., Edelman, M., and Sobolev, V. 2004. Importance of solvent accessibility and contact surfaces in modeling side-chain conformations in proteins. *J. Comp. Chem.* 25:712–724.

Fan, H., and Mark, A.E. 2004. Refinement of homology-based protein structures by molecular dynamics simulation techniques. *Protein Sci.* 13:211–220.

Felsenstein, J. 1981. Evolutionary trees from DNA sequences: A maximum likelihood approach. *J. Mol. Evol.* 17:368–376.

Fidelis, K., Stern, P.S., Bacon, D., and Moult, J. 1994. Comparison of systematic search and database methods for constructing segments of protein structure. *Protein Eng.* 7:953–960.

Fischer, A.J., Rockwell, N.C., Jang, A.Y., Ernst, L.A., Waggoner, A.S., Duan, Y., Lei, H., and Lagarias, J.C. 2005. Multiple roles of a conserved GAF domain tyrosine residue in cyanobacterial and plant phytochromes. *Biochemistry* 22:15203–15215.

Fiser, A., Gian Do, R., and Sali, A. 2000. Modeling of loops in protein structures. *Protein Sci.* 9:1753–1773.

Fiser, A., and Sali, A. 2003. Comparative protein structure modeling. In *Protein Structure* (D. Chasman, Ed.) New York, Dekker, pp. 167–206.

Fogolari, F., and Tosatto, S.C. 2005. Application of MM/PBSA colony free energy to loop decoy discrimination: Toward correlation between energy and root mean square deviation. *Protein Sci.* 14:889–901.

Francoijs, C.J., Klomp, J.P., and Knegtel, R.M. 2000. Sequence annotation of nuclear receptor ligand-binding domains by automated homology modeling. *Protein Eng.* 13:391–394.

Frishman, D., Goldstein, R.A., and Pollock, D.D. 2000. Protein evolution and structural genomics. *Pac. Symp. Biocomput.* 12:3–5.

Goldsmith-Fischman, S., and Honig, B. 2003. Structural genomics: Computational methods for structure analysis. *Protein Sci.* 12:1813–1821.

Gordon, D.B., and Mayo, S.L. 1999. Branch-and-terminate: A combinatorial optimization algorithm for protein desgin. *Structure Fold. Des.* 7:1089–1098.

Gotoh, O. 1982. An improved algorithm for matching biological sequences. *J. Mol. Biol.* 162:705–708.

Greer, J. 1980. Model for haptoglobin heavy chain based upon structural homology. *Proc. Natl. Acad. Sci. USA* 77:3393–3397.

Greer, J. 1981. Comparative model-building of the mammalian serine protease. *J. Mol. Biol.* 153:1027.

Gregoret, L.M., and Cohen, F.E. 1990. Novel method for the rapid evaluation of packing in protein structures. *J. Mol. Biol.* 211:959–974.

Harrison, R.W., Chatterjee, D., and Weber, I.T. 1995. Analysis of six protein structures predicted by comparative modeling techniques. *Proteins* 23:463–671.

Hassan, S.A., Gracia, L., Vasudevan, G., and Steinbach, P.J. 2005. Computer simulation of protein–ligand interactions: Challenges and applications. *Methods Mol. Biol.* 305:451–492.

Hassan, S.A., Guarnieri, F., and Mehler, E.L. 2000. A general treatment of solvent effects based on screened coulomb potentials *J. Phys. Chem. B* 104:6478.

Havel, T.F., and Snow, M.E. 1991. A new method for building protein conformations from sequence alignments with homologues of known structure. *J. Mol. Biol.* 217:1–7.

Holm, L., and Sander, C. 1992. Evaluation of protein models by atomic solvation preference. *J. Mol. Biol.* 225:93–105.

Huang, E.S., Koehl, P., Levitt, M., Pappu, R.V., and Ponder, J.W. 1998. Accuracy of side-chain prediction upon near-native protein backbones generated by ab initio folding methods. *Proteins* 33:204–217.

Huang, E., Subbiah, S., and Levitt, M. 1995. Recognizing native folds by the arrangement of hydrophobic and polar residues. *J. Mol. Biol.* 252:709–720.

Irving, J.A., Whisstock, J.C., and Lesk, A.M. 2001. Protein structural alignments and functional genomics. *Proteins* 42:378–382.

Ivanenkov, V.V., Meller, J., and Kirley, T.L. 2005. Characterization of disulfide bonds in human nucleoside triphosphate diphosphohydrolase 3 (NTPDase3): Implications for NTPDase structural modeling. *Biochemistry* 44:8998–9012.

Jacobson, M.P., Pincus, D.L., Rapp, C.S., Day, T.J.F., Honig, B., Shaw, D.E., and Friesner, R.A. 2004. A hierarchical approach to all-atom loop prediction. *Proteins: Struct. Funct. Genet.* 55:351–367.

Janardhan, A., and Vajda, S. 1998. Selecting near-native conformations in homology modeling: The role of molecular mechanics and solvation terms. *Protein Sci.* 7:1772–1780.

Jiang, L., Kuhlman, B., Kortemme, T.A., and Baker, D. 2005. A "solvated rotainer" approach to modeling water-mediated hydrogen bonds at protein–protein interfaces. *Proteins* 58:893–904.

Johnson, M.S., Srinivasan, N., Sowdhamini, R., and Blundell, T.L. 1994. Knowledge-based protein modeling. *Crit. Rev. Biochem. Mol. Biol.* 29:1–68.

Jung, J.W., An, J.H., Na, K.B., Kim, Y.S., and Lee, W. 2000. The active site and substrates binding mode of malonyl-CoA synthetase determined by transferred nuclear Overhauser effect spectroscopy, site-directed mutagenesis, and comparative modeling studies. *Protein Sci.* 9:1294–1303.

Karplus, K., Barrett, C., and Hughey, R. 1998. Hidden Markov models for detecting remote protein homologies. *Bioinformatics* 14:846–856.

Kelley, L.A., MacCallum, R.M., and Sternberg, M.J. 2000. Enhanced genome annotation using structural profiles in the program 3D-PSSM. *J. Mol. Biol.* 299:499–520.

Kim, D.E., Chivian, D., and Baker, D. 2004. Protein structure prediction and analysis using the Robetta server. *Nucleic Acids Res.* 32:W526–531.

Kim, C.G., Watts, J.A., and Watts, A. 2005. Ligand docking in the gastric H(+)/K(+)-ATPase: Homology modeling of reversible inhibitor binding sites. *J. Med. Chem.* 48:7145–7152.

Kingsford, C.L., Chazelle, B., and Singh, M. 2005. Solving and analyzing side-chain positioning problems using linear and integer programming. *Bioinformatics* 21:1028–1036.

Koelh, P., and Delarue, M. 1994. Application of a self-consistent mean field theory to predict protein side-chains conformation and estimate their conformational entropy. *J. Mol. Biol.* 239:249–275.

Koehl, P., and Delarue, M. 1996. Mean-field minimization methods for biological macromolecules. *Curr. Opin. Struct. Biol.* 6:222–226.

Krogh, A., Brown, M., Mian, I., Sjolander, K., and Haussler, D. 1994. Hidden Markov models in computational biology. Applications to protein modeling. *J. Mol. Biol.* 235:1501–1531.

Lazaridis, T., and Karplus, M. 1999a. Effective energy function for proteins in solution. *Proteins* 35:133–152.

Lazaridis, T., and Karplus, M. 1999b. Discrimination of the native from misfolded protein models with an energy function including implicit solvation. *J. Mol. Biol.* 288:477–487.

Lee, C. 1994. Predicting protein mutant energetics by self-consistent ensemble optimization. *J. Mol. Biol.* 236:918–939.

Lee, C., and Subbiah, S. 1991. Prediction of protein side-chain conformation by packing optimization. *J. Mol. Biol.* 217:373–388.

Lee, J., Liwo, A., Ripoll, D., Pillardy, J., and Scheraga, H. 1999. Calculation of protein conformation by global optimization of a potential energy function. *Proteins* 37 (Suppl. 3):204–208.

Lee, M.R., Baker, D., and Kollman, P.A. 2001. 2.1 and 1.8 Å average CαRMSD structure predictions on two small proteins, HP-36 and S15. *J. Am. Chem. Soc.* 123:1040–1046.

Lessel, U., and Schomburg, D. 1999. Importance of anchor group positioning in protein loop prediction. *Proteins* 37:56–64.

Levitt, M. 1992. Accurate modeling of protein coformation by automatic segment matching. *J. Mol. Biol.* 226:507.

Li, W., Liu, Z., and Lai, L. 1999. Protein loops on structurally similar scaffolds: Database and conformational analysis. *Biopolymers* 49:481–495.

Li, X., Jacobson, M.P., and Friesner, R.A. 2004. High resolution prediction of protein helix positions and orientations. *Proteins* 55:368–382.

Liang, S.D., and Grishin, N.V. 2002. Side-chain modeling with an optimized scoring function. *Protein Sci.* 11:322–331.

Liu, X., Fan, K., and Wang, W. 2004. The number of protein folds and their distribution over families in nature. *Proteins* 54:491–499.

Lolkema, J.S., and Slotboom, D.J. 1998. Estimation of structural similarity of membrane proteins by hydropathy profile alignment. *Mol. Membr. Biol.* 15:33–42.

Lovell, S.C., Word, J.M., Richardson, J.S., and Richardson, D.C. 2000. The penultimate rotamer library. *Proteins* 40:389–408.

Lu, H., and Skolnick, J., 2003. Application of statistical potentials to protein structure refinement from low resolution *ab initio* models. *Biopolymers* 70:575–584.

Maeyer, M.D., Desmet, J., and Lasters, I. 1997. All in one: A highly detailed rotamer library improves both accuracy and speed in the modelling of sidechains by dead-end elimination. *Fold. Des.* 2:53–66.

McDonald, I.K., and Thornton, J.M. 1994. Satisfying hydrogen bonding potential in proteins. *J. Mol. Biol.* 238:777–793.

McLachlan, A.D., and Shotton, D.M. 1971. Structural similarities between alpha-lytic protease of Myxobacter 495 and elastase. *Nat. New. Biol.* 229:202–205.

Mendes, J., Baptista, A., Carrondo, M., and Soares, C.M. 1999. Improvement of side-chain modeling in proteins with the self-consistent mean field theory method based on an analysis of the factors influencing prediction. *Biopolymers* 50:111–131.

Mishig-Ochiriin, T., Lee, C.H., Jeong, S.Y., Kim, B.J., Choi, C.H., Yim, H.S., and Kang, S.O. 2005. Calcium-induced conformational changes of the recombinant CBP3 protein from Dictyostelium discoideum. *Biochim. Biophys. Acta* 1748:157–164.

Mosimann, S., Meleshko, R., and James, M.N. 1995. A critical assessment of comparative molecular modeling of tertiary structures of proteins. *Proteins* 23:301–317.

Needleman, S.B., and Wunsch, C.D. 1970. A general method applicable to the search for similarities in the amino acid sequence of two proteins. *J. Mol. Biol.* 48:443–453.

Niv, M.Y., and Weinstein, H. 2005. Flexible docking procedure for the exploration of peptide binding selectivity to known structures and homology models of PDZ domains. *J. Am. Chem. Soc.* 127:14072–14079.

Nugiel, D.A., Voss, M.E., Brittelli, D.R., and Calabrese, J.C. 1995. An approach to the design of novel cognitive enhancers using molecular modeling and X-ray crystallography. *Drug Des. Discov.* 12:289–295.

Oakhill, J.S., Sutton, B.J., Gorringe, A.R., and Evans, R.W. 2005. Homology modelling of transferrin-binding protein A from Neisseria meningitidis. *Protein Eng. Des. Sel.* 18:221–228.

Orban, T., Kalafatis, M., and Gogonea, V. 2005. Completed three-dimensional model of human coagulation factor va. Molecular dynamics simulations and structural analyses. *Biochemistry* 44:13082–13090.

Petrey, D., and Honig, B. 2000. Free energy determinants of tertiary structure and the evaluation of protein models. *Protein Sci.* 9:2181–2191.

Petrey, D., and Honig, B. 2003. GRASP2: Visualization, surface properties, and electrosttics of macromolecular structures and sequences. *Methods Enzymol.* 374:492–509.

Petrey, D., Xiang, X., Tang, C.L., Xie, L., Gimpelev, M., Mitors, T., Soto, C.S., Goldsmith-Fischman, S., Kernytsky, A., Schlessinger, A., Koh, I.Y.Y., Alexov, E., and Honig, B. 2003. Using multiple structure alignments, fast model building, and energetic analysis in fold recognition and homology modeling. *Proteins Struct. Funct. Genet.* 53:430–435.

Pieper, U., Eswar, N., Ilyin, V.A., Stuart, A., and Sali, A. 2002. ModBase, a database of annotated comparative protein structure models. *Nucleic Acids Res.* 30:255–259.

Ponder, J.W., and Richard, F.M. 1987. Tertiary templates for proteins. Use of packing criteria in the enumeration of allowed sequence for different structure classes. *J. Mol. Biol.* 193:775–791.

Qian, B., Ortiz, A.R., and Baker, D. 2004. Improvement of comparative model accuracy by free-energy optimization along principal components of natural structural variation. *Proc. Natl. Acad. Sci. USA* 101:15346–15351.

Rapp, C.S., and Friesner, R.A. 1999. Prediction of loop geometries using a generalized Born model of solvation effects. *Proteins* 35:173–183.

Retief, J.D. 2000. Phylogenetic analysis using PHYLIP. *Methods Mol. Biol.* 132:243–258.

Sali, A., and Blundell, T.L. 1993. Comparative protein modelling by satisfaction of spatial restraints. *J. Mol. Biol.* 234:779–815.

Samudrala, R., Huang, E.S., Koehl, P., and Levitt, M. 2000. Constructing side chains on near-native main chains for *ab initio* protein structure prediction. *Protein Eng.* 7:453–457.

Samudrala, R., and Moult, J. 1998. Determinants of side chain conformational preferences in protein structures. *Protein Eng.* 11:991–997.

Sanchez, R., and Sali, A. 1997. Comparative protein structure modeling as an optimization problem. *J. Mol. Struct. (Theochem)* 398–399:489–496.

Sanchez, R., and Sali, A. 1998. Large-scale protein structure modeling of the Saccharomyces cerevisiae genome. *Proc. Natl. Acad. Sci. USA* 95:13597–13602.

Sauder, J.M., Arthur, J.W., and Dunbrack, R.L., Jr. 2000. Large-scale comparison of protein sequence alignment algorithms with structure alignments. *Proteins* 40:6–22.

Schindler, C., and Darnell, J.E., Jr. 1995. Transcriptional responses to polypeptide ligands: The JAK-STAT pathway. *Annu. Rev. Biochem.* 64:621–651.

Schwarzenbacher, R., Godzik, A., Grzechnik, S.K., and Jaroszewski, L. 2004. The importance of alignment accuracy for molecular replacement. *Acta Crystallogr. D Biol. Crystallogr.* 60(Pt. 7):1229–1236.

Schwede, T., Kopp, J., Guex, N., and Peitsch, M.C. 2003. SWISS-MODEL: An automated protein homology-modeling server. *Nucleic Acids Res.* 31:3381–3385.

Shakhnovich, B.E., Harvey, J.M., Comeau, S., Lorenz, D., DeLisi, C., and Shakhnovich, E. 2003. ELISA: Structure function inferences based on statistically significant and evolutionarily inspired observations. *BMC Bioinformatics* 4:34.

Shenkin, P.S., Yarmush, D.L., Fine, R.M., Wang, H.J., and Levinthal, C. 1987. Predicting antibody hypervariable loop conformation. I. Ensembles of random conformations for ringlike structures. *Biopolymers* 26:2053–2085.

Sippl, M. 1995. Knowledge-based potentials for proteins. *Curr. Opin. Struct. Biol.* 5:229–235.

Skolnick, J., Kolinski, A, and Ortiz, A.R. 1997. MONSSTER: A method for folding globular proteins with a small number of distance restraints. *J. Mol. Biol.* 265:217–241.

Smith, K.C., and Honig, B. 1994. Evaluation of the conformational free energies of loops in proteins. *Proteins* 18:119–132.

Smith, T.F., and Waterman, M.S. 1981. Identification of common molecular subsequences. *J. Mol. Biol.* 147:195–197.

Steinbach, P.J. 2004. Exploring peptide energy landscapes: A test of force fields and implicit solvent models. *Proteins* 57:665–677.

Still, W., Tempczyk, A., Hawley, R., and Hendrickson, T. 1990. Semi-analytical treatment of solvation for molecular mechanics and dynamics. *J. Am. Chem. Soc.* 112:6127–6129.

Sutcliffe, M.J., Haneef, I., Carney, D., and Blundell, T.L. 1987a. Knowledge based modelling of homologous proteins, Part I: Three-dimensional frameworks derived from the simultaneous superposition of multiple structures. *Protein Eng.* 1:377–384.

Sutcliffe, M.J., Hayes, F.R., and Blundell, T.L. 1987b. Knowledge based modelling of homologous proteins, Part II: Rules for the conformations of substituted sidechains. *Protein Eng.* 1:385–392.

Suyama, M., Matsuo, Y., and Nishikawa, K. 1997. Comparison of protein structures using 3D profile alignment. *J. Mol. Evol.* 44 (Suppl. 1):S163–173.

Talukdar, A.S., and Wilson, D.L. 1999. Modeling and optimization of rotational C-arm stereoscopic X-ray angiography. *IEEE Trans. Med. Imaging.* 18:604–616.

Taylor, W.R. 1986. Identification of protein sequence homology by consensus template alignment. *J. Mol. Biol.* 188:233–258.

Teichmann, S.A., Chothia, C., and Gerstein, M. 1999. Advances in structural genomics. *Curr. Opin. Struct. Biol.* 9:390–399.

Tometsko, A.M. 1970. Computer approaches to protein structure. II. Model building by computer. *Comput. Biomed. Res.* 3:690–698.

Topham, C.M., McLeod, A., Eisenmenger, F., Overington, J.P., Johnson, M.S., and Blundell, T.L. 1993. Fragment ranking in modelling of protein structure. Conformationally constrained environmental amino acid substitution tables. *J. Mol. Biol.* 229:194–220.

Tuffery, P., Etchebest, C., Hazout, S., and Lavery, R. 1993. A critical comparison of search algorithms applied to the optimization of protein side-chain conformations. *J. Comput. Chem.* 14:790–798.

Unger, R., Harel, D., Wherland, S., and Sussman, J.L. 1989. A 3-D building blocks approach to analyzing and predicting structure of proteins. *Proteins* 5:355–373.

Van Vlijmen, H.W., and Karplus, M. 1997. PDB-based protein loop prediction: Parameters for selection and methods for optimization *J. Mol. Biol.* 267:975–1001.

Vasquez, M. 1996. Modeling side-chain conformation. *Curr. Opin. Struct. Biol.* 6:217–221.

Vitkup, D., Melamud, E., Moult, J., and Sander, C. 2001. Completeness in structural genomics. *Nat. Struct. Biol.* 8:559–566.

Vorobjev, Y., Almagro, J., and Hermans, J. 1998. Discrimination between native and intentionally misfolded conformations of proteins: ES/IS, a new method for calculating conformational free energy that uses both dynamics simulations with an explicit solvent and an implicit solvent continuum model. *Proteins* 32:399–413.

Wallner, B., and Elofsson, A. 2005. All are not equal: A benchmark of different homology modeling programs. *Protein Sci.* 14:1315–1327.

Wang, M., and Hampson, D.R. 2005. An evaluation of automated in silico ligand docking of amino acid ligands to Family C G-protein coupled receptors. *Bioorg. Med. Chem.* 14:2030–2039.

Weber, I.T. 1990. Evaluation of homology modeling of HIV protease. *Proteins* 7:172–184.

Wojcik, J., Mornon, J.P., and Chomilier, J. 1999. New efficient statistical sequence-dependent structure prediction of short to medium-sized protein loops based on an exhaustive loop classification. *J. Mol. Biol.* 289:1469–1490.

Xiang, Z.X., Csoto, C., and Honig, B. 2002. Evaluating configurational free energies: The colony energy concept and its application to the problem of protein loop prediction. *Proc. Natl. Acad. Sci.* 99:7432–7437.

Xiang, Z.X., and Honig, B. 2001. Extending the accuracy limit of side-chain prediction. *J. Mol. Biol.* 311:421–430.

Xiang, Z., Steinbach, P., Jacobson, M.P., Friesner, R.A., and Honig, B. Prediction of side-chain conformations on protein surfaces (to be submitted).

Xu, W.P., Yuan, X.T., Xiang, Z.X., Mimnaugh, E., Marcu, M., and Neckers, L. 2005. Surface charge and hydrophobicity determine ErbB2 binding to the Hsp90 chaperone complex. *Nat. Struct. Mol. Biol.* 12:120–126.

Yang, A.S., and Honig, B. 1999. Sequence to structure alignment in comparative modeling using PrISM. *Proteins* 37(S3):66–72.

Yang, A.S., and Honig, B. 2000. An integrated approach to the analysis and modeling of protein sequences and structures. I. Protein structural alignment and a quantitative measure for protein structural distance. *J. Mol. Biol.* 301:665–678.

Zheng, Q., and Kyle, D.J. 1996. Accuracy and reliability of the scaling-relaxation method for loop closure: An evaluation based on extensive and multiple copy conformational samplings. *Proteins* 24:209–217.

Zhou, Y., and Johnson, M.E. 1999. Comparative molecular modeling analysis of 5-amidinoindole and benzamidine binding to thrombin and trypsin: Specific H-bond formation contributes to high 5-amidinoindole potency and selectivity for thrombin and factor Xa. *J. Mol. Recognit.* 12:235–241.

11 Modeling Protein Structures Based on Density Maps at Intermediate Resolutions

Jianpeng Ma

11.1 Introduction

Structural biology is now in a special era in which increasingly more complex biomolecules are being studied. For many of them, only low- or intermediate-resolution density maps (6–10 Å) can be obtained by, for instance, electron cryomicroscopy (cryo-EM) (Bottcher et al., 1997; Conway et al., 1997; DeRosier and Harrison, 1997; Kuhn et al., 2002; Li et al., 2002; Mancini et al., 2000; Zhang et al., 2000; Zhou et al., 2000, 2001a,b). In certain cases, analysis in terms of intermediate-resolution density maps is also inevitable in X-ray crystallography as exemplified in the lengthy process of structural determination of the 50S ribosomal subunit that incremented from 9 Å, 5 Å, to 2.4 Å (Ban et al., 1998, 1999, 2000). As a common feature in all these cases, it is usually impossible, with conventional methods, to construct reasonably accurate atomic models from density maps. However, for the purpose of structural analysis, it would still be very helpful if one can build some kind of pseudo-atomic models from the density maps because this will not only facilitate the structural determination to higher resolutions, but also assist further biochemical studies and functional interpretation. For example, significant insights into the architecture and organization of proteins can often be learned if one can roughly locate the major secondary structural elements such as α-helices and β-sheets. This rationale is supported by the fact that the knowledge of protein folds can be obtained primarily from the spatial arrangement of the secondary structural elements independent of the sequence identity of the proteins, as different sequences can have the same fold.

Toward this end, computational methods have recently been developed to identify α-helices (*helixhunter*) (Jiang et al., 2001) and β-sheets [*sheetminer* (Kong and Ma, 2003) and *sheettracer* (Kong et al., 2004)] in intermediate-resolution density maps. The outputs of *helixhunter*, *sheetminer*, and *sheettracer* outline the skeletons of secondary structures that describe the location of α-helices and β-strands. However, the skeletons do not contain any information of the directionality of the secondary structures and loop connectivity, i.e., the topology, or fold, of the protein remains unknown. To resolve this, we have recently developed an energetics-based procedure assisted by a complementary geometry-based analysis to effectively discriminate the native topology from the entire topology candidate pool.

Our study of topology determination supports an important hypothesis that, for a given protein skeleton, its native topology was the one chosen by evolution to accommodate the largest structural variation, not merely the one trapped in a deep, but narrow, energy well. Such a hypothesis led to the use of the average energy of an ensemble of structures, slightly randomized in the vicinity of native skeleton, as the parameter to rank the topology candidates. The ensemble-averaging scheme appears to be an effective way of compensating the inevitable errors in the artificially constructed structures and in empirical potential functions.

The contents of sections in this chapter are adopted from three seminal research papers (Kong and Ma, 2003; Kong et al., 2004; Wu et al., 2005a) with necessary modifications.

11.2 *Sheetminer*: Locating Sheets in Intermediate-Resolution Density Maps

Figure 11.1 shows the overall procedure of *sheetminer*, which does not rely on any 3D structure prediction methods. Rather, it is based on a morphological analysis of intermediate-resolution density maps, i.e., shape recognition in 3D space. One of the most important features of *sheetminer* is the flat density map on which most of the essential analyses are based. It allows one to maximally capture the elements of shape of the density maps without being severely influenced by the fluctuation of local density values. Based on their distance to the surface of the flat density map, the voxels in the flat density map are divided into two groups, surface voxels and kernel voxels. Then, for each kernel voxel, a condensation scheme is used to increase the contrast on the edge of density maps. After that, the identification of sheets is primarily achieved based on the ratio of two competing parameters, maximum disk inclusion number and minimum local thickness calculated for each kernel voxel. The identified sheet densities are then processed by a set of refinement steps before they are marked as the final output. The parameters used in *sheetminer* are chosen empirically based on exhaustive trials since there is no general rule in defining them.

This section is adapted from the original research article (Kong and Ma, 2003) from which interested readers can find more technical details.

11.2.1 Locating Sheets in Simulated Density Maps

The algorithm *sheetminer* was first tested on intermediate-resolution density maps simulated from high-resolution crystal structures. A total of 12 structurally unrelated proteins were chosen because, among them, the number, size, and shape of β-sheets vary widely and they are thus expected to reasonably represent a complete sampling of known β-sheet morphology. They are roughly split into three groups: group I protein contains a single β-sheet (Arnold and Rossmann, 1988; Hoover and Ludwig, 1997; Rees et al., 1983; Wittinghofer et al., 1991), group II contains

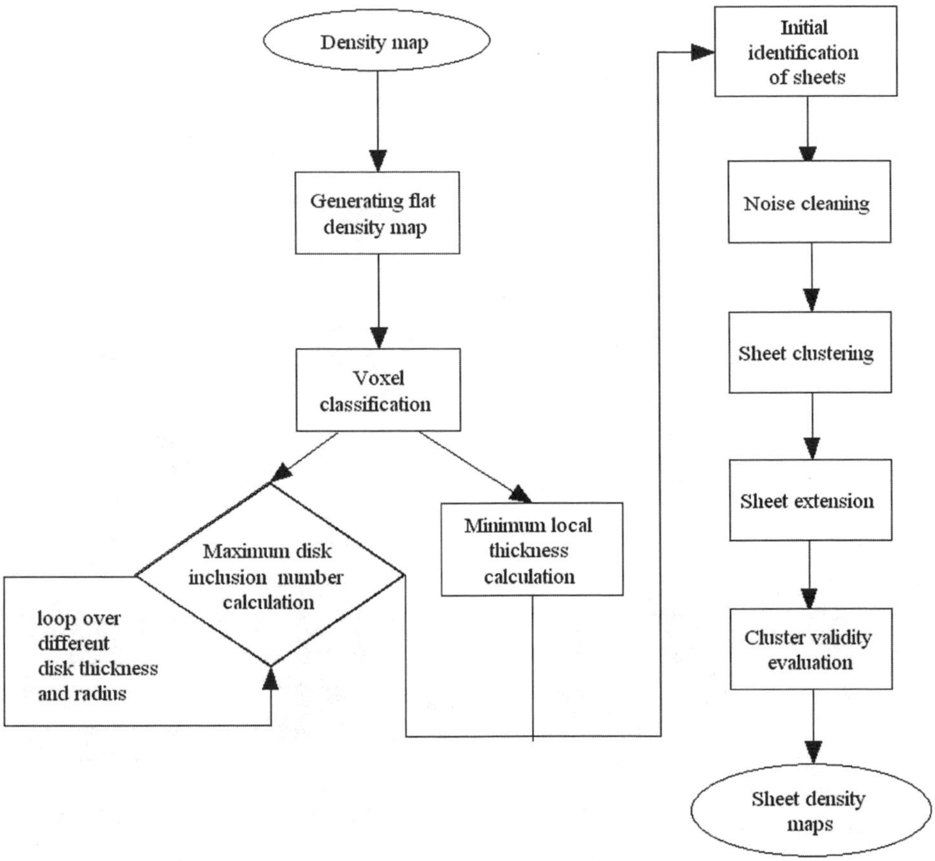

Fig. 11.1 Flowchart for the entire computational procedure of β-sheet identification in intermediate-resolution density maps implemented in *sheetminer*.

multiple independent β-sheets (Eklund et al., 1981; Khan et al., 2000; Mayer et al., 2002; Wang et al., 2000), and group III contains typical heavy β-motifs such as β-barrel and β-propeller (Gaudet et al., 1999; Steinbacher et al., 1994; Wilson et al., 1992; Zanotti et al., 1998).

11.2.1.1 Results at 8-Å Resolution

The selected PDB model was first blurred (Ludtke et al., 1999) to a resolution of 8 Å. At this resolution, visual identification of β-sheets is difficult, especially for the ones deeply buried inside proteins. Then *sheetminer* was used to identify sheet densities. In all 12 proteins tested, there were a total of 35 independent β-sheets and 34 of them were successfully located by *sheetminer* (Fig. 11.2, only three examples are shown here). One sheet was missed by *sheetminer* (circled in Fig. 11.3a) and two small regions were mistakenly identified as sheets (false positives; one such case

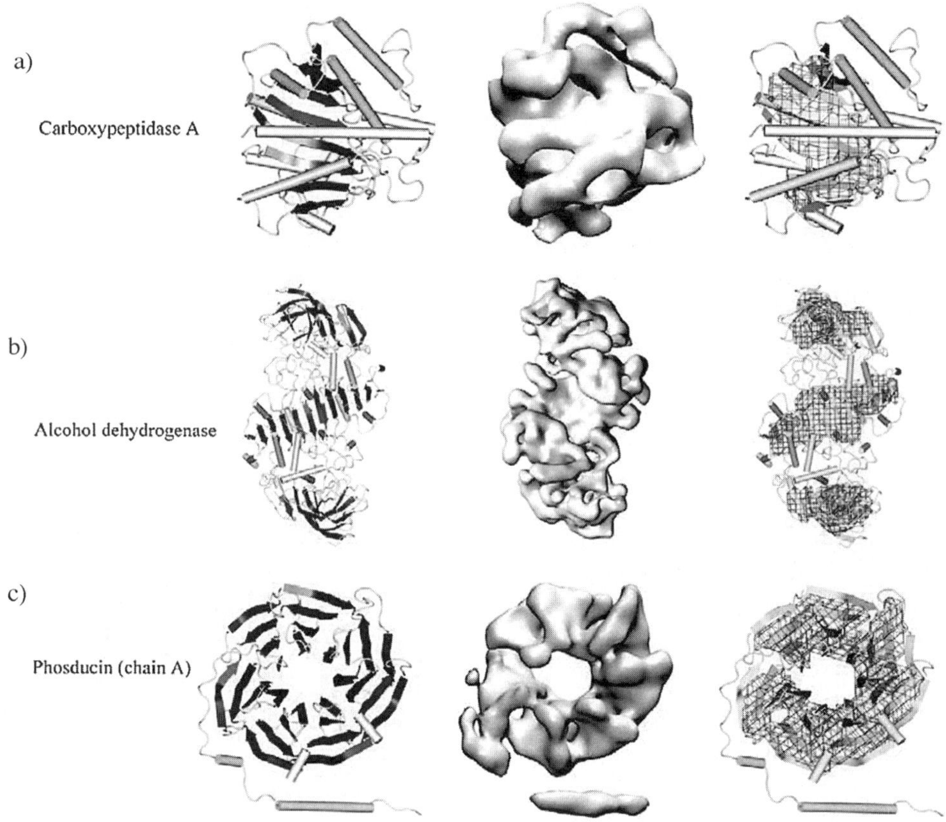

Fig. 11.2 Sheet-searching results based on simulated density maps from a total of 12 protein crystal structures. There are totally 35 independent β-sheets with a wide distribution of morphology. *Sheetminer* successfully located 34 of them and only missed one in MoFe protein of nitrogenase. The schematic ribbon diagrams on the left show the crystal structures with the β-sheets (in darker color). The middle diagrams show the blurred structures at 8-Å resolution. In diagrams on the right, the identified sheet regions are shown on top of the ribbon diagrams.

is shown in Fig. 11.3b). Therefore, *sheetminer* is very effective in defining sheet regions at this resolution.

In order to quantitatively investigate the accuracy of *sheetminer* in discerning sheet regions, we computed the values of *sensitivity* and *specificity*. Sensitivity is defined as the probability of a positive identification among voxels that are true sheet voxels, and specificity is defined as the probability of a true negative identification among the voxels that are not true sheet voxels. The average values of sensitivity and specificity are 87.1 and 73.3%, respectively. Thus, the agreement between the computationally searched sheet density maps and the real ones is very good at a resolution of 8 Å. The high value of sensitivity indicates that the method is reliable to outline the rough size of the sheet regions. The good value of specificity indicates that

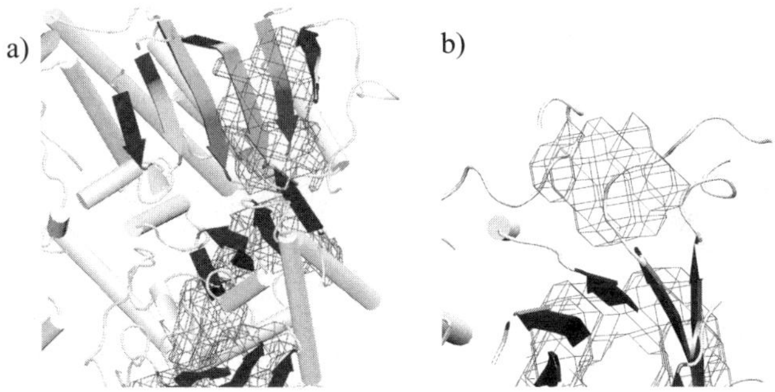

Fig. 11.3 Errors in sheet-searching results. (a) The β-sheet in MoFe protein of nitrogenase (PDB code lhl 1) that was missed by *sheetminer*. The region of the sheet is circled. This is the only one missed out of the total of 35 sheets in 12 proteins. (b) One of only two small regions that were mistakenly identified as β-sheets (false positives). The figure shows the one in aldose reductase (PDB code 1 ads).

the method seldomly predicts false positives, i.e., mistakenly identifying nonsheet voxels as sheet voxels. From a practical point of view, one would expect that the specificity would be somewhat more important than the sensitivity because it is far more important to correctly identify the overall locations of the sheets than to define the exact size of the sheets. The latter is also a variable quantity even between methods for assigning secondary structures at high resolutions.

11.2.1.2 Resolution Dependency

In our experience, *sheetminer* works best in the resolution range of 6 to 10 Å. However, the exact outcome also depends on the nature of systems: for large sheets, *sheetminer* can work to a lower resolution, but for small sheets, it is much harder with lower resolutions. The results for a typical five-stranded sheet in p21^{ras} (Wittinghofer et al., 1991) at resolutions of 6, 8, and 10 Å are shown in Fig. 11.4. At 6-Å resolution, the algorithm very accurately located not only the overall shape, but also the detailed edge of the sheet. At 8-Å resolution, the result is still satisfactory. At 10-Å resolution, the map is significantly fuzzier, but *sheetminer* was still able to find the rough location of the sheet, despite large errors on the edge. The sensitivity values are 87.2, 78.7, and 55.3%, while the specificity values are 92.5, 93.3, and 95.0%, for 6, 8, and 10 Å, respectively. The specificity seems to be much less resolution-dependent.

Certain degree of resolution dependence of the sensitivity is expected and should not undermine the applicability of *sheetminer*. The state-of-the-art cryo-EM techniques can now provide structures at intermediate resolutions and many of them are at or near resolutions of 6 to 8 Å (Zhou et al., 2001a). The program *helixhunter* (Jiang et al., 2001) also has a similar resolution dependency. Not surprisingly, the identification of β-sheets is more sensitive to resolution than is that of α-helices.

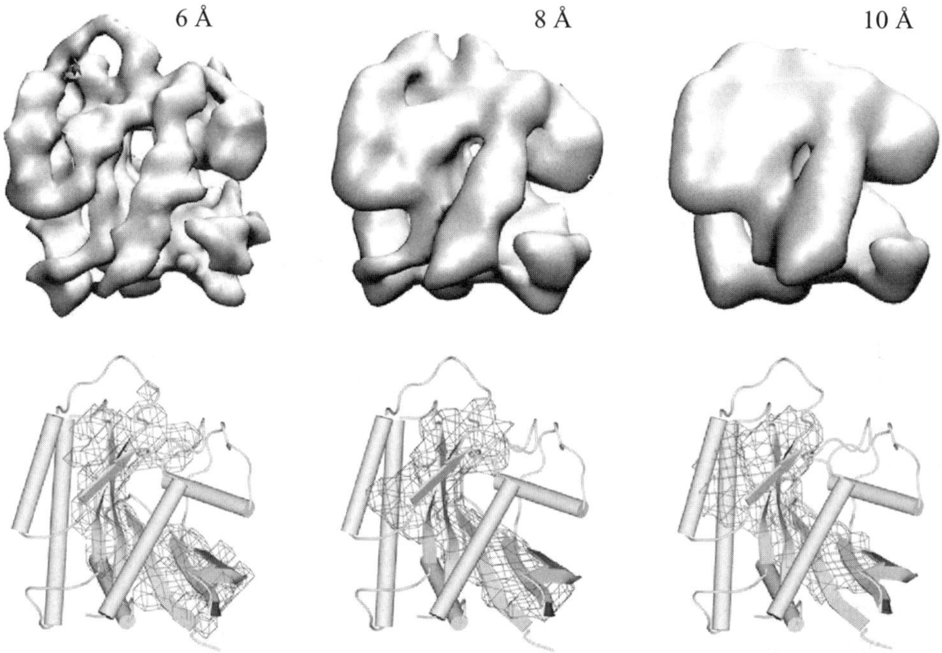

Fig. 11.4 Sheet-searching results of a typical five-stranded sheet in p21^{ras} (PDB code 121p) at resolutions of 6, 8, and 10 Å. The upper panels show the blurred structures at three resolutions and the lower panels show the corresponding results in which the found sheet density maps are superimposed on the ribbon representations of the crystal structures. At 6-Å resolution, *sheetminer* very accurately located not only the overall shape, but also the detailed edge of the sheet. At 8-Å resolution, the result is also satisfactory. At 10-Å resolution, the morphology of the density map is significantly fuzzier, but *sheetminer* was still able to identify the rough location of the sheet, despite large errors on the edges.

11.2.2 Application to Real Experimental Cryo-EM Density Maps

To test its applicability to real experimental density maps, *sheetminer* was also examined on the F41 fragment of bacterial flagellar filament that has both X-ray structure and intermediate-resolution cryo-EM structure available. The X-ray structure of the fragment of *Salmonella* flagellar filament was available at 2.0 Å (PDB code 1io1) (Samatey et al., 2001), and the cryo-EM structure of the same fragment was obtained from the 9-Å cryo-EM structure of the R-type straight flagellar filament (Mimori et al., 1995). *Sheetminer* successfully located two regions of β-sheets that encompass five out of the six β-sheets observed in the X-ray structure (Samatey et al., 2001) and missed only one isolated small two-stranded sheet (Fig. 11.5). It is worth pointing out that a 9-Å experimental cryo-EM density map is significantly noisier than a 9-Å simulated density map. Thus, the success in this case further confirmed the applicability of *sheetminer* in dealing with actual experimental data.

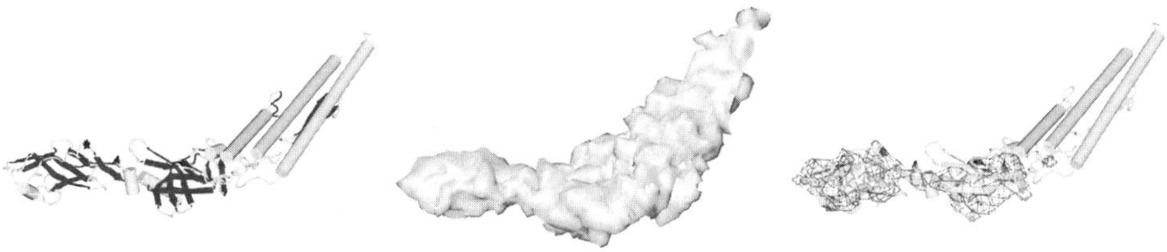

Fig. 11.5 Sheet-searching results for the F41 fragment of bacterial flagellar filament. The 2.0-Å X-ray structure of the F41 fragment of *Salmonella* flagellar is shown on the left (PDB code liol); the cryo-EM structure of the same fragment obtained from the 9-Å cryo-EM structure of the R-type straight flagellar filament is shown in the middle; and the sheet-searching results are shown on the right, superimposed on the ribbon diagram of the crystal structure. Five out of the six β-sheets observed in the crystal structure were successfully located. Only an isolated small two-stranded β-sheet was missed (behind the three longest helices).

11.2.3 Application to an 8-Å Experimental X-ray Density Map

Sheetminer was also tested on crystallographic electron density maps of equivalent resolution. An example is shown for flavodoxin (PDB code 1ag9). The X-ray electron density map was first generated from experimental diffraction data up to a resolution of 8 Å (the original structure has a resolution of 1.8 Å), and then *sheetminer* was applied to analyze the sheet density. The result is shown in Fig. 11.6 along with that from an 8-Å density map simulated based on the atomic coordinates. The overall results are similar in both cases. One important point is that, with *sheetminer*, the conventionally not-so-useful X-ray diffraction data in the resolution range of 4–8 Å can be used to extract meaningful structural information.

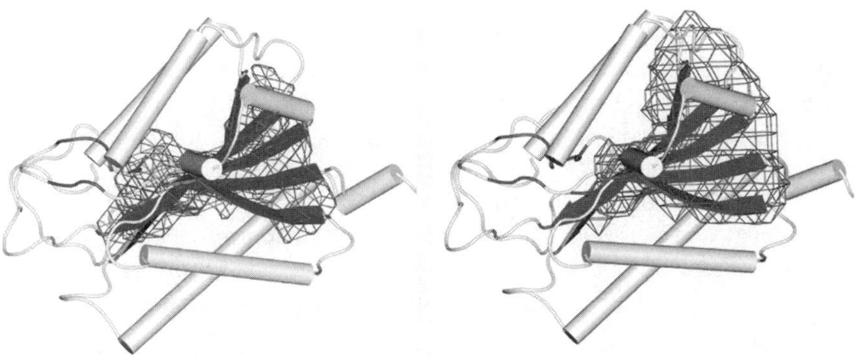

Fig. 11.6 Comparison of sheet-searching results for X-ray and simulated density maps of flavodoxin at 8 Å. The X-ray density map was generated using the experimental diffraction data up to a resolution of 8 Å (the resolution of the original X-ray structure, PDB code lag9, was 1.8 Å). The 8-Å simulated density map was obtained by artificial blurring of the X-ray structure. The sheet-searching results were superimposed on the ribbon diagrams of the crystal structure (left for simulated density map and right for X-ray density map).

11.2.4 Concluding Discussion

It is clear that *sheetminer* works better for large sheets than for smaller sheets. Usually, the locations of major sheets can be consistently found at intermediate resolutions, but the exact edges of the sheets can be fuzzy. Such an inaccuracy often makes it difficult to establish the exact length of strands even if their overall positions are well defined. Similar problems are also seen in the helix-hunting algorithm (Jiang et al., 2001). However, this should not be a severe problem in many regards because even the exact length of secondary structural elements in high-resolution X-ray structures can vary when using different assignment methods. More importantly, the identification of protein folds would be more sensitive to the overall spatial arrangement, rather than the exact length, of secondary structural elements.

The final outputs after the multistep processing of density maps by *sheetminer* are flat, but continuous, density maps corresponding to sheet regions. They could effectively narrow down the searching space for further model building into a pseudo 2D space. The algorithms for building pseudo-Cα-traces of β-sheets identified in density maps will be presented in the next section.

11.3 Sheettracer: Building Pseudo-traces for β-Strands in Intermediate-Resolution Density Maps

Sheettracer (Kong et al., 2004) is tightly coupled to the *sheetminer* method to trace individual β-strands based on the relatively thin, but continuous, sheet density maps output from *sheetminer* (Kong and Ma, 2003). Figure 11.7 shows the overall procedure of *sheettracer.* A deconvolution method was also developed to enhance the features of secondary structures in intermediate-resolution density maps.

The morphological analysis of density maps used by *sheettracer* is based on two observations: protein main-chain density is relatively higher in value than that of side chains and all neighboring β-strands are parallel or nearly parallel. The first observation enables the use of local peak-filtering to select backbone voxels, whose geometrical distribution helps define sheet morphology. The second observation facilitates local first principal component axis projection to condense the density without losing intrastrand connectivity. Differing from other thinning schemes that only consider the contacting neighbors, this local projection scheme reinforces the linear distribution of voxels but simultaneously increases the distance between voxels of different strands. This condensation results in a significantly increased efficiency in segments clustering.

We tested the methods on the simulated 6-Å density maps from 12 representative protein crystal structures, encompassing a wide range of sheet morphologies. *Sheettracer* successfully built pseudo-Cα models in the sheet densities output by *sheetminer*, with average values of 79.5%, 96.3%, and 1.54 Å for sensitivity, specificity, and rms deviations, respectively. For even lower-resolution (8 Å) simulated

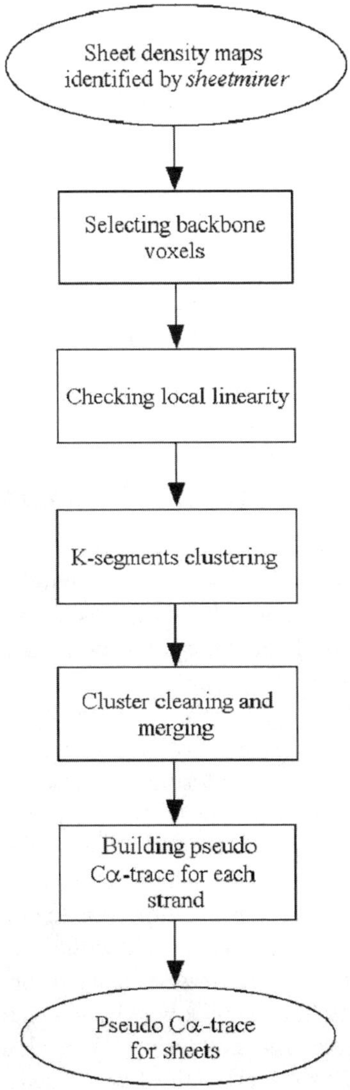

Fig. 11.7 Flowchart for the computational procedure of *sheettracer* in intermediate-resolution density maps.

data, a deconvolution method was used to permit *sheettracer* to build pseudo-Cα models with average values of 71.3%, 93.8%, and 1.77 Å for sensitivity, specificity, and rms deviations, respectively. Furthermore, *sheettracer* and the deconvolution method were also tested on experimental maps of the λ2 protein of reovirus at resolutions of 7.6 and 11.8 Å.

This section is adapted from the original research article (Kong et al., 2004) from which interested readers can find more technical details.

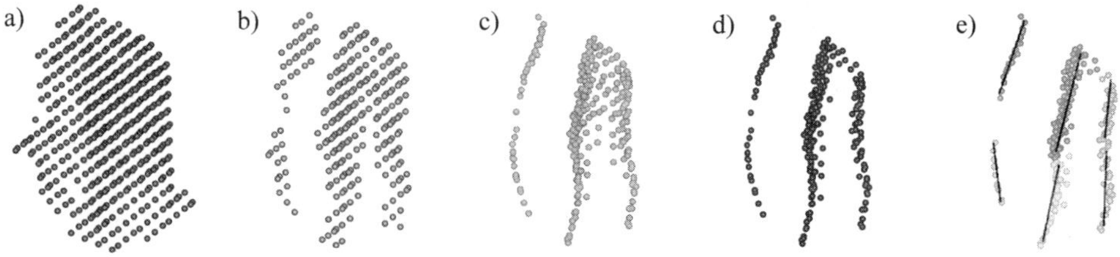

Fig. 11.8 Stepwise processing of sheet density maps to discern individual β-strands, using the sheet in the GroEL minichaperone as an example. (a) Sheet density identified by *sheetminer* shown in voxels. (b) Selected voxels by local peak filter. (c) Surviving voxels after local first principal component axis projection using the voxels in (b) as input. (d) Surviving voxels after local linearity filtering using the voxels in (c) as input. (e) Clustered backbone voxels after k-segments processing. The lines are the fitted segments (the first principal component axes).

11.3.1 Stepwise Discerning β-Strands on GroEL Minichaperone

Sheetminer (Kong and Ma, 2003) outputs clusters of voxels, each delineating a thin, but continuous volume of density representing a single β-sheet. *Sheettracer* then uses a multistep process to build pseudo-Cα-traces in each identified sheet. Here we first illustrate an example, a β-sheet of the apical domain of the molecular chaperonin GroEL, also known as the minichaperone (Wang et al., 2000) (PDB code 1fy9).

First, each cluster of voxels was processed by a local peak filter (Fig. 11.8a) to identify voxels that are most likely involved in forming the backbones of individual strands (Fig. 11.8b). The local peak-filtering algorithm enhances high *local* density values and thereby adjusts to variations in the magnitude of densities throughout the map, which permits effective selection of backbone voxels even in regions of relatively weak density. The next step was to condense the selected voxels using local first principal component axis projection. It is to enforce the voxel distribution along the longest axis that is meant to coincide with a strand backbone (Fig. 11.8c). The outcome was a significantly narrowed distribution of voxels that were then processed by a local linearity filter to pick backbone voxels with good local linearity (Fig. 11.8d). After that, k-segments clustering (Verbeek et al., 2002) was employed to group voxels into smaller subsets, each of which was to represent one part of a β-strand (Fig. 11.8e). Finally, all subsets belonging to the same strand were merged together so that each cluster of voxels represents an independent β-strand and a pseudo-Cα-trace was then built for each strand.

11.3.2 Discerning β-Strands and Building Pseudo-Cα-Traces in 12 Proteins

Sheettracer was further tested on simulated density maps of 11 other structurally unrelated proteins. They were the same set of proteins used to test *sheetminer* described

a) b) c)

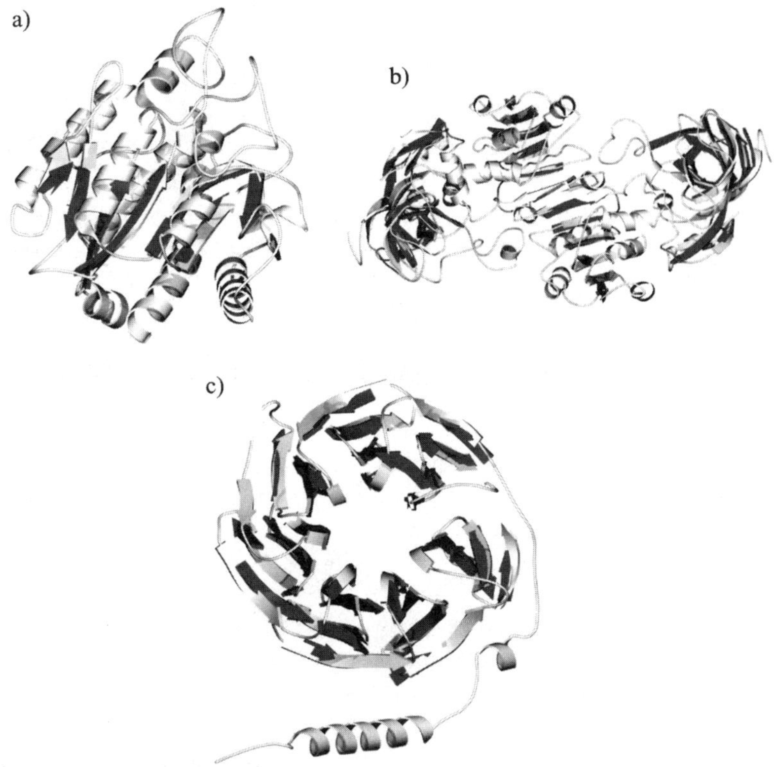

Fig. 11.9 Sheet-tracing results for all 12 proteins based on 6 Å simulated density maps. The pseudo-Cα-traces depicted in darker color are superimposed on the X-ray structures of the proteins shown in lighter color. Only one protein from each group is shown. They are (a) carboxypeptidase A; (b) horse liver alcohol dehydrogenase; (c) phosducin. The arrows in the pseudo-Cα-traces are artificially assigned based on the crystal structures.

in the previous section. Figure 11.9 shows the results with the built pseudo-Cα-traces superimposed on the crystal structures (only one example for each group is given). The results were statistically analyzed based on three separate measures: sensitivity, specificity, and rms deviations (Kong and Ma, 2003). The rms deviation was calculated as the average distance of each built pseudo-Cα-atom from its closest sheet Cα-atom in the superimposed crystal structure. The average sensitivity and specificity for the 12 proteins are 79.5 and 96.3%, respectively. The rms deviation is always smaller than 2.0 Å, with an average of 1.54 Å. Given the limited resolution, such statistical results of trace-building seem reasonable. Note that, in Fig. 11.9, strand directions were assigned according to the known X-ray structures since *sheettracer* was unable to specify them.

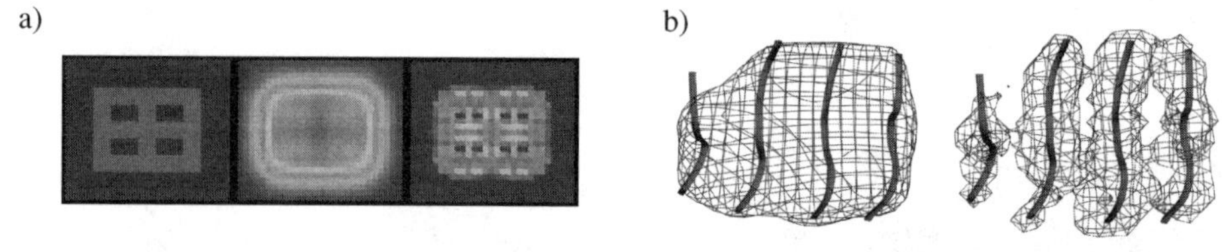

Fig. 11.10 A new deconvolution method. (a) A simple 2D example of deconvolution. The left is the original image, the middle is the image rendered with noises, and the right is the deconvoluted image. (b) A 3D example for deconvolution (right) on a piece of β-sheet density blurred to 8 Å (left). The Cα-traces of the sheet (red) are superimposed on the density.

11.3.3 A New Method for Deconvolution of Density Maps

In order to enhance the features of secondary structural elements in density maps for building pseudo-Cα-traces, we developed a new deconvolution method. An example is shown in Fig. 11.10a. A synthetic 2D geometrical object (left panel) was contaminated with a level of noise that nearly renders features in the original object indistinguishable (middle panel). Deconvolution resulted in a dramatic recovery of object features.

The method was then tested on a simulated 3D density map blurred to 8 Å (Fig. 11.10b, left). After deconvolution, strands are better resolved (Fig. 11.10b, right) and the subsequent building of pseudo-Cα-traces on the deconvoluted map became trivial.

The deconvolution method was then examined on experimental density maps of the λ2 protein of reovirus. In order to have a more systematic and self-consistent test, the cryo-EM structures of reovirus were purposely reconstructed to 7.6-Å resolution from 7939 single particle images (100%-particle structure) and to 11.8-Å resolution from a subset (12.5%) of the same particle images (12.5%-particle structure). Two helices that are distinct in the 100%-particle structure (Fig. 11.11a) are bridged by density that interconnects the helices in the 12.5%-particle structure (presumably owing to the higher level of noise) (Fig. 11.11b). The deconvolution of the 12.5%-particle structure yields a map with distinct densities for the helices (Fig. 11.11c).

11.3.4 Deconvolution and Trace Building in Simulated Density Maps of 12 Proteins

As demonstrated in previous sections, *sheettracer* can build pseudo-Cα-traces in simulated maps at resolutions as low as 6 Å. To increase its effectiveness, the deconvolution method was combined with *sheettracer* to trace strands at lower resolutions. Figure 11.12 shows the results of tracing with simulated maps of $p21^{ras}$ at 8 and 9 Å. The sensitivity, specificity, and rms deviations are 76.6%, 98.3%, and 1.65 Å and 70.2%, 96.7%, and 1.73 Å for 8 and 9 Å, respectively. The methods were then applied to simulated density maps of the other 11 proteins at 8 Å, and resulting average sensitivity, specificity, and rms deviations are 71.3%, 93.8%, and 1.77 Å, respectively.

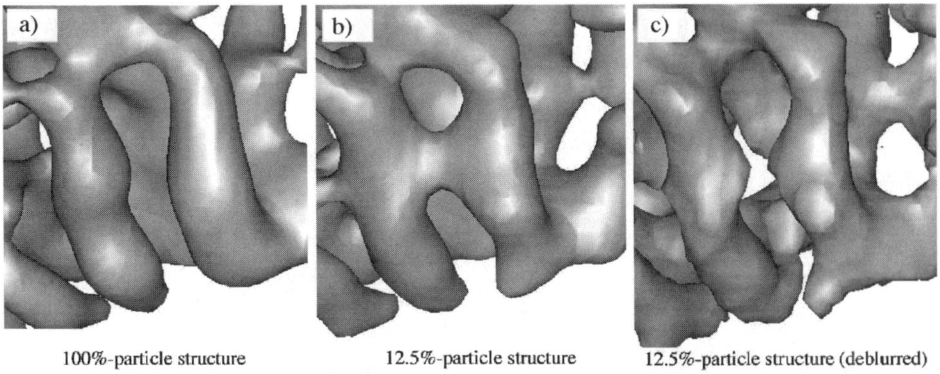

Fig. 11.11 The improved appearance of secondary structural elements in the experimental density map of the λ2 protein of reovirus by the deconvolution. (a) The cryo-EM structure generated using 100% particle images (100%-particle structure) highlighting the two well-separated helices. (b) The structure generated using 12.5% particle images (12.5%-particle structure) in which the two distinct helices are wrongfully connected. (c) The deconvolution procedure recovered the separation of these two helices in the 12.5%-particle structure.

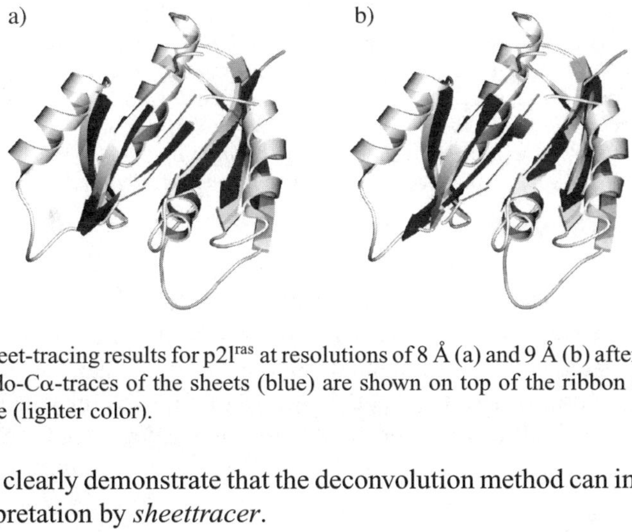

Fig. 11.12 Sheet-tracing results for p21^{ras} at resolutions of 8 Å (a) and 9 Å (b) after deconvolution. The built pseudo-Cα-traces of the sheets (blue) are shown on top of the ribbon diagrams of the crystal structure (lighter color).

These results clearly demonstrate that the deconvolution method can indeed enhance density interpretation by *sheettracer*.

11.3.5 Deconvolution and Trace Building in Experimental Maps of Reovirus λ2 Protein

To test *sheetracer* and the deconvolution method on real experimental data, we used the 7.6-Å cryo-EM structure of the λ2 protein of reovirus (Zhang et al., 2003), the crystal structure of which has been solved independently (Reinisch et al., 2000) (PDB code 1ej6) and could be used to validate the sheet-tracing results. The λ2 protein has 16 β-sheets, 12 of which contain three or more strands. The results of

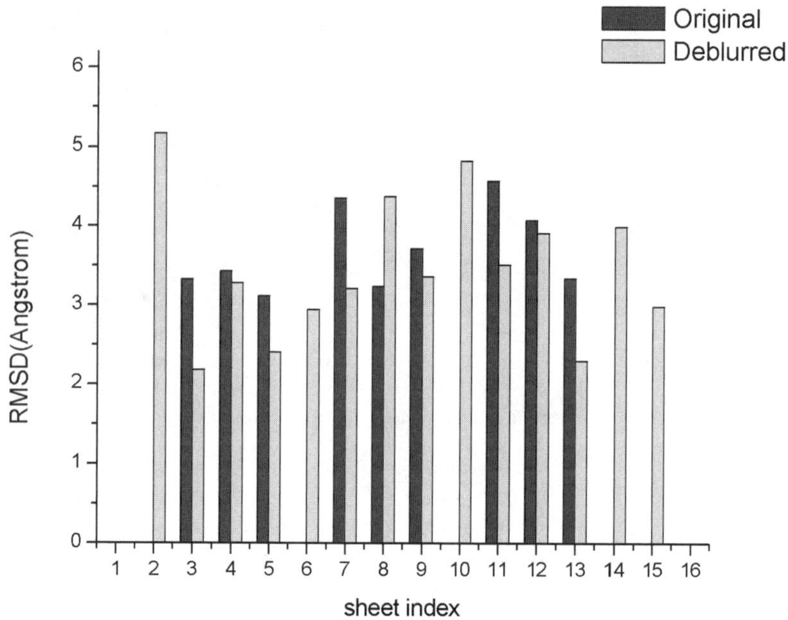

Fig. 11.13 Comparison of sheet-tracing results in the 7.6-Å density maps of the λ2 protein of reovirus with (yellow bar) and without (blue bar) deconvolution. There are a total of 16 β-sheets, 12 of which are large (three-stranded or more) and 4 are small (short two-stranded). In all cases except one (sheet 8), the deconvolution resulted in smaller rms deviations relative to the crystal structure than without. Moreover, the deconvolution brought up 5 additional β-sheets (sheets 2, 6, 10, 14, and 15) for which no pseudo-Cα-traces could be built on the original maps without deconvolution.

building pseudo-Cα-traces with and without deconvolution are shown in Fig. 11.13. Except sheet 8, the deconvoluted maps always have better rms deviations of pseudo-Cα-traces compared with those obtained from the original map. Deconvolution also improved five additional sheets (sheets 2, 6, 10, 14, and 15) for which pseudo-Cα-traces could not be built from the original maps before deconvolution.

11.3.6 Discussion

Not surprisingly, the accuracy of tracing generated by *sheettracer* depends at least in part on the reliability of *sheetminer* because the input to *sheettracer* consists of sheet density maps identified by *sheetminer* from raw density maps. Usually, the sensitivity of tracing is closely coupled to the performance of *sheetminer*, but the specificity of tracing is not and is always quite good. Moreover, similar to *sheetminer*, the size of β-sheets also affects the performance of *sheettracer*. *Sheetracer* naturally performs better when the sheets are large and the strands are long because errors tend to occur near the edges of β-sheets.

Our results have shown that the deconvolution method significantly enhances one's ability to build pseudo-Cα-traces for β-strands at relatively low resolutions. However, it is hard to objectively and quantitatively measure the improvement on the effective resolution brought by the deconvolution method.

Computational methods for identifying and tracing secondary structural elements in intermediate-resolution density maps should be valuable for several reasons. First, the pseudo-Cα-traces will facilitate more biochemical and functional studies and will help structure refinement at higher resolutions. Second, the combination of *sheettracer* with other related computational methods (Elofsson et al., 1996; Jiang et al., 2001; Lu et al., 2002; Miller et al., 1996; Skolnick et al., 2001) will eventually make it possible to reveal protein folds from data at intermediate or lower resolutions. Third, the secondary structural elements established by *sheettracer* and related methods (Jiang et al., 2001) can provide guiding landmarks for docking atomic models of sub-components or homology-derived models into intermediate-resolution density maps. Accuracy of rigid-body docking should be significantly improved if even just a few points *inside* a density map can be reliably identified (Rossmann, 2000).

With *sheetminer* and *sheettracer*, the secondary structural skeletons can be deduced from intermediate-resolution density maps, but the topology, or the fold, remains unknown. Topology determination is the topic of the next section.

11.4 Determining Protein Topology Based on Skeletons of Secondary Structures

The output from programs like *helixhunter* (Jiang et al., 2001), *sheetminer* (Kong and Ma, 2003), and *sheettracer* (Kong et al., 2004) gives the locations of α-helices and β-strands, i.e., the skeleton of secondary structures. But it does not contain any information of the directionality of the secondary structures and loop connectivity, i.e., the topology of structure is undetermined. The next question is naturally how to determine the topology, or fold, based on skeletons of secondary structures. This is a very difficult problem since there are a large number of ways to connect the secondary structural elements for a given skeleton (Fig. 11.14), among which only one is the native topology selected by evolution.

In order to discriminate the native topology from all other topology candidates, we developed an energetics-based procedure in which sequence information was first mapped onto the modeled Cα-traces and then a knowledge-based pairwise potential function (Bahar and Jernigan, 1997) was employed for energetic evaluation. To make the energetics-based procedure more effective, we also developed a complementary geometry-based analysis, based on knowledge extracted from high-resolution protein structure database, to improve the initial screening.

The empirical potential functions used in our study are very approximate. The structures constructed around the native skeleton evidently carry large errors regardless of the extensive optimization. This is particularly true for loops that were

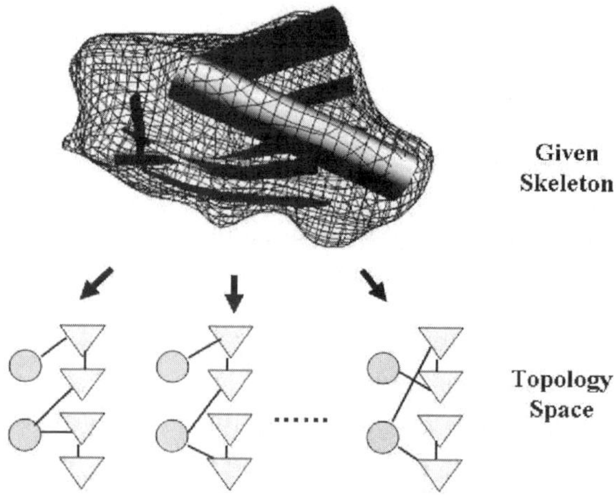

Fig. 11.14 Schematic representation of protein topology space. For a given secondary-structural skeleton, there are a large number of possible topology candidates associated with it. Together they form a topology space. In the figure, the skeleton is depicted in such a way that helices are drawn as cylinders and strands are drawn as ribbons. In the schematic diagrams, the circles are for α-helices and the triangles are for β-strands.

essentially built arbitrarily (although it was found that inclusion of loops was critical for covering a substantial portion of hydrophobic surfaces). Consequently, it is impossible to distinguish the native topology by the energy value of a single constructed structure because the energy of an individual structure of a nonnative topology can frequently be lower than that of an individual structure of the native topology. To solve this issue, we adopted a major working hypothesis that the native topology of a given protein skeleton is the one chosen by evolution to accommodate the largest structural variation, not merely the one trapped in a deep, but narrow, energy well. From such a hypothesis, one can deduce that the average energy of an ensemble of structures varying in the vicinity of the native skeleton should be the lowest, and the standard deviation of the average energy should be the smallest. Our results seem to support the hypothesis well. Another implication is that, in structural prediction, the ensemble-averaging scheme is an effective way for compensating the inevitable errors in the artificially constructed structures and in empirical potential functions.

We first examined the method on secondary-structural skeletons of 50 medium-sized single-domain proteins, among which 25 were all-helical proteins and 25 were sheet-containing proteins. We also tested the method on skeletons in which one or more secondary structures were purposely removed in order to examine the ability of the method to cope with the mismatches between secondary structdures extracted from density maps and those predicted from sequence. Finally, an eight-stranded skeleton obtained from an experimental 7.6-Å cryo-EM density map was also analyzed by the method. In most cases, the native topology was successfully identified as the most energetically favorable topology. Thus, our results suggest that it is indeed possible to derive protein native topology from secondary-structural skeletons.

This section is adapted from the original research article (Wu et al., 2005a) from which interested readers can find more technical details.

11.4.1 Secondary Structure Prediction Based on Protein Primary Sequence

The results of secondary structure prediction showed 12 out of 50 proteins with mismatches in the number of secondary structures between skeletons and assignments. Consensus evaluation had to be used for successful assignment in three cases (3icb, 1a1w, and 1d1l). The other nine proteins with mismatches are marked with asterisks in Tables 11.1 and 11.2.

For secondary structure assignment for an eight-stranded sheet of the λ2 protein of reovirus, PSIPRED was first employed and gave significant mismatch with the

Table 11.1 Results on 25 single-domain all-helical proteins

PDB ID	Total residues	N_α^1	Possible topologies[2]	Accessible topologies[3]	Native rank geometry (Method I)[4]	Native rank geometry (Method II)[5]	Topologies used for energetics	Native rank energetics (mean)[6]	Native rank energetics (median)[7]
1erc*	40	3	48	30	5th	5th	26	4th	4th
1mbg	40	3	48	20	1st	1st	20	1st	1st
2ezh	65	4	384	28	5th	4th	28	2nd	2nd
1a32	85	4	384	2	2nd	1st	2	1st	1st
1utg	70	4	384	6	5th	3rd	6	1st	1st
1mho	88	4	384	64	7th	26th	22	1st	1st
1nol	66	4	384	22	10th	14th	22	1st	1st
1i2t	61	4	384	4	1st	1st	4	1st	1st
1eo0	76	4	384	148	18th	25th	40	1st	1st
1lpe	144	5	3840	8	4th	4th	8	1st	1st
1vls	146	5	3840	12	2nd	1st	12	1st	1st
1aep	153	5	3840	15	15th	5th	15	1st	1st
1bz4	144	5	3840	4	2nd	2nd	4	1st	1st
1nkl	78	5	3840	49	5th	1st	26	1st	1st
3icb	75	5	3840	48	2nd	1st	24	1st	1st
2psr*	95	5	1920	960	2nd	4th	40	5th	7th
1l0i*	77	5	23040	58	6th	9th	34	8th	7th
2cro	65	5	3840	1544	184th	227th	200	12th	10th
2asr	142	5	3840	21	2nd	6th	21	1st	1st
1g7d	101	5	3840	15	6th	6th	15	1st	1st
1abv	105	6	46080	166	8th	8th	56	2nd	2nd
1a1w	83	6	46080	113	8th	9th	35	1st	3rd
1c15	94	6	46080	12	4th	4th	12	1st	1st
1ngr	74	6	46080	32	5th	5th	24	2nd	2nd
1bvc	153	8	10321920	14	1st	2nd	14	1st	1st

[1] N_α is the total number of α-helices in the crystal structures. [2] The total possible topologies. [3] The total accessible topologies were the number of topologies surviving through the initial screening. [4] Rank of the native topology among all accessible topologies using geometry analysis Method I. [5] Rank of the native topology among all accessible topologies using geometry analysis Method II. [6] Rank of the native topology among all accessible topologies by energetics approach and ranked according to arithmetic mean. [7] Rank of the native topology among all accessible topologies by energetics approach and ranked according to median.

Table 11.2 Results on 25 sheet-containing proteins

PDB ID	Total residues	N_α^1	N_β^2	Possible topologies[3]	Accessible topologies[4]	Native rank geometry Method I[5]	Native rank geometry Method II[6]	Native rank geometry Method III[7]	Topologies used for energetics	Native rank energetics (mean)[8]	Native rank energetics (median)[9]
1igd	61	1	4	768	22	2nd	3rd	3rd	22	1st	1st
1em7	56	1	4	768	40	1st	1st	2nd	13	1st	1st
1h0y	89	2	4	3072	36	2nd	2nd	2nd	17	1st	1st
1ctf*	68	3	3	4608	23	4th	2nd	2nd	23	15th	16th
1d1l	61	3	3	2304	6	3rd	1st	2nd	6	1st	2nd
1p1l	102	3	4	18432	7	1st	1st	1st	7	1st	2nd
1cm2	85	3	4	18432	18	1st	2nd	1st	5	1st	1st
1h75	76	3	4	18432	109	9th	13th	2nd	21	1st	1st
1lba	145	3	5	184320	640	27th	23rd	23rd	31	1st	1st
3fx2*	147	4	5	737280	3476	54th	300th	220th	44	1st	1st
1rlk	116	4	5	1474560	22	1st	3rd	3rd	12	1st	1st
1thx	108	4	5	1474560	97	7th	3rd	3rd	12	1st	1st
1rrb*	76	2	5	7680	1712	4th	1st	1st	14	1st	1st
1orc*	59	3	3	1152	8	1st	1st	1st	8	1st	1st
2hpr	87	3	4	18432	76	1st	1st	1st	5	2nd	2nd
1eof	100	5	4	1474560	28	6th	6th	16th	20	2nd	2nd
1ubq	76	2	5	30720	56	1st	1st	1st	7	3rd	2nd
1ck2	104	5	4	1474560	276	42nd	42nd	2nd	44	3rd	3rd
1aba	87	3	3	2304	640	87th	87th	53rd	88	6th	3rd
1e0n	27	0	3	48	16	1st	1st	1st	8	1st	1st
1mjc	69	0	5	3840	1872	9th	7th	7th	13	1st	1st
1fna	91	0	7	645120	74	7th	11th	12th	7	1st	1st
1tpm*	50	0	5	3840	192	18th	14th	9th	20	8th	9th
1ten	89	0	7	645120	1728	10th	10th	10th	19	1st	1st
3ait*	74		6	23040	2352	3rd	11th	23rd	40	1st	1st

There are a total of 25 sheet-containing proteins tested, the first 19 are alpha-beta-mixed proteins, and the last 6 are all-beta proteins. [1] N_α is the total number of α-helices in the crystal structures. [2] N_β is the total number of β-strands in the crystal structures. [3] All possible topologies. [4] Accessible topologies were the number of topologies surviving through the initial screening allowing a maximal 50% variation for α-helices and 33.3% for β-strands. [5] Rank of the native topology among all accessible topologies using geometry analysis Method I. [6] Rank of the native topology among all accessible topologies using geometry analysis Method II. [7] Rank of the native topology among all accessible topologies using geometry analysis Method III. [8] Rank of the native topology among all input accessible topologies by energetics approach and ranked according to arithmetic mean. [9] Rank of the native topology among all input accessible topologies by energetics approach and ranked according to median.

skeleton modeled based on an experimental 7.6-Å cryo-EM density map. Then, a consensus approach was employed. Among all of the methods, DSC (King and Sternberg, 1996) resulted in an assignment that matched with the skeleton from cryo-EM data. This assignment was used to align with the skeleton.

11.4.2 Packing Geometry of Two Consecutive Secondary Structures

To study the geometrical packing preference between two consecutive secondary structures (α-helices or β-strands), 1084 nonhomologous protein structures with resolutions better than 1.8 Å compiled in PISCES (Wang and Dunbrack, 2003) were examined. Three parameters, θ_1, θ_2, and φ, were employed to express the relative arrangement of two secondary structures and their connecting loop (Fig. 11.15a). Figure 11.15b shows the distribution of dihedral angle φ. It resembled a Gaussian distribution with a peak near zero, which suggests that the majority of two consecutive secondary structures are arranged in a plane with a *cis*-configuration. The ridge of the distribution of two packing angles, θ_1 and θ_2, was along the diagonal line from the lower-right corner to upper-left corner, with a sum $\Theta = \theta_1 + \theta_2$ of π (Fig. 11.15c). This indicates that the two consecutive secondary structures have a strong tendency to be antiparallel. When the distribution was plotted against Θ and φ (Fig. 11.15d), the dihedral angle was found to be centered at approximately zero and the sum of the two packing angles was centered around π. Two other methods were also used to analyze packing geometry, and similar statistics was obtained (Fig. 11.16). These statistical data serve as the basis of the geometry scoring function in geometry filter.

11.4.3 All-Helical Proteins

We examined 25 all-helical proteins (Table 11.1). They contain two major types of architecture of all-helical proteins with a single domain: up-down bundle and orthogonal bundle (Orengo et al., 1997), and represent 14 types of topology (three proteins do not have classified architecture and topology).

Geometry approach. In the geometry analysis, three proteins have their native topologies ranked as 1st, and 19 other proteins have their native topologies ranked within the top ten, and only one (PDB code: 2cro) has its native topology ranked as 184th (Table 11.1, sixth column). We further examined this particular protein and found that it has a highly globular structure and almost all of the helices are similar in length. These features resulted in a large number of accessible topology candidates that survived the initial screening. In sharp contrast, myoglobin (PDB code: 1bvc) has wider variations in the length of helices and loops, and, as a consequence, it dramatically narrowed down the accessible topology to 14 in the initial screening out of the total of 10^7 possible topologies.

Energetics approach. Table 11.1 (columns 9 and 10) illustrates the results of energetics analysis on these 25 proteins, which was performed after geometry filter. A

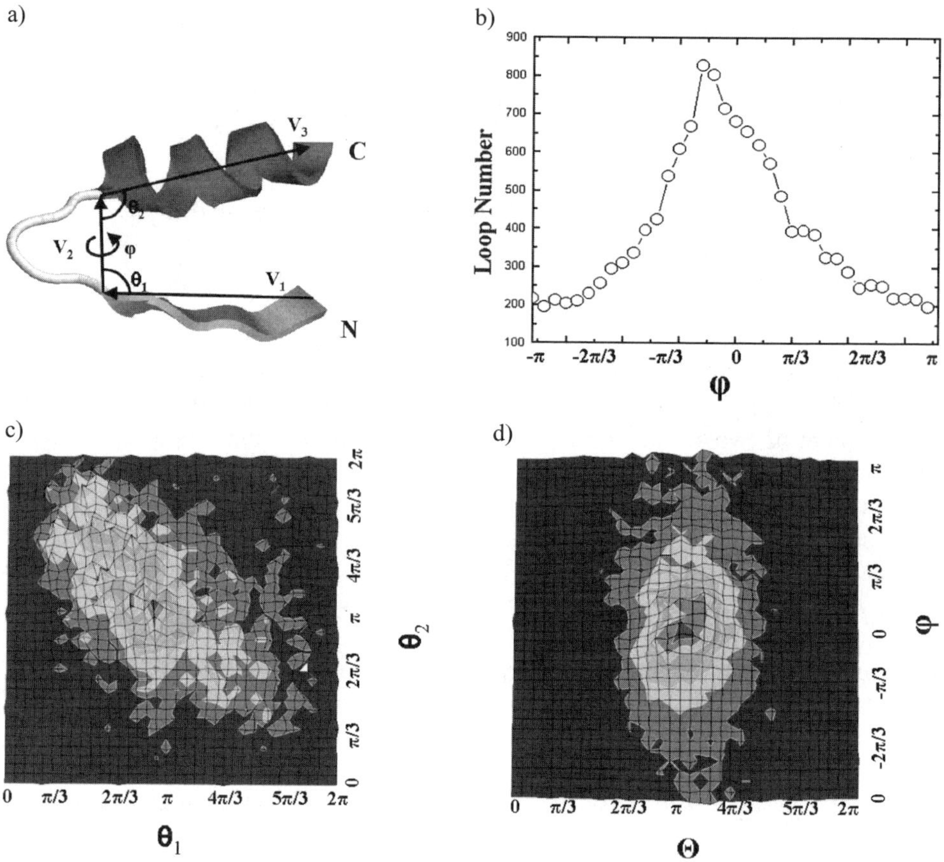

Fig. 11.15 Geometry of two consecutive secondary structures connected by a loop (Method I). (a) Three parameters, θ_1, θ_2, and φ, were used to describe the relative arrangement of the two consecutive secondary structures connected by a loop. For an α-helix, it is represented by a vector of the axis of the cylinder directed from the N-terminus to the C-terminus. For a loop or β-strand, the vector runs from the first Cα-atom to the last Cα-atom of the loop or strand. Based on these three vectors, we defined the packing angle θ_1 between vectors V1 and V2, packing angle θ_2 between vectors V2 and V3, and the dihedral angle φ formed by the three vectors. (b) The distribution of loops as a function of the dihedral angle φ. The curve resembles a Gaussian distribution with a peak near zero. (c) Two-dimensional contour representation of the distribution of angles θ_1 and θ_2. The ridge is along the diagonal line. The loops included in this calculation are within the dihedral values between $-\pi/6$ and $\pi/6$ around the peak of the Gaussian profile shown in Fig. 11.15b. (d) Two-dimensional contour representation of the distribution of Θ and φ. The dihedral angle is clearly centered at approximately zero and the sum of the two packing angles is centered at around 180°.

cutoff was used so that all topology candidates above the cutoff were used as input for energetic analysis. The native topologies of 18 proteins were successfully found to be of the lowest average energy (ranked as 1st), which is a reasonably high successful rate. It is worth pointing out that the number of randomly perturbed structures in

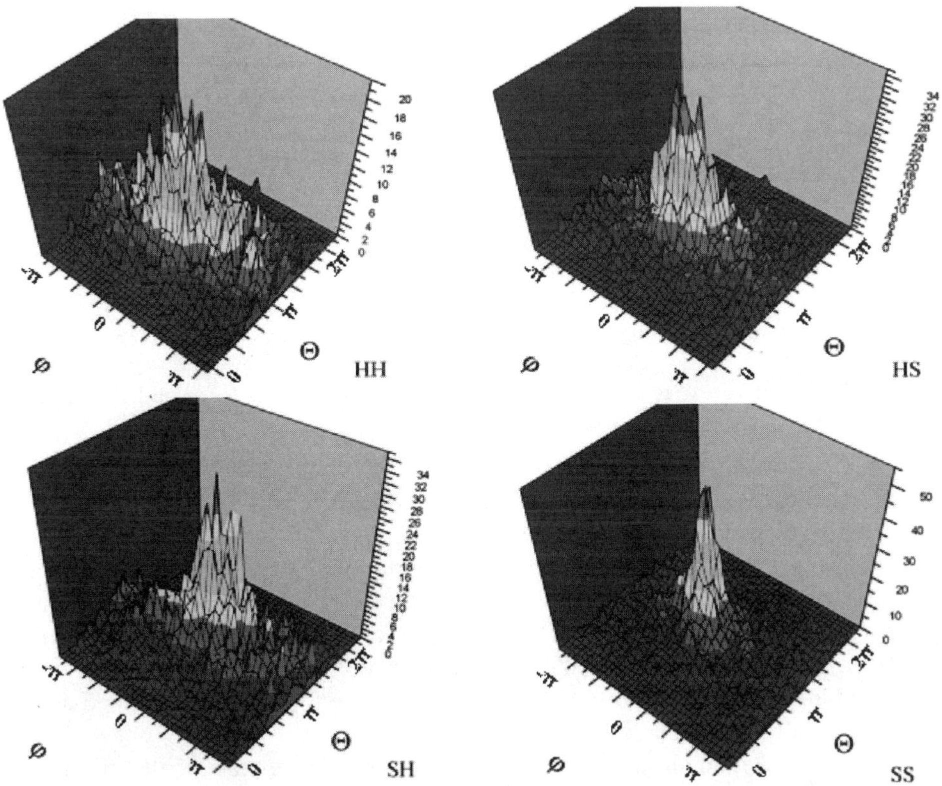

Fig. 11.16 Geometry of two consecutive secondary structures connected by a loop (Method II). The distributions of angles Θ and φ for loop motifs of helix–helix (HH), helix–strand (HS), strand–helix (SH), and strand–strand (SS) are shown separately.

the ensemble of each accessible topology differs significantly due to the fact that, for some of the energetically unfavorable topology candidates, it was much harder to generate perturbed structures around the given skeleton that satisfied all of the criteria. A correlation is found between the number of perturbed structures and the average energy.

For the remaining seven proteins, three (2ezh, 1abv, and 1ngr) have their native topology recognized as the 2nd lowest in average energy (ranked as 2nd in the ninth column of Table 11.1). The topology of the lowest energy (1st) is very similar to the native topology (2nd) in all cases. Figure 11.17 schematically illustrates the three lowest energy topologies for protein 2ezh. The difference between the 1st and the native topology (2nd) was a swap of two helices that are very similar in length and nearby in space.

The native topologies of two other proteins, 2psr and 1l0i, were ranked 5th and 8th, respectively. Mismatches did happen between the assignment and skeleton for both cases: two helical regions were predicted as one in the assignment of 2psr, and

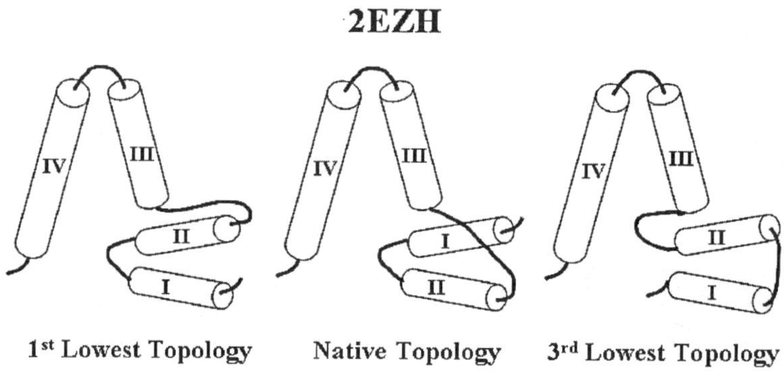

Fig. 11.17 Comparison of the three lowest-energy topology candidates for an all-helical protein 2ezh whose native topology was ranked the 2nd lowest.

one extrahelical region was predicted for 1l0i, which obviously influence the ranks of the native topologies of these two proteins.

In the case of 2cro, due to the length similarity of all five helices of this protein that gave rise to the largest number of accessible topology candidates in the initial screening, the native topology was ranked as the 12th lowest in average energy. However, despite the large error, the methods were very effective in narrowing down the searching space of possible topologies (the native topology was ranked as the 12th among all 1544 accessible candidates).

Finally, the median energy value of the ensemble, instead of the arithmetic mean, was computed to rank the topology candidates. This was to cross-validate the errors in our ranking procedure resulting from the non-Boltzmann random sampling in generating the perturbed structure ensemble. Mathematically, the median indicates a true average in the absence of a *priori* knowledge of data distribution. The results, shown in the 10th column of Table 11.1, are very consistent with those ranked according to arithmetic mean (9th column), indicating the fidelity of the ranking procedure.

11.4.4 Sheet-Containing Proteins

Geometry approach. We had a total of 19 alpha-beta-mixed proteins and 6 all-beta proteins. The 19 alpha-beta-mixed proteins contain three different types of architecture and seven types of topology [3 proteins do not have classified architecture or topology in CATH (Orengo et al., 1997)]. The 6 all-beta proteins contain three types of architecture and three types of topology (1 protein does not have classified architecture or topology). The seventh column of Table 11.2 shows the results of geometry analysis. In seven cases, the native topology was ranked as the lowest energy (1st) and in eight other cases for alpha-beta-mixed proteins was ranked within the top 10. In the all-beta cases, the native topology of one was ranked as the 1st and four others within the top 10.

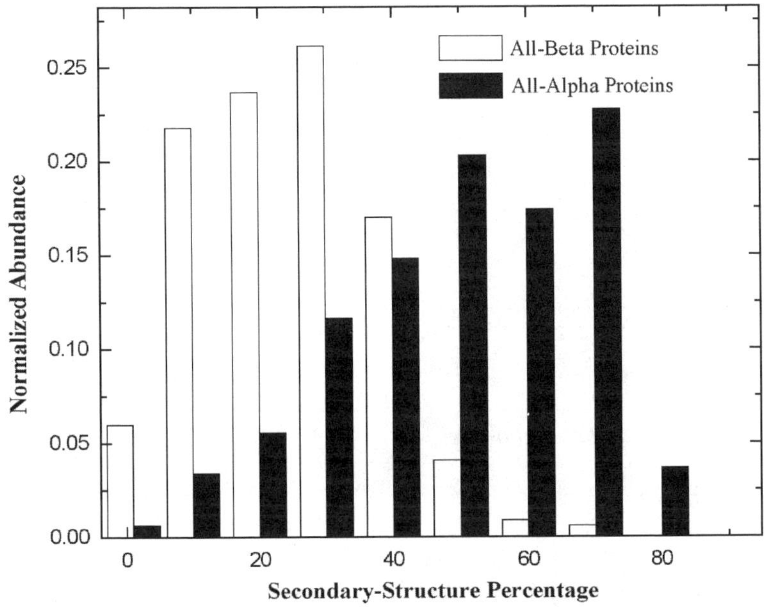

Fig. 11.18 Comparison of secondary-structural content in all-helical proteins versus all-beta proteins. It is clear that all-beta proteins have much lower secondary-structural content than all-helical proteins.

There are several reasons why the search for sheet-containing topology is more difficult. All-beta proteins have an overall lower percentage of secondary structures and higher percentage of loop regions compared with all-helical proteins (Fig. 11.18). Therefore, they have increased the complexity of the topology space, i.e., fewer topology candidates can be filtered out in the initial screening. Moreover, α-helices have more rigid structures with strong local interactions, while β-strands can bend and twist, and also involve long-range stabilizing interactions.

Energetics approach: Table 11.2 shows the results of energetics approach on the 25 sheet-containing proteins. All of the accessible topologies were initially screened by geometry analysis. Both the arithmetic mean and median of the energy were used as ranking criteria to avoid sampling bias. The final ranks of the 25 proteins are shown in columns 11 and 12, respectively, of Table 11.2. The results from the two ranking methods are quite similar. Totally, the native topologies of 18 out of 25 proteins have their average energy ranked the lowest. For the remaining 7 proteins, 2 have their native topologies ranked as the 2nd-lowest average energy and 2 others as 3rd. In all of these 4 cases, the difference between the lowest-energy topology (1st) and the native topology (2nd) was an exchange of two secondary structures with similar length and symmetric spatial location or the shift of the direction of certain strands in the skeleton. As an example, Fig. 11.19 compares the lowest-energy topology (1st) and the native topology (2nd) of protein 1eof.

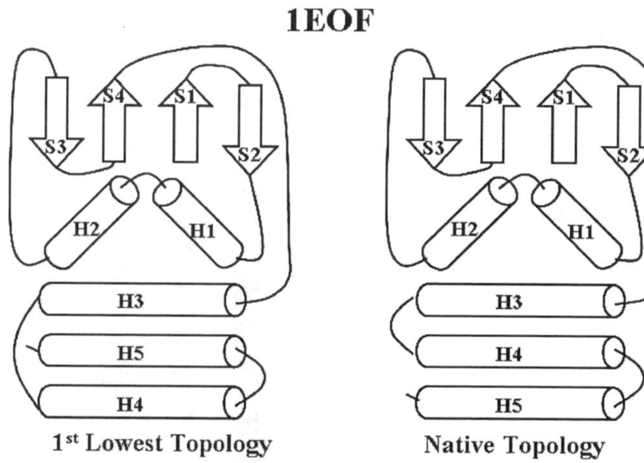

Fig. 11.19 Comparison of the lowest-energy topology with the native topology for a sheet-containing protein 1eof whose native topology was ranked the 2nd lowest.

11.4.5 Application to Incomplete Skeletons

In all previous test cases, the secondary structures in skeleton were assumed to be correct and they are used to judge the correctness of predicted assignment on sequence. In reality, however, it is very likely that skeletons from experimental maps have one or more secondary structures, especially short ones, missing. This issue of incomplete skeleton was tested on protein 1bvc that has eight α-helices. The skeletons of 1bvc purposely have one of the short helices H3 or H4 or both missing, which led to more accessible topology candidates being retained for the skeleton after initial screening. The geometry approach, however, consistently identified the native topology as the most favorable topology (1st) in all three cases. The employment of the energetics approach ranked the native topology 2nd, 3rd, and 1st when the missing component(s) was H3, H4, and both, respectively. This simple example suggested that our methods can tolerate small errors in skeleton.

It should be emphasized that, in general, the performance of the method does depend on the accuracy of secondary structures both in skeleton and in assignment. Usually, when the skeletons are correct (the normal assumption), the predicted assignment is judged based on that; when the skeletons have some small ones missing, the dependence on predicted assignment in sequence becomes stronger. In cases where both are drastically mistaken, the likelihood for the method to fail will be inevitably larger.

11.4.6 Application to Real Experimental Data

The λ2 protein of reovirion structure [solved to 7.6 Å by cryo-EM (Zhang et al., 2003)] has 16 β-sheets. One of them located at the tip of the structure was chosen to test our energetics procedure because of its continuity in sequence and comparable

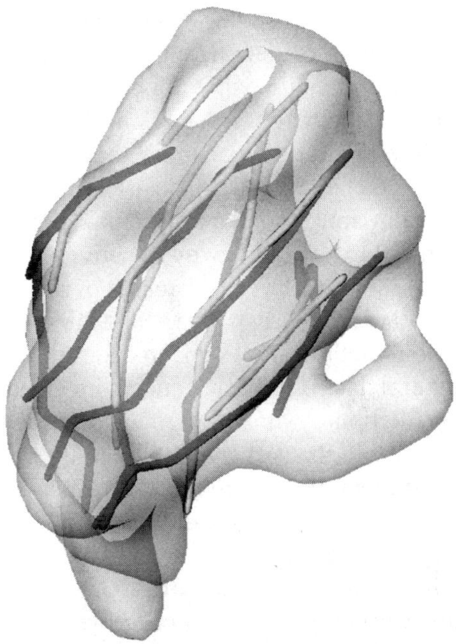

Fig. 11.20 Superposition of the secondary-structural skeleton modeled by *sheetminer* and *sheet-tracer* (yellow) based on an experimental 7.6-Å cryo-EM electron density maps (the transparent envelope) with that from the crystal structure (blue, PDB code: 1 ej6) of the λ2 protein of reovirus.

size to all other test cases. First, the skeleton of this β-sheet was generated by *sheetminer* (Kong and Ma, 2003) and *sheettracer* (Kong et al., 2004). All eight strands were successfully traced, as shown in Fig. 11.20 superimposed with the independently solved crystal structure (Reinisch et al., 2000). The secondary structure assignment was given by the algorithm DSC (King and Sternberg, 1996).

In the initial screening, a large number of accessible topologies were retained and geometry filtering ranked the native topology as 116th. By combining the sheet motif filter with the geometry filter, the native topology was ranked as 4th. When eight topology candidates were processed by energetic analysis, the native topology moved to 1st. In this case, despite the large deviations of the main chains of the traced β-strands from the crystal structure, our method was still able to correctly identify the native topology.

11.4.7 Concluding Discussion

Our computational method is fully applicable to determining topology for skeletons of unknown structures. The procedure is to first use the initial screening to remove any inaccessible topologies, then to use the geometry-based filter to rank all of the accessible topology candidates, and finally, with an appropriate cutoff, to select a fraction of accessible topologies for energetics analysis. This procedure does in many cases narrow down the native topology to be the most energetically favorable one on the final list. Moreover, in real applications, any additional knowledge about the structure can be used to filter the native topology. For example, if one knows the identity of one or a few secondary structures in the density maps, it should be enormously helpful for filtering out the nonnative topology candidates.

The method is not perfect at this stage. It suffers from the errors contained in both structural measurement and secondary structural prediction. There are cases where the method would fail to narrow down the native topology candidates as top choices, particularly in cases where severe mismatch of secondary structures occurs between the skeleton modeled from density maps and the assignment predicted from sequence. Nevertheless, our method allows one to determine native protein topology from fairly limited structural data. The basic concept involved in this study may also be useful in structure prediction by allowing effective discrimination of nonnative topology (fold) candidates from the native topology in the vast topology space.

Finally, the successful use of the ensemble average energy of randomly perturbed structures for evaluating topology candidates may also have an important implication for threading research (Elofsson et al., 1996; Jones et al., 1995; Jones and Thornton, 1996; Kihara et al., 2001; Lu et al., 2002; Miller et al., 1996; Skolnick et al., 2001). One could in principle get a better answer in evaluating decoys if effective structural variations and averaging around the given template are taken into account.

11.5 Future Perspectives

In the coming years, as the field of structure biology continues to deal with larger and more complex systems, it is inevitable that the resolutions for some of them are lower, experimental information available for structural modeling is more meager, and thus computational modeling aided by partial experimental data is increasingly more important. Other examples already in the literature include recent development of computational methods that utilize small-angle X-ray scattering (SAXS) for assisting low-resolution structural determination, in which the one-dimensional X-ray scattering profile is used as a constraint for deriving three-dimensional structures of small globular proteins (Wu et al., 2005) and large complexes (Costenaro et al., 2005b; Davies et al., 2005b; Svergun et al., 2001; Svergun and Koch, 2002). Incomplete experimental data were also used to derive biological structures in NMR-related fields.

In all of these cases, a common feature is that the effective resolutions of the structures were significantly improved with the assistance of powerful computational methods. It is expected that, in the near future, more and more new modeling algorithms will be developed to effectively make use of those either incomplete or low-resolution experimental data, from which no structural models can be built by any conventional methods.

Acknowledgment

The author acknowledges support from the National Institutes of Health (R01-GM067801).

References

Arnold, E., and Rossmann, M. G. 1988. The use of molecular-replacement phases for the refinement of the human rhinovirus 14 structure. *Acta Crystallogr. A* 44(Pt. 3):270–282.

Bahar, I., and Jernigan, R. L. 1997. Inter-residue potentials in globular proteins and the dominance of highly specific hydrophilic interactions at close separation. *J. Mol. Biol.* 266:195–214.

Ban, N., Freeborn, B., Nissen, P., Penczek, P., Grassucci, R. A., Sweet, R., Frank, J., Moore, P. B., and Steitz, T. A. 1998. A 9 Å resolution X-ray crystallographic map of the large ribosomal subunit. *Cell* 93:1105–1115.

Ban, N., Nissen, P., Hansen, J., Capel, M., Moore, P. B., and Steitz, T. A. 1999. Placement of protein and RNA structures into a 5 Å-resolution map of the 50S ribosomal subunit. *Nature* 400:841–847.

Ban, N., Nissen, P., Hansen, J., Moore, P. B., and Steitz, T. A. 2000. The complete atomic structure of the large ribosomal subunit at 2.4 Å resolution. *Science* 289:905–920.

Bottcher, B., Wynne, S. A., and Crowther, R. A. 1997. Determination of the fold of the core protein of hepatitis B virus by electron cryomicroscopy. *Nature* 386:88–91.

Conway, J. F., Cheng, N., Zlotnick, A., Wingfield, P. T., Stahl, S. J., and Steven, A. C. 1997. Visualization of a 4-helix bundle in the hepatitis B virus capsid by cryo-electron microscopy. *Nature* 386:91–94.

Costenaro, L., Grossmann, J. G., Ebel, C., and Maxwell, A. 2005. Small-angle X-ray scattering reveals the solution structure of the full-length DNA gyrase a subunit. *Structure* 13:287–296.

Davies, J. M., Tsuruta, H., May, A. P., and Weis, W. I. 2005. Conformational changes of p97 during nucleotide hydrolysis determined by small-angle X-ray scattering. *Structure* 13:183–195.

DeRosier, D. J., and Harrison, S. C. 1997. Macromolecular assemblages. Sizing things up. *Curr. Opin. Struct. Biol.* 7:237–238.

Eklund, H., Samma, J. P., Wallen, L., Branden, C. I., Akeson, A., and Jones, T. A. 1981. Structure of a triclinic ternary complex of horse liver alcohol dehydrogenase at 2.9 Å resolution. *J. Mol. Biol.* 146:561–587.

Elofsson, A., Fischer, D., Rice, D. W., Le Grand, S. M., and Eisenberg, D. 1996. A study of combined structure/sequence profiles. *Fold. Des.* 1:451–461.

Gaudet, R., Savage, J. R., McLaughlin, J. N., Willardson, B. M., and Sigler, P. B. 1999. A molecular mechanism for the phosphorylation-dependent regulation of heterotrimeric G proteins by phosducin. *Mol. Cell* 3:649–660.

Hoover, D. M., and Ludwig, M. L. 1997. A flavodoxin that is required for enzyme activation: The structure of oxidized flavodoxin from Escherichia coli at 1.8 Å resolution. *Protein Sci.* 6:2525–2537.

Jiang, W., Baker, M. L., Ludtke, S. J., and Chiu, W. 2001. Bridging the information gap: Computational tools for intermediate resolution structure interpretation. *J. Mol. Biol.* 308:1033–1044.

Jones, D. T., Miller, R. T., and Thornton, J. M. 1995. Successful protein fold recognition by optimal sequence threading validated by rigorous blind testing. *Proteins* 23:387–397.

Jones, D. T., and Thornton, J. M. 1996. Potential energy functions for threading. *Curr. Opin. Struct. Biol.* 6:210–216.

Khan, A. R., Baker, B. M., Ghosh, P., Biddison, W. E., and Wiley, D. C. 2000. The structure and stability of an HLA-A*0201/octameric tax peptide complex with an empty conserved peptide-N-terminal binding site. *J. Immunol.* 164:6398–6405.

Kihara, D., Lu, H., Kolinski, A., and Skolnick, J. 2001. TOUCHSTONE: An ab initio protein structure prediction method that uses threading-based tertiary restraints. *Proc. Natl. Acad. Sci.* USA 98:10125–10130.

King, R. D., and Sternberg, M. J. E. 1996. Identification and application of the concepts important for accurate and reliable protein secondary structure prediction. *Protein Sci.* 5:2298–2310.

Kong, Y., and Ma, J. 2003. A structural-informatics approach for mining β-sheets: Locating sheets in intermediate-resolution density maps. *J. Mol. Biol.* 332:399–413.

Kong, Y., Zhang, X., Baker, T. S., and Ma, J. 2004. A structural-informatics approach for tracing β-sheets: Building pseudo-Cα traces for β-strands in intermediate-resolution density maps. *J. Mol. Biol.* 339:117–130.

Kuhn, R. J., Zhang, W., Rossmann, M. G., Pletnev, S. V., Corver, J., Lenches, E., Jones, C. T., Mukhopadhyay, S., Chipman, P. R., Strauss, E. G., Baker, T. S., and Strauss, J. H. 2002. Structure of dengue virus: Implications for flavivirus organization, maturation, and fusion. *Cell* 108:717–725.

Li, H., DeRosier, D., Nicholson, W., Nogales, E., and Downing, K. 2002. Microtubule structure at 8 Å resolution. *Structure* 10:1317.

Lu, L., Lu, H., and Skolnick, J. 2002. MULTIPROSPECTOR: An algorithm for the prediction of protein–protein interactions by multimeric threading. *Proteins* 49:350–364.

Ludtke, S. J., Baldwin, P. R., and Chiu, W. 1999. EMAN: Semiautomated software for high-resolution single-particle reconstructions. *J. Struct. Biol.* 128:82–97.

Mancini, E. J., Clarke, M., Gowen, B. E., Rutten, T., and Fuller, S. D. 2000. Cryo-electron microscopy reveals the functional organization of an enveloped virus, Semliki Forest virus. *Mol. Cell* 5:255–266.

Mayer, S. M., Gormal, C. A., Smith, B. E., and Lawson, D. M. 2002. Crystallographic analysis of the MoFe protein of nitrogenase from a nifV mutant of Klebsiella pneumoniae identifies citrate as a ligand to the molybdenum of iron molybdenum cofactor (FeMoco). *J. Biol. Chem.* 277:35263–35266.

Miller, R. T., Jones, D. T., and Thornton, J. M. 1996. Protein fold recognition by sequence threading: Tools and assessment techniques. *Faseb J.* 10:171–178.

Mimori, Y., Yamashita, I., Murata, K., Fujiyoshi, Y., Yonekura, K., Toyoshima, C., and Namba, K. 1995. The structure of the R-type straight flagellar filament of Salmonella at 9 Å resolution by electron cryomicroscopy. *J. Mol. Biol.* 249:69–87.

Orengo, C. A., Michie, A. D., Jones, S., Jones, D. T., Swindells, M. B., and Thornton, J. M. 1997. CATH—A hierarchic classification of protein domain structures. *Structure* 5:1093–1109.

Rees, D. C., Lewis, M., and Lipscomb, W. N. 1983. Refined crystal structure of carboxypeptidase A at 1.54 Å resolution. *J. Mol. Biol.* 168:367–387.

Reinisch, K. M., Nibert, M. L., and Harrison, S. C. 2000. Structure of the reovirus core at 3.6 Å resolution. *Nature* 404:960–967.

Rossmann, M. G. 2000. Fitting atomic models into electron-microscopy maps. *Acta Crystallogr. D Biol. Crystallogr.* 56(Pt.10):1341–1349.

Samatey, F. A., Imada, K., Nagashima, S., Vonderviszt, F., Kumasaka, T., Yamamoto, M., and Namba, K. 2001. Structure of the bacterial flagellar protofilament and implications for a switch for supercoiling. *Nature* 410:331–337.

Skolnick, J., Kolinski, A., Kihara, D., Betancourt, M., Rotkiewicz, P., and Boniecki, M. 2001. Ab initio protein structure prediction via a combination of threading, lattice folding, clustering, and structure refinement. *Proteins* Suppl. 5:149–156.

Steinbacher, S., Seckler, R., Miller, S., Steipe, B., Huber, R., and Reinemer, P. 1994. Crystal structure of P22 tailspike protein: Interdigitated subunits in a thermostable trimer. *Science* 265:383–386.

Svergun, D. I., and Koch, M. H. 2002. Advances in structure analysis using small-angle scattering in solution. *Curr. Opin. Struct. Biol.* 12:654–660.

Svergun, D. I., Petoukhov, M. V., and Koch, M. H. 2001. Determination of domain structure of proteins from X-ray solution scattering. *Biophys. J.* 80:2946–2953.

Verbeek, J. J., Vlassis, N., and Krose, B. 2002. A k-segments algorithm for finding principal curves. *Pattern Recognition Lett.* 23:1009–1017.

Wang, G., and Dunbrack, R. L., Jr. 2003. PISCES: A protein sequence culling server. *Bioinformatics* 19:1589–1591.

Wang, Q., Buckle, A. M., and Fersht, A. R. 2000. Stabilization of GroEL minichaperones by core and surface mutations. *J. Mol. Biol.* 298:917–926.

Wilson, D. K., Bohren, K. M., Gabbay, K. H., and Quiocho, F. A. 1992. An unlikely sugar substrate site in the 1.65 Å structure of the human aldose reductase holoenzyme implicated in diabetic complications. *Science* 257:81–84.

Wittinghofer, F., Krengel, U., John, J., Kabsch, W., and Pai, E. F. 1991. Three-dimensional structure of p21 in the active conformation and analysis of an oncogenic mutant. *Environ. Health Perspect.* 93:11–15.

Wu, Y., Chen, M., Lu, M., Wang, Q., and Ma, J. 2005a. Determining protein topology from skeletons of secondary structures. *J. Mol. Biol.* 350:571–586.

Wu, Y., Tian, X., Lu, M., Chen, M., Wang, Q., and Ma, J. 2005b. Folding of small helical proteins assisted by small-angle x-ray scattering profiles. *Structure* 13:1587–1597.

Zanotti, G., Panzalorto, M., Marcato, A., Malpeli, G., Folli, C., and Berni, R. 1998. Structure of pig plasma retinal-binding protein at 1.65 Å resolution. *Acta Crystallogr. D* 54:1049–1052.

Zhang, X., Shaw, A., Bates, P. A., Newman, R. H., Gowen, B., Orlova, E., Gorman, M. A., Kondo, H., Dokurno, P., Lally, J., Leonard, G., Meyer, H., van Heel, M., and Freemont, P. S. 2000. Structure of the AAA ATPase p97. *Mol. Cell* 6:1473–1484.

Zhang, X., Walker, S. B., Chipman, P. R., Nibert, M. L., and Baker, T. S. 2003. Reovirus polymerase lambda 3 localized by cryo-electron microscopy of virions at a resolution of 7.6 Å. *Nat. Struct. Biol.* 10:1011–1018.

Zhou, Z. H., Baker, M. L., Jiang, W., Dougherty, M., Jakana, J., Dong, G., Lu, G., and Chiu, W. 2001a. Electron cryomicroscopy and bioinformatics suggest protein fold models for rice dwarf virus. *Nat. Struct. Biol.* 8:868–873.

Zhou, Z. H., Dougherty, M., Jakana, J., He, J., Rixon, F. J., and Chiu, W. 2000. Seeing the herpesvirus capsid at 8.5 Å. *Science* 288:877–880.

Zhou, Z. H., Liao, W., Cheng, R. H., Lawson, J. E., McCarthy, D. B., Reed, L. J., and Stoops, J. K. 2001b. Direct evidence for the size and conformational variability of the pyruvate dehydrogenase complex revealed by three-dimensional electron microscopy. The "breathing" core and its functional relationship to protein dynamics. *J. Biol. Chem.* 276:21704–21713.

Index

(*continued from page ii*)

Physics of the Human Body: A Physical View of Physiology Herman, I.P., 2006

Intermediate Physics for Medicine and Biology Hobbie, R.K., Roth, B., 2006

Computational Methods for Protein Structure Prediction and Modeling (2 volume set) Xu, Y., Xu, D., Liang, J. (Eds.) 2006

Artificial Sight: Basic Research, Biomedical Engineering, and Clinical Advances, Humayun, Weiland, Greenbaum., 2006

Physics and Energy Landscapes of Proteins Fraunfelder, H., Austin, R., Chan. S., 2006

Biological Membrane Ion Channels Chung, S.H., Anderson, O.S., Krishnamurthy, V.V., 2006

Cell Motility, Lenz, P., 2007

Applications of Physics in Radiation Oncology, Goitein, M., 2007

Statistical Physics of Macromolecules, Khokhlov, A., Grosberg, A.Y., Pande, V.S. 2007

Biological Physics, Benedek, G., Villars, F., 2007

Protein Structure Protein Modeling, Kurochikina, N., 2007

Three-Dimensional Shape Perception, Zaidi, Q., 2007

Structural Approaches to Sequence Evolution, Bastolla, U.

Radiobiologically Optimized Radiation Therapy, Brahme, A.

Biological Optical Microscopy, Cheng, P.

Microscopic Imaging, Gu, M.

Deciphering Complex Signatures: Applications in Life Sciences, Morfill, G.

Biomedical Opto-Acoustics, Oraevsky, A.A.

Mathematical Methods in Biology: Mathematics for Ecology and Environmental Sciences, Takeuchi, Y.

Mathematical Methods in Biology: Mathematics for Life Science and Medicine, Takeuchi, Y.

In Vivo Optical Biopsy of the Human Retina, Drexler, W., Fujimoto, J.

Tissue Engineering: Scaffold Material, Design and Fabrication Principles, Hutmacher, D.W.

Ion Beam Therapy, Kraft, G.H.

Biomaterials Engineering: Implants, Materials and Tissues, Helsen, J.A.

Printed in the United States of America.